Bridge Management

Bridge Management

Bojidar Yanev

JOHN WILEY & SONS, INC.

Published by John Wiley & Sons, Inc., Hoboken, New Jersey
Published simultaneously in Canada

For general information about our other products and services, please contact our Customer Care Department within the United States at (800) 762-2974, outside the United States at (317) 572-3993 or fax (317) 572-4002.

Wiley also publishes its books in a variety of electronic formats. Some content that appears in print may not be available in electronic books. For more information about Wiley products, visit our web site at www.wiley.com.

Library of Congress Cataloging-in-Publication Data:

Yanev, Bojidar, 1947–
      Bridge management/by Bojidar Yanev.
            p. cm.
      Includes index.
      ISBN-13 978-0-471-69162-4 (cloth)
      ISBN 0-471-69162-3 (cloth)
   1. Bridges—Maintenance and repair.   2. Bridge failures—Prevention.   3. Bridges—Inspection.   I. Title.
TG315.V36 2006
624.2028′8—dc22

                                                          2006013363

Printed in the United States of America
10 9 8 7 6 5 4 3 2 1

Contents

Preface

> In general, it is a sign of the man who knows and of the man who does not know, that the former can teach, and therefore we think art more truly knowledge than experience, for artists can teach, and men of mere experience cannot.
> Aristotle, *Metaphysics* (Book 1, 981b, 1941)

Aristotle believed that the acquisition of knowledge resulted in a state opposite to the one at the origin of the inquiry (*Metaphysics,* 983a). Twenty-four centuries after his prediction, art is produced by individualistic and therefore unique talents, whereas experience is studied and reproduced by people and machines. The tangible advantages of training are increasingly preferred to the less specific benefits of education. Hence, the demand for teaching bridge management is exceeded by the interest in the tools for practicing it. The subject is presumed to be premature for beginners and superfluous for experts. The basics are presented in many excellent publications on structural analysis, design, construction, maintenance, operation, inspection, rehabilitation, economics, computer science, and information theory. The latest developments are reported in conference proceedings, government agency directives, the National Cooperative Highway Research Program (NCHRP), manuals, handbooks, and periodicals and on the Internet.

The Organization for Economic Cooperation and Development (OECD, 1992, p. 17) attempted to define bridge management as follows:

> The term *bridge management* encompasses a broad range of activities aimed at ensuring the safety and functionality of bridges. An efficient bridge management scheme is necessary to support the highway and bridge agencies' organizational set-up and their administrative and technical functions. . . . Above all, adequate qualified and well-trained manpower with clear responsibilities and duties must be ensured. The personnel must be aware of its role in the different management processes."

The *scheme* in the preceding description is a bridge management support system (BMS) for agencies and for specific operations. "Adequately qualified and well-trained" personnel supporting the BMS "must be ensured," but not necessarily by a supply of authoritative instruction. Visionary strategists have no detailed knowledge of individual projects, whereas hands-on managers cannot afford a global network vision. A text can never be definitive, but it risks becoming obsolete or redundant even before publication. Rigorous treatments of specific aspects do not capture the scope of the subject. General descriptions run the risk of becoming all-inclusive and rambling.

Noteworthy exceptions include the texts written and coauthored by W. R. Hudson, such as NCHRP Report 300 (1987) and Hudson et al. (1997) (the latter of which the author personally provided to this writer shortly before the completion of the present manuscript). NCHRP Report 300 (1987) described the purpose of the BMS as follows (p. 1): "A Bridge Management System should assist decision-makers to select optimum cost-effective alternatives needed to achieve desired levels of service within the allocated funds and to identify future funding requirements. . . . A BMS provides benefits to administrators, engineers, and managers at all levels within a transportation agency."

The demand for competent administration, engineering, and management is growing in the following areas:

- Large infrastructure networks comprising technologically diverse structures and facilities mostly ranging in age between 0 and 150 years
- Dynamic global economy, dependent on the transportation infrastructure
- Rapidly growing capabilities in all areas related to bridge management, including analysis, construction, data acquisition, and processing

Concluding a three-year mandate as part of the Federal Infrastructure Strategy (FIS) program, the Committee on Measuring and Improving Infrastructure Performance (NRC, 1995) described the state of the art as follows (p. 19):

> Increasingly powerful and cost-effective computer-based forecasting and simulation methods and new technology for measuring and monitoring system conditions have made more sophisticated approaches to assessing system performance widely available. Remote sensing, real-time monitoring, and network analysis and simulation models provide powerful new capabilities for measuring system-wide conditions and evaluating system changes. These tools will support more meaningful multi-jurisdictional and multi-modal cooperation. Despite the availability of such new tools, there remain many impediments to infrastructure performance measurement and management.

Einstein (1950) anticipated the arising risks as follows: "The perfection of means and confusion of goals seem to characterize our age." Under the assault of technical advances and social change, expertise and its products are equally maintenance intensive. Professionals struggle to keep up with the latest information and technology. Professions compete for managerial responsibilities. In order to assert itself as a discipline, management tends to impose an accelerated pace on structures intended to span over generations of users. David Brooks commented (*New York Times,* July 20, 2004, op-ed, p. A19, column 1) that "too many universities have become professional information transmission systems." In the dynamic exchange between academic education and market demands, the risk of knowledge obsolescence is matched by the hazards of indiscriminate innovation. According to Shenk (1997, p. 91), "the proliferation of expert opinion has ushered in a virtual anarchy of expertise."

Bridge management has resisted anarchy by polarizing into *top-down* and *ground-up* schools. In the past, managers rose from the ground up as engineers, builders, or craftsmen, educated in design, construction, and the related trades (including the military). The recent advances in statistics, probability, quantitative analysis, systems design, nondestructive testing, data processing, and software design are steering engineers toward decision support and implementation rather than to executive positions. In the meanwhile economics, business management, administration, law, architecture, urban planning, and, as ever, politics are claiming their shares of the top-down decision-making process. The overarching scope of *asset* management has absorbed the specific concerns of bridge management along with the interest of top-down managers.

Since electronic data searches supply all information (except the one of immediate relevance) and since engineers and managers by definition should be able solve their own problems, the presented text attempts to do neither. Rather, it can serve as a tour guide to the relevant tasks and around the common pitfalls in the field. In order to avoid catastrophic obsolescence, the material is structured into partially redundant modules of general discussion, appendices derived from essential sources, and illustrative examples derived from practice. References draw from the archaic, the recent and the latest, as a reminder that the subject is as transient as it is permanent. Direct contact with the original authors is encouraged.

Analysis advances from design to construction and to management of the produced assets. Practice has followed the opposite path: management demands service, builders construct the facilities, and designers give form to complex functions. The subject of bridge management should therefore be accessible, as it is practiced, from different points of entry, for different purposes, and with different background. Topics recur, reflecting the cyclic nature of the process and the redundancy that serves it particularly well in matters of safety. The added risk of losing the reader's interest remains to be evaluated.

Acknowledgments

The author was able to assemble this material largely as a result of the education and advice he received during 30 years of studies, research, and teaching at Columbia University, a 2-year postdoctoral engagement at the University of California at Berkeley, 10 years of practice in engineering consulting companies in New York City, and 18 years of managing the office of Bridge Inspection & Management, which he established at the New York City Department of Transportation. (Some of these activities were concurrent.) Personal experience has strengthened the appreciation of the talent and knowledge invested in this subject by the many sources quoted or described in the text. Visits with colleagues in the United States, Europe, Japan, China, Taiwan, Korea, Indonesia, and Australia have helped to expand the view beyond local interests (even if those local interests are quite diverse). The encouragement of R. Ratay and J. Harper of John Wiley & Sons gave life to the project.

INTRODUCTION

ENGINEERING
AND MANAGEMENT

Chapter 1

Engineering and Management: The Dynamic Equilibrium

Professional specialization seems to have put an end to the long and distinguished succession of master builders. The descendant engineers and managers are cultivating their own tastes, activities, competence, and terminology. Their purpose is not always the same. Still shared are the professional tools and the possibly disputed field of application. It is tempting to lament the split but more practical to renegotiate the terms of coexistence, because, so long as they remain distinct, engineering and management will be entirely complementary. The same is true of the categories, definitions, and disciplines comprising or informing the two professions. A few examples are listed below:

Art	Science
Knowledge	Information
Information	Data
Quality	Quantity
Function	Form
Form	Content
Content	Strength
Shape	Stability
Process	Product
Discovery	Invention
Analogy	Analysis
Structural analysis	Design
Design	Construction
Evaluation	Inspection
Uncertainty	Probability
Safety	Reliability
Value	Money
Leadership	Management
Management	Engineering

The list can be extended considerably, merely corroborating the predictable conclusion that engineering and management as well as their sources, methods and objectives, processes, and products are dialectically enjoined. A philosophical treatment would therefore be appropriate and, if true to tradition, should raise more questions than it could answer. In the meanwhile, engineering and management must deliver services for the present and structures for the future by all the means of art and science.

1.1   ASSETS, ACTIVITIES, STATICS, AND DYNAMICS

Engineering produces the material assets of the future and preserves those of the past. Management optimizes the use of physical, financial, and human resources, present and anticipated. The bonds and tensions between the two are at the source of human creativity. Nowhere are they more conspicuously linked than in constructing and operating bridges.

Change and, hence, life are expressed in terms of energy in the physical world, information in the realm of ideas, and money in commerce. Philosophy perceives human understanding as a synthesis of dichotomous ideas. Physics models natural phenomena as energy equilibria. A static equilibrium describes sufficiently many structural problems and designers can grow used to it. In contrast, the forces separating, uniting, reorganizing, and redesigning society are dynamic, and so is leadership. Bridges similarly carry vehicles not only in space but also over time. They serve the community as ideas propagate thought—by providing more or less temporary shortcuts. No matter how durable a structure may appear, its life depends on the dynamic forces acting in the social brain that created it. A neglected bridge is as worthless as an abandoned vehicle or a forgotten idea.

Engineering management must adjust bridges to the social dynamics as well as to the physical ones. When this is accomplished, engineering and management blend inseparably into masterpieces; at other times they lack the energy to come to terms and appear distinct, even opposed. Art and science continually reshape them, combining to various degrees experience and understanding, and application and abstraction.

1.2   ART, SCIENCE, EMPIRICISM, AND ABSTRACTION

In his *Nicomachean Ethics* (1941, Book VI, 1139a, 1140a), Aristotle (384–322 B.C.) (Fig. 1.1) defined art and science as follows: "The object of scientific knowledge is of necessity, it is eternal. . . . Art is a creative state concerned with making, involving a true course of reasoning."

Aristotle's examples of science and art are medicine and architecture, respectively. In his *Metaphysics* (1941, 981a) he added:

> With a view to action, experience seems in no respect inferior to art, and men of experience succeed even better than those who have theory without experience. The reason is that experience is knowledge of individuals and art of universals, and actions and productions are all concerned with the individual. But we think that *knowledge* and *understanding* belong to art rather than to experience, and we suppose artists to be wiser than men of experience, because the former know the causes, but the latter do not.

By this standard, set a century before him, Archimedes (287–212 B.C.) was an artist of great experience. By modern standards, his boundless daring in theory and application

Figure 1.1 Aristotle, marble, Roman, first to second centuries A.D., the Louvre, Paris.

define the protoengineering spirit. It is a fitting legend that, after defending Syracuse with his practical inventions, he should have lost his life protecting his abstractions.

Philosophy and science drifted apart during the Middle Ages, but crafts attained artistic quality. Armytage (1976) considered the empirical achievements, evidenced by the cathedrals of the period and by the works of the leading scholars, notably Roger Bacon (1219–1292), as the precursors of the scientific revolution that followed. Heyman (1996, p. 1) pointed out that, if theory was not up to modern standards yet, building practices had already been regulated in various forms since 600 B.C. By A.D. 1400 Mignot restated Aristotle: "Practice is nothing without theory."

Stark (2005) argued that the scientific revolution of the Renaissance and the following Industrial Revolution were nurtured by faith. During the Renaissance art and science blended as organically as science and philosophy had done in antiquity. On the walls of the Vatican's Sala della Signatura Raphael (1483–1520) portrayed the ancient Greek philosophers with a reverence appropriate for the apostles. The School of Athens (Fig. 1.2) brilliantly summarizes the major philosophical and scientific trends, including the harmonics of Pythagoras, the geometry of Euclid, the astronomy of Ptolemy and Zoro-

Figure 1.2 The School of Athens, Raphael, Sala della Signatura, the Vatican, Rome. *Bottom:* Pythagoras, left; Heraclitus (bearing the features of Michelangelo, writing), center left; Diogenes (reclining), center right; Euclid, bending to explain, right; Ptolemy (with a globe) and Zoroaster (with a stellar sphere), far right. *Top:* Socrates (in green toga, counting on his fingers), sixth from the left; Plato (bearing the features of Leonardo), center left; Aristotle, center right. The building suggests a design by Bramante reminiscent of Saint Peter's Cathedral (Fig. 1.3).

aster, the pessimism of Heraclitus, the cynicism of Diogenes, the rhetoric of Socrates, culminating in a synthesis of Plato's idealism and Aristotle's materialism.

The domes of Brunelleschi (1377–1446) in Florence and Michelangelo (1475–1564) (Fig. 1.4) in Rome unify mechanics and sculpture. Leonardo da Vinci (1452–1519) sought employment as artist and designer of fortifications. Niccolo Machiavelli (1469–1527) proposed rational management principles to the same employer, Cesare Borgia. Galileo Galilei (1564–1642) was professor of physics and military engineering. He used mathematics to model the orbits of the planets and the deflections of cantilever beams. Nicolaus Copernicus (1473–1543) set the sun at the center of the planetary system. René Descartes (1596–1650) placed thinking at the core of human existence. "Standing on the shoulders of giants" (of antiquity, but perhaps also of the more recent past), Isaac Newton (1642–1727) perceived the mechanics of the universe. He and Leibnitz (1646–1716) formulated the calculus that described it.

Figure 1.3 Saint Peter's Cathedral, Rome.

Calculus, however, was soon to venture into the unimaginable. Voltaire (1694–1778) (Fig. 1.5) called it "the art of measuring and numbering exactly a Thing whose Existence cannot be conceived." Kline (1953, p. 232) concluded: "It was a very fortunate circumstance that mathematics and science were closely linked in the Newtonian era and that physical reasoning could guide the mathematicians and keep them on the right track. Because the results they obtained were useful and sound in application, they maintained confidence in their methods and the courage to proceed farther."

In 1678 Robert Hooke (1635–1703) observed that iron extended proportionally to the applied tension. Expanded in 1776 to solids by Coulomb (1736–1808) and quantified by Young's (1773–1829) modulus in 1807, elasticity formed the theoretical backbone of engineering. In practice, George Stephenson's (1781–1848) steam engine ushered in the

Figure 1.4   Michelangelo, bronze, Daniele da Volterra, 1509–1566, the Louvre, Paris.

Industrial Revolution. Billington (Gans, 1991, p. 4) places the "the origins of modern engineering . . . in the west Midlands of Britain in the late eighteenth century."

Discoveries and inventions increasingly competed on the marketplaces rather than on the battlefields. Mathematics grew abstract, sciences specialized, arts and crafts became professions and trades. The management of assets in times of peace shifted attention from the capacity to sustain and inflict destruction to the efficiency of production and the profitability of transaction. Adam Smith (1723–1790) quantified wealth and value for purposes of state. A volatile mix of theory, empiricism, ethics, and speculation began shaping into political economy.

Eiffel stated that "the advancements of science, metallurgy and the art of engineering distinguish the latter part of the [nineteenth] century" (Harriss, 1975, p. 81). The interaction of these ingredients depended on the historic backgrounds. Once structural demands exceeded the competence of the "trades," engineering took shape as a profession in England, a business in the United States, and a science in France. Alexis de Tocqueville (1805–1859) observed (2000, p. 550) that "in aristocracies . . . the notion of profit

Figure 1.5 Voltaire, marble, J.-B. Pigalle, 1776, the Louvre, Paris.

remains distinct from that of work . . . but in democratic societies the two notions are always visibly united." Gordon (1978, Chapter 4) described the difference in approaches as a contrast of "French theory versus British pragmatism." Anglo-Saxon and American engineering showed a taste for empirical evidence in the tradition of Roger Bacon. France championed the analytic method of Descartes (Fig. 1.6).

Despite his highest esteem for the "art" of mastering theoretical principles, Aristotle noted that practical experience could sell better in the marketplace. In decentralized Eng-

Figure 1.6   University René Descartes, Paris.

land theoretically minded engineers were said to belong "on tap, but not on top." In highly centralized France *genie civil* became a subject fit for leaders because of the implied analytical competence. The English language, with its Germanic, Celtic, and Italic influences, may be more conducive to a laissez faire adaptability, whereas the Romance French is more prescriptive and exacting. The terms *engine* in the former case and *genius* in the latter (arguably) imply different roles.

De Tocqueville (2000, Vol. 2, Chapter 10) explained "why the Americans are more concerned with the applications than with the theory of science" as follows (p. 459): "Those in democracies who study sciences are always afraid of getting lost in utopias.

They mistrust systems and like to stick very close to the facts and study them for themselves. . . . They penetrate, as far as they can, into the main parts of the subject that interests them, and they like to expound them in popular language. Scientific pursuits thus follow a freer and safer course but a less lofty one."

Thus the difference between "the lust to make use of knowledge and a pure desire to know" (Tocqueville, 2000, p. 461):

> If Pascal had had nothing in view beyond some great gain, or even if he had been stimulated by the love of fame alone, I cannot conceive that he would have been able, as able he was, to rally all the powers of his mind to discover the most hidden secrets of the Creator. . . .
>
> The future will show whether such rare, creative passions come to grow as easily in democracies as in aristocratic communities.

The application of ideas also differs, somewhat analogously to the *top-down* and *ground-up* approaches (Tocqueville, 2000, p. 641): "Since there is no link binding the inhabitants of democracies to each other, each man has to be convinced separately, whereas in aristocracies it is enough to influence the views of certain individuals, and the rest will follow."

One hundred and twenty years later Jacques Barzun (1959, p. 11), Dean of Faculties and Provost of Columbia University (Fig. 1.7), would write:

> Men have come to believe that they can link knowledge and action without a regular gradation of intellects to harbor and diffuse ideas, or a common concern about the welfare of Intellect as an institution. They trust to the exceeding fineness and particularity of the information stored in print and made foolproof by formulas. Yet at the very time when the sum of fact on all subjects begins to seem adequate to the demand, a silent panic

Figure 1.7 Columbia University, New York City.

overtakes both thinkers and doers: "the problem of communication", Babel [has] become an everyday experience.

As communication is the domain of language, different cultures conduct it on their own terms. In France, *ouvrages d'art* were inspected by government decree since 1665. The task is currently managed by the highly qualified graduates of École Nationale des Ponts et Chaussées (Fig. 1.8), admitted to the school through intensely competitive ex-

Figure 1.8   École Nationale des Ponts et Chaussées, Paris.

aminations. In the United States many highly qualified universities produce graduates who spend their professional careers competing with each other over the financial and structural management of the *assets.*

Karnakis (1997) compared the interaction of theory and empiricism in the design and construction of suspension bridges in the Anglo-Saxon and French traditions. Judge James Finley (1762–1839) patented serviceable chain suspension bridges long before large-deflection theory had been formulated. Navier's (1785–1836) suspension bridge over the Seine had faulty anchorages, but his treatise on the subject became a lasting theoretical contribution. A fellow graduate of École Politechnique, Auguste Comte (1798–1857) founded positivism in philosophy. Both theoreticians were duly recognized. Navier's marble bust stands in the lobby of École Nationale des Ponts et Chaussées along with Saint-Venant's and those of other illustrious colleagues (Fig. 1.9*a*). Comte's monument looks down on Place de la Sorbonne (Fig. 1.9*b*).

Anecdotal evidence gravitates to established stereotypes. The pragmatist supposedly builds for a dollar what others would accomplish for ten. A theorist (allegedly at École Normale Superieure) argued that "just because something works in practice, it does not necessarily work in theory."

(*a*)   (*b*)

Figure 1.9   (*a*) Navier, marble, École des Ponts et Chaussées, Paris. (*b*) Auguste Comte, Place de la Sorbonne, Paris.

After De Tocqueville's dream of replacing the old aristocratic society by a new democratic one became institutionalized, the stereotypical attitudes conformed. Parkinson (1957, p. 16) speculated that, whereas the British House of Commons was structured on a bilateral premise, with each team considering itself right and the opposite one wrong, the semicircular chamber of the French Senat, with "a multitude of teams facing in all directions, . . . allows for subtle distinctions between the various degrees of rightness and wrongness." The author mischievously acknowledged that, sitting arrangements notwithstanding, language and the traditional sense of gamesmanship also contributed to the difference. The American Senate, also semicircular yet staunchly bipartisan, corroborates the latter rather than the former conjecture.

Regardless of the prevalent cultural traditions, engineering owes more to art than the present state of the practice indicates. According to Picon (1992, p. 47), the Bridge Corps of France headed by Peronnet in 1747 had a "more descriptive than analytic approach." Aubry expressed the philosophy of the early École Nationale des Ponts et Chaussées as follows: "Speculative knowledge, even if based on solitary observations, can lead to many errors if the latter are not related to experiments conducted in natural conditions."

The professionals of the period are characterized as *engineers-artists* and *hydraulic architects*. According to Picon (1992, p. 230), professional relations were "much less tenuous between bridge engineers and architects than between military and civil engineers." "Mathematisation" was soon to be introduced, notably by Coulomb, Carnot, Meusnier, Borda, and Dubuat.

In the *American Railroad Journal and Mechanics Magazine* of April 1, 1841, No. 379, Vol. XII, John Roebling (1806–1869) argued in favor of securing suspension bridges dynamically by experience and analysis (at a time when both were lacking):

> To ensure the successful introduction of cable bridges into the United States, their erection, and especially the construction of the first specimen, should not be left to mere mechanics. No modern improvement has profited more by the aid of science than the system of suspension bridges. And we see that all the noble and bold structures of this kind which have been put up in Europe, were planned and executed under the immediate superintendence of the most eminent Engineers, whose practical judgment was aided by a rich store of scientific knowledge.

The most prominent feature shared by the masters of the art appears to be the ability to blend empiricism and theory seamlessly. Marveling at the "total beauty" of medieval churches, Paul Valéry (1871–1945) speculated (1945, p. 144): "The builders of the great periods apparently always conceived their structures in one single stroke, rather than in two mental phases or in two series of operations, one relating to form and the other to matter. . . . they thought in materials."

Harriss (1975, p. 52) applied the quote to Eiffel as well. By 1917 the French surrealist Marcel Duchamp (1887–1968) rated plumbing and bridges highest among the achievements of American art.

Billington (1983) showed that the leading bridge engineers have always absorbed cultural and scientific influences in order to outgrow and reshape them. Thomas Telford (1757–1834) and Francois Hennebique (1843–1921) shared an empirical background in stonemasonry. Theoretical lucidity was common to the artistically daring Isambard Kingdom Brunel (1806–1859) and Gustave Eiffel (1832–1923). Gustav Lindenthal (1850–1935), New York Bridge Commissioner, studied in Dresden but admired I. K. Brunel and R. Stephenson. Carl Culmann (1821–1881) explored civil engineering at its professional source in Britain and in the United States before establishing a school in Zurich. Othmar Ammann (1879–1966) studied with him and built his bridges in New York. David Steinman (1887–1971) translated Joseph Melan's (1854–1941) work on large deflections of suspension bridges from the German. Leon Moisseiff (1872–1943) translated Considere (1903) on reinforced concrete from the French.

Boller (1885, p. 43), designer of the landmark Macomb's Dam bridge in New York City (Fig. 1.10) among many others, denounced the "purely empirical structures, [whose] construction should under no circumstances be permitted." He particularly objected (p. 48) to "the crude conclusions, drawn from very meagre experiments, made more than a dozen years since in England [that] have been a sort of blind guide for engineers." Boller stated his professional position as follows: "It is bordering on criminality to build any structure on a plan that no human being can tell definitely any thing about, when there are so many plans that we thoroughly understand."

During the twentieth century information became widely accessible and professional expertise gained precision and narrowed in scope. Barzun (1959, p. 11) lamented as follows:

> The same abundance of information has turned into a barrier between one man and the next. They are mutually incommunicado, because each believes that his subject and his language cannot and should not be understood by the other. This is the vice we weakly deplore as specialization. It is thought of, once again, as external and compelling, though it comes in fact from within, a tacit denial of Intellect. It is a denial because it rests on the superstition that understanding is identical with professional skill.

Deplored or not, professional specialization is inexorably enforced by both competitive markets and competing expertise. The Transportation Research Board (TRB, 2001, p. 8) dedicated a paragraph to "eliminating art and introducing science" (from and to pavement management). Through refinement of their analytic and computational tools,

Figure 1.10 Macombs Dam Bridge, Harlem River, New York City.

mathematics, physics, and biology constantly subdivide their fields of application. Economics is distinct from financial management and business administration. Architecture claims exclusive rights over structural form. Every structural type requires specialized designers and contractors.

Responding to different demands, engineering and management, as well as architecture and economics, have emerged as professions from the ancient and thoroughly crafty arts of war, building, and leadership. But professions never quite outgrow the need for sound judgment and good taste. Theories hold true within (not always defined) boundaries and beyond them must revert to the intuitive and analogue methods more typical of art. Before analysis can optimize the forms appropriate for new functions, subjective choices and personal visions perfect them by trials and errors.

1.3   ENGINEERS AS MANAGERS

Management and engineering evolved jointly during the early stages of organized warfare. Sun-Tzu (1994, p. 184) defined the art of war as a synthesis of measurement, estimation, calculation, weighing, and victory, distinctly engineering tasks. The advent and increasingly destructive power of artillery created a corresponding demand for engineered fortifications. Under Louis XIV of France (1638–1715), Vauban (1633–1707) was marechal of the army, commissioner general of fortifications, and superintendent of buildings, arts, and manufactures. As a young officer, Napoleon demonstrated the advantages of artillery but eventually acknowledged that "an army marches on its stomach." After "civil"-izing in the nineteenth century, the profession retained some quasi-military aspects in practice and education. Apprenticeship under a famous practitioner was considered the best education in England. In Germany, John Roebling studied with Hegel. Amman studied with Culmann in Switzerland. At Rensselaer Polytechnic Institute (1854–1857) Washington Roebling studied nearly a hundred subjects ranging from calculus of variations and qualitative and quantitative analysis to rhetorical criticism, ethical philosophy, and French literature. He was one of 12 to graduate from the 65 who started in his class (Steinman, 1945).

Unique lifetime experience completes the image of the bridge manager as a mythical character, defying formalized education and transcending the professional boundaries. Biotechnical and historical accounts, for instance by Gies (1963) and Petroski (1993), as well as autobiographic accounts, such as Freyssinet (1993), confirm that impression. Certain engineers are associated with a particular bridge type, others have championed a variety of designs. In every case the ability to satisfy existing demands and to generate new ones is outstanding. All exemplars show a firm grasp of theory *and* a relentless drive toward application (imagination and practicality would be another way to describe the winning combination, as in all creative achievements).

Jean Rodolphe Peronnet (1708–1794) is credited as the father of the Corps of Bridge Engineers in France. Lacking the funds required for admission in the Corps of Fortifications, Peronnet worked for the first architect of the City of Paris, before entering the Bridge Corps. A friend of King Louis XV of France, he was the first director of École des Ponts et Chaussées, designed the Pont de Neuilly across the Seine in Paris among many others, and established the first comprehensive database of roads in France (Picon, 1992).

James Finley (1762–1839), the first suspension bridge patent holder, was a judge, a landowner, and a businessman (Karnakis, 1997).

Thomas Telford (1757–1834) is considered by Billington (Gans, 1991, p. 5) as the "first great structural artist" who designed for "light weight, low cost and structural appeal." The suspension bridge across the Menai Straits is perhaps the most famous monument of his achievement. Telford saw engineers as the new master builders. He was the first president of the Institution of Civil Engineers in London (the first engineering society ever founded) until the end of his life.

Sir Marc Isambard Brunel (1769–1849), of French lineage, became chief engineer of New York, then moved to England. There he excelled in ship building for the war against Napoleon and built the first tunnel under the Thames (1822–1843). His importance was such that the Duke of Wellington prevented his departure to Russia on the invitation of Tsar Alexander (Billington, 1983).

Figure 1.11   James Buchanan Eads, Hall of Fame, the Bronx, New York City.

Marc's son Isambard Kingdom Brunel (1806–1859) designed and built some of the most imaginative and daring bridges of the time but also launched the largest and most innovative steamships, introduced the 7-ft (2135-mm) gauge to railways, and authored numerous other inventions. An artist of engineering, Brunel succumbed to the demands of managing the construction of his most ambitious steamship, *The Great Eastern.*

Ferdinand de Lesseps (1805–1895) built the Suez Canal but could not repeat that feat at Panama. McCullough (1977) traces both outcomes to the unique combination of engineering and managerial talents demonstrated in these two endeavors, successful in French Africa but inapplicable in American Panama. George Goethals (1858–1928) completed the latter project on behalf of the United States by different management and engineering methods.

James Buchanan Eads (1820–1887) (Fig. 1.11) designed and built the Great St. Louis Bridge over the Mississippi River. He was captain and manager in salvage operation on the river, organized numerous construction projects, and obtained their international financial backing. McCullough (1972, p. 181) added the following colorful detail: "Eads was the sort of person who liked to play chess with two or three others at a time and in a recent (circa 1870) weight-lifting contest among some of his blacksmiths, he had come in second."

John Roebling (1806–1869) is recognized as one of the greatest bridge designers of all time, primarily in view of the Brooklyn Bridge (Example 1); however, he was many other things as well. Beyond studying philosophy under Hegel and designing and building the longest and most brilliant hybrid suspension and stay cable spans of the nineteenth century, he established an agrarian community in Pennsylvania, founded the Roebling high-strength wire company, and mobilized the political and financial backing needed for the merger of Manhattan and Brooklyn into a New York for the twentieth century.

Example 1. New York City Bridge History

The 1908 annual City Bridge Report listed 45 bridges. Included were the four new record-breaking East River crossings, an assortment of movable bridges, as well as some timber structures. Lindenthal, Wadell, Boller, Buck, Ammann, Steinman, Mojeski, Moisseiff, and many other distinguished engineers contributed to the bridge network in the so-called tristate area (New York, New Jersey, and Connecticut).

In 2004 the city was responsible for 790 structures, including 6 tunnels and over 100 pedestrian bridges. The total number of spans approached 5000 and the average age was approximately 80 years. New York State operates a similar number of arterial bridges in the same geographic area with an average age of 40 years. Authorities manage tolled vehicular and rail bridges and tunnels. The total number of New York City bridges surpasses 2200.

Table E1.1 shows a list of 12 long span bridges in the tristate area, 5 of which have set world records. The longest bridge on the city inventory is the Gowanus Expressway, 3.52 miles (5634 m), 1,745,606 ft² (162,385 m²), 322 spans. Figures E1.1–E1.11 illustrate the close proximity of different structural types and services.

Table E1.1 Long-Span Bridges in New York City

Year	Bridge	Maximum Span Length	Structure	Service	Owner[a]
1883	Brooklyn (J. Roebling)	487 m[b] main; 284 m side	Suspension/stay	6 lanes	NYC
1888	Washington (Schneider/McAlpine/Hutton)	155 m	Multisteel arch	6 lanes	NYC
1903	Williamsburg (L. L. Buck)	488 m[b] main	Suspension	8 lanes, 2 tracks	NYC MTA
1908	Manhattan (R. Mojeski/L. Moisseiff)	449 m main; 222 m side	Suspension	7 lanes, 4 tracks	NYC MTA
1912	Queensboro (G. Lindenthal)	361 m main; 1136 m total	Cantilever truss	10 lanes	NYC
1912	Hell Gate (G. Lindenthal)	310 m main	Steel arch	4 tracks	AMTRAK
1931	Bayonne (O. Ammann)	511 m[b] main	Steel arch	8 lanes	PA NY&NJ
1931	George Washington (O. Ammann)	1067 m[b] main	Suspension	14 lanes	PA NY&NJ
1936	Triboro (O. Ammann)	421 m main	Suspension	7 lanes	MTA
1936	Henry Hudson (D. Steinman)	244 m main	Fixed solid arch	6 lanes	MTA
1937	Whitestone (O. Ammann)	701.5 m main	Suspension	6 lanes	MTA
1961	Throg's Neck (O. Ammann)	549 m main	Suspension	6 lanes	MTA
1964	Verrazano (O. Ammann)	1300 m[b] main	Suspension	12 lanes	MTA

[a] NYC, New York City; MTA, Metropolitan Transit Authority; PA NY&NJ, Port Authority of New York and New Jersey.
[b] World record.

19

Figure E1.1 Brooklyn and Manhattan Bridges, East River.

Figure E1.2 Williamsburg Bridge, East River.

Figure E1.3 Queensboro Bridge, East River.

Figure E1.4 Triboro and Hell Gate Bridges, Harlem River.

Figure E1.5   Bayonne Bridge.

Figure E1.6   George Washington Bridge.

Figure E1.7 Henry Hudson Bridge.

Figure E1.8 Bronx–Whitestone and Throg's Neck Bridges.

Figure E1.9 Verrazano Bridge.

Figure E1.10   Alexander Hamilton and Washington Bridges, Harlem River.

Figure E1.11   The 9th Street lift bridge and elevated subway line across Gowanus Canal.

Roebling's advice to engineers was to become engaged in manufacturing in order to gain better pay than engineering offered (Billington, 1983). A believer in the power of his will, Roebling felt qualified to treat his leg injury sustained at the Brooklyn Bridge construction site (McCullough, 1972) and dismissed the doctors. That managerial failure to delegate responsibility proved fatal.

Colonel Washington Roebling (1837–1926) graduated as an engineer from Rensselaer Polytechnic Institute and fought in the Civil War under the direct orders of General Ulysses S. Grant. To test the safety in the Brooklyn Bridge tower caisson, he fired his pistol in the chamber with no one present. Crippled by the work but assisted by his wife

Emily, he remained in charge of the project until its successful completion in 1883 (Fig. 1.12), despite efforts of both Manhattan and Brooklyn mayors to depose him (McCullough, 1972).

A graduate of Ecole Centrale des Arts et Manufactures in Paris, Gustave Eiffel (1832–1923) (Fig. 1.13) demonstrated mastery of steel design with the railroad arch bridges over the Douro River at Opporto (1876) and over the Truyere River at Garabit (1884). At 400 ft (122 m) the latter was the tallest arch in the world. The subsequent design and construction of the tallest tower in the world however proved Eiffel not only a brilliant designer but a masterful manager.

Harriss (1975) refers to the "production line" efficiency of the construction under Eiffel's supervision. The selection and design of the tower's elevators (a matter exceeding Eiffel's technical qualifications) are an instructive display of managerial skills. The process included sometimes tenuous but ultimately productive negotiations with a fellow giant of engineering accomplishment, the American elevator manufacturer Otis. A represen-

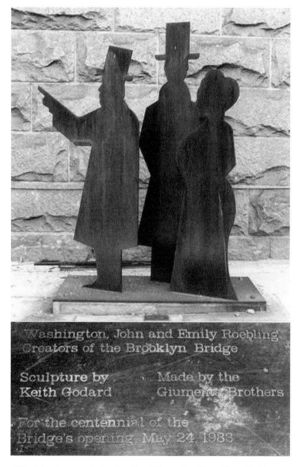

Figure 1.12 John, Washington, and Emily Roebling, the Brooklyn Bridge.

Figure 1.13 Gustave Eiffel, the Eiffel Tower, Paris.

tative of the preceding generation, Elisha Graves Otis (1811–1861) had made high-rise structures possible by inventing construction hoists in 1854 and the elevator in 1857.

Francois Hennebique (1843–1921) developed his ideas and field experience in reinforced concrete into thousands of structures and an international business (Billington, 1983).

The father of prestressed concrete, Eugene Freyssinet (1879–1962) wrote (1993, p. 30) that on the construction sites of his *ouvrages d'art* he was supervisor, contractor, and designer, in his own words "absolute master who received no orders and no advice from anyone."

Gustav Lindenthal (1850–1935) designed the landmark Hell Gate Bridge and supervised the design of the Queensboro Bridge in New York City (Example 1), as well as many others, with a strong preference for eye-bars (Fig. 1.14). He served as bridge commissioner during the greatest expansion of New York City at the turn of the twentieth century. Both O. Ammann and D. Steinman apprenticed under Lindenthal.

Othmar Ammann (1879–1966), (Fig. 1.15), chief engineer of the New York and New Jersey Port Authority, designed the Triboro, George Washington, Bayonne, Whitestone, Throg's Neck, and Verrazano Bridges in New York City (Example 1). Ammann investigated the collapse of the Quebec bridge in 1907 and was consulted on many others, including the Golden Gate in San Francisco (with Chief Engineer Joseph B. Strauss).

Despite his outstanding accomplishments in design, Ammann was considered a "political entrepreneur," "infinitely abler as an executive" than his mentor, the mercurial G. Lindenthal (Tobin, 2001, Chapter 6). Ammann expressed his priorities in bridge design

Figure 1.14 The Smithfield Bridge, Pittsburgh, PA.

Figure 1.15 Othmar Ammann, Port Authority, George Washington Bridge, New York City.

as "economics and utility, but also aesthetics" (Tobin, 2001, pp. 188–225). His design for the future George Washington Bridge won over Lindenthal's because of its (surprisingly low) bid of U.S. $25 million. In 1931 the George Washington span was the first to exceed 1 km with eight lanes of vehicular traffic. On the opening of the lower bridge deck in 1962, *The New York Times* wrote (Reier, 1977, p. 107): "(Ammann) was the dreamer, he was the artist, he was the solid and reliable planner who made this beautiful structure possible and durable."

As the Brooklyn Bridge and the Eiffel Tower, Ammann's spans, still admired for their record size, have accumulated an impressive service record. Forty years after opening to traffic, the Verrazano remains the longest span in the United States. Seventy-four years after opening to traffic, the George Washington is the busiest of the long spans, with daily toll collection of approximately U.S. $1 million (Example 2). Ammann's bridges (even the less successful ones) demonstrate that sound management is a feature of the original structural design.

Example 2. George Washington Bridge, New York City

The George Washington Bridge (Figs. E2.1 and E1.6) was built across the Hudson River after a traffic study dating from 1925 foresaw that a U.S. 50-cent car toll would generate annual revenue of U.S. $5.25 million by 1932 (Reier, 1977). The bridge was designed by Othmar Ammann, chief engineer of the New York and New Jersey Port Authority with a top deck able to accommodate 8 traffic lanes and a lower deck that could carry either trains or 6 traffic lanes. Construction began on October 21, 1927. Four lanes were opened to traffic in October 1931, ahead of schedule and, at a construction cost of U.S. $59 million, under budget. Four more lanes were opened to traffic in 1943. The lower deck was completed with 6 lanes in 1963 at a cost of U.S. $76 million, nearly doubling the traffic capacity and providing a welcome stiffening against torsion. Capital rehabilitation was conducted in 1992–1993 at a cost of U.S. $36.2 million.

The 3500-ft (1067-m) 14-lane double-deck main span held the world record for length until 1937, when it was surpassed by the 4200-ft (1280-m) span of the Golden Gate carrying 6 traffic lanes between San Francisco and Marin County (Figs. 4.41 and 4.42). In 1964 Ammann pushed the record to 4260 ft (1300 m) with the 12-lane double-deck span of the Verrazano Bridge between Brooklyn and Staten Island (Fig. E1.9). The Humber Bridge in Wales extended the record to 4623 ft (1410 m). The Akashi-Kaikyo Bridge between the islands of Honsu and Shikoku, Japan (Fig. 1.39), holds the current record with a 6-lane main span of 6528 ft (1991 m).

Akashi-Kaikyo, George Washington, Verrazano, and Golden Gate share a design relying on stiffening trusses (Figs. 1.40*a* and 1.42). In contrast, the Humber, the

Figure E2.1   George Washington Bridge, Hudson River. Painting in progress at West Tower.

Figure E2.2   Box girder of Great Belt Bridge.

Figure E2.3   Tsing Ma Bridge, Hong Kong.

Table E2.1 Tolls, Annual User Counts, and Revenues at George Washington Bridge

Year	Annual East-Bound Vehicles	Average Toll (U.S.$)	Annual Revenue (million U.S.$)
1932	10,500,000	0.50	5.25
1991	47,952,700	4.30	207.78
1992	47,764,900	4.70	223.76

Great Belt in Denmark (5325 ft, 1624 m) (Figs. 1.41 and E2.2), Tsing Ma in Hong Kong (4515 ft, 1377 m) (Fig. E2.3) and many other modern suspension spans are box girders.

Tolls and user numbers evolved at the George Washington Bridge, as shown in Table E2.1.

A comprehensive list would have to include other countries and cultures. Much inspiration and expertise have been imported from the shipbuilding, automotive, and aircraft industries. The complexities of great structures increasingly force engineers to focus their competitive efforts on one of the many levels of design and management. Van Der Zee (1986) described the tenuous collaboration of L. Moisseiff, the inspired artist, Charles A. Ellis (1876–1949), the meticulous designer, and Joseph B. Strauss (1870–1938), the ambitious manager, on the creation of the Golden Gate Bridge (Fig. 1.16).

Figure 1.16 Joseph Strauss, the Golden Gate Bridge, San Francisco.

Competence and dedication have often brought less than the desired success at higher than the anticipated price. Individual talents consistently adjust to the diverse demands for and of transportation. Bridge managers remain self-made, at their own risk.

1.4   ENGINEERS AND MANAGERS

A competitive distinction between engineers and managers is implied in the Grecian myth of the judgment of Paris at the beauty contest between Hera, queen of Olympus (Fig. 1.17), Athena, beloved brainchild of Zeus (Fig. 1.18), and Aphrodite, goddess of love

Figure 1.17   Hera (Juno, Roman) marble, second century A.D., the Louvre, Paris.

Figure 1.18   Athena Pallas, marble, first century A.D., the Louvre, Paris.

(Fig. 1.19). Provoked by Eris, goddess of discord, the contestants entice the arbiter with mutually exclusive offers of political power, sagacity in war and peace, and beauty, respectively. Paris, banished prince of Troy (Fig. 1.20), modestly opts for the last, causing the outbreak of the Trojan War.

According to Graves (1992, p. 631) the ensuing bloodshed was instigated by Zeus, king of Olympus, and Themis, goddess of order and vengeance, "for reasons that shall remain obscure." During the 10-year siege, Hera and Athena join forces against Aphrodite and Apollo, patron of the arts. Homer's (Fig. 1.21) *Iliad* and *Odyssey* depict the events as a confrontation between relentlessly inventive aggression and conservative traditional methods (e.g., the "proactive" vs. the "reactive" approach).

Historic evidence corroborates the myth. Leaders equally successful in war and peace have been even rarer than engineers rising to the top in management. It can be speculated that Alexander the Great (356–323 B.C.) (Fig. 1.22) was moved to tears (and possibly to an early death) by the prospect of managing the world he had handily conquered.

Attempts to integrate politics, management, and science have proven dangerously unrewarding for the perpetrators. Aristotle (*Politics,* II 8, 1267a) related that in 433 B.C. Pericles (495–429 B.C.) appointed Hippodamus of Miletus to lay out Piraeus. Not content with merely planning a town for 10,000 inhabitants, the latter also wrote its constitution and regulated the judicial system. This eccentricity, according to Armytage (1976), drove the overreaching urbanist into exile.

Plutarch (ca. A.D. 150–125) concluded that, although Marcus Licinius Crassus (115–53 B.C.), the parsimonious developer, defeated Spartacus, and Cneius Pompey (106–46 B.C.), the popular conqueror of Spain, judiciously arbitrated peaceful disputes, neither could match the thorough competence of Caius Julius Caesar (102–44 B.C.), the relentlessly ambitious all-purpose manager. Caesar's exceptional versatility is reflected in the diverse artistic portrayals inspired by his biography over the centuries (Figs. 1.23 and 1.24).

On the day he was to be declared emperor, Caesar was slain in the Roman Senate, one more victim of political backstabbing. His grandnephew Octavian Augustus (63 B.C.–A.D. 14) (Fig. 1.25) cultivated the image of "first citizen" and became the first Roman emperor. During his 41-year reign, Augustus built the administration and the infrastructure. He took pride in transforming Rome from a city of bricks to one of marble.

Occupational hazards have been somewhat different for engineers and managers. Machiavelli's constructive writings on management earned him jail and torture (but not the outright assassination awaiting his dangerous bosses, the Borgias). In order to survive, Galilei had to renounce his conclusions regarding the universal order, a matter deemed beyond his competence. Copernicus and Descartes (the former foreseeing Galilei's fate, the latter learning from it) published their more controversial scientific conclusions posthumously. Sir Isaac Newton sat uneventfully in the House of Lords but reportedly objected only once, to a draughty window.

Antoine Laurent Lavoisier (1743–1794) entered l'Academie Francaise at 25 for his contributions to chemistry and soon became its director for his management skills. For his efforts in economics, particularly as tax collector and commissioner of the treasury, he was guillotined during the Terror. His plea for a stay of execution that would have allowed him to complete essential research was rejected by the brilliant military engineer Lazare Carnot (1753–1823), war minister under the French Revolution, later exiled.

Figure 1.19 Aphrodite, holding the apple of discord, marble, first century A.D., the Louvre, Paris.

Figure 1.20   Paris, marble, A. Canova (1757–1822), the Metropolitan Museum of Art, New York City.

Following the Panama Canal scandal, Ferdinand de Lesseps was spared jail due to advanced age, but his son Charles served a sentence. Gustave Eiffel avoided a similar fate on a technicality. John and Washington Roebling sacrificed life, limb, and fortune at the Brooklyn Bridge. Strauss, no stranger to nervous and financial breakdown, died within a year of the Golden Gate Bridge opening.

The evidence of precedents and the growing complexity of technical expertise are discouraging to generalists. The successful record of the few who have survived the test of time appears to be edited accordingly. Benjamin Franklin's (1706–1790) contributions

Figure 1.21 Imaginary head of Homer, marble, second century A.D., the Louvre, Paris.

to science, for instance, would seem to detract rather than add to his stature as a founding father of the United States. They are omitted from the otherwise detailed inscription under his monument at Park Row in New York City (Fig. 1.26). Franklin's historic image, perhaps more complex than Caesar's, alternates between the public figure (Fig. 1.27), the statesman (Fig. 1.28), and the savant, but rarely the empiricist.

Thomas Paine (1737–1809) (Fig. 1.29) promoted the design and construction of iron bridges in the young United States and in England, but he is known as the author of *The Age of Reason, The Rights of Man,* and *Common Sense.*

Thomas Jefferson (1743–1826) earned his place on the U.S. 5-cent coin by writing the Declaration of Independence and by serving as the nation's third president. The

Figure 1.22 Alexander the Great, marble, Roman, second century A.D., the Louvre, Paris.

reference to the design of Monticello on the tail side of the nickel can be considered as coincidental. Reissues of the coin clearly seek to supplant it with worthier accomplishments, such as the Louisiana Purchase, the expedition of Lewis and Clark, and the exploration of the Western Territories (Fig. 1.30a). At Columbia University (Fig. 1.30b), Jefferson stands much more comfortably in front of the School of Journalism than he might have in front of the School of Architecture.

In contrast, Herbert Hoover (1874–1964), the thirty-first president of the United States, is never denied due credit for his accomplishments as a mining engineer and as

Figure 1.23   Julius Caesar, marble, first century A.D., the Louvre, Paris.

an international manager of "the first massive foreign aid operation in history" (Drucker, 1973, p. 25).

It is hard to perceive simultaneously the scientific significance of Thomas Edison's (1847–1931) (Fig. 1.31) inventions, the number of his patents, and the economic importance of his companies. The legacy of Edison's contemporary and friend, Henry Ford (1863–1947), is similar. He designed some of the first automobiles and, more importantly, their mass production; however, most biographies focus on his legacy as a major em-

Figure 1.24   Julius Caesar, marble, A. Ferrucci (1465–1526), the Metropolitan Museum of Art, New York City.

ployer and industrialist. The name of Eleuthere Irenee Du Pont (1771–1834), pupil of Lavoisier and leading early American chemical engineer (Armytage, 1976), is associated with the industrial complex he launched.

Howard Hughes (1905–1976), a one-time Caltech student, daring designer, and manager, is known mainly for personal idiosyncrasies, adventures in the film industry, and unorthodox financial practices. Jay (1994, p. 37) compared the ouster of Hughes from the management of TWA, the airline company he had created, to the execution of King Charles I of England (1600–1649). The author saw the industrial shift from family ownerships to professional management as analogous to the earlier political transition from personalized dynasties to professional meritocracies. Phillips (2004), in contrast, found

Figure 1.25   Octavian Augustus, marble, Roman, 20 B.C., the Louvre, Paris.

examples of hereditary dynasties still competing successfully against intellectual elitism in politics and economics at the turn of the twenty-first century.

Even if engineering and management share long-term goals, their immediate priorities can be directly opposed. Engineers cultivate and exploit natural resources; managers supervise the transactions of resources between social groups. Engineering pursues enduring results by equally enduring and therefore objective and reproducible reasoning. Management satisfies the urgent and transient priorities of the moment by the most effective and preferably unique means. Plutarch (p. 879) believed Caesar gifted "above all men with the faculty of making the right use of everything in war, and most especially of seizing the right moment."

Berlin (1996, pp. 40–53) dismissed the *social engineers,* who attempt to resolve the mysteries of government by reducing them to social statics and dynamics, as well as the

Figure 1.26   Benjamin Franklin, inscription, Park Row, New York City.

messianic preachers, such as Saint-Simon (1760–1825), Fourier (1772–1837), and Comte (Fig. 1.9). In his view, scientists seek the universally valid aspects of specific problems, whereas politicians must grasp the uniqueness of common situations. The latter imperative is shared by original architects, such as F. L. Wright (1869–1959), whose stated goal was to create structures uniquely suited for the specific environment and functions. In matters of law O. W. Holmes, Jr. similarly insisted that experience rather than logic must guide (Appendix 1). Engineering, on the other hand, is not satisfied until logic and experience are reconciled. That is not always possible. Despite the rigorous engineering methods and well-meaning intentions of Hippodamus, Aristotle (*Ethics,* 1941, 1268b) criticized the latter's attempt to divide the population into the classes of artists, "husbandmen," and warriors, concluding: "There is surely a great confusion in all of this."

Twenty-one hundred years later Marechal Vauban ruined his illustrious career of a soldier and a builder by venturing beyond his presumed competence and proposing a tax reform to his patron, Louis XIV.

Figure 1.27 Benjamin Franklin, Park Row, New York City.

Figure 1.28 Benjamin Franklin and George Washington, terracotta, Houdon, the Louvre, Paris.

Berlin distinguished between the theoretical and pragmatic (or objectivist and subjectivist) knowledge of successful scientists and politicians. He argued that overreliance on analytic principles is fatal in political judgment (and vice versa) (1996):

> Utopianism, lack of realism, bad judgment here consist not in failing to apply the methods of natural science, but, on the contrary, in over-applying them. . . . What is rational in a scientist is therefore often Utopian in a historian or a politician (that is, it systematically fails to obtain the desired result), and vice versa. . . . Should scientists be put in authority, as Plato or Saint-Simon or H. G. Wells wanted? . . . Most of the suspicion of intellectuals in politics springs from the belief, not entirely false, that, owing to a strong desire to see life in some simple, symmetrical fashion, they put too much faith in the beneficient results of applying directly to life conclusions obtained by operations in some theoretical sphere.

The public debate between "political pragmatists" and "social engineers" remains lively (and sometimes deadly). Antoine Caritat, marquis de Condorcet (1743–1794), and Pierre-Simon, marquis de Laplace (1749–1827), updated the efforts of Hippodamus to regulate the judicial system by applying statistical analysis to court decisions. Both were designated "immortals," i.e., members of the French Academy. Condorcet (Fig. 1.32), mathematician, philosopher, economist, president of the Legislative Assembly, and believer in the ultimate perfection of society, poisoned himself in order to avoid the guillotine during the Terror. Laplace confined his efforts to mathematics, physics, and astronomy and died of natural causes. In 2005 the voters of France rejected the European Constitution proposed by Valéry Giscard d'Estaign, graduate of École Politechnique and former president of the Republic, immortalized in 2004.

Figure 1.29 Thomas Paine, Hall of Fame, the Bronx, New York City.

1.5 DEMAND AND SUPPLY

Once engineering had evolved into a profession, its practitioners became a resource that management must allocate cost-effectively. With commerce gradually replacing warfare as the principal means of transferring wealth between nations, the economical constraints imposed on engineering assume the role of the political ones. Commerce is "so like war," stated Jay (1994, p. 201), "that every stage of military history has its parallel in industrial and commercial warfare."

The supply–demand flowchart of Fig. 1.33a is bilateral. Engineering delivers specialized services, management provides resources. Suspension bridge patent holder John Finley (Section 1.3) may have dealt with his clients in this manner. A more complex chain of supply and demand develops between the public and the political management it elects and subsidizes. In the past, the great artists of bridge design, for instance those analyzed by Billington (1983), found ways to interact directly with the public and with

(a)

(b)

Figure 1.30 (a) The "nickel," bearing the profile of Thomas Jefferson on the head side and alternative designs for the tail side. (b) Thomas Jefferson, School of Journalism, Columbia University, New York.

Figure 1.31   Thomas A. Edison, Hall of Fame, the Bronx, New York City.

the highest levels of political management. Such an interaction is becoming increasingly difficult. Yao and Roesset (2001) noted that "an increasing number of intermediate layers have been established between the engineers and the decision-makers." Fig. 1.33*b* shows different management levels communicating with the users and the technical experts and between each other.

Whereas human activities diversify, their best products, such as structural master-pieces, integrate the results. No engineering achievements ever gain significance without becoming managerial feats. The duke of Wellington (1769–1852) reportedly dismissed the steam engine, designed by George Stephenson (1781–1848), as a "means for the lower classes to wander aimlessly about." The demand for that means allowed Stephen-son's son Robert (1803–1859) to build the monumental Britannia Bridge that survived until 1970 (Billington, 1983).

To the extent the two professions (or their historic precursors) are distinct, managers supply to the public but demand from the technical support staff that might include engineers. When building is elevated to an art, the master-builders are personal favorites of the leaders.

Figure 1.32 Condorcet, Quai de Conti, Paris.

The demands shaping the pyramids of Egypt, the Great Wall of China (Fig. 1.34), and the Roman arches (Figs. 1.35 and 1.36) were both utilitarian and spiritual. A victory over time was a primary design requirement. Supplying maximum construction effort was the proof of excellence. Having outlived their utility by millennia, these ancient structures still set the human standard for monumentality.

Transportation would not advance very far on monumental and record-setting spans alone. The many structures using timber, rope, and other perishable materials satisfy equally valid management considerations. A rope suspension bridge, a stone slab, or a tree trunk may have been the best or the only options that could have bridged the gap to be crossed. Despite the less permanent nature of some of these structures, they too may be masterpieces of engineering and management in terms of cost and service.

By the twentieth century social and transportation dynamics had turned timeless bridges into a luxury. John Roebling's bridges demonstrate the difficulty of surviving as a geographic landmark in a competitive market. The magnificent crossings at Niagara Falls and at Pittsburgh have vanished along with their serviceability. The Cincinnati–Covington Bridge (Fig. 1.37a) survives as an international landmark and has been strengthened to carry two lanes of traffic.

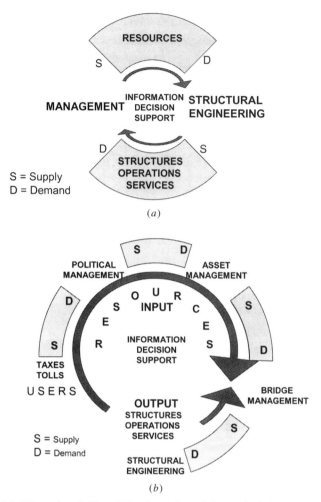

Figure 1.33   (*a*) Bilateral and (*b*) multilevel supply-and-demand relationships.

The landmark suspension bridge across the Danube at Budapest was partially de-molished during World War II and subsequently restored to meet both symbolic and serviceability demands (Fig. 1.37*b*).

One hundred and twenty-three years after the opening of the Brooklyn Bridge (Ex-amples 1 and 3) it is impossible to separate its monumental stature from its utility (up to 100,000 passengers use the structure for daily commutes between Manhattan and Brooklyn). That combination appears to have been achieved by design. Reier (1977, p. 11) quoted from Roebling's letter to the New York Bridge Company (1867) as follows: "The contemplated work, when constructed in accordance with my design, will not only be the greatest bridge in existence, but it will be the greatest engineering work of this continent, and of the age. Its most conspicuous features, the towers, will serve as land-

Figure 1.34 The Great Wall of China.

marks to the adjoining cities, and they will be entitled to be ranked as national monuments."

Eiffel promoted his controversial tower project with very similar arguments: "Besides its soul-inspiring aspects, the Tower will have varied applications for our defense as well as in the domain of science" (Harriss, 1975, p. 101).

The implied research value was in the nascent field of aeronautics. A century later, the Eiffel Tower (Fig. 1.38) is a tourist attraction, a television antenna, a national symbol, and the most profitable enterprise in Paris.

In contrast, Eiffel's first railroad bridge in Bordeaux is scheduled for demolition in 2008, despite its recognized historic significance. *Le Monde* (February 2, 2006, p. 26) quoted a representative of the French railroads (*RFF*) as follows: "*RFF* has no calling to preserve a structure on which trains no longer travel."

Figure 1.35   The Colosseum, Rome, begun in A.D. 72.

Figure 1.36   Ponte Rotto, Rome, collapsed in 1598.

(a)

(b)

Figure 1.37 (a) Cincinnati–Covington Bridge. (b) The Danube Bridge at Budapest.

Figure 1.38 The Eiffel Tower, Paris.

Figure 1.39 Mid-Hudson railroad bridge.

Attempts to preserve the bridge for pedestrians and cyclists are encountering the usual financial difficulties. Whereas the new railroad bridge is estimated to cost EU 117 million, the costs and benefits associated with the conservation and future use of the old one appear less clear. (See also Section 11.5.)

The Mid-Hudson railroad bridge in New York State (Fig. 1.39) presents a similar dilemma. Completed in 1888, the bridge has outlasted train service as a pedestrian crossing; however, that service is unable to fund the costly maintenance needs.

The "tension between business and structural art" (Billington, 1983), both stimulating and stifling, has taken on the contemporary form of interaction between management and engineering. To that tension must be added the one between individualistic design and popular consensus. Engineers specializing in design and construction champion a "supply-side" bridge management, emphasizing innovative projects. That position was expressed by N. Olsson in *Civil Engineering* (April 1993, pp. 57–59) regarding a proposed replacement of the Williamsburg Bridge (Example 3).

Example 3. Williamsburg Bridge, New York City

The Williamsburg Bridge (Figs. E1.2, E3.1, and E3.2) opened to traffic between Brooklyn and Manhattan on December 19, 1903. The chief engineer was Leffert L. Buck. G. Lindenthal was the bridge commissioner. Suspension bridges were still finding their stride. Measuring 1600 ft (488 m), the main span was the world's longest suspended structure; however, the cantilever truss at the Firth of Forth was longer. The side spans are not suspended. The bridge originally carried six train tracks, four automobile and carriage lanes, and two pedestrian walkways. In 1924

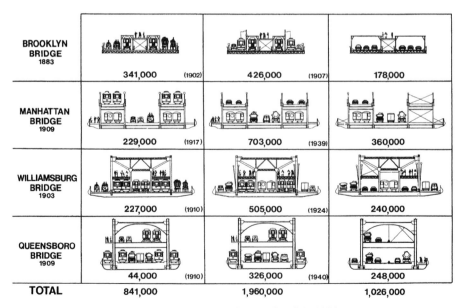

Figure E3.1 East River bridges, New York City, NYC DOT (1988).

Figure E3.2 Williamsburg Bridge, East River.

average daily users peaked at 505,000. In the post–World War II years 4 train tracks were replaced by roadways. As the automobile became the principal mode of transportation, train tracks were eliminated from the bridges and the number of daily users was reduced by half (see Fig. E3.1), with trucks transporting goods.

Each of the four suspension cables contains 7696 wires grouped into 37 strands. The high-strength parallel wires have typical diameter (0.189 in., 5 mm) and yield strength (220 ksi, 1517 mPa). Atypically, they are not galvanized. It was assumed that wires must be adequately waterproofed and consequently did not need the protection of the zinc coating. Another noteworthy feature is the cantilevered floor beams supporting two traffic lanes (Fig. E3.3).

NCHRP Synthesis 330 (2004, p. 3) illustrated the importance of bridge maintenance with the following reference: "Neglect of maintenance can have dramatic consequences. A routine inspection of New York's now notorious Williamsburg bridge in 1988 discovered extensive deterioration of the steel girders—altogether some 400 areas where structural conditions required immediate attention. The bridge was closed first to mass transit trains and then to all traffic for 3 months as emergency repairs were made, making news headlines and extensively disrupting the city's commerce."

The importance of the Williamsburg Bridge commanded public and media attention. On February 24, 1988, R. Levine wrote in *The New York Times* (p. E24):

Figure E3.3 Corroded cantilever floor beam-to-truss connection.

Officials say that about half the city's (846) bridges are rated in poor or fair condition, 31 have had to be at least partly closed off and 100 need to be replaced. . . . For decades bridges have been subject to the vicissitudes of tight city budgets, a status that has shortened the structures' lifespans and caused many to age prematurely.

The eye bars anchoring the cables of the Manhattan Bridge [ ] are deteriorating [ ]. The steel stringers on the Queensboro Bridge are corroded. So are the steel columns of the Queens Boulevard Bridge. The concrete deck and supporting floor beams have deteriorated on the Washington Bridge over Harlem River. Cracks mar the grating of the Madison Avenue Bridge. Last spring a gap the size of a Buick appeared in the pavement of a lane of the Pulaski Bridge.

Depending on the results of an inspection effort that began nearly a year ago, the cables that hold up the Williamsburg Bridge may have to be repaired or replaced—a task that could cost $250 million or even more. . . .

[Transportation Commissioner Ross] Sandler said that over the next 10 years his department is planning to spend $1.37 billion, including Federal and state funds. Still, he insists he could use $600 million more, not counting money for regular maintenance.

Although the city has hired an additional 30 bridge inspectors, it has spent only about $6 million annually on maintenance in recent years. . . . The amount is about a sixth of what officials say is needed.

In contrast, the Triborough Bridge and Tunnel Authority—which, not coincidentally charges tolls—spent $20 million in 1985 to maintain its seven bridges and two tunnels.

To fix the bridges, the bureaucracy must be also changed. . . . The remedy, [Commissioner Sandler] maintained, is the creation of a bureau, department or authority solely for bridges. . . . Since the collapse of Connecticut's Mianus River Bridge in 1983 and the Schoharie Creek Bridge in upstate New York last year, which together claimed 13 lives, no one can accuse him of playing Chicken Little.

The bridge commission responsible for the construction of the East River bridges had been absorbed by other agencies decades earlier. The long spans built between the 1920s and 1960s were managed by authorities and funded by tolls. A bureau of bridges was established at the NYC DOT later in 1988. The function of the bureau is illustrated in Appendix 18.

The temporary closure was recommended after the regular biennial inspection found advanced deterioration in the cantilever floor beams (Fig. E3.3) and the suspension cables (Fig. E3.4) (Williamsburg Bridge Technical Advisory Committee,

Figure E3.4   Broken suspension cable wires.

1988). Analysis of the cables concluded that the cables' "safety factor" had declined from 4.1 to 2.3.

The conclusion was based on two key assumptions:

- The number of broken wires at any point along the length of the cable other than the anchorages can be extrapolated from the number of broken wires observed on the surface of the cable.

- The capacity of the cable is reduced proportionally to the number of *all* broken wires.

An alternative model obtained a reduced safety factor of 3.6 as a result of the following differences with the former study:

- Broken wires are counted throughout the cable cross section. Access is gained by driving wedges into the cable (Fig. E3.5) at representative locations.

- Broken wires are assumed to regain their functionality through friction over a length approximately equal to the distance between two to three cable bands (40–60 ft, 12–18 m).

The former study recommended replacement of the cables and consequently the bridge. A number of qualified consultants proposed cable-stayed and suspension bridge designs for a replacement in the same or different alignments. Three conceptual designs were recognized for their excellence (Figs. E3.6a,b,c).

The influence of the Brooklyn Bridge is manifest in the hybrid cable-stayed and suspension bridge, proposed by J. Schlaich and R. Walther, as well as in the towers of the suspension bridge proposed by Steimman, Boynton, Gronquist, and Birdsall.

Figure E3.5   Suspension cable inspection.

(a)

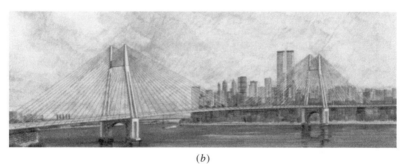

(b)

(c)

Figure E3.6   Replacement concepts; (a) J. Schlaich and R. Walther; (b) T. Y. Lin and N. H. Bettigole; (c) Steinman, Boynton, Gronquist, and Birdsall.

Nonetheless, the latter engineers recommended rehabilitation of the existing bridge. Rehabilitation was selected for two main reasons:

- The cost of rehabilitation was estimated at roughly U.S. $400 million, compared to U.S. $950 million for a new bridge (1988).
- Constructing a new bridge in the same alignment implied a traffic interruption of three to four years.

The decision to rehabilitate the Williamsburg Bridge was intensely debated by the entire community and in professional circles. Against the rehabilitation, it was argued that:

- Substandard features of the existing bridge, such as traffic lane width, cannot be corrected.
- The useful life of the rehabilitated bridge cannot be estimated reliably.

Olsson (*Civil Engineering,* April 1993, p. 59) wrote: "Anyone with any knowledge of this poorly treated workhorse understands that to renovate instead of rebuild is impossible. The main bridge, excluding approaches, is in such an advanced state of decay that there is not a healthy member to which a new element can be attached. . . . The bridge has terminal cancer; the only fix is to rebuild."

Ultimately, only the approaches were entirely rebuilt. Once it was determined that the main cables could be saved, rehabilitation became feasible. To the arguments in its favor, the following can be added retrospectively:

- The user costs caused by the traffic interruption would have been destructive to the community. The effect would have been exacerbated by the concurrent rehabilitation of the Manhattan Bridge, providing the only alternate truck route across the East River (Fig. E1.1).
- Actual cost overruns should not be compared with the initial cost estimate of a new bridge construction, because the latter would have been subject to changes as well.
- The maintenance needs of the existing bridge are better known than would have been those of a new structure.
- A 75- to 100-year life extension of the existing bridge permits a future replacement by a structure that cannot be anticipated at present.

Robison (1988, pp. 75–78) summarized the "rehab after all" option as follows: "Construction costs, estimated in 1987 dollars, would be about $378 million. Escalation at 6%, however, could boost the total to more than $600 million by 1999. Basic maintenance costs were estimated as $150 million for the entire 21st century. In 100 years, there would be one deck replacement at $14 million and 12 painting cycles at $65 million on the suspension spans. The approaches would need two new decks for $31 million and 12 paint jobs for $40 million."

The problems to be resolved at the Williamsburg were typical of bridge management and, also typically, exceeded the scope of engineering solutions. The numerical data resulting from technical investigations were critical but supported

Table E3.1 Rehabilitation Cost Estimates for Selected Years for East River Bridges, New York City (million U.S.$)

	1990	1996	2000	2004	1990 ($\times 1.04^{15}$)
Brooklyn	231.32	321.29	351.26	464.07	416.60
Manhattan	316.20	611.30	702.20	788.70	569.46
Williamsburg	398.53	697.21	748.51	989.56	717.73
Queensboro	337.60	447.70	516.40	741.02	608.00

Table E3.2 Rehabilitation Contracts and Costs at Williamsburg Bridge, 1990–2006 (NYC DOT Annual Bridge Report, 2004)

Rehabilitation Items	Estimated Cost (millions U.S.$)[a]
Replace main span outer roadway (1983)	11.20
Replace one-third of suspenders (1984)	3.20
Repair pier 20E foundation and replace bulkhead (1986)	2.30
Paint side spans and towers (1985)	1.10
Paint main and approach spans (1989)	4.24
Emergency interim repairs (1989)	10.00
Install temporary hand-rope system on main cables (1990)	0.63
Main cable preservation (field test, oiling) (1991)	0.44
Main cable strand splicing at Manhattan anchorage (1991)	0.29
Interim pedestrian walkway (1994)	1.05
Component repairs of flag conditions on the north outer roadway and north inner roadway (1994)	4.12
Rehabilitate main cables and new redundant suspender system (1996)	88.30
Demolish existing building under approaches (1993)	1.50
Testing program for bored-in piles (1993)	0.74
Demolish DOS and DOH buildings, replace entire south outer roadway approach structures, rehabilitate south outer roadway deck and south inner roadway deck of main bridge, and replace south inner roadway substructure of approaches (1998)	198.00
Portion of Contract #6 BMT track structure work transferred to Contract #5 south approach roadway reconstruction work	65.00
Paint main and intermediate towers (2001)	14.90[b]
Reconstruct BMT Subway structure; install new signals, tracks, and communication system (2000)	166.65
Miscellaneous rehabilitation work; rehabilitation of towers, replace bearings, travelers, architectural work, painting of north and south trusses, suspender adjustment, tower jacking, construction of colonnades	172.90[c]
Replace north approach structures (Manhattan/Brooklyn) and rehabilitate north half of bridge (2002)	233.00
Seismic retrofit	10.00[d]
Total	989.56

[a] Construction complete except as noted.

[b] Painting suspended in 1996 pending publication of Environmental Impact Statement (EIS) in 1998. Painting resumed under a new schedule in 1999 and was completed in 2001.

[c] In construction.

[d] In design.

Figure E3.7   Installation of new floor beams and orthotropic deck.

decisions only to a point. The investigation combined project and network considerations on a citywide and, owing to the federal involvement, a national level. The final choice could not be weighed quantitatively against the rejected alternatives. As in the case of the San Francisco–Oakland East Bay Bridge, regional and urban considerations were decisive.

Work on the project has taken 16 years (as planned), at a cost of U.S. $1.0 billion, compared to the 1988 estimate of U.S. $378 million. At a 6% inflation rate, the $378 million of 1988 would grow to $1.018 billion by 2005. Table E3.1 illustrates the evolution of the costs from estimates in 1991 to actual in 2004.

The costs cited in Table E3.1 have escalated over the reported period for a number of reasons, most important among which are the following:

- Over a reconstruction period of 15–20 years new items are likely to be added. As an example, the scope of rehabilitation work at the Williamsburg Bridge is listed by contracts in Table E3.2. The first item, the replacement of the outer roadway, had to be repeated, because the relatively new floor beams did not meet the alignment of the orthotropic deck installed 17 years later (Fig. E3.7).

- Traffic must be maintained during reconstructions as much as possible.

- Costs must be adjusted for inflation over the period, as illustrated in the last column of Table E3.1.

Table E3.2 lists the rehabilitation items and their respective costs. Rehabilitation of major bridges proved costlier and more complex than new construction. Assessment and redesign are new types of expertise to be mastered by the profession. In contrast to new construction, they are inevitable, because bridges, once built, become an integral and irreplaceable part of the urban infrastructure. The value of maintenance was underscored, and the main tasks of management were refocused from new construction to conservation of existing assets.

Absorbed in the beauty of their designs, engineers can underestimate construction costs and overrate expected benefits as well as public support. Plato attributed to Socrates a depiction of this tendency (quoted in the following section). More recent examples are numerous.

The first tunnel under the Thames (1822–1846) was a magnificent feat, popular with pedestrians. Yet it was losing money and was eventually taken over by the underground trains (Gramet, 1966). Some of the greatest twentieth-century feats of transportation engineering, including the Channel Tunnel, the Concord supersonic airplane, and even the Apollo landings on the moon, have been criticized on a management level for their cost ineffectiveness. The innovative designs of the Eiffel Tower and the George Washington Bridge, on the other hand, receive extra credit because of their highly profitable long-term performance.

New champions in the world of bridges could not be constructed if they were not both essential and architecturally imposing. The term *signature bridge* (clearly implied in Roebling's preceding quote) has come to signify such a synthesis of qualities. Examples include the Akashi-Kaikyo (Fig. 1.40*a*) and the Tatara (Fig. 1.40*b*) in Japan, the Humber in Wales, the Great Belt in Denmark (Fig. 1.41), the Stonecutter Bridge in Hong Kong, Pont de Normandie and Viaduct de Millau in France, and the proposed crossing at the Messina Straits. These structures transform the landscape and capture the imagination, but crossing them must still be worth the tolls that support their maintenance and repay their concessions. Lacking a visual impact evocative of its technical accomplishment, the Channel Tunnel is associated primarily with high construction costs and disappointing revenues.

The San Francisco–Oakland Bay area bridges (Fig. 1.42) perform a dual function as well: They are symbols of the U.S. West Coast and major traffic arteries.

When the "signature" status becomes a primary design objective, rather than a consequence, costs surpass quantifiable benefits. The East Bay crossing between San Francisco and Oakland is an essential truss bridge too costly and too vulnerable to retrofit seismically (Fig. 1.43*a*). A unique self-anchored suspension bridge (Camo, 2004) was proposed as a replacement (Fig. 1.43*b*). The proposal satisfied the demand for a signature structure worthy of the Bay area but was also described as "the most expensive bridge in the world" (Mladjov, 2004). These contrasting evaluations earned the design an approval, a subsequent rejection, and further review. An article in *The New York Times* of April 17, 2005 (p. 20) discussed the "back and forth on bridge reconstruction."

On the feasibility of the original crossing, Waddell (1921, p. 8) commented as follows:

> There is in contemplation a project for building a long, high, and exceedingly expensive bridge across San Francisco Harbor, so as to connect the city of San Francisco with the

(*a*)

Figure 1.40 (*a*) Akashi-Kaikyo, Japan.

cities of Oakland, Berkeley, and its other suburbs. This project has been a dream for at least a decade; but it is not an idle dream, because some day in some manner or other it is certain to be realized. . . . The communities interested, however, have taken as yet no sensible step towards making a thorough study of the question.

In October 2005 California Governor A. Schwartzenegger consulted the administration of Millau, France, on the method of selecting the multispan cable-stayed design for the Viaduct de Millau, celebrated as the tallest bridge in the world (Fig. 1.44).

Creativity may be less constrained on bridges of lesser importance and cost. Figure 1.45 shows a highly imaginative pedestrian bridge which, despite its relatively minor scale and limited purpose, can claim signature status.

The cost–benefit assessment of designs anticipating and generating new demands reaches beyond the quantifiable *utility* optimized by Von Neumann and Morgenstern (1964) or Cornell (Appendix 2). During his 17 years as prefect of the Seine under the Second Empire, Baron Georges Haussmann (1809–1891) transformed Paris from a medieval town to the City of Light according to a personal vision rather than popular demand. From the 1920s to the 1960s Robert Moses (1888–1981) shaped New York City for an urban life he envisioned, rich in parks and highways (Fig. 1.46).

At that level of innovation, management relies on engineering in order to interface with the public and the political leadership. More than a century after Haussmann's work, the Parisian boulevards are emblematic of Paris (Fig. 1.47). The legacy of New York master builder and powerbroker R. Moses is still debated on the strength of continually reinterpreted evidence. While his controversial expressways and parkways are filled be-

(b)

Figure 1.40   *(b)* Tatara Bridge, Japan.

yond capacity, his acclaimed parks and swimming pools (Fig. 1.48) are often deserted and undermaintained. Caro's (1974, p. 1162) comprehensive biography of Moses ends with a question: "Why weren't they grateful?"

1.6   KNOWLEDGE AND INFORMATION

Supply and demand shape the interactions of theory, application, information and knowledge. Socrates (470–399 B.C.) (Fig. 1.49a) repeatedly discussed the role of memory in acquiring knowledge from information.

Figure 1.41   The Great Belt, Denmark.

Armytage (1976, p. 7) quoted the argument of Socrates against recorded information, as recounted by Plato (427–347 B.C., Fig. 1.49*b*) in *Phaedrus* 275. (*The Great Dialogues,* 1956). In B. Jowett's translation (1942, p. 442), Thamus, king of Egypt and hence God, judged the alphabet of Theuth as follows:

> O most ingenious Theuth, he who has the gift of invention is not always the best judge of the utility or inutility of his own inventions to the users of them. And in this instance a paternal love of your own child has led you to say what is not the fact; for this invention of yours will create forgetfulness in the learners' souls, because they will not use their memories, they will trust to the external written characters and not remember of them-

Figure 1.42 Golden Gate and San Francisco–Oakland Bay Bridges.

selves. You have found a specific, not for memory but for reminiscence, and you give your disciples only the pretence of wisdom; they will be hearers of many things and will have learned nothing. They will appear to be omniscient and will generally know nothing; they will be tiresome, having the reputation of knowledge without the reality.

In the interest of productivity and safety, management theory, design specifications, and knowledge-based systems deliberately pursue the same effect—they provide written guidance (if not true wisdom) to those who, in a perfect world, should not need it. As the analytic method supersedes the analogue in every maturing form of human understanding, the ancient caution that there is either more or less to knowledge than just memory remains relevant.

Upon graduating from RPI in 1857, Washington Roebling vowed to forget "the heterogeneous mass of knowledge that [he] could only memorize, not really digest" (Steinman, 1945). Information that could not be digested into knowledge appears to be implied in the preceding quote.

The need for a distinction between knowledge and information became critical with the advent of the indiscriminate electronic memory. Asimov (1990) anticipated the resulting moral and technological difficulties from a precautious computer age (Appendix 3). With a typically masterful mix of brilliant imagination, scientific understanding, and sly sense of humor, the author demonstrated how the collaboration of humans and machines can lead to the escapist options of either flipping a coin or trusting a black box. The usefulness of the former method for real management decisions has been considered

(a)

(b)

Figure 1.43 (a) San-Francisco–Oakland East Bay Bridge. (b) Proposed self-anchored suspension crossing, San-Francisco–Oakland East Bay. Courtesy of Weidlinger Assoc.

Figure 1.44 Viaduc de Millau. Courtesy J. Stubler, Freyssinet.

and remains unlikely (Appendix 2). Despite being morally equivalent, the latter method has not been entirely ruled out. In "2001" Arthur Clarke foresaw that once the computer had joined the experts, it would be a matter of time before it attempted to take over, as everyone else does (some engineers excepted).

The databases at the core of all management systems could play such a role. They are central to all operations and all experts communicate through them. They have made

Figure 1.45 Pedestrian bridge, Paris.

Figure 1.46 Robert Moses, Fordham University, New York City.

possible complex statistical modeling and forecasting. Expert systems provide decision support, combining subjective evaluations and quantitative data. A process can be designed such that empiricists supply data, theorists design self-learning databases, and managers implement the computer's findings. Decisions, once suspect because of their subjectivity, would instead emanate from a black box of guaranteed impersonality.

As a result of its newly gained abundance, information has become identified with intelligence, or with the ability to comprehend and anticipate input. Converting information into knowledge remains both sought after and questioned in every profession, including engineering and management (and even "intelligence"). Aristotle (*Ethics,* Book VI, 1140b, 1141b) considered wisdom "the most finished of the forms of knowledge."

Figure 1.47 Aerial view of Paris, the Louvre and the Tuilleries at the center.

He defined it as "intuitive reason combined with scientific knowledge." Aristotle praised philosophical, practical, and political wisdom but considered none of them independently sufficient.

Drucker (1968, Part IV) considered knowledge "central to contemporary society," precisely because information has become so abundant and accessible. Later (1995, p. 115) the author urged "to measure, not count." Confusing the quantity of precision in electronic computations with the quality of accuracy is an occupational hazard for all who use computers, perhaps greater for engineers than for managers. One explanation might be that managers value their subjective expertise more highly, whereas engineers, objective to a fault, place greater trust in information and hence in the computer. Furthermore, as descendants of the most ingenious Theuth, engineers trust the computer because it is their own creation and tool. Managers recognize in it another competitor.

1.7 BENEFITS AND COSTS

Practices, supplying according to demand, become professions with theoretical and financial support. The former manages the transactions between knowledge and information, the latter balances costs and benefits. Engineering and management became professions as a result of the Industrial Revolution when energy was harnessed by physics and production—by economics. During the subsequent information age the focus in-

Figure 1.48 Aerial view of Queens, New York City, with Flushing Meadows–Corona Park and Long Island Expressway in the forefront.

creasingly shifted to the processing of data (and the new science of informatics). Engineering designs products and has a primary interest in supply and demand. Management of processes, in contrast, focuses on costs and benefits. The detachment of process from product has given rise to operations management.

Professional management originated with the first economists, namely, Adam Smith (1723–1790) and David Ricardo (1772–1823). Drucker (1973, p. 21) attributed the term *entrepreneur* to J. B. Say (1767–1832) of France. Management developments followed those of industry and organizations, accelerating in Europe, the United States and, particularly, Japan, after World War II. Galbraith (1967) and Servan-Schreiber (1967) predicted a world domination of professional "asset management" in the industrial state. Drucker (1973, p. 15) found that they "had begun to sound naive by 1970." By then Galbraith (1973, Chapter XVII) was already announcing *transnational systems*. Toffler (1980, Chapter 19) heralded the replacement of the Cartesian method by new "system" thinking. He anticipated adaptable fast-forming networks capable of assuming dual forms, somewhat reminiscent of the indeterminate state of subatomic particles.

Figure 1.49 (*a*) Socrates, marble, Roman, first to second centuries A.D., the Louvre, Paris. (*b*) Plato, marble, The Vatican Museum, Rome.

Bell (1973, p. 14) envisioned a postindustrial society characterized by the following "five dimensions":

- Economic sector: the change from a goods-producing to a service economy
- Occupations distribution: the preeminence of the professional and technical class
- Axial principle: the centrality of theoretical knowledge as the source of innovation and of policy formulation for the society
- Future orientation: the control of technology and technological assessment
- Decision making: the criterion of a new "intellectual technology"

Fukuyama (1999, p. 3) summarized the period as the arrival of the *information age*. Drucker (1973, p. 17) wrestled with the vagueness of management as a professional *practice* as follows: "Management is a practice rather than a science. In this, it is comparable to medicine, law, and engineering. It is not knowledge but performance. Furthermore, it is not the application of common sense, or leadership, let alone financial manipulation. Its practice is based both on knowledge and on responsibility."

By 2005, according to Fukuyama (1992), history should have been a thing of the past. Still on the same authority (Fukuyama, 1999), a great social disruption is underway.

The globalization of organizations and interests, envisioned by D. Bell, J. K. Galbraith, J.-J. Servan-Schreiber, and many statesmen of the post–World War II period, is a central and much debated topic. Management and engineering are, as ever, joined in a dynamic equilibrium, under the watchful eye of accounting.

1.8 DETERMINISM, UNCERTAINTY, AND FAITH IN THEORY AND APPLICATION

Aristotle based his *Metaphysics* on the following reasoning:

All men by nature have a desire to know. [980]

Evidently, we have to acquire knowledge of the original causes. [982b]

The possession of it might be justly regarded as beyond human power . . . Yet the acquisition of it must in a sense end in something which is the opposite of our original inquiries. [983a]

In this infinite pursuit, Aristotle adopted the method of mathematical logic (prior analytics). Resuming the search for the appropriate analytic method two millennia later, Descartes (Fig. 1.50*a*) dismissed the interim philosophical endeavors tactfully but summarily (1976, p. 98): "I shall say nothing of philosophy, except that it has been cultivated by the most excellent minds that ever lived over the centuries, and that there still remains to be found something that is not disputed, and consequently doubtful."

Descartes found people's opinions diverse "not because some are more reasonable than others, but only because their thinking follows different routes and does not consider the same things." He proposed to clarify reason by the rigorously deductive method of mathematics. Galilei believed that nature spoke the language of mathematics. John Locke (1632–1704) expected logic to be mathematically substantiated. Kline (1953, p. 263) called him a "rational supernaturalist."

Newton rejected hypothesizing and described, in his own mathematical language, a universe uniquely defined by mechanics. In the penultimate paragraph of his *Principia,* he stated: "I frame no hypotheses; for whatever is not deduced from the phaenomena is to be called an hypothesis; and hypotheses, whether metaphysical or physical, whether of occult qualities or mechanical, have no place in experimental philosophy. In this philosophy particular propositions are inferred from the phaenomena, and afterwards rendered general by induction."

Immanuel Kant (1724–1804) hypothesized convincingly (along with Laplace) on the origin of the solar system. According to Kant (1965, p. 41), all our knowledge *begins* with experience but does not necessarily *arise* from it. He separated knowledge into *pure* and *empirical,* or *a priori* and *a posteriori*. A priori knowledge was characterized by necessity and universality. Sciences contained *analytic* and *synthetic* judgments, namely, identities and derivations. Mathematics was purely synthetic. "In any particular theory there is only as much real science as there is mathematics."

In his *Science of Logic* (1831), G. W. Friedrich Hegel (1770–1831) criticized Kant's references to motion and gravity [*Critique of Pure Reason,* (1965) 1781] and raised

(a)

(b)

Figure 1.50 (a) René Descartes, the Louvre, Paris. (b) Blaise Pascal, the Louvre, Paris.

questions regarding the quality and quantity of matter and force that relativity and quantum mechanics had yet to address (Hegel, 1989).

Scientific determinism fitted comfortably with monotheistic religion but not always with scientific discovery. Thomas Bayes (1702–1761) set out to prove the existence of God mathematically. The result defined *conditional probability* (Appendix 4). In his *Theorie Analytique des Probabilités,* Laplace stated: "Given for one instant an intelligence which could comprehend all the forces by which nature is animated and the respective positions of the beings which compose it . . . nothing would be uncertain." (Kline, 1980, p. 67)

Lacking such intelligence, Laplace applied the Bayesian equation to educated guesses. Guessing was considered an appropriate starting point for estimating the otherwise unpredictable social behavior. In nature, however, physics was encountering similarly unpredictable, unstable, and indeterminate phenomena. In logic, philosophy and mathematics were confronted by undecidable propositions. Certainty had been taken to its limits (Kline, 1980). Beyond those limits, the new field of statistics delivered the best approximation of rigor.

Mathematicians diverged on the appropriate treatment of uncertainty in nature, in society, and in logic, just as general thinkers debated the nature of common sense. J. R.

Saul (2004) offers an interesting overview of the latter subject. Whereas Descartes derided common sense as "the best shared quality," Pascal (1623–1662) (Fig. 1.50*b*) argued that "the heart had reasons reason did not know." He found Descartes "largely true," but "useless and uncertain," "mechanical," and "painful" [*Pensées,* Section 84 (79)]. Siding with Descartes, Voltaire treated common sense as the median of the intellectual range, situated "midway between stupidity and intelligence." For Thomas Paine it implied the memory, judgment and imagination of democratic society, as opposed to autocratic government (Paine, 1945). The same traits were to become the trademarks of sound management. Einstein's (1941) summary of "the whole of science" as "nothing more than a refinement of everyday thinking" is usually quoted as praise of the commonsensical nature of reason, but it also affirms the exclusivity of rigorous deduction.

To explain the outcomes of games of chance, Pascal formulated probability but eventually concluded that the product of odds and benefits favored faith. (Certainty is commonly associated with indifference.) Despite his commitment to rigorous deduction, Descartes admitted a priori knowledge.

D'Alembert (1717–1783) assured students of calculus that faith would come to them (Kline, 1953, p. 232). Over the centuries, faith and deduction became inseparable.

Frequentists, or *objectivists,* counted outcomes and identified patterns in the behavior of large populations. James Maxwell (1831–1879) applied objectivist statistics to the kinetic theory of gasses. *Probabilists,* or *subjectivists,* allowed, as Augustus de Morgan (1806–1871) that "probability is a feeling of the mind, not an inherent property of a set of circumstances" (McNeill and Freiberger, 1994, p. 179). Arthur Schopenhauer (1788–1860) examined "the world as will and as representation" and defined pessimism as the only correct forecasting method (or attitude). Soren Kierkegaard (1811–1855) (Fig. 1.51) (1849, 1980, p. 35) shared the pessimistic view but deplored the failure of "fatalists and determinists" to "raise higher than the miasma of probability." His answer to despair was faith, attained by imagination. Kierkegaard's terminology ("necessity is the constraint in relation to possibility") sounds like a precursor of modern optimization.

Eventually theory moderated its self-assurance in explaining all empirical evidence while faith adjusted to explanations it had once banished. Einstein would still have liked to believe that "God does not play dice with the Universe." Nevertheless, he conceded (1921): "So far as the laws of mathematics refer to reality, they are not certain; and so far as they are certain, they do not refer to reality."

Barrow (1991, p. 31) quoted Alan Turing (1921–1954) to the effect that "science is a differential equation, religion is a boundary condition." The uncertainty, or indeterminacy, principle formulated by Werner Heisenberg (1901–1976) placed quantum physics outside the deterministic Newtonian frame (a development almost anticipated by the last words of Newton's *Principia*). In the 1930s Kurt Godel (1906–1978) demonstrated that "what is intuitively certain extends beyond mathematical proof" (Kline, 1980, p. 263). Godel's incompleteness theorem demonstrated undecidable propositions in mathematical logic. Thus both mechanics and rational decision making and, hence, engineering and management, once unlimited, appeared to be exceeded.

Heisenberg (1958, Chapter XI) feared that after the "dissolution of the rigid frame of 19th century concepts due to empirical advances and mathematical modeling," the certainty offered by irrational beliefs (e.g., fanaticism) might prove attractive. Dewey (1929, Chapter VIII) did not consider the natural randomness discovered by Heisenberg

Figure 1.51 Sören Kierkegaard, Copenhagen.

as an end of the quest for moral certainty. He found uncertainty to be 'primarily a practical matter' (p. 223).

The father of statistical nuclear physics, Max Born (1968, Chapter 2) resolved the "ancient problem of necessity and freedom [ ] in everyday life" by drawing an analogy with the *complimentarity* introduced by Niels Bohr (1885–1962) in quantum mechanics (p. 107):

Only two possibilities seem to exist: either one must believe in determinism and regard free will as a subjective illusion, or one must become a mystic and regard the discovery of natural laws as a meaningless intellectual game. Metaphysicians of the old schools have proclaimed one or the other of these doctrines, but ordinary people have always accepted the dual nature of the world. Bohr's idea of complimentarity is a justification of the common people's attitude, because it directs attention to the fact that even an exact science like physics has reconciled itself to the use of complimentary descriptions, which provide a true image of the world only when they are combined.

In 1950 Jean-Jacques Servan-Schreiber (1991, p. 399) announced from the front page of *Le Monde* that, in contrast to the deterministic totalitarianism, democracy was a translation of fundamental natural uncertainties into human terms. Galbraith (1977) entitled his history of economic ideas and their consequences *The Age of Uncertainty.* In it he expressed concern about the management of the essentially deterministic aspects of urban infrastructure in advanced democratic societies. Drucker (1995) dedicated a chapter to management and planning under uncertainty (more precisely, unpredictability). On the op-ed page of the *New York Times* (April 8, 2005), Columbia University professor and expert on String Theory Brian Green celebrated "One Hundred Years of Uncertainty," counting from the publication of Einstein's papers on relativity and quantum mechanics.

In 1963 the Caltech professor Richard Feynman (1918–1988), who was to win the Nobel prize for physics in 1965, lectured (1998, pp. 26–27): "Scientists are used to dealing with doubt and uncertainty. . . . What we call scientific knowledge today is a body of statements of varying degrees of certainty. Some of them are most unsure; some of them are nearly certain; but none is absolutely certain."

1.9 VAGUENESS, IGNORANCE, AND RANDOMNESS

During the reign of certainty, Leonard Euler (1707–1783) had recognized the need to relax it into perceptual, demonstrative, and moral (Casti, 1990, p. 23). After uncertainty became generally accepted, it had to be similarly diversified into ignorance, randomness, and vagueness (Appendix 5). Each category had to be treated by specifically suitable methods, championed by dedicated schools of thought.

Bertrand Russell (1985) warned: "Everything is vague to a degree you do not realize until you have tried to make it precise."

"Everything" in this case appears to target particularly deterministic assertions. Distilling vagueness into precision seems to be implied in Einstein's (1941) aphorism to the effect that "the whole of science is nothing more than a refinement of everyday thinking." In contrast, professions such as engineering, economics, and medicine apply scientific principles to everyday life. Von Neumann and Morgenstern (1964) developed game theory as a tool for economic decisions, departing from the following reasoning (p. 4):

The empirical background of economic science is definitely inadequate. . . . Our knowledge of relevant facts of economics is considerably smaller than that commanded in physics at the time when mathematization of that subject was achieved. Indeed the decisive break which came in physics in the seventeenth century, specifically in the field of mechanics, was possible only because of . . . several millennia of systematic, scientific, astronomical observation, culminating in an observer of unparalleled caliber, Tycho de Brahe. Nothing of this sort has occurred in economic science.

In order to limit the ignorance and vagueness of economic forecasts, the authors assumed that individuals rationally maximize a numerically quantifiable *utility,* serving as a measure of satisfaction or preference (which de Tocqueville considered typically democratic, see Section 1.2). The approach is based on the following (frequentist) argument: "It is a well known phenomenon in many branches of the exact and physical sciences that very great numbers are often easier to handle than those of medium size. This is, of course, due to the excellent possibility of applying the laws of statistics and probabilities in the first case."

"Quantitative" methods for the treatment of large data samples proliferated, although not all of them were purely frequentist. Appendix 6 lists some of the most widely used ones.

Leonardo da Vinci, admitted in his notebooks (1935): "All knowledge is based on opinion." Laplace combined that approach with the application of the Bayesian theorem. De Finetti (1974, p. XI) wrote: "Probabilistic reasoning is completely unrelated to general philosophical controversies, such as Determinism versus Indeterminism." In his view probability is a measure of belief (e.g., if not faith, an educated opinion). Opinions matter because they differ. Hays (1994, p. 47) stated: "The Bayes's theorem was applied inappropriately and yielded some rather ridiculous results . . . casting many perfectly proper applications of the theorem into dispute."

The Bayesian reliability approach (Appendices 2 and 4) is more inclusive than the deterministic and the statistical methods. Nonetheless, it assumes knowledge of *all pertinent* (random) properties and phenomena. Consequently, the cumulative probability of all outcomes must be unity. Although the frequentist game theory of Von Neumann and Morgenstern (1964) differs from Bayesian structural reliability, the assumed knowledge of all outcomes in the latter parallels the hypothesis of rational behavior of all players in the former. The two also share a dependence on quantifiable "utility" and an inability to address unique situations.

The Bayesian assumptions could become too rigid under vaguely defined conditions, just as frequency statistics had been too constraining to opinion-based probability. Vagueness and its nuances, such as ambiguity, nonspecificity, and confusion (McNeill and Freiberger, 1994, p. 189), were modeled by fuzzy logic. The authors point out (p. 182) that while fuzzy logic may be more general and less constrained, it is also less specific. Critics compare it to "ad hoc engineering" or to allowing "people with no engineering background to perform control." Fuzzy logic is also compared to "managing by committee." Since these are realistic possibilities, fuzziness is clearly part of reality. In keeping with the advice to speak softly and carry a big stick, logic grows fuzzier, as it draws on ever-increasing data processing power. The unknown has been found to contain the unknowable and the indescribable. To model the full range of uncertainties, mathematics eventually had to adopt the inconclusive language of philosophy.

1.10. OBJECTIVITY AND SUBJECTIVITY; QUANTITY AND QUALITY

A poet educated in mathematics, Valéry (1941, p. 9) noted (ca. 1910) that theory and practice were married in a reciprocal redesign. He perceived thought as an intermediary between analysis and application, producing a *utilizable value,* within the *realm of pos-*

sibilities, subject to *constraints.* A poetically inclined philosopher, George Santayana (1863–1952) considered (1928, p. 20) that "science is a half-way house between private sensation and universal vision." He expected science to "articulate experience and reveal its skeleton" (p. 72).

Engineering practice has grown to expect that theory will eventually quantify or reject all of its empirical models. Einstein and Infeld (1942) stated: "One of the most important characteristics of modern physics is that the conclusions drawn from initial clues are not only qualitative but also quantitative. We wish to be able to predict events and to determine by experiment whether observation confirms these predictions and thus the initial assumptions. To draw qualitative conclusions we must use the language of mathematics."

Mathematics could therefore be used for nuanced interpretations.

Dewey (1929, p. 178) speculated: "Theory of knowledge or epistemology would have been very different if instead of the word *data* or *givens,* it had happened to start with calling the qualities in question *takens.*" Theoreticians are inclined to regard data as *given,* practitioners are more likely to *take* the data they need.

"God was an engineer, not a scientist," professed Terence Sejnowski of the Laboratory of Computational Neurobiology during a conference on brain neurology at the Salle Institute in Cambridge, Massachusetts, in 1988 (Yanev, 1989). The past tense may have been intended as a reference to the Act of Creation, of which He was also The Manager. Closer to the source, the Old Testament tells: "A false balance is abomination to the Lord, but a just weight is His delight" (Proverb 11–1, King James Version, 1611).

Equilibrium is thereby not a given but a design option.

For engineering design, Cornell (Freudenthal, 1972, p. 48) recommended "a highly refined technical pragmatism." He explained (p. 47) the combination of determinism, objectivism, and subjectivism resulting in Bayesian statistical decision theory for reliability-based design as follows:

> If it is accepted that our purpose is the engineering one of design rather than the scientific one of description, then a number of important qualitative implications follow immediately. . . . If one seeks to design rather than to describe, a number of deep unresolved arguments within science about the fundamental nature of probability become unimportant. It no longer matters whether nature (e.g. the motion of molecules or the maximum wind velocity) is basically deterministic or whether probability is just a convenient tool to describe phenomena too complex to treat with the present level of science. It only matters that from an engineering design point of view it is useful (i.e. it produces more economical structures) to represent particular quantities, such as loads and resistances, as random variables.

Treating all variables as random, which is to say probabilistically, but updating the probabilities according to improved knowledge, is the pragmatic method, essentially similar to all learning, described in Appendix 2. Adopting this method, the deterministic *safety* of engineering design yielded to probabilistic *reliability,* quantifiable and negotiable in mechanical and financial terms (Appendix 7). The *prescriptive* American Association of State Highway Transportation Officials (AASHTO) bridge design specifications were superseded by the load and resistance factor design specifications (LRFD, 1998a), described as *performance based* and *probabilistic* (Section 4.2.1). The formulation is not unique. The Eurocode for bridge design is called *semiprobabilistic* in recognition of the inevitable measure of determinism.

The reliability of engineering structures and infrastructure assets could be quantified by the methods of operations research and decision-making analysis (Appendix 8). Some subjectivity, however, is inevitable. According to Cornell (Freudenthal, 1972, p. 47), "the distinction between subjective and objective probabilities becomes in fact vague and unnecessary." A more extensive excerpt is presented in Appendix 2.

In sum, engineers and managers must base decisions on quantified measurements, calculated odds, estimated probabilities, simulated vague patterns, and personal opinions (Appendix 9). Responsible management must allocate exactly quantified funding on the basis of statements combining qualitative and quantitative assessments, such as the following: "More than 40% of the nation's bridges are structurally deficient or functionally obsolete" [Chase et al., in TRB Circular (TRC) 498, 2000, C-6, see also Section 10.3].

Contemporary engineering management translates qualities and quantities bilaterally and the outcomes are not uniquely defined. Axioms, assumptions, and tests contribute in varying degrees. Ignorance of the actual state of bridge elements can be treated statistically if the scatter of the properties is adequately explored. The randomness of phenomena can be modeled by probability distributions based on informed expectations. Quantification appears to improve the reliability of opinions, but it suffers from ignorance and therefore indecision. Opinions appear decisive, but their basis is vague. Ignorance, randomness, and vagueness are complementary and not always separable. By their respective means, determinists, frequentists, probabilists, and fuzzy logicians should converge upon acceptable translations of measured quantities into defined qualities.

The diverse methods used in various combinations by engineering and management allow for equally rational but superficially opposite professional approaches. One can select a quality (e.g., the form) and seek the quantities (e.g., the content) that will satisfy it or arrange the quantities to fit the constraints, letting the quality define itself. In Section 5.5 the probabilist and frequentist methods are (vaguely) compared to the kynematic (upper bound) and static (lower bound) methods for plastic analysis of frames (Appendix 10). In a broader analogy, subjectivism is comparable to the freedom of interpretation practiced in the arts and objectivism to the scientific reliance on verifiable data. Both art and science, however, notoriously borrow each other's methods. J. Bailey (2003) argued for a "parallel" dynamic integration of ideas, drawing on all valid analogies in contrast to the governing "sequentially fixed" analytic reasoning.

Although engineering and management operate in compatible mathematical dialects, the former is informed by the (mostly deterministic) mechanics, the latter by the largely statistical economics. Consequently, engineers determine reliability, whereas managers rely on determination.

At their best, the guidelines adopted by the practitioners of complementary skills should converge to the desired results. At their worst, professional attitudes are mummified into mutually exclusive stereotypes. Judges are decisive, jurists argumentative, doctors sympathetic, executives dynamic, architects creative, and educators inspiring. After educating engineers at Columbia University in New York for half of the twentieth century, Professor Mario Salvadori found them lacking the ambition of scientists and the imagination of architects, reliant on axioms, and "socially and politically conservative" (Gans, 1991, pp. xiii–xiv). This apparent conformism can also be seen as a professional sacrifice in the name of reliability. Engineers prefer to be judged by their work and the best work attracts the least publicity. Inventors, most prominently represented by Thomas A. Edison, break out of this mold by their creativity, particularly if they manage to corner the market on their invention.

Outstanding designs are perceived as discoveries, preexisting in the abstract, rather than as inventions. In that view, the improvisational aspect of management would appear incompatible with the rigorous scientific methods of engineering. The engineering output is packaged with an emphasis on quantification, for example, as estimates, assessments, design calculations, work schedules, drawings, or, most extremely, proposals and recommendations (always backed up by numbers).

Estimating quantities amounts to decision support; decisions of any genuine consequence are qualitative. Neither design nor management can fully retreat behind quantification. Be they *prescriptive* or *performance based,* design specifications are based on qualitative interpretations of phenomena (Section 4.3.1). Their use requires yet another qualitative interpretation. Innovative designs always exceed the scope (if not the spirit) of specifications.

Genuinely new information clashes with accumulated expertise because it does not fit the operating formal models. In 1992 the Bridge Design Committee of the American Society of Civil Engineers (ASCE) rejected a proposal to establish a bridge management subcommittee on the grounds that management was beyond the scope of design. A year later the Subcommittee for Bridge Maintenance voted to change its name to Bridge Management, Maintenance and Inspection. Soon thereafter it obtained full committee status. Engineering and management continually renegotiate their common ground.

1.11 COMPETENCE AND QUALIFICATIONS

The great bridge designer and manager Othmar Ammann (Section 1.3) was chief engineer of the New York and New Jersey Port Authority. That and other autonomous public organizations were created by Robert Moses, the urban and transportation master builder. A multispan suspension bridge between downtown Manhattan and Brooklyn was proposed by Moses and might have been designed by Ammann. Instead, it was overruled in favor of the Battery Tunnel on grounds of national security by President Franklin D. Roosevelt (Fig. 1.52). The ambitious multispan cable-stayed bridge Moses proposed across Long Island Sound was rejected by New York State Governor Nelson Rockefeller, because it proved unpopular with the local communities. The priorities of societies invariably exceed the needs of the bridges that generate and support their activities. No matter how accomplished the bridge manager might be, a transportation manager is likely to have a higher (and still not final) responsibility.

Drucker (1995) defined the broader range of ability required for higher levels of management as *competence,* surpassing the narrower *qualifications*. Both have their uses. Bridges are in the pontiff's competence, but Rome maintains a technically qualified transportation department as well.

The demands for specialized competence and qualifications have cultivated a formal and a technical type of management. *Top-down* managers administrate the process. *Ground-up* managers deliver the product. Budget managers increasingly serve as intermediaries between the two, as in Fig. 1.33*b*. Architects similarly serve as intermediaries between structures, art, and public taste.

Figure 1.52 Franklin D. Roosevelt, bronze by Neil Estern, F. D. R. Memorial, Washington, DC.

According to John Kenneth Galbraith, specialization is "the parent of boredom, irrelevance and error" (Frank, 2005). At least since Ford's conveyor belt, however, maximizing talent has been discredited as a means of attaining production efficiency. Instead, processes are relentlessly streamlined and specialized.

Yao and Roesset (2001) express concern that "future civil engineers could be relegated to the status of clerks, available to only provide numerical data or to perform computations, but without any major say in the planning and management process." To reverse this professional decline, the authors propose the following curriculum in infrastructure management:

- Mathematics and basic science
- Engineering science
- Technical aspects of infrastructure systems
- Principles of uncertainty and risk analysis
- Decision analysis in the face of uncertainty
- Management and business principles
- Societal needs, ethics, public policy, and political science
- Communication skills

The course work ends with "communication skills," at the point where leadership begins. According to Drucker (1954, p. 159) managerial education can go no further: "Three thousand years of study, exhortation, injunction and advice do not seem to have increased the supply of leaders to any appreciable extent nor enabled people to learn how to become leaders."

Drucker (1995) considered "executives" rather than "managers" essential in time of change (particularly the present).

The distinction between management and leadership is as vague as the one between engineering and management. Henri Fayol (1841–1925) formulated the division of services under the unity of command, which can be perceived as a combination of leadership and management. The activities distinguishing this function include forecasting, planning, organization, commanding (or directing), motivation, and communication.

Northouse (1997, p. 9) defined management as a sort of "organization engineering," related but opposed to leadership, as follows:

Management	Leadership
Produces order and consistency	Produces change and movement
Planning/budgeting	Vision building/strategizing
Organizing/staffing	Aligning people/communicating
Controlling/problem solving	Motivating/inspiring

In this type-casting exercise leadership (industrial, economic, and political) innovates, whereas management optimizes the use of human and other resources. Good leadership should be effective, whereas management must be efficient (e.g., cost effective).

The situational leadership model proposed by Hersey (1985) represents leadership behavior as a combination of relationship and task-oriented functions, consisting primarily of delegating, participating, selling, and telling. For the purposes of construction management, Cooke and Williams (2004, Chapter 2) structure Fayol's seven activities into two basic models described as theories X and Y, respectively autocratic and democratic. The autocratic theory X emphasizes obtaining the product (as in Fig. 1.33*a*). The democratic theory Y optimizes the process according to the demands of the many interests involved, as in Fig. 1.33*b*.

Example 4 illustrates the determination of leadership to live up to its "reinventive" image. Leaders give crowds a sense of direction or define already shaping directions in the timeliest and most eloquent manner. Engineers supply the means to advance. Once management is charged with rigorously defined and quantifiable tasks, it is distanced from leadership as engineering has been from management (and from architecture).

Example 4. Reengineering of Government

Fig. E4.1 shows a letter by two New York City deputy mayors urging the managers of publicly owned facilities to identify tasks suitable for outsourcing while improving

THE CITY OF NEW YORK
OFFICE OF THE MAYOR
NEW YORK, N.Y. 10007

MEMORANDUM

TO: Agency Heads

FROM: Adam L. Barsky
               Robert M. Harding

DATE: October 30, 1998

SUBJECT: Reinventing Government

One of the Mayor's goals is to reduce the size of government to its most essential and necessary components. As agency heads and managers, we are challenged to differentiate between those core services that must be provided by City work forces and services that can be 1) discontinued; 2) consolidated; 3) privatized or 4) outsourced. It is also our responsibility to ensure that the services we do provide are the most effective and cost-efficient, meeting our customers' needs while maximizing the use of our scarce resources.

We must critically review City services and periodically coordinate our efforts to ensure that we are all advancing together toward the goal of a more efficient, streamlined and "reinvented" government. To this end, we ask that you submit to our offices your ten (or more) highest priority proposals to eliminate, consolidate, privatize, or outsource services as well as to reengineer, streamline or otherwise dramatically improve the efficiency or effectiveness of a service. Where possible, you should also identify surplus assets that can be sold.

Your proposal package should be delivered, on hard copy and disk, to each of us by Wednesday, November 25th. It should include a project name or title; a brief description of the initiative, including dollar savings or revenue and performance improvements; a timeline for major milestones; and, a point person at your agency who will be specifically responsible for implementing the initiative. We will be scheduling meetings for agency heads to present their plans shortly after the due date for submission. We look forward to your response.

Figure E4.1   Letter by New York City Deputy Mayors A. L. Barsky and R. M. Harding on reinventing government by outsourcing tasks, October 30, 1998.

McCall rips DOT for consultant work

.Y NEWS • Tuesday, March 27, 2001

ALBANY — The state Transportation Department wastes millions of dollars by hiring outside consultants for routine construction projects that should be handled by its own engineers, state Controller Carl McCall charged yesterday.

McCall warned that he might halt payment on consulting contracts if transportation officials cannot provide sufficient reasons for farming out the work to consultants.

He said the state could have saved $137 million had the department reduced spending on consultants by 25% from 1991 to 1999.

"These tax dollars could have been better spent repairing roads and bridges across New York," said McCall, a Democrat who has his sights set on becoming governor next year.

The department said it is moving to reduce its reliance on consultants and use them more wisely.

A spokesman also returned fire at McCall, blasting the controller for being interested in electioneering.

"They [the controller's office] didn't want to know the facts about what we are already doing," said spokesman Michael Fleischer.

Joe Mahoney

Figure E4.2 Article in *Daily News,* March 27, 2001, quoting the position of New York State Controller C. McCall against excessive outsourcing.

the quality of services. Figure E4.2 shows an article by the New York State controller arguing that outsourcing is costlier and less productive.

C. E. Harris, Jr., professor of philosophy at Texas A&M (*New York Times,* June 11, 2006, p. WK 3, col. 2) concluded that engineers lack the "professional identity" of physicians and lawyers. Whereas the latter see themselves as members of their respective professions first and employees second, engineers function according to their contractual tasks, such as design or management.

The separation of the socially dynamic and static functions of any process owes a lot to personal preferences and is, to that extent, inevitable. Salvadori (Gans, 1991, pp. xiii–xv) found architects and engineers to be "two entirely different breeds of the human species." In agreement with Drucker, he considers it "impossible to train geniuses closing entirely the gap between engineering and architecture." As Yao and Roesset (2001), Salvadori recommended a minimum of shared competence. In the counterproductive alternative, leadership would become the exclusive domain of public relations experts, while managers would be reduced to accountants, engineers to technicians, and architects to decorators.

The National Commission Report (2004) on the terrorist attacks perpetrated upon the United States on September 11, 2001, cited a "failure of imagination, policy, capability and management." Management is subdivided into *operational* and *institutional* (Chapter 11, "Foresight—and Hindsight," pp. 339–360). An alternative would be to consider the absence of imagination, policy, and capability as failures of management. Such seems to be the view taken by Newman and Summer (1961) in Chapter 21, entitled "The Place of Leadership in Management."

In his op-ed article related to the reconstruction of Ground Zero in Manhattan (*The New York Times,* October 23, 2005, p. 12), Sudjic argued that an "edifice complex" has motivated the "architectural patronage" of many political leaders from the time of the pharaos to the present. The author considers this form of motivation effective against a "state of inertia."

Engineers and managers face the personal choice of limiting their competence to recommendations or extending it to implementation. Organizations must navigate a course that combines decisions by consensus and creative individualistic visions. In a simplistic separation of competence, personal style would define ground-up management, whereas top-down management would be restricted to formalized systems. Alternatively, both projects and networks could benefit from a share of the ground-up and top-down approaches.

Regardless of the terminology, talents do not fit neatly into prescribed professional categories. Excessive conformism produces stereotypical "talking heads," "bean counters," "number crunchers," and "nuts-and-bolts" operators. As soon as an academic curriculum is defined, those who transcend it make the difference. The highest authority takes charge of human resources, not material assets. Plutarch (transl. Dreyden, p. 864) attributed Caesar's victories to a supreme mastery "of the good-will and hearty service of his soldiers." Technical experts obtain top performance from natural resources, but they deliver their best efforts under inspiring leadership, not always their own.

Leadership is the management of human resources without recourse to dictatorial power tools. The technocratic master builders of the past may have shared a common deterministic language with their contemporaneous autocratic rulers. The supply-and-demand relationship between the two could be fully described by the scheme shown in Fig. 1.33a. In modern democracies various forms of checks and balances separate authority into legislative, executive, and judiciary branches. Most organized social activities follow a similar model: Leadership popularizes, management optimizes, engineering obtains. These three levels of competence negotiate bridge management decisions in terms of budget constraints, production capabilities, community interests, and visions of the future. The process, illustrated in Fig. 1.33b, must be transparent and systematic, therefore engineered and managed.

PART I

DEMAND: FROM STRUCTURES TO SYSTEM

Chapter 2

Objectives, Constraints, Needs, and Priorities

The objective of the builder is to produce the best new structure. The manager, in contrast, must use an existing structure to the best advantage, according to the governing standard of the moment. As structures evolve from concepts to utilities, the designer's view is increasingly outweighed by the manager's. The diverging viewpoints should nonetheless remain consistent with the general purpose of engineering as it is defined by Newmark and Rosenblueth (1971, p. 443): "The aim of any purposeful activity is optimization of the outcome . . . *the purpose of engineering design is to maximize the utility to be derived from the system produced*" (italic in original).

Maximizing utility under constraints was defined by Von Neumann and Morgenstern (1964, Section 1.9) as optimization. Diwekar (2003) attributed the origin of rigorous optimization to John Bernoulli (1654–1705). Heyman (1996) recounted the contributions of James (1667–1748) and Daniel Bernoulli (1700–1782) to structural analysis. At the urging of the latter, Euler applied his calculus of variations to the problem of elastic deformation and obtained the "elastica" in 1744. In the spirit of proverb 11-1 of the Old Testament (Section 1.8 herein), Euler believed that God not only delighted in equilibrium but also maximized it: "Since the fabric of the universe is most perfect, and is the work of a most wise Creator, nothing whatsoever takes place in the universe in which some form of maximum and minimum does not appear."

Mechanical systems assume shapes minimizing their potential energy. More broadly, the transportation infrastructure systems of the present are shaped by the changing supply and demand driving engineers and managers over the last 150 years.

As engineering activities are increasingly optimized not only in the physical but also in the social domain, the once absolute constraints and objectives become interchangeable and negotiable. The transition is illustrated in Fig. 1.33a and 1.33b).

2.1 MAXIMIZING UTILITY

Bridge engineering more or less continually responds to changes in the modes of transportation, material availability, labor costs, and social and cultural demands. The resulting bridges, since the advent of the steam engine, have undergone the following evolution:

Railroads over timber, iron, and steel structures in the nineteenth century

Vehicular and/or train traffic over steel and concrete bridges in the twentieth century

High-speed trains and vehicles over new structural types and materials at the beginning of the twenty-first century

The maximum span length grew from 2×1710 ft (521 m) at the Forth rail cantilever truss bridge in 1889 to 6529 ft (1991 m) at the Akashi-Kaikyo vehicular suspension bridge in 1998. A 3.3-km suspension span carrying automobile and rail traffic across the Messina Straights is under design.

Railroad bridges are regulated by the American Railway Engineering and Maintenance Association (AREMA), originally AREA, first published in 1905. The American Association of State Highway (Transportation) Officials (AASHTO) was founded in 1921 (as AASHO) and the first specifications appeared in 1931.

AASHTO (1999a) begins as follows: "When highway departments were first formed in the early 20th century, they were seen as construction organizations. Design was a minor responsibility."

The New York City bridge construction history, briefly highlighted in Example 1, illustrates both a qualitative and a quantitative expansion responding to urban transportation demands. The national trend is similar. At the turn of the twenty-first century there are approximately 650,000 vehicular bridges and culverts (Appendix 11), 101,000 railroad bridges, and a comparable number of pedestrian bridges in the United States.

During the early period of rapid network growth material and labor cost estimates and construction practices were of primary interest. Waddell (1916) dedicated volume I of his comprehensive *Bridge Engineering* to design (primarily of railroad trusses) and volume II to construction, its cost estimates, and optimization.

Personal tastes for materials and construction methods shaped today's infrastructure in equal measure with utilitarian considerations. In his professional engineering dissertation at Columbia University, Steinman (1909) analyzed the feasibility of steel and concrete arch alternatives for a proposed Henry Hudson Bridge over Harlem River in New York City. He completed his steel fixed-arch version in 1928 (Fig. E1.7). Steinman's (1949) text on suspension bridges pays equal attention to "design, construction and erection."

Waddell (1921, Chapter XIII) defended the feasibility of cantilever trusses for long-span railroad bridges against Steinman's conclusion in favor of suspension bridges. Despite his lifetime dedication to steel, however, Waddell (1921, p. 66) deemed it "impossible to come to any fixed or reliable conclusion concerning the relative economics of steel and reinforced-concrete bridges."

Lindenthal found Waddell's text "valuable and authoritative only so far as it deals with the engineering of bridges of ordinary spans and type that have become more or less standardized" (Petroski, 1995, p. 200).

Comparing Waddell's Goethals Bridge (Fig. 2.1) with Lindenthal's Hell Gate (Fig. E1.4) and Smithfield (Fig. 1.14) Bridges suggests that the beginning of the twentieth century offered ample opportunities for both standard and unique bridge solutions on a large scale. Less imaginative but well-executed designs have served transportation reliably as have inspired unique structures.

The construction boom at the turn of the twentieth century did not detract designers from the objective of durability. Boller (1885) recommended changes in design that would extend bridge life under the maintenance practices of his period. Waddell (1921, p. 6)

Figure 2.1   Goethals Bridge, Staten Island–New Jersey.

defined economics as "the science of obtaining a desired result with the ultimate minimum expenditure of effort, money and material," adding a caution against limiting economic analysis to first cost (Appendix 12). He wrote (1921, p. 430):

> The preservation of bridges against rapid deterioration is just as important a matter as
> ensuring that they are properly proportioned and constructed—yes, even more important,
> for what behooveth it the owner of a steel structure to take the utmost care in its designing
> and building, if he neglect to protect it effectively against the ravages of rust? The life
> of a metal bridge that is scientifically designed, honestly and carefully built, and not
> seriously overloaded, if properly maintained, is indefinitely long, but if badly neglected
> is often quite short, especially when it is exposed to acid fumes, such as those contained
> in the smoke from locomotives passing through or beneath.

Despite earlier arguments in favor of life-cycle management policies, Kohoutek (in
Ireson and Coombs, 1988, p. 3.7) traced life-cycle costing (LCC) to investigations by
the Logistics Management Institute of the Assistant Secretary of Defense in the early
1960s. Appendix 11 highlights the milestones of vehicular bridge management history
in the United States since that time. The interest of the owner and not that of the designer
emerges as the driving force in this development.

The highway design model (HDM) was developed by the World Bank in 1981 and
is still in use under the management of the World Road Association (PIARC). The
National Cooperative Highway Research Program (NCHRP Synthesis 330, 2004) also

quoted MicroPAVER, developed by the U.S. Army Construction Engineering Research Laboratory in cooperation with the American Public Works Association (APWA). NCHRP Report 285 (1986) recommended life-cycle cost analysis (LCCA) for pavements and bridges, pointing out the much greater complexity of the latter.

Roadways seem to preoccupy contemporary bridge management, because they distinguish the more recent vehicular structures from the older railroad ones and because they support a still growing transportation mode. Shaking down its problems since the nineteenth century, railroad bridge management has become self-effacing. Nonetheless, its precedents remain relevant.

The transportation leap from horse-drawn carriages to steam-driven trains was much bolder than the one from trains to internal combustion engines. Railroad bridges are highly susceptible to fatigue fractures (Section 4.2.4) and track irregularities. Their design and maintenance practices are influenced by well-publicized catastrophic failures.

Bridge ownership, crucial to management policies, is clearer in railroads than in the multiple overlapping highway administrations. Whereas a railroad can only be perceived as a network, the management of vehicular transportation networks and corridors is occasionally presented as a revelation.

The deterioration of roadway surfaces and decks is more gradual and obvious than that of rail bridge floor beams. The automobile alleviated live loads and relaxed geometric constraints. Moreover, vehicular loads must be estimated statistically, whereas train loads are nearly deterministic. The relatively fuzzier interaction between vehicles, their bridges, and the numerous responsible owners has permitted a broad range of presumably feasible maintenance and management strategies (Chapters 11 and 12) based on a spectrum of deterioration models (Chapters 9 and 10) and priorities. Once the management of vehicular bridges has adjusted to middle age, it is likely to become as low key as railway bridge management already is. High-speed and magnetic levitation trains might supply the next high-profile problems.

2.2 OPTIMIZATION OF OBJECTIVES

Maximizing the benefits of bridges would be appropriate if they were perceived as individual projects. As vital links in a complex transportation network, they can be (at best) *optimized*.

Ang and Tang (1975, p. 12) point out that safety might be maximized by assuming "consistently worst conditions," but so would be the cost. The authors develop procedures for obtaining solutions that are "optimal, in the sense of minimum cost and/or maximum benefits" (Section 5.2).

Since neither the constraints nor the benefits are readily quantifiable, the optimization cannot be entirely rigorous. It involves a subjective interaction between two levels of management, one on the transportation network level, the other on the level of structures or projects. Network managers must allocate available funding according to negotiable sets of criteria. Project managers must justify budget requests and conduct bridge-related activities. If the two participants in Fig. 1.33a were to optimize unilaterally the utility of their assets, management would focus on budgets and engineering on structures. The objective of each would become the constraint of the other, as in a negotiated transaction or a "game." Money is often the only representation of "utility" conveying the same meaning to all negotiators. In the extreme cases when money becomes the only pertinent information to be managed, accounting assumes a key position.

In contrast, when management seeks to optimize the product, which may consist of service, it stimulates cost-effective engineering solutions. Small et al. (in TRC 498, 2000, A-1) present the network- and project-level bridge management system philosophies as "top-down" and "bottom-up" approaches. The former implements policies, the latter manages expenditures. Both comply with existing standards.

In recent decades truck traffic has increased in volume and weight, while transportation infrastructure networks have grown denser and older. TRB (1990) described the continuing evolution of truck loads on the highways of the United States. Management priorities gradually shifted from new construction to maximizing the benefits of aging assets. In the meanwhile science, engineering, and society in general have achieved new levels of integration owing to the digital computer. Bridge utility is no longer *maximized* in isolation, nor is it within the competence of bridge designers alone. The supply–demand relationship of Fig. 1.33*a* has been expanded to include a broader range of interests, as shown in Fig. 1.33*b*.

Example 3 highlights the debate whether to rehabilitate or replace a major mixed-use 90-year-old bridge in an urban environment (e.g., the Williamsburg Bridge in New York City). The San Francisco–Oakland East Bay Bridge (Section 1.5) similarly integrates structural, economic, and political considerations.

As many profound transitions, the priority shift from expanding a new system to maintaining an aging one was most acutely perceived at the surface. The roadway surface provides measurable transportation benefits. NCHRP Report 300 (1987, p. 1) stated: "Bridge management is not business as usual. . . . Bridge management is a relatively new concept that was adapted from successful application of systems concepts to pavement management functions."

The noteworthy departures from "business as usual" include network planning, advanced data acquisition, and processing. The intent is to distinguish the rigorously formalized optimization developed by operational research from management as a general practice. Nonetheless, the result is a definition of what bridge management *is not,* as in Drucker's general definition quoted in Section 1.8. Beale (1988, p. 1) similarly emphasized: "Before discussing what operations research involves, we will spend a few minutes distinguishing it from statistics."

NCHRP Report 300 (1987), O'Connor and Hyman (1989), and the Organization for Economic Cooperation and Development (OECD, 1992) define the bridge management tools and tasks and describe their potential applications.

NCHRP Report 300 (1987) summarized bridge management as follows (p. 1):

> The objective . . . is to develop a form of effective bridge management at the network level (that is, dealing with a group of bridges rather than with a single bridge) that will ensure the effective use of available funds and identify the effects of various funding levels. Bridge management . . . requires a practical, objective, and systematic consideration of the problem with a set of economic and technical tools not previously combined to solve the problem. Specifically, a bridge management system (BMS) is a rational and systematic approach to organizing and carrying out the activities related to planning,

designing, constructing, maintaining rehabilitating, and replacing bridges vital to the transportation infrastructure.

Beyond that point, the report discusses BMS. OECD (1992, p. 14) takes a similar view:

> Bridge management addresses all bridge-related activities from the moment of construction to the replacement of a particular structure. Bridge managers require an instrument that enables them to carry out their work efficiently, and guarantees the functionality and safety of bridges in the most economic way. In order to improve the ways and means of managing bridges which so far have been founded on an individual project-by-project approach, the sheer size of the existing bridge stock and capital invested has recently led member countries to search for a coherent and cost-effective network approach to bridge management. Bridge Management Systems are therefore being developed to provide guidelines for managers, which identify and clarify the implications and requirements of the system as to the technical and organizational structures.

The preceding quotes illustrate the struggle to define bridge management. As in Von Neumann's and Morgenstern's game theory (Section 1.4), the need for objectively quantifiable methods is emphasized. A consistent effort to avoid *prescriptive* in favor of *descriptive* (or in recent terminology *performance-based*) definitions is evident. Later publications, such as the National Research Council (NRC, 1994) and BRIME (2002), report on established bridge management practices and on the service record of BMS, thus contributing new empirical data.

NCHRP Report 300 (1987) addressed the decision support for pertinent activities on a network level, as shown in Appendix 13. Shirole at al. [in Transportation Research Record (TRR) 423, 1994, pp. 27–34] tabulated the decision support capabilities bridge management requires on network and project levels. Some of the bridge management flowcharts discussed in Section 6, such as BRIME (2002) (Fig. A16.5), clearly delineate project- and network-level operations. Network-level management addresses the process (e.g., transportation), whereas site-specific tasks must generate the product (e.g., the structures). Section 1.8 argued that the priorities of financial and bridge managers should be complementary rather than contradictory. The network manager demands transportation and supplies funding. The project-level engineer demands funding and supplies structural performance, as in Fig. 1.33a.

2.3 PRIORITIZING ACTIONS

As the scope of management narrows down from the national to the local network and to project levels, attention to details overwhelms a shrinking executive license. In Example 5, the president of the United States was sympathetic to the objective of overall social and economic development. In a follow-up letter, Chief Highway Administrator (later Secretary of Transportation) R. Slater more specifically defined the constraints as follows: "*Needs* far exceed the available funding."

Example 5. Views of High-Level Network Managers

In his letter of October 6, 1963, (Fig. E5.1) President W. J. Clinton stated that America must address its vast network of bridges and highways in order to remain a strong nation in the twenty-first century.

In a prompt follow-up letter (October 23, 1993, Fig. E5.2), the chief administrator of the FHWA and, later, Secretary of Transportation R. Slater emphasized that, despite the best intentions, money is "by its nature limited."

THE WHITE HOUSE

WASHINGTON

October 6, 1993

Mr. Bojidar S. Yanev
Assistant Commissioner
Bridge Inspection/Research
    and Development
New York Department of Transportation
Fourth Floor
2 Rector Street
New York, New York   10006

Dear Mr. Yanev:

    Thank you for your thoughtful letter
regarding New York City's infrastructure. I
agree with you that America must address the
problems of its vast network of bridges and
highways if we are to remain a strong nation
in the next century, and I have forwarded your
letter to the Department of Transportation for
futher review. I will keep your ideas in mind
as I face the great challenges ahead.

Sincerely,

Bill Clinton

Figure E5.1   Letter by W. J. Clinton, president of the United States.

U.S. Department
of Transportation

**Federal Highway
Administration**

Office of the Administrator

400 Seventh St. S W
Washington D C 20590

October 21, 1993
Refer to: HPD-1

Bolidar S. Yanev, P.E.
Assistant Commissioner
New York City Department
   of Transportation
New York, New York 10006

Dear Mr. Yanev:

This is in further reply to your August 27 letter to the President regarding the condition of New York City's major bridges. After replying on October 6, the President asked us to review your concerns.

The President mentioned that our network of highways and bridges is vital if we are to remain a strong Nation in the next century. The Intermodal Surface Transportation Efficiency Act of 1991 (ISTEA), which I am sure you are familiar with, is making it possible for us to address our highway and bridge needs while we develop a National Intermodal Transportation Network that will serve the Nation in an energy efficient, environmentally sound manner.

At the same time, the Federal Government faces some of the same problems that afflict New York City when it comes to our infrastructure. Needs typically exceed the means available to address them. At the national level, ISTEA has allowed us to set new funding records for highway and bridge projects, but still, needs far exceed the available funding. Resources, therefore, must be distributed as equitably as possible.

Under current apportionment formulas for the Highway Bridge Replacement and Rehabilitation Program, New York State receives more funds each year than all but one State, namely Pennsylvania, and has received more than Pennsylvania at times. (The State's share in fiscal year 1994 is $254,496,994.)

In addition, the State has actively sought discretionary bridge funds. Out of discretionary bridge-allocations totalling $2.7 billion through fiscal year 1994, New York State received $382,710,361 or 14.1 percent of the total. Of the amount allocated to the State, a total of $322,414,703 was for bridges in New York City (the Brooklyn, East Tremont, Manhattan, Queensboro, and University Heights Bridges). Thus, New York City bridge projects received nearly 12 percent of all discretionary bridge funds allocated nationwide during that period. For fiscal year 1994, we recently announced an allocation of $12,776,000 for rehabilitation and widening of Ramp D of the Brooklyn Bridge.

Clearly, if funds were unlimited, we would do more. However, we are proud of our role in helping New York City preserve its major bridges as important transportation links and as a valuable historic legacy for generations to come.

Sincerely yours,

Rodney E. Slater
Administrator

Figure E5.2   Letter by R. Slater, chief administrator, Federal Highways.

For the local networks and the individual projects (where the quoted correspondence originated), the options may be limited to closing unsafe bridges or spending the limited resources on hazard mitigation (Section 10.3). NCHRP Synthesis 331 (2004) elaborated:

> Key factors that drive the priority of projects on the STIP (Statewide Transportation Improvement Program) include safety, level of traffic (e.g., average daily traffic), consistency with long-range plans, cost-effectiveness, and condition of the existing facility. Development of the STIP is closely tied to the project development process with respect to the status of design completion. However, there are many different ways in which SHAs (State Highway Agencies) select specific projects for inclusion in their letting programs. There appear to be some dominant factors that influence this decision such as delivery status of the project, and project priority established by SHA district, regions, and divisions, with input from the FHWA.

With projects managed by engineers and budgets negotiated by politicians, bridge management is expected to provide objective assessments supporting structural engineering and transportation management decisions. On a national level, the assessment of infrastructure needs has significant implications. According to the National Bridge Inventory (NBI) 40% of the nearly 650,000 highway bridges and culverts in the United States are structurally deficient or functionally obsolete (Appendix 14). One of three bridge crossings is over a deficient bridge. Average annual rehabilitation and replacement expenditures amount to U.S. $7 billion nationwide (United States). BRIME (2002) reported on the bridges included in the national highway networks within the European Union as follows:

> The value of bridges on the national networks . . . is estimated at 12 billion Euros in France, 23 billion in the UK, 4.1 billion in Spain and 30 billion in Germany. . . . The high usage is causing congestion the costs of which are currently estimated at 120 billion Euros annually. . . . The annual expenditure on maintenance and repair on national bridges in England is of the order 180 million Euros, in France the figure is 50 million, in Norway 30 million and in Spain 13 million Euros.

Such critical assessments must employ the latest analytic and technological tools. However, the National Commission Report (2004, p. 339) recommended that management must also use imagination in reconciling local needs and global policies. When the constraints are too restrictive for rigorous optimization, a methodical alternative to compromise is needed to allocate the limited resources. *Optimization* takes the form of *prioritization.*

All resource allocation is optimized under conflicted constraints. Picon (1992, p. 38) pointed out that the transportation infrastructure engineering of France during the eighteenth century managed the territory and consequently had to reconcile the interests of governing administrations and land owners. Regional policies and priorities remain diverse. Appendix 11 outlines the steps taken by the Federal Highway Administration (FHWA) in order to accommodate bridge management priorities in the United States.

Appendix 15 describes the terminology and trends of multimodal transportation asset management. Project-level bridge management quickly teaches that all estimates of "remaining useful life," "level of service," and infrastructure "deterioration" are fraught with uncertainty (see Chapter 10). Considerable latitude and intensive further research are needed to accomplish the asset management goals.

Engineering societies and associations worldwide have responded to the demands of infrastructure management by conducting numerous congresses, such as ASCE (1993, 1997), PIARC (1996), and the International Association for Bridge and Structural Engineering (IABSE, 2000). Conferences and symposia are organized periodically on the subject, notably by the International Association for Bridge Maintenance and Safety (IABMAS). Other examples include École National des Pont et Chaussées (ENPC, 1994), Harding et al. (1990, 1993, 1997, 2000a, 2005), Forde (1999), Vincentsen and Jensen (1998), Das et al. (1999), Frangopol (1998, 1999a, b), Frangopol and Furuta (2001), Miyamoto and Frangopol (2001), Miyamoto et al. (2005), and TRC (2000, 2003). The NCHRP and the TRB of the FHWA have produced the most comprehensive and continually expanding library of reports addressing every significant development in the field.

Bridge design texts, such as Xanthakos (1994), Barker and Puckett (1997), and Taly (1998), discuss bridge management or its components, including inspection, maintenance, and cost–benefit analysis in introductory or concluding chapters. The bridge engineering manuals by Chen and Duan (1999) and Ryall et al. (2000) dedicate sections to the subject. White et al. (1992) discuss bridge maintenance, inspection, and evaluation. Xanthakos (1996) and Calgaro and Lacroix (1997) address strengthening and rehabilitation of existing bridges. Texts directly addressing bridge management, such as Troitsky (1994) and Ryall (2001), demonstrate, among other points, the impossibility of a comprehensive discourse on the subject. Hudson et al. (1997) achieve an encyclopedic overview of infrastructure management.

Software companies offer customized BMS. States, including North Carolina, Louisiana, Pennsylvania (TRR 1083, 1986, pp. 25–34), and New York, have developed their own BMS. Representative BMS flowcharts are shown in Appendix 16. All of them consist of database, input, and output modules intended to provide decision support for the responsible bridge owners at various management levels. The activities included in the flowcharts predate bridge management; however, their systematic treatment is expected to improve management qualitatively.

Bridge management is appropriately reluctant to call itself *management*, because, as shown in Fig. 1.33b, it does not determine its resources and long-term policies. Arner et al. (in TRR 1083, 1986, pp. 25–34) reported on the Pennsylvania Bridge *Maintenance Management System*. All activities managed by the system were assigned a priority ranking. Repairs were prioritized as follows:

1. Emergency—within 6 months

2. Emergency—within 12 months

3. Priority—within 2 years

4. Routine structural—can be delayed until funds are available

5. Routine nonstructural—can be delayed until programmed

Maintenance activities were ranked from A (highest priority) to E (lowest priority) "based on their generalized relative importance to the current structural stability of the bridge."

As of this writing the Pennsylvania Department of Transportation (PennDOT) is updating the BMS to reflect the evolving needs and conditions.

For its 221.2-km bridge network in the area of Kobe and Osaka in Japan, the Hanshin Expressway Public Corporation has introduced a Maintenance Information Management System. The report by Hearn et al. (FHWA, 2005b) describes predominantly maintenance management systems currently used in Europe, as does BRIME (2002). In his comprehensive chapter on the subject, Hearn (in Frangopol, 2002, p. 208) concluded: "Existing bridge management systems are maintenance management systems, but work in progress can extend these systems into evaluation of bridge strength and safety."

2.4 MINIMIZING RISK

Design practices have been strongly influenced by past failure, such as les Invalides (1827), Dee (1847), Tay (1879), Quebec (1907), and Tacoma (1940). All of these events have occurred during construction or early use. In response, design specifications, such as AREMA (formerly AREA, first published in 1905) and AASHTO (originally AASHO, 1928) have evolved from prescribing safety margins to modeling the levels of reliability in design and construction, as in AASHTO LRFD and Eurocode. The result should be safe structures in "as-built" condition. The safety of structures long in existence depends on the method of operation, for example, on the management and use of the assets, neither of which can be specified or controlled to the same extent as design and construction.

In recent decades the bridge useful life, once considered indefinite, became a debated design parameter. The AASHTO has set a default value of 75 years; however Chase et al. reported an average bridge life of 42 years (in TRC 498, 2000, C-6). Even vital link bridges whose performance should be above average are affected. *The New York Times* (January 17, 2006, p. B1) reported the 50th anniversary of the Tappan Zee Bridge (Fig. 2.2) as follows: "The Tappan Zee turns 50, a risky age for its kind."

The loss of bridge services at relatively early structural ages has suggested deficient funding, maintenance, and ultimately management. Interest in risk analysis and inspection was stimulated. Since these tasks cannot be codified as rigorously as are analysis and design, a number of *recommendations, guidelines, manuals, instructions,* and *advisories* have proliferated in lieu of *specifications.* Such are AASHTO (1983, 2000a and b) on maintenance, FHWA (1986, 1995a, 2002a and b) on inspections, and AASHTO (1989, 2000, 2003) on condition evaluation.

All of these texts are in some degree motivated by specific events. Risk and its complementary safety are perceptions rather than objectively quantifiable parameters. If a structure is declared unsafe at the midpoint of its expected useful life, the resulting economic loss may be perceived as a management failure, whereas the averted collapse could be claimed as an engineering success. In contrast, if a bridge collapses during active service, the failure affects the users, who blame summarily design and management. The *Engineering News Record* (ENR) issue of September 12, 2005 reported the break of the levee system around New Orleans during the hurricane Katrina as a "breach of faith." The same sentiments were expressed after the fatal collapse of a ceiling panel

Figure 2.2   Tappan Zee Bridge, Hudson River.

in a tunnel that is part of the "Big Dig" vehicular network in Boston on July 10, 2006. On that occasion the media stated that "the infrastructure is an extension of the administration." The loss of public and professional confidence combined with a demonstrable loss of structural integrity remains the driving force in management. BMSs provide support by quantifying the consequences of management decisions and by estimating the likelihood of various outcomes, most significantly that of failures.

Chapter 3

Failures

Because of their professional commitment to succeed, engineering and management diverge from art and science in their attitude toward failure. Valéry (1941, p. 95) proposed "to call *Science* only the ensemble of recipes that always succeed; the rest is literature." Descartes and Newton, however, raised science above success. Rather than hypothesize what might be, they investigated what is (Section 1.8). Rudyard Kipling (1865–1936) recommended treating "Triumph and Disaster . . . those two impostors just the same." Success, ever the impostor, depends on uncertain circumstances and perceptions. Santayana (1928, p. 60) viewed it as a "brief erratic experiment made in living." Contemporary science has come to terms with uncertainty (Section 1.9), whereas art embraces risk. Hence, engineering and management must ensure success without firm theoretical support and guaranteed artistic approval.

Daring spans (perhaps even more than tall towers, because of their greater utility) embody success. Although evident and durable, however, that success is not necessarily reproducible. The simplicity of the final product usually conceals important successive approximations and fortuitous coincidences. A single outcome does not prove infallibility, because it may be the result of an unstable process. A reliable routine precludes innovation and fosters complacency. The boundaries within which engineering systems remain successful (or safe) cannot be strictly defined, because the relevant properties, the operating conditions, and the demands cannot be fully modeled, much less foreseen.

Failures, in contrast, can be reconstructed and systematically eradicated. Thus the pursuit of success assumes the form of avoiding failure. "Success is foreseeing failure," stated Petroski (1992, p. 53). That foresight is sought in theory, in practice, in extratechnical disciplines, and even in the occult. Godfrain (2003) has woven the many controversies and spectacular ultimate success of the Viaduc de Millau (Fig. 1.44) into a highly entertaining plot engaging art, architecture, politics, love, religion, and superstition. In a couple of edifying personal appearances, Lucifer explains that bridges "revolt against their creators" when the latter ignore the mystical and potentially destructive powers taking a keen interest in such creations. His examples include the collapses at Quebec and Tacoma and the temporary closures of the Millennium Bridge in London (Fig. 3.1) and the Passerelle de Solferino in Paris (Fig. 3.2). In contrast, roughly 300 bridges survive in France under the name Pont du Diable, most prominently the landmark one in Cahors. All resemblance with real characters is disclaimed on page 1, and engineers are appropriately absent from the text.

In the technical domain, Farrar, Lieven, and Bement (in Inman et al., 2005) cite the Aloha Airlines aircraft fuselage failure in 1988 as a trigger for the Aging Aircraft Pro-

Figure 3.1   Millennium Bridge, London.

(*a*)

Figure 3.2   Passerelle de Solferino, Paris.

(b)

Figure 3.2   Passerelle de Solferino, Paris.

gram instituted by the Federal Aviation Administration (Chapter 1.3, Motivation for damage control prognosis). A direct analogy can be drawn between that reaction and the consequences of several noteworthy bridge malfunctions. Foremost among them is the collapse of the Silver Bridge across the Ohio River (Example 6), but other examples have also enriched the "failure database."

Example 6. Silver Bridge, Point Pleasant

On December 15, 1967 the "Silver" Bridge carrying U.S. Route 36 over the Ohio River at Point Pleasant collapsed. Forty-six motorists perished, setting the twentieth-century record in the United States, triggering the developments highlighted in Appendix 11.

The evolving viewpoint of contemporary ENR reports (Ross, 1984, pp. 239–253) aptly illustrates post-collapse speculations. The initial assessment (p. 241) was that "the collapse may never be solved" because of the (not entirely independent) contributions of "age, corrosion, fatigue and overload."

The forensic investigation of the Silver Bridge identified several factors contributing to the failure. Despite a deep stiffening truss, the bridge relied on a globally nonredundant pin and eye-bar suspension system. In contrast to other multi-eye-bar systems (such as those shown in Figs. E1.3 and E1.14), the failed one consisted of only two eye-bars. Thus the globally nonredundant suspension was matched by the

Figure E6.1 Cracked eyebar from the Pasco-Kenniwick Bridge, Washington. Courtesy of G. Washer and M. Lozev.

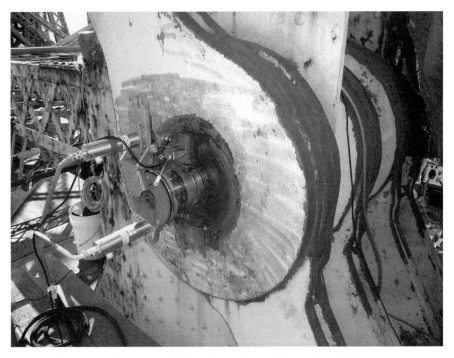

Figure E6.2 Ultrasonic testing of pins and eye-bars at Queensboro Bridge.

internal nonredundancy of the main members. Local material imperfections were subsequently detected. Ambient temperature may have been around $-23°C$, further embrittling the material. The numerous fracture-critical details could not have been inspected visually at safe intervals. Fatigue life estimates would only have guessed at the actual stress levels and amplitudes at critical locations. No preventive maintenance measures could have been effective. In retrospect, every stage of the Silver Bridge life-cycle placed a greater burden of responsibility on the following stages.

(Rolfe and Barsom, 1987, p. 4) quote Bennett and Mindlin (*Journal of Testing and Evaluation,* March 1973, pp. 152–161) on the metallurgical aspects of the failure of the Point Pleasant Bridge:

> The fracture resulted from a combination of factors; in the absence of any of these it probably would not have occurred: (a) the high hardness of the steel which rendered it susceptible to stress-corrosion cracking; (b) the close spacing of the components in the joint which made it impossible to apply paint to the most highly stressed region of the eye, yet provided a crevice in this region, where water could collect; (c) the high design load of the eyebar chain, which resulted in a local stress at the inside of the eye greater than the yield strength of the steel; and (d) the low fracture toughness of the steel which permitted the initiation of complete fracture from the slowly propagating stress-corrosion crack when it had reached a depth of only 0.12 in (3.0 mm).

Once the vulnerability of the two eye-bar suspension system was identified, a number of bridges with that structural feature were decommissioned or closely investigated. Figure E6.1 shows a crack in an eye-bar from the Pasco-Kenniwick Bridge in Washington State. The condition of multi-eye-bar systems is not considered as critical, because of their redundancy; however, ultrasonic testing is recommended, as shown in Fig. E6.2.

On June 28, 1983, a span carrying Interstate 95 collapsed over the Mianus River in Connecticut as a result of a failed pin and hanger assembly. The FHWA published the manual for inspection of fracture-critical bridge members in September of 1986.

On April 3, 1987, a New York State Thruway Authority bridge fell in the Schoharie Creek as a result of scour. The Surface Transportation and Uniform Relocation Assistance Act of 1987 expanded bridge inspection programs to include special procedures for the inspection of fracture-critical members and underwater components.

On October 17, 1989, the Loma-Prieta 7.1-magnitude earthquake caused an estimated U.S. $6 billion damage and 42 deaths in the San Francisco–Oakland Bay area (Housner, 1990). In 1990 Presidential Order No. 12699 mandated seismic design provisions for federally funded projects.

The Williamsburg Bridge closure in 1988 (Example 3 herein) coalesced the New York City bridge management effort. After the failure of the Sungsu bridge in Seoul in 1994, bridge inspections in Korea were mandated by law in 1996.

3.1 UNDERSTANDING AND/OR AVOIDING

Aristotle (Section 1.2) appears to have credited science with avoiding failures by virtue of experience and art with understanding their causes. The terminology has since undergone the complete cycle he predicted for learning in general—modern science theorizes and art improvises. Nonetheless, Aristotle's conclusion that experience can "succeed even better than . . . theory" remains key to avoiding failures whose causes may not be fully understood. Leonardo da Vinci (1935) wrote in his notebooks: "Experience is never at fault, it is our interpretation that is in error."

Under the title "Falling Down Is Part of Growing Up," Petroski (1992, Chapter 2) presented failure as an inevitable part of the creative empirical process, which advances by trial and error. The author argued that scientific researchers can reject hypothesizing (as did Newton) because they discover existing truths, however engineers cannot, because they create new entities. According to Karnakis (1997), the analysis of failure causes evolved concurrently with and sometimes independently from the empirical elimination of incalculable risk.

Aversion to risk is the subject of De Finetti (1974, Section 3.2). His examples, mostly derived from economics, point to a strong preference for limiting uncertain gains in exchange for reducing the likelihood of equally uncertain losses. In the domain of statesmanship, the U.S. Declaration of Independence (July 4, 1776) stated: "Prudence, indeed, will dictate that governments long established should not be changed for light and transient causes; and accordingly all experience hath shewn, that mankind are more disposed to suffer, while evils are sufferable, than to right themselves by abolishing the forms to which they are accustomed."

Paraphrased as "if it ain't broke, don't fix it," this principle often guides not only social development but also design and management. Its conservatism perpetuates existing theory and accepted practice. On the other hand, things do wear out eventually, and at that point engineering is expected to have a superior alternative ready for use, whereas management is held responsible for ensuring a smooth transition. Change therefore must be engineered and managed so as to maximize the benefits of the transitions from old to new operating structures, mechanical or otherwise, without access to all the pertinent information.

Sibly and Walker (1977) and Petroski (1993) detected a cyclic trend in the failures at the Dee Bridge in England (1847), the Tay bridge in Scotland (1879), the Quebec Bridge (1907), and the Tacoma Narrows Bridge (1940). The cycles begin with cautiously successful (partly empirical) designs and culminate by overextending the practice beyond the validity of the model. The failure of the Dee Bridge at Chester (Petroski, 1994) is presented as an example of the "success syndrome," essentially an error of complacency. Structural failures of this type result from applying known routines beyond their valid range.

The relative contributions of theory and experience or those of design and management to the successful use of a bridge over the course of centuries cannot be quantified and compared. Nevertheless, engineering is usually blamed for theoretical oversights causing failures of the product. The mismanagement of the process is customarily attributed to inexperience. The refinement of engineering product and process has produced distinctions such as nonperformance, malfunction, local, partial, and catastrophic failures, to be kept under close scrutiny in theory and avoided in practice.

3.2 CATASTROPHIC FAILURES

Catastrophic failures command public attention, spur legislation, and shape professional practice. They are studied in the greatest detail and contribute the most enduring lessons.

Farrar, Lieven, and Bement (in Inman et al., 2005, Chapter 1) state: "In the civil engineering field, the driver for prognosis is largely governed by large-scale discrete events rather than more continuous degradation."

Catastrophic failures are discussed in the introductory chapters of design textbooks, such as Taly (1998) and Barker and Puckett (1997). Feld (1968) and Ross (1984) refer to construction failures; however, both authors examine every stage of the structural life. Ryall (2001, p. 416) lists 22 bridge failures of the last 30 years, noteworthy because of their causes and magnitude. Design, construction, maintenance, collision, and scour are among the main causes. The Millennium footbridge in London (Fig. 3.1) stands out from the list, as it suffered no damage but had to be closed and dampened (Dallard et al., 2001). This is a "non-conformity with design expectations", as defined by Feld (1968, p. 2), in this case exacerbated by the even higher expectations of the general public. The point is made that a failure to perform as designed is still a failure. The "nonperformance" is highlighted by the prominence of the structure. The Passerelle de Solferino in Paris (Fig. 3.2) suffered a similarly conspicuous opening delay. Although affecting only the serviceability of the structures and not their integrity, these failures may have been catastrophic to the managers involved.

Catastrophic failures in civil structures are atypical and eventually acquire a mythical aspect. Related details are exaggerated, embellished, or otherwise distorted. Partial and near failures are much more abundant but easier to overlook and misinterpret.

3.3 PARTIAL FAILURES, NEAR FAILURES, AND OVERDESIGN

The errors, identified by Levy and Salvadori (1992) as potentially fatal, are occasionally in evidence at ("safely") surviving structures. After the Point Pleasant Bridge collapse a similar structure crossing the Ohio River at St. Mary's, West Virginia, was promptly decommissioned without incident (Taly, 1998). In engineering terms, that structure had failed, albeit not catastrophically.

"Overdesign" can be considered a failure of both design and management because it is cost ineffective (e.g., wasteful). What is excessive by one standard, however, may be routine according to another. As shown in Chapter 5, the terms *reserve, safety,* and *reliability* stem from different analytic modeling of reality, a wide variety of objectives, and may be differently defined. Overdesign is an easy target for elimination in hypothetical structures, but it is not so readily credited for averting any hypothetical failures of enduring actual structures. Moreover, the elimination of overdesign is invariably justified by an utterly unlikely "overmaintenance."

The role of structural reserve is uniquely demonstrated during partial or local failures. Failures remain limited or partial because of load redistribution. When anticipated by design, they are no longer perceived as failures at all. The forming of plastic hinges in steel or ductile behavior in concrete are examples. The service limits of structures are most accurately demonstrated under partial rather than total failures. Properly investigated, partial failures can point to the structural assets or attributes that have prevented a total collapse or failure.

The bracing in Fig. 3.3 yielded during the Loma-Prieta earthquake in 1989 and was replaced with the stronger bracing shown in Fig. 3.4. Yet it performed adequately during the earthquake.

Figure 3.3   Bracing, yielding during the Loma-Prieta earthquake, San Francisco, 1989.

At the Williamsburg Bridge (Example 3), corrosion had reduced the estimated reserve of the main suspension cables from a factor of 4.2 to 3.6, where it could be maintained after an intensive rehabilitation. Thus the cable overdesign typical of the late nineteenth century, but unlikely in the late twentieth century gave the bridge managers the option to maintain it in service. This and similar incidents prompted NCHRP Project 10-57 to investigate the safety of suspension bridge cables in general.

The stay rupture at the Brooklyn Bridge in 1981 proved fatal to one pedestrian, but it did not propagate through the structure. The entire system of vertical suspenders and diagonal stays had suffered advanced corrosion and was beginning to fail. The suspension, however, was multidegree redundant (Fig. 3.5). Suspenders were spaced 8 ft (2.4 m) apart, the stiffening truss was substantial, and diagonal stays provided additional stiffening. J. Roebling (in Reier, 1977, p. 16) had stated that, if one of these three primary

Figure 3.4   Replaced bracing of Fig. 3.3.

systems failed, the bridge "would sag but not fail." That design philosophy not only averted failure but allowed the owner to replace all stays and suspenders under traffic.

Contemporary cable-stayed and suspension bridge design anticipates the failure of individual stays and suspenders. A suspender broke at the first Bosporus bridge in 2004, but traffic was not interrupted. Traffic was maintained at the Pont de Tancarville after one strand of the main cable broke.

Attention to strength reserve is evident throughout Roebling's design. Figure 3.6 shows a truss span on the Brooklyn Bridge approach. Given the age of the structure (opened in 1883), load rating must assume that its material is iron with 10 ksi (69 mPa) ultimate strength. Under that assumption, the bottom chords of the trusses would be overstressed. Furthermore, the pinned eye-bar connections showed indications of bending, rather than pure shear. The truss had to be supported by arches as shown, but traffic was never interrupted. Continuous monitoring did not reveal any distress during the three-

Figure 3.5   Hybrid suspension and stay system, Brooklyn Bridge.

Figure 3.6   Frankin Square trusses and supporting arches, Brooklyn Bridge approach.

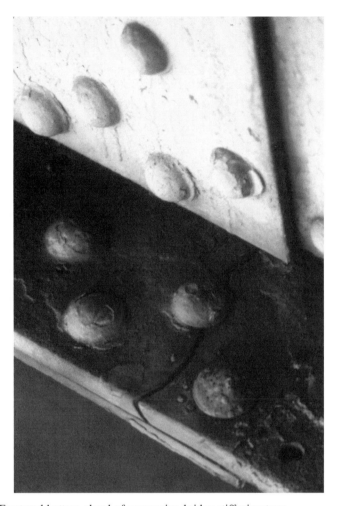

Figure 3.7 Fractured bottom chord of suspension bridge stiffening truss.

year period between the engineering determination and the completion of the project. A test of a material sample indicated much greater strength than the specified value. Strain-gauge monitoring suggested that live load caused much lower stresses than anticipated in the primary members.

Figure 3.7 shows a fracture of the bottom chord of a suspension bridge stiffening truss that did not have consequences to the structure.

Figure 3.8 shows a column failed at the base. A nonredundant structure could have failed as a result of this condition.

Bearings and other load transfer and release devices often fail or perform below expectations, as in Example 7, without global consequences, because of redundancy, adequate strength reserve, and geometric tolerance.

Figure 3.8 Column disconnected at base by corrosion.

Example 7. Malfunctioning Linkages

A. TIE-DOWNS

The tie-downs shown in Fig. E7.1 are superficially reminiscent of pin-and-hanger assemblies (Chapter 14) and shear release linkages (Section B below); however, their intended function is entirely different.

The bridge girders shown in Fig. E7.1 were designed for a minimum of 50 kips (22.5 tons) tension in the tie-downs at the ends of the girders. That force would counteract live loads and reduce bending moment midspan, allowing the girder depth to vary as shown. The resulting moment diagram is shown in Fig. E7.2a.

Vulnerable in this case was the assumption that the relatively crude detailing of the pinned rigid tie-downs and the rocker bearings located under the expansion joints could sustain the required tension under service conditions. That this was unrealistic is demonstrated by computing the elastic extension of the tie-downs (Fig. E7.2b) under the prescribed load of 50 kips (22.5 tons) as follows:

$$\Delta_{el} = \frac{50.0L}{EA} = \frac{50.0 \times 90.00}{29,000 \times 24} = 0.006 \text{ in.}$$

or

$$\Delta_{el} = \frac{22.5L}{EA} = \frac{22.5 \times 2286}{21 \times 15484} = 0.16 \text{ mm}$$

Figure E7.1 Girders with tie-downs.

where E = 21 tons/mm^2 (29,000 ksi)
$\quad\quad A$ = 15484 mm^2 (24 in.^{2}), cross-sectional area
$\quad\quad L$ = 2286 mm (90 in.), length

Visual inspections (Fig. E7.3) could not have detected a reduction of the tension in the tie-downs.

Fifty years of debris accumulation, corrosion, and thermal effects not only eliminated but also reversed the calculated Δ_{el}. Some of the tie-downs became loose, others buckled in compression, thus increasing the midspan positive bending moment, as shown in Fig. E7.2a. Attempts to restore the tension by cutting and rewelding the tie-downs (Fig. E7.1) proved ineffective.

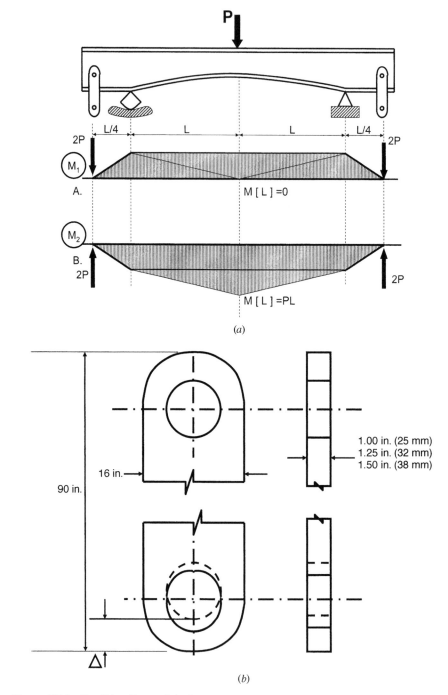

Figure E7.2 Possible effects of tie-downs.

Figure E7.3   Hands-on inspection of tie-downs. Courtesy of S. Teaw.

The total failure of the tie-downs was compensated by the strength reserve of the girders. The bridge remained in operation for more than a decade after the condition was first observed. The eventual rehabilitation retained the original girders, adding new ones in the longest spans. The tie-downs were retrofitted as shown in Fig. E7.4, allowing for adjustments in their tension. The rocker bearings were replaced with pot bearings.

Figure E7.4   Retrofitted tie-downs.

B. LINKAGE FOR LATERAL LOAD RELEASE

The 11-ft (3.35-m) "linkage" in Fig. E7.5 was intended to allow lateral displacement between the top of a 90-ft (27-m) pier and the bridge above it. Over a period of 70 years the linkage "froze." Seasonal and daily variations in the ambient temperature dragged the pier (Fig. E7.6) laterally so that the load of the superstructure was

Figure E7.5   Hands-on inspection of linkage.

Figure E7.6   Tilted 90-ft (27-m) pier.

supported by only two of the four pier legs. A source of concern was the integrity of the stone masonry (Fig. E7.7) supporting the steel pier. It proved adequate for the eccentric load over the time of the rehabilitation. The two legs could carry the full load safely until the frozen linkage was replaced with a different detail (Fig. E7.8).

Figure E7.7   Uplift at base of pier in Fig. E7.6.

Figure E7.8   Roller bearing replacing linkage of Fig. E7.5.

Partial failures motivate two opposite types of developments. One is to eliminate nonperforming details, the other to improve structural reliability, for instance by redundancy. In many cases both options are exercised concurrently. The tie-downs and bearings shown in Example 7A were replaced, but new girders were also added to the superstructure. The pier in Example 7B had been added to the bridge after it had already been in service for more than a decade. The malfunctioning linkage was eventually replaced with a modern nickel alloy steel roller, while the pier was strengthened.

Whereas partial failures receive full professional and public attention, "near" failures are easily overlooked until they fully develop. Many are the examples of potentially hazardous practices that remain common until they fail. It is therefore essential to closely monitor the "close calls" and the "near misses."

The overextended sliding bearing discussed in Section 4.2.3 was observed by a timely and competent inspection. Although no accident occurred, the condition was followed by emergency repairs, investigation, and capital rehabilitation. Approximately 10 years elapsed before the rebuilt structure carried train traffic again.

Construction safety procedures consider close calls as failures, requiring investigation and appropriate action. A similar policy benefits the entire bridge management process, including analysis, design, construction, maintenance, inspection, and operation (Section 4.1.3).

The proverbial assumption that "anything that can go wrong will," known as Murphy's law (optimistic version), may serve as an inspection wake-up call (it is attributed to a World War II aircraft inspector) but amounts to "hands-off" forecasting. Petroski (1992, p. 28) appreciated it as "a joke." The law evades criticism by admitting that it, too, will go wrong at the first opportunity, which it does. By its logic, any uneventful performance proves that nothing could have gone wrong. That conclusion is false. In his minority report to the Space Shuttle Challenger Inquiry in 1986, Feynman (1999) debunked such "elliptic" reasoning as follows (p. 155): "The fact that this danger did not lead to a catastrophe before is no guarantee that it will not the next time, unless it is completely understood. When playing Russian roulette, the fact that the first shot got off safely is little comfort for the next."

During the Northridge earthquake in 1994 structures that had survived earlier earthquakes suffered considerable damage. During the Loma-Prieta earthquake in 1989, the Embarcadero elevated highway in downtown San Francisco failed (Fig. 3.9), but not catastrophically. That performance demonstrated the critical accumulation of causes leading to the collapse of the Cypress Avenue Viaduct in Oakland (Fig. 3.10) during the same event (Nims et al., 1989; Housner et al., 1990). The Viaduct carrying Route I-880 failed because of the discontinuity between the lower and upper level frames, the poorly confined column reinforcement, *and* the deep layer of "bay mud" underneath (see Section 3.5).

Near failures and overdesign reveal an engineering contradiction. Whereas analysis strives to understand and eliminate failure causes from the designed product, management is wise to assume that knowledge is incomplete and structures must survive despite a level of ignorance accompanying every stage of the imperfect process.

Figure 3.9 Embarcadero elevated highway, San Francisco, October 29, 1989.

3.4 CAUSES

Levy and Salvadori (1992) concluded that "structures fall down" because of errors. That position reflects the Cartesian judgment that "all error is a fault" (*toute erreur est une faute*).

Consistently with Leonardo da Vinci (Section 3.1), Schopenhauer (1942, p. II-105) defined error as "making a wrong inference, that is, ascribing a given effect to something that did not cause it." On the subject of experience, he pointed out (p. II-104): "Perhaps in no form of knowledge is personal experience so indispensable as in learning to see that all things are unstable and transitory in this world."

Figure 3.10 Cypress Avenue Viaduct (Route I-880), Oakland, October 29, 1989.

Forensic investigations of failure causes reconstruct errors in the process and flaws in the product, expanding the professional experience.

Thoft-Christensen and Baker (1982, p. 241) attributed structural failures to *gross errors* that may be found at all stages of the building *process*. The latter term encompasses design, analysis, construction, inspection, and use which form the structural life cycle. The following primary deficiencies were identified:

- Formal qualifications
- Education
- Experience (most frequently)
- Communication
- Authority
- Competence
- Negligence and "sharp" practices

The same deficiencies are sorted out according to a number of different possible criteria (p. 242):

(a) Nature (or source) of the error, for example, design, analysis, construction, inspection, and use

(b) Type of failure: (A) in a mode as designed and (B) in a mode not as designed

(c) Consequences

(d) Responsibility (original and oversight)

The distinction of subgroup (b) can be compared to the difference between vagueness and ignorance, two of the three main types of uncertainty (Appendix 5). The authors then focus on the third one (e.g., randomness).

Whereas Thoft-Christensen and Baker (1982), Melchers (1987), and Schneider (1997) seek analytic flaws, Brown (1995), Ratay (2000), and Neale (2001) find them in design, fabrication, construction, and management by forensic analysis. Neale (2001, p. 83) lists the following operational causes:

- Poor quality of build, pointing to design, materials, and workmanship
- Inappropriate methods of build, including loadings imposed during the process
- Delivering more efficient structures within a competitive environment and perhaps novel features (which are inadequate in some way)
- Inadequate monitoring of existing facilities to deal with changing conditions, including loadings
- Ineffective use and maintenance strategies by owners and/or controllers of facilities
- Not realizing or understanding the consequences of particular actions or inactions
- Inadequate knowledge and experience of those involved
- Ineffective management of knowledge, including lack of ready access to information

Failures are thus attacked on both the analytic and applied fronts. Theory determines the ultimate capacity of models; actual failures shape management practice. Strictly within the domain of structural behavior, Gourmelon (1988) recommended both "materialist" and "phenomenological" lines of investigation, the former focusing on mechanics, the latter on load history. The periodical *Engineering Failure Analysis* (Pergamon, Elsevier Science) offers rigorous discussions of structural failures.

3.5 COMPOUNDED EFFECTS AND COINCIDENCES

Forensic investigations tend to reveal decision patterns or chains, rather than single causes, precipitating global structural failures. This is so by design. As much as possible, engineering design relies on redundancy to improve reliability in both the product and the process. When that is the case, total failures are the cumulative effect of several partial failures. Neale (2001, p. 51) stated: "In practice it is extremely rare for a failure to have a single cause. The common situation is that a number of factors occur simultaneously or sequentially, and the critical combination gives rise to the incident."

Numerous examples point to coincidences, rather than single causes of failures. The Point Pleasant failure (Example 6) proved particularly instructive because it had been just as likely to occur elsewhere for a number of similar reasons.

After the failure of the covered pedestrian bridges at the Charles de Gaulle Airport on May 23, 2004, (Fig. 3.11), *The New York Times* (May, 25, 2004, p. A10) reported:

Figure 3.11   Failed shell structure at Charles de Gaulle Airport, Paris.

"It is too early to say whether a design flaw, engineering error or construction mistake caused the collapse."

Distinctions are vaguely implied between *design flaw, engineering error,* and *construction mistake. Le Monde* (May 25, 2004, p. 18) phrases the question more specifically: "Is the collapse . . . due to a design flaw? Error in the materials? Or perhaps a construction too fast and poorly controlled?"

The *Engineering News Record* (ENR, May 31, 2004, p. 11) described construction delays and repairs of supporting columns with external fiber reinforcement prior to the collapse of the shell superstructure. In *The New York Times* of July 7, 2004 (p. A3), C. Smith reported that the collapsed shell structure was of "great complexity" due to its asymmetry, had suffered punctures during construction, and "folded like a wallet." The latter statement invites speculation about the formation of a collapse mechanism. That hypothesis is supported by evidence of steel struts puncturing the concrete shell structure (ENR, July 12, 2004, p. 10). Subsequent reports (ENR, Sept. 27, 2004, p. 16) focus on the quality of the concrete and the reinforcement. Preliminary results of the investigation released on February 15, 2005, cite concrete fatigue (Section 4.3.1) induced by thermal expansion and contraction. As in the case of the Silver Bridge (Example 6), the contribution of the individually noncritical vulnerabilities to a compounded catastrophic effect was not anticipated and is hard to evaluate in retrospect.

Design, construction, maintenance, and service were summarily questioned following the collapse of the suspended truss span of the Sungsu Bridge across the Han River in Korea (1994).

In reference to the effects of Hurricane Katrina, Professor R. G. Bea of University of California, Berkeley, is quoted by *New York Times* (June 11, 2006, p. WK 3, col. 2) to the effect that big failures "generally develop over long periods of time, involve large numbers of people and different organizations, and involve a multitude of breakdowns or malfunctions."

Uniqueness of engineering masterpieces is readily recognized, but failures are expected to follow predictable scenarios. Contrary to the popular expression, neither triumphs nor disasters endow survivors with perfect hindsight. The contributions of coincidences to failures are as important and elusive as they are to successes.

3.6 LESSONS

Seneca (A.D. 4–65) considered knowledge "not justified unless shared with others." After the Charles De Gaulle Airport collapse, an ENR editorial (May 31, 2004) appealed: "We hope that the worldwide industry studies this event intensely to squeeze every last lesson out of it."

The current litigation will delay the publication of detailed information beyond the attention span of the initial media outburst.

Feld (1968, p. 2) expressed hope that his study of construction failures would "loosen up the files and records of private investigations for the benefit of the entire industry." Ross (1984) summarized *Engineering News Record* reports of noteworthy earlier failures.

Teaching reduces knowledge into significant information, whereas learning absorbs information into knowledge. Mach (1838–1916) observed (1956, p. 1787): "It is the object of science to replace, or *save*, experiences, by the reproduction and anticipation of facts in thought. Memory is handier than experience and often answers the same purpose."

Information is more manageable than knowledge. Whereas experts are deriving knowledge from information (such as specifications) and experience (such as failures), "knowledge-based" systems attempt to express that knowledge in terms of transferable information. Thus, after evolving from experience, knowledge is parsed into information and retested under new circumstances by inexperienced users. A perpetual cycle of intended high points and unintended low ones evolves. Successes are expected to add to the growing structure of scientific knowledge. Errors and faults become targets of elimination, an activity typically associated with art. Each marks the professional and public memory in its specifically biased way.

According to Gordon (1978, p. 63), "a deep intuitive appreciation of the inherent cussedness of materials and structures is one of the most valuable accomplishments an engineer can have." This oblique inherent cussedness of structures must be managed by combinations of the two available methods: analysis and empiricism (e.g., demystification and accommodation).

Not every construction feat advances the professional practice of design. The presumed lessons of successes are shared (reluctantly or effusively) by practitioners who are not always fully aware of their implications. Successful empirical solutions do not teach an enduring lesson until theory has explained them. Roebling's Brooklyn Bridge was followed by less brilliant but theoretically clearer designs. Without rigorous analysis, failures too are no more than cautionary tales—either exaggerated or forgotten. Post-failure reactions owe more to spontaneity than to design. After the collapse of the Quebec Bridge in 1907, the design of the Queensboro Bridge in New York City (an entirely different, although still "cantilever," truss) was publicly questioned, reviewed, and found conservative (Reier, 1977; Petroski, 1995).

The lessons of bridge failures must be deduced from unforeseen and mostly unobserved events (the filming of the Tacoma Bridge collapse remains as unique as was the event). The few planned demolitions have yielded inconclusive but very interesting findings, recommending more such investigations in the future. Testing to failure (proof load testing) is required when feasible, as in determining the ultimate strength of wire ropes. Even when successful, such tests advance primarily a specific design.

Deriving benefits from disasters is an extreme test for engineering management. Among their many accomplishments, Navier and L. Moisseiff are credited for their willingness to analyze the failures of their own bridges, at Les Invalides in Paris and Tacoma Narrows in Washington State, respectively.

The following sequence can be selectively extracted from the history of suspension bridges. In 1823 Navier published the seminal *Memoire sur les Ponts Suspendus,* showing a particular admiration for the designs of Sir Marc Isambard Brunel (Section 1.8.) On a visit to Navier's Pont des Invalides in 1826, Sir Marc prevented his son Isambard Kingdom Brunel from setting foot on the bridge for fear that he would end up in the river (Hopkins, 1970, p. 203). With the anchorages rapidly deforming, Navier had to demolish the bridge in 1827 (Karnakis, 1997). Before his early death, he published a detailed analysis of the failure. I. K. Brunel's favorite suspension bridge at Clifton, spanning 702 ft (214 m) was completed after his untimely death and still survives (in a rehabilitated version).

Was Sir Marc prescient or merely cautious? Was Navier overly confident in theory? Did Isambard learn more from his and his father's experience at les Invalides or from Navier's analysis? The questions are particularly instructive, because they cannot be answered definitely. Failure is an edifying but not necessarily a scientific experience. The generally applicable information extracted from failures does not add up to professional knowledge. Specifications codify design, but practitioners must have matching knowledge, in order to produce structures of quality.

Shepherd and Frost (1995) show that the mere enumeration of failures and their identified causes does not necessarily advance practice. Foreseeing failures is an analytic success, but it is up to management to avert them. Shortly before the German invasion of France during World War II, general Gamelin, commander in chief of the French army, accurately predicted it to Romains (1940). The author marveled at this futile "dreamlike lucidity" as follows (p. 102): "A dreamer, for instance, if he were an architect, dreams in great detail of the admirable buildings that he would like to construct; but he neglects one last detail—petty, tedious, and rich in unforeseen annoyances—which is, to build."

Despite its merits, the study of failures remains "reactive." The general who passively envisions defeat is more vulnerable than the designer who does not build, because the army is an already committed asset. This is also true of the built infrastructure; which is to say that it is already performing under the attack of loads (anticipated and unexpected).

The "proactive" alternative is to identify and eliminate vulnerable features of the product and the process. The task can be a two-step perpetual cycle, consisting of (1) gathering and assessing information and (2) taking and implementing decisions.

Chapter 4

Vulnerabilities in Product and Process

Vulnerabilities (past, present, and future) must be sought in the following tasks and constraints:

Practices and outcomes

Conditions and resources

Needs

Credible hazards

Alternative options

The demands are contradictory and require the typical engineering optimization, intended to maximize performance and minimize expenditures (Section 2.2). Analysis must model structures simply but accurately. Design should be economical but safe and consistent with guidelines but innovative. Management seeks to improve the quality of life but at minimum cost. The infrastructure consists of durable static assets that must be managed with dynamic adaptability. Bridges must last for centuries, but the service demands and management policies change.

The opposed demands increasingly polarize competence, particularly in the following specialized areas:

Supply	Demand
Benefit	Cost
Structure	Operation
Product	Process
Project	Network
Service	Performance
Condition	Need
Target	Task
Ground up	Top down
Hazards	Penalties
Engineering	Management

These categories can be broadly grouped as shown in Table 4.1.

Table 4.1   Demand and Supply and Product and Process in Management and Engineering

Competence	Demand/Product	Supply/Process
Management	Services (transportation), functioning organization, decision and engineering support: funding, time and material, equipment	Decisions, implementation, budgeting, planning, reporting, communications, coordination, negotiations, supervision, quality assurance and quality control (QA&QC)
Engineering	Structures (bridges, roadways), standards, specifications, estimates, assessments, plans, calculations, drawings, schedules, reports	Analysis, design, construction, maintenance, inspection, operation, research and development, QA&QC

Whereas management and engineering can appear at the opposite sides of a supply-and-demand relationship, their objectives are singular structures in continuous services. The challenge is to balance the priorities of the disparate components. Breaking down the process and the product into discrete tasks and elements focuses and streamlines the scope of the required expertise, but the fragmentation of competence is artificial and counterproductive. Reservations notwithstanding, Beale (1988, Appendix 8) recommended a similar distinction between statistics and operational research:

> If we make this distinction we must add that a practical statistician must then spend some of his time practicing operational research, and that a practical operational research(er) must spend some of his time practicing statistics. . . . The quality of the solution to any practical problem may be impaired by dividing the problem up in this way. However such arbitrary divisions of problems into manageable components are often necessary initial steps towards finding any solutions at all.

Two possible sets of manageable components contain bridge-related tasks and their outputs (e.g., the process and the product). The product can be the structure, its model, or its service, all of which can be represented as systems of elements linked in series or parallel. In Table 4.2 the structure is presented as a product, changing from one stage of the life cycle to the next. In Table 4.3 the process of the structural life cycle is considered as a sequence of tasks grouped into stages.

This "initial" expedient is itself vulnerable to becoming the only recognized method. The "arbitrary divisions" become final. Statisticians diversify from operational researchers as do engineers from managers and their intended integration never follows. Provisional recommendations solidify into prescriptions that govern practice long after their validity has expired.

Forensic investigations focus independently on the product and the process and then correlate the findings. Investigating the product can be perceived as a lower bound or static approach (e.g., the form is fixed, the quantities are checked). Reviewing the process is comparable to the kinematic or upper bound approach, which identifies collapse mechanisms of the adopted form (or, in this case, method). If the bridge management process consists of consecutive stages, it can be viewed as a system of "failure elements," according to the definition adopted by Cremona (2003; Appendix 17 herein). Whereas

Table 4.2 Product: Partial List of Bridge Component Needs over Their Life Cycle

Bridge Components/ Elements	Bridge Life-Cycle Stages					
	Concept	*Proposal*	*Project*	*Asset/Liability*	*Inspection and Evaluation*	*Management*
Primary/ secondary member, pier	Model, stress/strain distribution, compression/ bending/shear/ torsion, stability, fatigue, redundancy, loads, short- and long-term material properties, failure modes, load ratings, load postings	Drawings, specifications, material properties, connections, constructability, maintainability, useful life, vulnerability, repair/ replacement, strengthening, budget	QA: alignment, scaffoldings, formwork, stability (permanent/ temporary), connections: welds, bolts, pins, linkages, splices, reinforcement, concrete, waterproofing, drainage, schedule, budget	Budget, schedule, personnel, underdeck protection, deicing, equipment, access, lights, wearing surface, traffic signs, movable bridge operation, traffic control, alternate routes, retrofits, postevent emergency measures, scheduled rehabilitations, environmental protection, community outreach	Access, corrosion, alkali–silica reaction, errors in analysis, design, construction, use, fatigue, scour, seepage, erosion, environmental hazards, fracture-critical details, temporary and permanent repairs	Services (transportation), functioning organization, decision and engineering support: time and material, equipment, funding

Deck, wearing surface		Deck type, thickness, strength, detailing of reinforcement, corrosion protection, rebar cover, admixtures, orthotropic, grid deck detailing, waterproofing, wearing surface, useful life/ replacement		
Bearings/ pedestals	Load, load transfer, articulation, damping, isolation	Material, detailing, maintainability, useful life, method of replacement	Material, details, alignment, waterproofing	Waterproofing, fillers
Joint	Articulation, continuity	Displacement capacity, connection to deck, maintenance needs		
Drainage		Discharge capacity, maintenance needs		
Footings/ piles abutment/ approach	Soil, channel properties, scour, flood, impact, earthquakes		Piles, excavation, backfill	Settlement, scour, undermining, liquefaction, collision
Paint		Paint, repaint type and cycle	QA	Useful life, toxicity, corrosion protection
Utilities	Type of service, ownership, impact on structure, maintenance/ replacement needs	Connections, safety	Installation	Replacement

Table 4.3 Process: Partial List of Bridge-Related Tasks

Bridge Component/ Element	Bridge Tasks					
	(Re)analysis	Design	Construction and Rehabilitation	Maintenance and Operation	Inspection and Evaluation	Management
Primary/ secondary member, pier	Structural and material modeling, selecting loading conditions/ combinations, forecasting hazards, interface with design	Design specifications, type of service, estimate first and life-cycle costs, scheduling, repair/ replacement, strengthening, interface with analysis, construction, maintenance, inspections	QC: testing, fabrication, transportation, installation, demolition, rehabilitation, alignment, welding, bolting, reinforcement, casting, curing, waterproofing, resurfacing, traffic control, schedule, interface with design, maintenance, inspections, construction specifications, community outreach, budget	Manage budget, sweeping, cleaning, washing, hazard mitigation, emergency and scheduled repairs, underdeck protection, anti-icing, interface with design, construction, inspection, operations, operate movable bridges, control traffic, emergency planning, hazard estimates, retrofits, postevent emergency measures, interface with other agencies, owners, community	In-depth, essential completion, regular, special, movable, underwater, hazards: structural, safety, environmental, fracture-critical details; health monitoring, inventory management, interfacing with analysis, design, construction, maintenance and operations	Decision, implementation, planning, reporting, communication, coordination, negotiations, supervision, QA&QC

132

Component				
Deck, wearing surface	Select structural model, load distribution, continuity, composite action, contributing width, span, shear lag, deflection, fatigue	Adopt specifications, select materials, detailing, reinforcement, corrosion protection, waterproofing, orthotropic, grid deck detailing, wearing surface, useful life/replacement		
Bearings/pedestals	Select load, displacement, damping, isolation	Select material, detailing, maintenance, inspection, useful life, method of replacement	Verify material, detailing, installation	Clean, replace
Joint	Estimate type of release	Select type, displacement capacity, connection to deck		
Drainage		Discharge capacity		
Footings/piles abutment/approach	Determine soil, channel properties, scour, flood, seismic, impact hazards		Drive piles, excavate, backfill, compact	Clean channel, install pier protection
Painting		Establish painting protocol: specification, coatings, surface preparation, method of application, paint cycle	QC: surface preparation, application, environment protection	Lead paint removal, repainting, spot painting, washing
Utilities	Impact on structure, maintenance needs	Verify connections, safety	Install, test	Replacement, ownership

structural system identification generally determines the parameters modeling the service modes of systems, the identification of vulnerabilities models their failure modes.

Design generally follows the static or lower bound approach. The objective is to meet the demand of the loads in terms of strength, and stiffness. Thus content is supplied for a selected form. In contrast, the kinematic or upper bound approach anticipates the shape of failures and may identify the need for a different form. The recently introduced requirements for ductility reflect a kinematic or upper bound approach. Model parameters (e.g., quantities) are relatively easier to optimize than model shapes (e.g., qualities). Failure criteria are much better defined for physical than for organizational structures, allowing for greater tolerances in processes than in their products. New operational models are adopted and processes are overhauled by qualitative managerial decisions.

The interest in the process is not new. Picon (1992, p. 77) observed: "Peronnet (see Section 1.3) may not have been the only one to perceive the structure (l'ouvrage d'art) in terms of a process, but he pursued to the limit the harmonious production sequence."

Prescriptive specifications of the product and performance indicators of the process were eventually found unable to fully exploit the available material and operational strengths (see Section 4.2.1). Performance-based specifications were recommended in design and benchmarking in management. Their implementation, however, requires broader and continually updated knowledge of the subject. The emphasis on performance is associated with risk mitigation. Picon (1992, p. 76) reported: "Peronnet related performance to risk (danger). According to him, modern bridges were comparable to church colonnades, which *would necessarily collapse in large parts if one little support or one column fell*."

The columns of Table 4.4 enumerate typical bridge-related tasks to be managed. The rows correspond to the stages in the bridge life cycle. The result is a matrix representation of the information depicted graphically in Figs. 4.4*a, b*. The term *matrix* is used in a purely formal sense. Mintzberg (1979, p. 168) has described the use of the matrix method for management purposes. Cleland and Cokaoglu (1981, p. 32) apply it to engineering projects. For a "total highway management system," Sinha and Fwa (1987) proposed a matrix structure with three dimensions, for facilities, functions, and objectives.

Applied to organizational structures, the method separates engineering and managerial expertise but assigns equal importance to both. The National Aeronautics and Space Administration (NASA) and IBM were early proponents. One result has been the establishment of distinct and comparably remunerated chief engineer and chief executive positions. The method can also be useful in public agencies where relatively constant technical responsibilities must interact with administrative authority subject to quadrennial changes.

The matrix method allows for redundancy and consequently a possible ambiguity (or even conflict) in the chain of responsibility. Hassab (1997) recognizes the following three basic organization structures:

- Hierarchical (top-down or circular chains of responsibility)
- Network (trees with added cycles or multiple superiors possible)
- Relational (matrices allowing dynamic relationships to be established according to the demand)

Table 4.4 Stages in Life Cycle of Structure and Process

Structure	Process					
	Analysis	Design	(Re-)construction	Maintenance and Operation	Inspection and Evaluation	Management
Proposed structure	Soil, channel, environmental conditions, extreme-event forecasts, needs, existing structure assessments	Final design, material properties, design loads, costs: first, life cycle, schedules, recommended maintenance	Letting	Review of maintenance and operation manual, maintainability, inspectability		Needs (network, community), budget, request for proposal, design review selection
Project	Feasibility of alternative designs	Peer review, field verification	Construction schedule, progress evaluations, QA&QC			Contract documents, construction supervision, QA&QC, budget
As built	As-built drawings, essential completion/final acceptance inspections, liability resolution, inventory input					
Present condition	Inventory, operating ratings	Repairs, component replacement		Implementation	Potential hazards, condition assessment, diagnostics	Supervision, interfacing with other agencies, community
Future condition	Serviceability forecasts			Hazard mitigation	Structural condition forecasts	Vulnerability forecasts
Needs estimates	Repairs, strengthening, rehabilitation				Needs forecasts	10–20-year capital plans, budget requests

Computer databases assisting the management process evolve concurrently and are structured similarly (Appendix 18).

The "matrix" shown in Table 4.5 superimposes the main stages of the process and the corresponding product. The matrix representation is an attempt to maximize the insights contained in information from established sources. The knowledge accumulated during one stage of the cycle, for instance analysis, becomes new information at the following stages, such as design, construction, maintenance, inspection, and back again. The two-dimensional matrix provides some independence to the management of the process and the product at network and project levels. Process and product designers can check one another. Engineering knowledge may be new information to managers and vice versa. Quality control examines the product. Quality assurance examines the process. Recent NCHRP syntheses summarize best *values* (presumably of products) and best *practices*. Organizational matrices must be adaptable to the changing needs of the process and the product.

Along the main diagonal of Table 4.5, vulnerabilities are internal to the respective life-cycle stages and tasks, namely, analysis, design, construction, maintenance and operation, and inspection. Nontrivial off-diagonal terms can be interpreted as potential vulnerabilities resulting from poor transitions between the discrete stages and tasks. Examples include analysis insufficient for design, design, misinterpreting analysis, construction at variance with design, inability to maintain the structure, and inconclusive inspections. If no off-diagonal terms exist, the process reduces to a vector or a chain, each tasks delivering a QC certified product to the next.

If the objective is to expedite a process, the matrix can be reduced to one or several concurrent vectors. Flowcharts can identify the critical path and auxiliary branches and adequately model construction project tasks. Toffler (1980, p. 280) considered the matrix structure too restrictive for the society of the future and (so far correctly) anticipated mutable flat networks supplanting centralized "top-down" systems.

The tasks associated with both individual structures and networks can be modeled in a matrix configuration. The inventory of structures is relatively static and the needs evolve over the life cycle. Large network inventories are more dynamic, whereas needs may tend toward an average relatively steady level. As a result, a project must anticipate future conditions and needs (Chapters 10 and 11). During the design of a bridge, provisions must be made for construction, maintenance, inspection, operation, and rehabilitation. As project needs and benefits recede into the future, they lose urgency and gain uncertainty. In contrast, large networks face simultaneously conditions and choices typical of every life-cycle stage. The recurring demand for all types of resources is more uniform and predictable at the network level. Project (*ground-up*) and network (*top-down*) priorities interact as do products and processes. Project management can contribute the knowledge of details (with a possible emphasis on the product). Network management must engineer the equilibrium between service demand and the network life cycle.

Separating the product from the process parallels the separation of management and engineering. The best engineering creations are unique blends of form and content with the process evolving to suit the specific product. If the process is responsible for the product, management is ultimately the source of all vulnerabilities. Neale (2001) stated (p. 83): "The findings show that there is more than one reason for a failure, and usually many reasons. . . . The management of the process can thus also come into scrutiny."

Table 4.5 Assessments of Product and Process

Product (Output) (Table 4.2)	Process (Table 4.3)					
	Analysis	Design	(Re)construction	Maintenance and Operation	Inspection and Evaluation	Management
Transportation needs, loads, structural model, response, impact	Peer review	Review of load conditions	Prequalifying	Review of maintenance needs	Structural health monitoring, in-depth inspections	Supervision, performance review (QA&QC), resources (budget, personnel, material, database), safety assurance, audits, emergency
Design calculations, drawings, inventory, rehabiltation, repair	Analysis of design alternatives	Peer review, value engineering	Review of shop drawings, construction schedules			
Budget, schedule, project, structure		Field verification of drawings	Resident engineer, construction support services, QA&QC	Inspection of completed project		
Service, repairs		Maintenance specifications		Field supervision	QA&QC	
Condition, needs evaluations, inventory updates	Load rating		Inventory of new and rehabilitated structures		Periodic inspection	

Table 4.5 (Continued)

Product (Output) (Table 4.2)	Analysis	Design	(Re)construction	Maintenance and Operation	Inspection and Evaluation	Management
				Process (Table 4.3)		
Life-cycle plans, transportation modes, budget, staff, data management (BMS)	Cost/benefit analysis					Asset management, assessments (risk, needs), internal audits, interfacing with external sources

4.1 MANAGEMENT

Vulnerabilities are likely to recur at all stages of the life cycle unless management is scrutinized as one of them. Rather than question its own validity, however, management is likelier to seek isolated weak links in the chain of individual tasks under its purview. Laplace (1749–1827) commented that human reason has "less difficulty making progress than investigating itself" (Kline, 1980, p. 335).

As a task, serving any particular professional practice, such as engineering or accounting, management is guided by the specifications and standards of that practice. As an independent profession, management is much more flexible and, hence, vulnerable. Purely managerial flaws have been repeatedly blamed for affecting the performance of transportation facilities and networks. Gibble (1986) presented case histories of mismanagement on various levels. All of them include administrative failures.

4.1.1 Administration

In order to optimize (or prioritize) operations involving human, financial, and engineering resources, management must define its own medium to match or exceed the level of detail relevant to personnel, financial accounting, engineering design, construction, and maintenance. Administration serves that purpose and, consequently, it must be conversant in the abstractions adopted by engineering and economics, particularly forces and money. No engineering project or organization can function without effective administration. The agencies or authorities responsible for unique structures and networks are designed along with their respective facilities. Effective administrations are inconspicuous (as are forces in equilibrium or demands exceeded by existing supplies). Failing ones are "bureaucracies."

Administrations become bureaucracies by succumbing to the following most common vulnerabilities.

Inexperience/Incompetence

The subjectivity of the operating models is more prevalent in economics than in mechanics and even stronger in administration. As a result, administrative competence is acquired primarily through direct experience and applies mostly to the specific circumstances where it has originated. Administrative procedures (including recording practices known as "paper trails" and tools termed "boiler plates") adapt to transient demands and constraints, tempting many managers, particularly engineers, to underestimate their importance, assuming an attitude above such trivia. A common counterproductive consequence is the development of an organization-specific administrative language consisting of acronyms incomprehensible to anyone but the entrenched insiders. It persists until the next generation of administrators replaces it with its own equally opaque but essential terminology.

Expertise in fiscal and contracting procedures is required on all management levels. Section 1.11 pointed out that specialized business management is only one aspect of infrastructure management. The successful courses in business administration provide useful but limited guidance, because in engineering the process and the product cannot be treated independently. Many of the great designers discussed in Section 1.3, notably

Eiffel and Ammann, were outstanding administrators. The high demand for the few experts who have supplemented their engineering education with business and law degrees signifies a need for engineers who are competent administrators.

Formalism

Essential as it is, the administration of infrastructure management is particularly vulnerable to self-absorption. Administrative tools assume the importance of products and methods become the objectives, because the assets appear deceptively tolerant to formalism (e.g., "red tape"). Structural life cycles are longer than those observed in economics and politics. Structures enjoy greater stability than businesses. Ineffective administrations can execute their habitual functions long after they have become obsolete without apparent consequences to the services in their charge. The governing equilibrium grows increasingly unstable, culminating with a structural or operational collapse under more or less extreme conditions. Post facto analysts invariably discover that "a disaster has been waiting to happen."

Purely formalistic restrictions can be imposed upon an administration as a reaction to the mismanagement of individual projects or overregulation by preceding ineffectual administrations. Optimizing and maintaining the equilibrium between administrative and physical actions is the most difficult task of any management. Appendix 19 points out that the latter difficulty is an important argument in favor of privatization, which in turn requires a different but no less onerous administrative management. Appendices 20 and 21 and Example 8 illustrate the complexity of administrating the relationship between public ownership of the assets and private contractors.

> **Example 8. Contract Letting Procedures for Bridge Management Agency in Urban Setting**
>
> The responsible local owner of several hundred bridges in an urban environment has adopted the following sequence of steps for competitive contract letting:
>
> 1. Project engineer/contacts engineer in charge of specifications (engineer in charge, e.g., EIC of specs) must arrange a guidance meeting at about 60% completion of final design.
> 2. The EIC of specs arranges the meeting and sends a copy of guidelines for preparation of proposal book to be read by the consultant before coming to the meeting.
> 3. Guidance meeting.
> 4. Ninety percent submission (review time two to three weeks), including list of items, solicitation provisions (SPs), and special specifications.
> 5. Communicate comments to consultant at meeting arranged by project group for federally aided contracts. Any special specifications (SSs)* should first be submitted to owner for review. After owner's review, SSs should be approved by state DOT and assigned new item numbers. For new SSs for locally funded contracts, owner will issue item number after the comments resulting from the review are addressed.

6. One hundred percent submission (review time two to three weeks).

    a. Consultant/project engineer submits draft presolicitation review (PSR) to EIC for review.

    b. Comments on the draft PSR are sent to the project engineer and a revised PSR is obtained from the project engineer.

    c. Submission of PSR package to the agency chief contracting officer (ACCO) by specification section (usually takes one week for approval).

    d. Receipt of PSR approval from ACCO.

7. Communicate comments to consultant at meeting arranged by project group.

8. Final submission (review time two weeks).

9. Final review and preparation of book for law department approval incentive/disincentive clause for a project must be approved by the mayor's office of construction and the office of management and budget (OMB). A memo with a justification package should be sent by the project group to the mayor's office and OMB for their review and approval. The final submission consists of original unbound and one bound (screw-posted) copy. Any new specifications should have been approved before preparing final submission.

10. Submission of specification books to DOT legal department (usually takes two to three weeks for approval).

11. Receipt of approval from law department. Comply with comments for law department final review and prepare book for printing.

12. Send advertisement for bid to contract/purchasing.

13. Approve specifications. Books will be sent to consultant for printing along with the printing requirements. Consultant will deliver three copies of the books to 2 Rector Street two days prior to start to mass printing for final review. Specification books will be sent to contracts/purchase one day prior (by 3:00 p.m.) to start of advertisement by the consultant.

14. Publish advertisement in newspaper by ACCO.

15. Prebid meeting (two to three weeks after start of advertisement).

16. Issue addenda as required. Addendum will be prepared by consultant as per instruction from the owner's specification section which will be reviewed and approved by specification section and transmitted by the project group to ACCO for approval. If required, approval must be obtained from owner's law department before submitting. If the project is part of the Federally Aided Urban System (FAUS), the addendum should be approved by FHWA, owner's Structures Division, state DOT headquarters, and regional offices. Addendum is necessary if bid opening date must be changed.

17. Bid opening by contract manager. After bid opening the specification section will need 7 or 9 contract books for contract award and distribution. These copies must include any addendum issued, 7 copies of specification books for 100% city funded, and 11 copies of books for FAUS-funded project.

18. Specs section will fill out procedures checklist form (specs portion only) and send it to project group.

19. Procedures checklist will be sent to contract registration office by project group for processing of contract award.

20. Receive 7 or 9 sets of books (including addenda) from project engineer for numbering)

21. Number the books from 1 to 7. Send books 2–9 to the project engineer. Keep book 1 for specification section. Project engineer will send these books to contracts/purchasing for execution of contract.

22. Project group will send recommendation for contract award signed by the deputy chief engineer (DCE) to deputy agency chief of contracts (DACCO) and copy to concerned persons.

23. Director of contract management will issue contract award letter.

24. Contract registration.

25. Notice to proceed (NP) to the contractor.

The preceding list is provided in XLS file form with samples of all required letters.

*SPECIAL SPECIFICATIONS

An urban construction project must conform with SSs other agencies or owners with overlapping responsibilities in the following alphabetical order:

Coast Guard. If the bridge is on the waterways, U.S. Coast Guard requirements and Army Corps of Engineers permits are to be included. Consultant should obtain latest version of U.S. Coast Guard requirements and Army Corps permits.

Electrical Specifications. All electrical specifications must be reviewed and approved by the electrical section of the owner's department for engineering review and support (ERS).

Department of Environmental Protection (DEP). Specifications related to sewers, water mains, water and air pollution, noise control, and other items under the jurisdiction of the DEP must be reviewed and approved by DEP.

Fire Department (FD). Consultant shall obtain the approval of FD for any specifications related to FD. A new item number is assigned following the approval.

Machinery of Movable Bridges. All machinery specifications must be reviewed and approved by the mechanical section of the owner's ERS.

Parks and Recreation. Any SS related to areas under the jurisdiction of a department of parks (DOP) must be approved and signed by DOP. Consultant should send the new special specifications to DOP. The DOP approval is assigned a new item number.

Private Utilities (PUs). All private utility SSs must be reviewed and commented on by the owner's specifications section, primarily for conformance with the owner's standard format. Technical aspects must be reviewed by the concerned private utilities. Consultant shall check for any conflict with the owner's requirements and resolve with PUs. After comments are addressed, the owner shall assign item numbers to new private utility specifications.

Railroad (RR). RR requirements/protective shields provisions must be reviewed and approved by the concerned railroads.

Railroad Insurance (RRI). RRI for the project must be obtained by the consultant from the concerned RR. Requirements must be updated as necessary.

Street Lighting/Traffic Signals. All street lighting specifications must be approved by the signal division. In the absence of local special specifications, the consultant will prepare new special specifications in conformance with state specifications.

In addition to the SS requirements, approvals must be obtained by any agency with relevant jurisdiction, such as the Landmark Commission.

Accountability/Liability

Accountability accompanies any position of trust, such as those of administrators and designers. Discussing structural failures in general and the New Orleans levee system failure during Hurricane Katrina respectively, Feld (1968, p. 5) and J. Schwartz (*New York Times,* June 11, 2006, p. WK 3, column 1) cite the Code of Hammurabi (1730–1685 B.C.), which is very specific regarding the responsibilities and liabilities associated with construction. Over the centuries the liability has varied along with the social systems. In his article, entitled "Too Bad Hippocrates Wasn't an Engineer," J. Schwartz quoted E. Wenk, Jr., science advisor to Presidents Kennedy, Johnson, and Nixon, to the effect that "safety is a social judgment."

Administrations vulnerable to tort liability dedicate key resources, and ultimately their best effort, to legal defense. For administrative purposes, Banks (2002, Chapter 2.2.3) outlined the differences between *ministerial functions* of public agencies, involving clearly defined tasks (and hence tort liability), and *discretionary functions,* allowing for independent judgment (and therefore exempt from tort liability). Government is liable for an unsafe condition if it has received *notice* of its existence. *Constructive notice* implies management should have known about the condition.

Following the fire in the Mont Blanc tunnel that caused 39 deaths on March 24, 1999, the president of the administrative council of the French Concessionary for the tunnel was fined EU 15,000 and received (along with several responsible managers) a two-year suspended sentence. The court found that by the time of the catastrophe "the president had been in charge for three years, and therefore, had had the opportunity to change the conditions that had certainly preceded him, but instead, he had submitted to them" (*Le Monde,* July 29, 2005, p. 7, column 2).

4.1.2 Personnel

Personnel adds a third (and therefore easy-to-overlook) dimension to the product/process matrix of Table 4.5. Separating the management of personnel from that of bridges is a byproduct of the general trend for specialization, which distances engineering from management. Personnel management is particularly vulnerable, because it requires both administrative and technical competence.

Outsourcing tasks considerably increases the options and hence the complexity of personnel management. The supervision of contracts and in-house operations requires different competence. The resulting organization and, ultimately, the infrastructure reflect that difference.

O. Ammann designed the George Washington Bridge as Chief Engineer of the Port Authority of New York and New Jersey, but the Whitestone, Throg's Neck, and Verrazano bridges as a consultant.

The procurement of services requires legal and fiscal expertise. Public agencies award contracts following due process, for instance as illustrated in Appendices 20 and 21. Delayed projects lapse from one fiscal year to the next and can lose their funding. Structures remain in service despite their poor condition.

The lowest (first cost) and the most qualified bidders are not always the same. Appendices 20 and 21 describe methods to resolve possible contradiction between these two constraints.

Responsible owners can easily grow dependent on outside professional help for any bridge-related task unless they exercise rigorous QC, which implies some duplication of effort. Over the long periods typical of bridge life cycles, only the owner can provide continuity in management and expertise.

Maintaining competent in-house expertise for the tasks of Table 4.3 is a continuing challenge for bridge owners. As with structural maintenance deferral, the lapses in personnel maintenance take hold slowly and are very costly to repair. A highly effective engineering unit may require 5–10 years to build. Turnovers and upgrading must be perpetual. Budget constraints and the lack of incentives can gradually reverse the process, reducing the same unit to a liability.

Personnel is vulnerable to inexperience and complacency (or "burnout") resulting from monotonous experience. FHWA (1986) cautions against inspecting bridges as "old friends" and missing important clues of impending failures. Analysis, design, construction, and maintenance are similarly susceptible to routine. Many bridge owners counteract these trends by continuing education and training. The FHWA conducts refresher and introduction courses on inspection, fracture critical structural elements, bridge management systems, seismic evaluation, security, and so on. Annual budgets must anticipate personnel upgrades. The cost is negligible, compared to that of restarting a lapsed organization. A relevant example is quoted in Section 4.1.7.

4.1.3 Emergency Management/Damage Control

In order to fulfill its two primary tasks (to avoid emergencies and to minimize their consequences), management may cultivate entirely different organizational structures. The causes of potential emergencies are anticipated as "extreme events" (Section 4.2.4) and mitigated by design in a civil mode of operation. In actual events engineering reverts to its military origins under the various designations of crisis management, damage control, post-disaster mitigation, and emergency response. Teams or squads are assembled according to availability and expertise and charged with tasks to be delivered daily or even hourly. If the emergency is related to a structure, the tasks usually include ensuring the safety of the public in its vicinity and the design and execution of emergency repairs. The usually distinct functions of inspection, analysis, design, construction, planning, and communication (Tables 4.3–4.5) become concurrent within the same project. The benefits of expertise are thus maximized. In a major city numerous infrastructure networks are managed by respective specialized agencies, such as the Departments of Buildings, Housing Preservation, Transportation, Design and Construction, Parks, Environmental Protection, Sanitation, Fire, and Police in New York City. Emergencies, such as the water main break shown in Fig. 4.1, the retaining wall collapse shown in Fig. 4.2, or, ultimately, the clean-up after the collapse of the World Trade Center towers (Fig. 4.3), required the coordination of all expertise by the Office of Emergency Management (OEM) appointed for that purpose.

Emergency management succeeds tactically. Long-term benefits elude it, because the strategic goal is to eliminate its causes. In his *Common Sense,* Tom Paine (1945, p. 38) wrote: "Immediate necessity makes many things convenient, which if continued would grow into oppressions. Expedience and right are different things." While the operational latitude gained by temporarily relaxing administrative and engineering procedures underscores the inefficiency of standard practices, administrative illiteracy and engineering incompetence can also sneak in. Management can take engineering decisions without all the necessary theoretical backing, whereas engineering has no time to justify all of its recommendations by all the normally available tools. Extraordinary competence is required in both domains, usually of the same individuals, exacerbating the already high risk. Lord Baker (1978), whose academic and common sense credentials were of the highest order, recounted the design and construction of bomb shelters during the WW II air raids over London under the revealing title "Enterprise Versus Bureaucracy."

Prioritization by Worst First and by Triage

As a prioritization, "worst first" is a step above "first come, first served," that is, addressing all conditions in the order in which they are reported. Scheduling conditions in

Figure 4.1   Water main break, 5th Avenue, 19th Street, New York City.

Figure 4.2   Retaining wall collapse, West Side Highway at 178th Street, New York City.

Figure 4.3   Cleanup at Ground Zero, New York City, September 13, 2001.

the order of descending urgency requires considerably better optimization capabilities than the method can usually afford. Example 9 shows that the number of conditions slipping into the worst category is bound to become overwhelming with time.

Example 9. Cost Estimates Based on Condition Ratings

In the 1980s the New York City bridges were rated according to the NYS DOT rating system (see Appendix 40) and a local system consisting of four levels. In 1991 the two rating systems were correlated, as shown in Table E9.1.

The following example shows that estimates of the immediate needs of the bridge stock based on the condition assessments in Table E9.1 can lead to diverging conclusions.

Total bridge deck area $\approx$ 1,500,000 m^2
Average bridge condition rating $\approx$ 4.4
Average recommended cost of maintenance $\approx$ U.S. \$65 m^2 (U.S. \$6 ft^2)
Average estimated cost of rehabilitation[a] $\approx$ U.S. \$4,840 m^2 (U.S. \$450 ft^2)

---

[a] In 2005 the estimated cost of rehabilitation was adjusted to U.S. \$600 ft^2 (U.S. \$6454 m^2)

The rehabilitation needs estimates are as follows:

(a) based on the proportion of the poor bridges of the total deck area:

Rehabilitation 1,500,000 (7.3% $\times$ U.S. \$4840) = U.S. \$529,700,000
Maintenance   1,500,000 (92.7% $\times$ U.S. \$65) = U.S. \$90,382,500
   Total   U.S. \$620,082,500

(b) based on the proportion of all spans on poor bridges:

Rehabilitation 1,500,000 (22.3% $\times$ U.S. \$4840) = U.S. \$1,619,000,000
Maintenance   1,500,000 (77.7% $\times$ U.S. \$65) = U.S.   \$75,757,500
   Total   U.S. \$1,694,757,500

The large differences between the two estimates tend to discredit both of them. Estimate (a) errs by assuming that all bridges are of the same size. Estimate (b)

Table E9.1 Bridge Condition Ratings, Annual Report (NYC DOT, 1992)

Condition	Rating	Bridges	%	Spans	%
Poor	1 −3.0	64	7.3	1,110	22.3
Fair	3.01–4.5	430	49.1	2,163	43.4
Good	4.51–6.0	335	38.3	1,412	28.3
Very good	6.01–7.0	46	5.3	301	6.0
Total		875	100	4,986	100

assumes that all spans are comparable in size and condition throughout a given bridge.

If a hypothetical worst span defines the bridge condition, the estimate should be conservative; for example, a multispan bridge may be scheduled for rehabilitation because of a single poor span or a number of scattered poor elements. For a more realistic assessment, the NYS DOT introduced a span condition rating (Appendix 41). Despite the sound reasoning of that approach, it produced rehabilitation need estimates similar to those based on assumption (b). The most likely explanation is that the need to rehabilitate an isolated span on an otherwise perfectly serviceable multispan bridge is not very realistic.

By the year 2004, the condition ratings of the bridges were distributed along the condition rating scale shown in Fig. E9.1. The nearly normal distribution is consistent with the central limit theorem (Appendix 9). The observed distribution is not purely random, however. It is influenced by the following significant trends in the condition rating process and the rehabilitation policies:

Inspectors rate more frequently toward the center of the scale.

The age and condition rating of rehabilitated older bridges are often adjusted to the median of the scale (see Example 12).

The sample of spans is 6 times larger and therefore should fit a normal distribution even better. As in Table E9.1, however, that distribution (Fig. E9.2) shows a stronger skew toward the lower ratings. The overall average condition rating is approximately 4.5 for the 4986 spans and 5.0 for the 750 bridges. The average age is between 70 and 75 years.

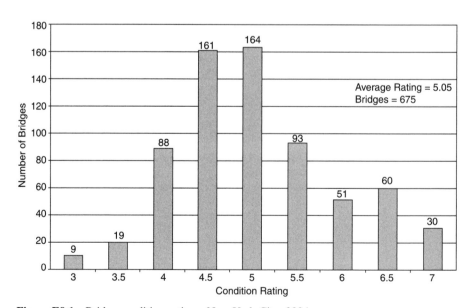

Figure E9.1   Bridge condition ratings, New York City, 2004.

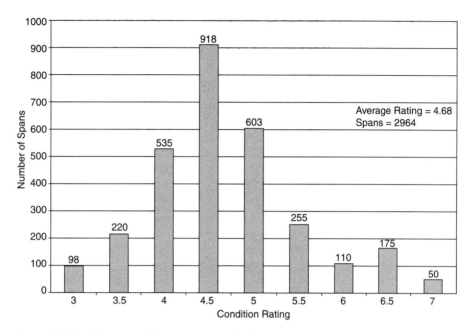

Figure E9.2 Bridge condition ratings weighted according to number of spans, New York City, 2004.

The span condition ratings in Fig. E9.2, however, are not obtained independently. Rather, each span is assigned the rating of the bridge of which it is part. Hence the distribution merely confirms that long-span bridges take longer to rehabilitate.

Estimating needs according to the deck area falling within prescribed ratings is a crude from of averaging. The rehabilitation cost on the East River crossings (Example 3) exceeded U.S. $20,000/m². The scope of rehabilitation work (Sections 11.3 and 13.3) is estimated more reliably by in-depth inspections (Section 14.3). The 10-year rehabilitation plan reported by NYC DOT (1992, p. 17) was to address 260 bridges at an estimated cost of U.S. $3.4 billion (considerably exceeding the preceding high estimate).

Fig. E9.3 shows essentially contrasting distributions of the same bridges with respect to the "condition" and "sufficiency" rating scales (described in Appendix 41). The sufficiency rating, reflecting serviceability, declines insignificantly so long as structural conditions are above average. For below average structural conditions serviceability declines precipitously and is the first to fail (as in closed or obsolescent bridges).

Although much more realistic, the Bridge Rehabilitation Proposal Reports (BRPR) resulting from in-depth inspections still do not take all needs into account. NYC DOT (1992, p. 23) concluded that *flags* (e.g., potential hazards) are another important indicator of bridge conditions (Appendix 46). The cost of the temporary or permanent flag repairs ranges from U.S. $10,000 to $15,000 in direct costs, usually borne by the expense budget. When flag repairs (see Section 10.3) cannot be imple-

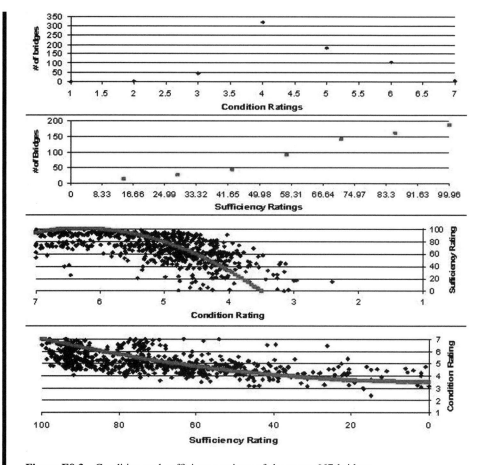

Figure E9.3   Condition and sufficiency ratings of the same 667 bridges.

mented within the prescribed time, bridges must be closed, incurring user costs and still requiring repairs. NYC DOT (1992, Appendix C2) reported 16 closed, 12 partially closed, and 23 posted bridges. A system in such condition requires "demand maintenance," consisting of emergency inspections, repairs, possible closures, traffic management, and community outreach programs.

DEICING VERSUS ANTI-ICING

For a large urban center in moderate climate, the significant strategic bridge maintenance decision may be (a bit simplistically) reduced to a choice between deicing (with chlorides) and anti-icing (with noncorrosive substitutes). Circa 2005–2006, it has been reported by the mayor's office that New York City spends approximately C = U.S. \$1 million/in. of snowfall in rock salt application and plowing. The city Department of Sanitation applies 300,000 tons of salt during an average winter. This effort saves the city motorists an amount of time denoted by T_{salt} and accidents numbering A_{salt} (compared to no treatment). Estimates suggest that chlorides are reducing the bridge useful life by roughly 50%. Thus, the bridge-related consequences of replacing deicing salt with a noncorrosive anti-icer can be evaluated by comparing the following simplified costs and benefits:

	Cost	Benefit
Deicing	$C_{salt} + R$	$T_{salt} \times CT + A_{salt} \times CA$
Anti-icing	$c \times C_{salt} + R/2$	$T_{sub} \times CT + A_{sub} \times CA$

where $R \approx 550$ million = bridge reconstruction/rehabilitation costs (U.S.\$/year)

$\qquad T_{salt}, T_{sub}$ = time saved by motorists if salt and a substitute are used, respectively (hrs)

$\qquad CT \approx 10 \div 20$ = average cost of motorist time (U.S.\$/hr)

$\qquad A_{salt}, A_{sub}$ = number of accidents averted by salt and substitute application, respectively

$\qquad CA$ = average accident costs (U.S.\$/accident)

$\qquad C_{salt} \approx 50$ million = cost of deicing with salt (U.S.\$/year)

$\qquad C_{sub} = c \times C_{salt}$ = costs of anti-icing with substitute

It is commonly assumed that $c = 3 \div 10$.

If $A_{salt} = A_{sub}$ (an assumption likely to be disputed in court after any accident occurring during the use of a salt substitute), the expected losses in motorist time caused by winter weather should not exceed the following limit:

$$(T_{salt} - T_{sub}) \times CT < R/2 - C_{salt}(c - 1) \qquad (E9.1)$$

Assuming $CT =$ U.S. \$10/h and $c = 3$, salt substitutes would appear cost effective if they do not cause delays (relative to the use of salt) exceeding

$$\frac{550/2 - 2 \times 50}{10} = 17.5 \text{ million hours/year}$$

At a user cost $CT =$ U.S. \$20/h, the acceptable time loss would be 8.75 million hours/year. For $c > 6.5$, according to Eq. E9.1, the salt substitute would have to yield a *net gain* of user time in order to be considered as cost effective.

The comparison is elementary and the numerical values are highly uncertain; however, several valid reasons emerge in support of salt as the deicer of choice:

Losses to the motorists are perceived as more urgent than the expenditures of the owner.

The savings of the owner appear relatively fixed compared to the volatile traffic demands.

The owners expect to benefit from a "no-salt" policy after years of implementation, whereas overhauling the operation immediately increases the operating and user costs.

Traffic-related losses do not accrue linearly with time but according to the perception and circumstances of the users (in the extreme cases of stalled ambulances or fire trucks, they may amount to lost lives).

> Salt is less expensive and more effective than the noncorrosive substitutes.
>
> In favor of potential salt substitutes, it can be argued that the entire infrastructure, including all lifelines, roadways, and vehicles, suffers from the salt-induced corrosion. The benefits of noncorrosive anti-icers must be evaluated for all the assets. In the meanwhile salt has been eliminated from use on select, particularly sensitive bridges.

When resources fail to satisfy safety requirements by the worst-first method, management is forced to prioritize work by triage. The term *triage* is derived from the French verb *trier,* meaning selecting or screening. The assets are divided into three categories according to their condition, and the available funding is spent on keeping the middle group in service. In the case of bridges, it is assumed that the better group will survive without urgent assistance. In extreme conditions, the worst group can be "made safe" by closure.

Triage is an expedient emergency response, because it concentrates all capabilities on manageable targets. If extended to a long-term strategy, it allows the assets to decline. Resources spent on emergency measures typically do not improve the condition of the assets. Determining the boundaries between the three categories becomes critically important, even though such a determination is uncertain and inaccurate and can be biased. Safety is easily compromised. Hence, life-cycle management must eliminate the need for triage.

Relying on safety imperatives to supersede procedural constraints is an abdication of management. As a moral choice, Kierkegaard (1849, 1980, p. 41) attributed this recourse to "fatalists and determinists who have sufficient imagination to despair of possibility." As a deliberate professional choice, such practice dissipates the technical and managerial acuities and resources, and, if identified, can be the subject of legal actions.

Technical experts grow used to the relatively generous expenditures lavished on duplicated tasks and turn the practice into a routine. Management considers such efforts as penalties for past lapses and seeks to recover the costs (usually by budget cuts) as soon as the primary distress is mitigated. The very need for intensive cross-unit collaboration is acknowledged only in emergencies, just as management comes under scrutiny mainly after failures. In such instances it is invariably concluded that some level of duplication must be a part of the process, for instance as QC&QA (Chapter 14).

NCHRP Report 525 (Appendix 22) represents a nationwide effort to replace the exceptional nature of emergency transportation operations by systematic management. The report is clearly inspired by the events of September 11, 2001. Hurricanes Katrina and Rita of September 2005 further demonstrate that the development and implementation of emergency management procedures are indeed an emergency task in itself.

4.1.4   Ignorance/Miscommunication

Ignorance can be regarded as a source of *gross errors* in the terminology of Thoft-Christensen and Baker (1982) (Section 3.4). Alternatively, it can be viewed as a gross error, divided into types A and B, meaning lack and inadequate use of information, respectively. Ignorance of type A, the insufficiency of knowledge, is always present and cannot be entirely remedied by communication. Ignorance of type B, inadequate use of knowledge, can be reduced by overlapping qualifications of the personnel engaged in technical and managerial activities and by reliable communication between these activities.

A propos of the shell collapse at the Charles De Gaulle Airport, ENR (May 31, 2004) editorialized: "Construction's curse and joy is that virtually every project is unique. . . . Some clients like to push the envelope of earlier accomplishments. And some in the industry are probably not lying when they say that they can pretty much design and build anything, given sufficient time and money. And the lack of these two elements generally is where projects sour."

In Fig. 1.33*a* management is the supplier of resources, such as time and money. Engineering is responsible for undertaking an undersupplied assignment. Structural and budget management optimize different life cycles. The former envisions (say) a 75-year life cycle, the latter responds to immediate demands. The engineering estimates of structural needs are uncertain but not negotiable. In contrast, budget allocations are negotiated at the highest levels of management and then become a firm operational constraint.

The resulting gap invites the "intermediary layers separating managers from engineers" noted by Yao and Roesset (Section 1.5 herein) and the loss of professional identity observed by Harris (Section 1.11 herein). That gap is reduced most effectively under the pressure of emergencies (Section 4.1.3).

In a matrix configuration, the flow of resources and responsibility can be represented as follows:

Responsibility

Resources	Engineering $\rightarrow$ Engineering	Engineering $\rightarrow$ Management
	Management $\rightarrow$ Engineering	Management $\rightarrow$ Management

The terms on the main diagonal of this elementary matrix represent performance within the separate units. The lack of knowledge in those relationships amounts to incompetence. Communicating highly specialized technical information is highly vulnerable. Although Von Karman's findings on the stability of aircraft wings were published before the Tacoma Bridge failure in 1940, bridge design did not benefit from that information. The knowledge gained subsequently has led to a redundant system of elaborate analytic modeling and wind tunnel investigations of scaled physical models. The synchronized pedestrian-induced excitations had been already investigated (Fujino et al., 1993) when they occurred at the bridges shown in Figs. 3.1 and 3.2.

The various forms of QC and peer review (Chapter 13) are intended to detect and limit vulnerabilities in the specialized technical expertise. Managerial performance, on the other hand, has proven easier to replace than to enhance. Rather than influence management, decision support systems (e.g., BMS) can easily become technical tools operated and understood only by their specialized experts.

The off-diagonal terms of the matrix show the need for communication between areas of different competence. Management compensates for the inevitable incompleteness of knowledge (e.g., ignorance) by the transmittal of significant information (as in Section 3.6). Since no amount of information can transmit the specialized knowledge between different operation levels, that function is assigned to indicators capable of conveying the quantity and quality of performance in a manner suitable for assessment on any level (see Chapter 13).

When data allow, production and performance indicators are derived statistically. The *frequentist* and *probabilist* approaches to statistical analysis are discussed in Sections 1.3–1.5. Hays and Abernathy (Steiner et al., 1982, p. 117) wrote: "American managers have increasingly relied on principles which prize analytical detachment and methodological elegance over insight, based on experience. As a result, maximum short-term financial returns have become the overriding criteria for many companies."

Projects and operations are initially managed with direct hands-on knowledge. As networks expand, essential information replaces direct knowledge in the form of performance indicators. Hence, management must balance the contradictory needs for direct knowledge of the tasks on the project level and for meaningful indicators on the network level (Chapter 12). Operations typically fail when they neglect specific needs and rely excessively on generic models. Past experiences offer abundant "lessons for the future" but are somehow inapplicable to the immediate present. Retrospective evaluations (of which this text offers numerous examples) draw on incomplete information and apply current views occasionally beyond their valid range. The resulting conclusions do not contribute directly to the quality of projects currently underway. Traditional indicators do not reflect new vulnerabilities. In order to remain meaningful, performance indicators must be continually reevaluated.

Feynman (1999, p. 169) found "an almost incredible lack of communication between [managers] and their working engineers" at NASA. Yet the managers in question were highly qualified engineers expected not only to communicate but also to know. Feynman attributed the Challenger failure in 1986 to a top-down method of management and design (p. 169) (also quoted in Chapter 5).

Top-Down/Ground-Up, Network/Project Level

According to Feynman (1999), the top-down method envisions the product in its entirety e.g., as a series or a nonredundant system. A feature of the method that might serve it well under some but not all circumstances is that success is envisioned as part of the process. As a result, the process is assumed to be a success in itself and redesign is most frequently limited to the product. The drawback of the top-down method is that failures are global and require complete redesign. The opposite would be a parallel system of redundant (even competing) ground-up modules, possibly by different management procedures. In system analysis and design, Mittra (1988) recommends a *structured* approach as a *quantitative* improvement of the top-down method (Section 4.1.7).

In this view, the terms top down and ground up describe centralized and decentralized management methods, respectively. An analogy can be drawn to the *hierarchical* and *relational* organizational structures (Appendix 18). Top down and ground up are therefore more coincidental than synonymous with *network* and *project,* used to describe the level and scope of responsibilities (Section 2.2). Shirole et al. (in TRC 423, 1994, pp. 30–33) assembled a list of network- and project-level tasks as perceived by 53 bridge owners. Many of the tasks appear at both levels with different scope, complexity, and responsibility. Project considerations dominated early bridge management. As the infrastructure started forming coherent networks, their priorities increasingly influenced local developments.

NCHRP Report 300 (1987) listed the main activities of network- and project-level management as shown in Appendix 13. The differences reflect the larger scope and broader considerations of the former and the greater significance of technical detail in the latter. In practice, network management prioritizes projects, whereas project management creates networks of tasks. The networks differ in size, diversity, and importance.

The polarized two-tier approach may be effective during early managerial efforts. If it becomes institutionalized over longer periods, it can foster a discontinuous and even conflicted process. Project-level management supplies services and demands funding, whereas the network level provides funding and demands accountability. If this specialization is maximized, engineers and managers are confined to their respective technical and financial expertise, and no amount of communication can compensate for the relative ignorance at each end.

The regimentation of competence into project and network levels is overcome by the *project manager, project engineer,* or *engineer in charge* (*EIC*). The position is indispensable to any large construction project (discussed also in Chapter 13). It requires both engineering and managerial expertise. The project is managed as a network of concurrent operations (e.g., ground up), but overall responsibility is maintained, as in the top-down approach. The efficiency and the competence required of the project manager/engineer have been honed over millennia of construction. As modern life-cycle bridge management develops, similar standards may eventually be adopted for "network" engineers and managers.

Network management must decide on a national level to what extent it should be top down. The level of centralization can vary. The federal structure of the United States allows for a clear interaction between central and local governments. The critical task then becomes to maintain a relatively equitable funding allocation. Example 5 illustrates the difficulties encountered by top managers in performing this task at the level of the national infrastructure network.

Funds are allocated proportionally to needs, such as traffic, and conditions (e.g., structural deterioration and obsolescence). The mayors of major cities in the United States have campaigned for federal funding allocation that would take into account the density and the criticality of the urban networks. Example 5 shows a federal-level response to a similar argument advanced by the author. The perception of these parameters on the project, local network, and central network levels is different even if the data are stored in a compatible format. In recent years the FHWA has conducted investigations and audits aimed at reconciling condition evaluations and rehabilitation needs in different states. At the same time, the traditionally centralized France is delegating the bridge management of local bridges to its Regions.

The quality of the future structures depends predominantly on the selection and supervision of consultants and contractors (see Section 13.3). These essentially project-level tasks are funded on the network level. So constrained, projects may not always achieve the quality projected by network managers.

Discontinuity

The terms *break down, disruption,* and currently *disconnect* are often used to describe critical discontinuities in the chain of technical and nontechnical managerial functions.

Discontinuities in any process imply communication failures (discussed earlier); however, they are not necessarily the results of oversight or ignorance. Rather, they may result from the deliberate elimination of all duplication of effort.

According to Fig. 4.4*a* bridges can be designed and inspected independently. Inspectors rate the structure's condition and designers analyze the structure's model, each communicating with the database. The flow of information depends on the data management center. That is where the typically managerial failures of *miscommunication* can occur.

The dynamic continuity of the bridge management process emerges if Fig. 4.4*a* is considered as a cross section and viewed in plan, as in Fig. 4.4*b*. So represented, design and inspection are not opposed but complementary functions. The bridge life cycle can be managed and optimized as it evolves from design to project, to asset, and finally to liability. Although Fig. 4.4*b* shows the stages of the bridge life cycle as overlapping, concurrent tasks do not guarantee a continuous process. Tables 4.2 and 4.3 illustrate the difficulty of transitioning smoothly from one stage of the bridge lifecycle to the next. UK Bridges Board (2005, p. 102, fig. 3.8) depicts a similar asset lifecycle, noting that "most highway structures need to be maintained in perpetuity".

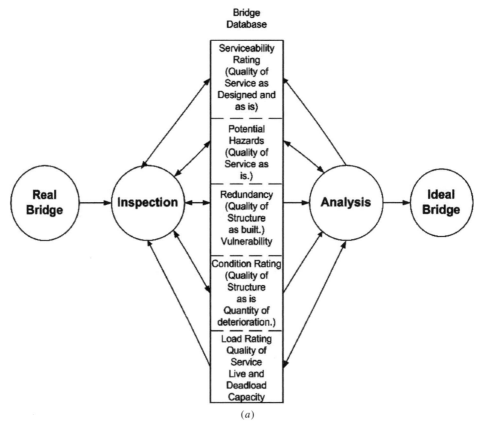

(*a*)

Figure 4.4   (*a*) Input and output of bridge condition database.

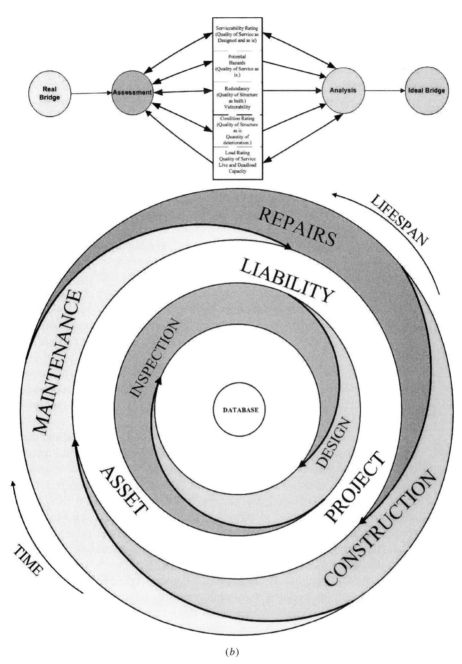

(*b*)

Figure 4.4   (*b*) Bridge life cycle and database.

Discontinuities can be anticipated between nonconsecutive life-cycle stages, as for instance analysis and construction or design and maintenance. Consecutive stages, particularly construction/maintenance, maintenance/inspection, or design/construction, can also lose touch with each other. Designers expect contractors to follow their drawings, whereas contractors revise them to improve constructability.

Because of its high productivity, regimentation of technical expertise into specialized units is hard to avoid. Managers of larger infrastructure networks are likely to streamline activities along the columns of Table 4.2. Even if concurrent, the activities along the rows of Table 4.2 are most likely to remain independent. Different constraints and priorities may govern the separate tasks. It is not uncommon to favor a particular design because of earlier experience or to supersede a design in favor of a preferred construction method. A bridge originally designed with bolted connections can end up fully welded during fabrication. Maintenance and inspections often object to choices made by design and construction.

An analogy can be drawn between organizational and structural discontinuities, such as articulation devices (e.g., expansion joints and bearings; Section 4.2.3). They ensure the partial independence of separate structural elements. If they fail, the elements either fall apart or unduly stress each other. The production of such devices is as discontinuous as is their function. They are assumed during analysis, specified by design, fabricated by a manufacturer, installed by a contractor, and often remain inaccessible for inspection and/or maintenance. Predictably, both product and process remain consistently vulnerable.

Discontinuities cannot be entirely eliminated from either structures or processes. Drucker (1968) observed growing discontinuities in technology, economics, politics, and learning. In that discontinuous environment, management must provide the one unbroken link.

The discontinuity between the fixed civil engineering facilities and the services they provide is fundamental to transportation. A further, even more critical discontinuity exists between the engineering of bridges and that of the rest of the transportation network. The construction and maintenance costs of highways and bridges relate roughly as $1:6$ and can reach $1:10$ for special structures. These ratios reflect the complexity of the engineering tasks involved in the respective design and construction. The ratio of respective criticality can be even higher. As a result, bridges command the attention of structural experts. In contrast, the length of highway networks is predominant and transportation system managers focus on it. Banks (2002) reported on measures, reducing the fragmentation of multimodal transportation systems (see Section 12.1), but divided civil engineering activities related to transportation into *physical civil engineering* and *system engineering* (p. 19). The former includes design, construction, and maintenance of fixed transportation facilities, including bridges. The latter consists of transportation planning, including analysis of the demand, the system capacity, and operational characteristics and design of traffic control and operating strategies. The text is concerned with the system aspects of transportation. An integrated approach is provided by Hudson et al. (1997).

If management fails to establish an equilibrium between services and assets, highways absorb the funding dedicated to transportation until bridge failures reverse the trend (see Example 3) and roadways begin to decline. Recent efforts in asset management (Section 12.1) emphasize the need for integration of multimodal transportation databases.

Administratively, executives in charge of engineering, personnel, and budget must find a balance. The alternative of consolidating management in a single position (the leader as manager/engineer) eliminates the risks of managerial miscommunication and duplication of effort but creates a potentially critical nonredundancy. Since, as Drucker (1954, p. 159) observed, "management cannot create leaders," it may have a tendency to distrust them, the more so in traditionally democratic environments, where "each man has to be convinced separately" (Tocqueville, 2000, see Section 1.2). Each case is resolved by competition between authoritative managers and more or less "micro"-managing leaders.

AASHTO (1999a) pointed out (p. 1-1) that business management principles have informed maintenance since statistical quality control introduced "management by objectives." The "level of service" approach advanced by NCHRP Report 273 (1984) is credited for the elimination of unnecessary duplication and the identification of contract opportunities. On the downside, this approach can be blamed for micromanagement, loss of authority, low morale, excessive cost reduction, and employment fluctuations.

Duplication of Effort and Critical Path

Redundant structures are not necessarily inefficient—they allow load redistribution. Similarly, it is a recommended engineering practice to confirm analytic conclusions by independent methods, when available. In contrast, a redundant process appears unproductive and is commonly targeted for downsizing. In a well-functioning organization management too can be perceived as redundant. To demonstrate its effectiveness, management typically targets any duplication of effort for elimination. The immediate savings are demonstrable. The potential disadvantages take longer to surface.

Eliminating all operational overlaps sets the life cycle on a critical path. Where a product lends itself to modular production, the method can be cost effective without loss of quality. Airbus aircraft modules are completed and certified by QC in different countries, then directly assembled in Toulouse. If that method were applied to bridges, the structural features listed in Table 4.2 would receive independent and self-sufficient treatment during each of the stages shown in Table 4.3.

The critical path method (CPM) is a popular project management technique. Halpin and Riggs (1992, p. 269) point out that it underestimates project durations. Whatever its limitations as a tactic, CPM was never intended as a long-term transportation strategy. Throughout their life, the structural components listed in Table 4.2 continually depend on all the tasks of Table 4.3. This is concisely expressed by the New York State Commission of Investigation for the Schoharie Bridge collapse (1987): "While the bridge's design and construction may well have been deficient, with proper inspection and maintenance, the bridge would not have collapsed."

The question arises: What is the likelihood of superior maintenance following deficient design and construction? If deficiencies are likely at all stages, every task should be able to overperform, as in the concept of safety factors. Better maintenance and inspections might have indeed averted the collapse, but reliability could have been improved only by a different combination of design, construction, maintenance, and inspection.

Bridge operations duplicate some functions both within the owner's organization and in interactions with other entities.

Internally, the process becomes critical if the activities of Table 4.3 are conducted by independent staff with no duplication. Short-term costs are minimized, and the critical path is achieved. For a large asset network seeking optimal life-cycle cost–benefit performance, this strategy fails in the long term. Eventually accidents expose the losses to reliability and serviceability declines. Annual reports of safety hazards on the New York City bridges escalated from less than a hundred in the mid-1980s to 3000 in 1993. The average bridge life declined from 60 to 30 years, and the annual rate of bridge rehabilitation doubled.

Diverse organizations (e.g., private and public) regularly interact during the "outsourcing" of specific tasks. The variety and scope of bridge-related work makes contracting the only option for most bridge owners. Example 3 lists the costs of capital rehabilitation on the East River bridges in New York City over a period of 20 years. Example 9 describes typical annual expenditures on the city bridges. The predominant capital portion of these expenditures is channeled through contracts, performed by scores of consultants and contractors. The FHWA provides a substantial share of the funding and oversees the progress of the contracts.

The federally mandated biennial bridge inspections are conducted by consultants under contracts to state departments of transportation (DOTs). Without in-house interpretation and field verification by the responsible owner, however, the resulting inspection reports (e.g., the product) cannot be used.

As in structural redundancy (Section 5.3), task duplication improves the reliability of a process only if the activities are internally reliable as well. A formal duplication of functions provides many opportunities for a discontinuity. The dual ownership of structures carrying different modes of transportation is one example. Visual bridge inspection and analytic load rating (Chapter 10) similarly can overly depend on each other.

Figures 4.5 and 4.6 show an example of a structure which was rebuilt with the particular intent to separate vehicular-and train–supporting approaches in order to simplify their management.

Projects and Services

Financial management prefers projects of finite duration over perpetual services. Projects are clearly defined by their budgets, funding sources, schedules, lines of responsibility, and deliverables. Both culturally and economically, outsourcing of tasks in the form of projects currently holds sway over continuous in-house services. The Transportation Equity Act for the twenty-first century (U.S. Congress, 1998) defined life-cycle cost analysis as "a process of evaluating the total economic worth of a usable project segment by analyzing initial costs, such as maintenance, user, reconstruction, rehabilitation, restoring, and resurfacing costs, over the life of the project segment."

Nonetheless, the breaking of typically continuous tasks into discrete projects can also be counterproductive. The correspondence in Example 4 illustrates opposed views on the subject among top political management. Appendix 19 outlines fundamental views on the applicability of privatization as a management tactic. Maintenance (Section 4.4) is vulnerable primarily because its tasks do not fit comfortably into distinct projects with

Figure 4.5 Bridge pier supporting vehicular and train traffic.

Figure 4.6 Multiple piers, each supporting specific traffic mode.

tangible deliverables. Smaller versus larger can be a solution. Yanev and Testa (2001) concluded that painting may have been neglected for decades by some bridge owners because it was managed as a routine maintenance task, rather than as rehabilitation projects (Examples 23 and 24). Since Intermodal Surface Transportation Efficiency Act (IS-TEA) 91 (Appendix 11), the FHWA concurs and funds painting as capital improvement.

4.1.5   Economy and Economics

De Gramo et al. (1979, p. 19) consider engineering economy distinct from other economic studies because it is not governed by monetary considerations alone. Applying the methods of economics developed for private interests to publicly owned infrastructure facilities entails the following difficulties (p. 373):

- The financial effectiveness of nonprofit organizations cannot be measured in terms of profit (e.g., benefits do not equal profits).
- Benefits cannot be easily quantified in monetary terms.
- Publicly owned projects are disconnected from the public.
- Critically important political considerations can be unknown to engineers or biased toward short-time effects. Benefits do not equal interests.
- Indefinite motivation of personnel.
- Legal restrictions.

Cost–benefit analysis focuses on quantities, mostly expressed in terms of time and money. Since transportation networks are publicly owned assets, the services they provide are viewed as social gains, whereas reductions are perceived as penalties sustained by the users. For the purposes of transportation, a monetary expression of traffic volumes is determined. That determination is invariably approximate and highly controversial.

On page 17 of *Le Monde* (January 6, 2006) the economists P. Kopp and R. Prud'homme argue that the restrictions imposed by the administration of Paris on the use of passenger cars are costing the motorists 60 million hours annually at a rate of EU 9/hour. An additional 6 million hours of truck delays are valued at EU 30/hour. The authors estimate the corresponding benefit in air pollution reduction at EU 70 million. The highly approximate arithmetic translates the vehicular service reductions into ("maximized") user losses of roughly EU 830 million/year. On the same page, the mayor of Paris, B. Delanoë, rejects the findings as subjective and uninformed in matters of public health and the environment. He vows to "break with the old logic of *all automobile*" by developing public transportation and reducing congestion.

The costs and benefits of deicing city bridges with rock salt can be roughly estimated by similarly approximate calculations, illustrated in Example 9. Monetary expressions of "quality-of-life" issues, such as comfort, health hazards, historic, and aesthetic values must reflect public perception. Typically prescient, Eiffel described his requirements for elevator safety as follows: "The system is not only safe, but appears safe, a most desirable feature in elevators traveling such heights and carrying the general public" (Harriss, 1975, p. 95).

User costs are further discussed in Section 11.2.1. Example 10 illustrates the difference between cost–benefit estimates based on annualized and discounted life-cycle costs.

In Example 11, the consideration of user costs reverses the outcome of the life-cycle cost analysis (LCCA).

Example 10. Discounted Present Worth (PW)

Present worth analysis (PWA) compares alternative investment strategies by discounting future costs and benefits at an assumed annual rate i. Discount rates model the loss of interest in benefits retarded from the present to the future. In contrast to inflation rates, discount rates are subjective and apply to specific investments rather than to the overall economy. Hudson et al. (1997, p. 304) wrote: "The discount rate selected by most agencies is a policy decision, but usually it is the difference between the interest rate for borrowing money and the inflation rate."

Hawk pointed out in NCHRP Report 483 (2003) that discount rates generally comprise three components as follows:

$$i = (1 + cc)(1 + fr)(1 + pi) - 1 \qquad \text{(E10.1)}$$

where cc = "real" opportunity cost of capital
  fr = required premium for financial risk associated with considered investments
  pi = anticipated rate of price inflation

Neglecting the higher order terms is justified by the relatively small values involved and reduces Eq. E10.1 to

$$i = cc + fr + pi \qquad \text{(E10.1a)}$$

The present worth of an amount A occurring N years into the future is reduced by the factor $1/(1 + i)^N$. If amounts a occur annually, their present worth for a period extending over N years from the present is equal to

$$a \sum_{n=1}^{N} \frac{1}{(1 + i)^n} = a\left(1 + \frac{1}{i}\right)\left(1 - \frac{1}{(1 + i)^N}\right) \qquad \text{(E10.2)}$$

$$\lim_{N \to \infty} a \sum_{n=1}^{N} \frac{1}{1 + i^n} = a\left(1 + \frac{1}{i}\right) \qquad \text{(E10.3)}$$

Equation E10.3 implies that discounting transforms the nonconvergent infinite series $(a, a, a, \ldots)$ into a convergent one with boundary determined by i. Figure E10.1 (Yanev, in TRC 423, 1994, p. 132) shows the curves defined by Eq. E10.3 at different values of i within the practical range of PWA. Since they tend to finite limits with the increase of N (measured in years), analysis can only discern significant results for a limited time. The ratio of the sum obtained at a finite value of N to the infinite sum is equal to

$$\frac{a(1 + 1/i)[1 - 1/(1 + i)^N]}{a(1 + 1/i)} = \left(1 - \frac{1}{(1 + i)^N}\right) \qquad \text{(E10.4)}$$

Given i, N can be selected such that the error ε of substituting it for infinity would not exceed an acceptable threshold as follows:

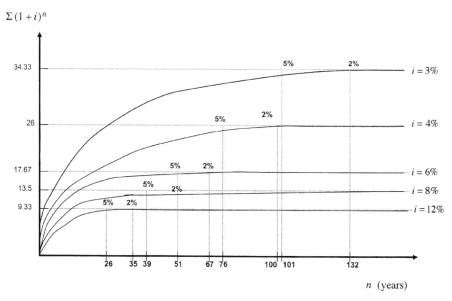

Figure E10.1 Cumulative present worth of future moneys at different discount rates.

$$N = -\frac{\ln \varepsilon}{\ln(1 + i)} \qquad (E10.5)$$

Representative numerical values obtained from Eq. E10.4 are listed in Table E10.1.

It is well known that the significant number of years N declines with the increase of i. High values of i imply that returns on an investment must be expected only in a relatively short term. An average value assumed for discounting is $i = 4\%$. In order to minimize the effect of discounting without completely eliminating it from life-cycle cost analysis, a low discount rate $i = 2\%$ is typically assumed.

Leeming (in Harding et al. 1993, p. 576) presented a graph similar to the one in Fig. E10.1. As in the present example, he concluded that "any costs beyond 30 or 40 years have a negligible influence on the outcome." Stopping short of rejecting discounting for maintenance expenditures, the author concluded: "If maintenance of

Table E10.1 Error ε Associated with Ignoring Annual Inputs Beyond Period of N Years at Different Discount Rates i

i (%)	2			3			4			5			6			7		
$1+1/i$	51			34.3			26			21			17.6			15.3		
ε (%)	2	5	10	2	5	10	2	5	10	2	5	10	2	5	10	2	5	10
N years	198	151	116	132	101	78	100	76	59	80	61	47	67	51	40	58	44	34

our bridge stock is to remain a fixed percentage of the total governmental expenditure on construction then there is an argument for a zero discount rate in calculating the net present value of maintenance."

For infrastructure facilities requiring periodic maintenance and replacements, De Gramo et al. (1973, p. 93) recommend *perpetuity,* i.e., a uniform series of indefinitely running payments. In order to provide for annualized payments X, a principal P must be set aside at annual interest in% (interest $\neq$ discount rate), such that P in $= X$. If the payments are not annual but arise at k periods, the relationship becomes

$$X = P\,[(1 + in)^k - 1] \qquad\qquad \text{(E10.6)}$$

where P is the *capitalized* value of X.

On the feasibility of capitalized annual maintenance expenditures, Leeming (in Harding et al. 1993, p. 576) commented: "Governments do not usually put aside sums of money for future expenditure, but maintain out of income from the taxes we pay. . . . It would be necessary to invest at 8% compound [interest] in order to keep pace with the increase in the road construction price index [UK]."

First versus Life-Cycle Costs

Cost and benefit estimates (Section 10.2.1) are both critical and highly vulnerable. After scour undermined a pier of the Schoharie Creek Bridge and caused its collapse in 1987, an earlier rejected proposal for a costlier longer span received nationwide attention. Yet, despite arguments against it, lowest first cost remains central to project selection.

Minimizing first costs to the detriment of life-cycle ones is a concern of long standing. Boller (1885, p. 8) admonished: "Ordinarily, the cheapest proposal wins the day, simply because to the average committeeman one iron bridge is as good as another, no matter from what source its plan emanates. . . . First-class bridge-builders find themselves underbid by ignorant or unscrupulous builders, who have no other ambition than that of getting work."

Calgaro (2004), chief engineer of bridges and highways, France, quoted a letter submitted by Marechal Vauban, commissioner general of fortifications, to King Louis XIV on July 17, 1683, as follows: "Enough, Monseigneur, to show you the imperfection of this practice (e.g. underbidding): abandon it for God's sake, reestablish good faith, give the structures their price and do not refuse the entrepreneur who will acquit himself of his duty his honest salary; this will always be the best bargain you can find."

Waddell (1916, p. 1182) updated that position to the early twentieth century (Appendix 12).

Billington (1983) pointed out, on the other hand, that economically competitive designs have also been among the most innovative. The difficulty is in drawing a distinction between cost effective and cheap. The application of *value engineering* to bridges does not always clarify that distinction. De Gramo et al. (1979, Chapter 16) attribute the concept to General Electric in the 1940s and the Bureau of Ships in 1954. It consisted of reviewing a design for a potential substitution of materials with less expensive ones without reducing quality. A number of cost-effective improvements resulted. In the case of bridges, value engineering may modify the method of construction or even the structural scheme. Life-cycle consequences can become apparent much later. *Signature bridges* emphasizing appearance are a possible (and occasionally extreme) reaction to *value* engineering.

Discounted Present Worth of New and Existing Bridges

Estimates of the initial (or construction) costs associated with the infrastructure are uncertain, and they grow increasingly unreliable when extended over the life cycle of the assets. The present worth (PW) of future costs and benefits is obtained by discounting. According to the Asset Management Office of the FHWA:

> Discounting is an economic method of accounting for the time value of an investment. The calculations of discounting are identical to those of compound interest. Lifecycle cost analysis (LCCA) uses discounting to convert anticipated future costs to present dollar values so that the lifetime costs of different alternatives can be directly compared. Because the level of service provided by each project alternative in the analysis is assumed to be the same, LCCA allows transportation agencies to evaluate alternatives on the basis of their lifecycle costs.

Discounting is occasionally referred to as the "time value of money" (Example 10). It is particularly convenient in comparing costs and benefits over different anticipated periods. It has also been blamed for obliterating any effect beyond its horizon. The following fundamental limitations should be considered.

The present value of an existing structure is not equal to its adjusted construction cost. The benefits (or urgency) of replacing a bridge already long in service are not the same as those of a new service, even if the projected traffic volumes are comparable.

The performance of proposed and existing structures cannot be evaluated under the same constraints. New bridges must respond to present and projected demands. Their structural and administrative characteristics are determined by their intended service. Existing bridges are part of the infrastructure fabric and the regional geography (Examples 1–3; Fig. 1.42). Reluctantly or willingly, some have become local and international landmarks, as for example the Brooklyn (Example 1), the Cincinnati-Covington (Fig. 1.37a), and the Smithfield (Fig. 1.14). Others aspire for "signature" status, as the proposed San Francisco–Oakland East Bay crossing (Fig. 1.44). Such bridges may be managed for maximized life span at the level of service they can best provide.

The estimated costs and benefits associated with proposed bridges, although highly uncertain, must compete with alternative investment opportunities. Rejecting a new bridge proposal is always a realistic option. The George Washington Bridge (Example 2) was the only one of many proposals materializing across the Hudson River in New York City. The multi-span cable stayed bridge across Long Island Sound was never realized (see also Section 1.11).

For bridges in service, the options are limited to maintenance (ranging from zero to optimal), repair and replacement but rarely include scrapping the entire service. The bridge shown in Fig. 4.7 was built as a replacement for the one in Fig. 4.8. The com-

Figure 4.7 Cable-stayed bridge across Delaware canal.

munity, however opposed the demolition of the old bridge, declared it a landmark. Traffic capability has substantially increased but not the maintenance funding.

The maintenance costs of established networks are relatively constant over time and proportional to the size. The operating costs are known and the benefits can be estimated by accepted models, based on measurable service indicators. If the network is large enough and its condition is uniformly (or normally) distributed across the governing condition rating scale, reconstructions become a routine expenditure, independent of discounting (Section 11.5). The maintenance practices of large asset networks are most revealingly expressed by the ratio of annual maintenance expenditures to the estimated replacement cost (see Section 11.3). Typical ratios of annual maintenance to estimated replacement costs were quoted by BRIME (2002) and Bieniek et al. (1989).

In contrast with discounted costs, annualized life-cycle costs represent an annual average of all expenditures over the useful life of an asset. Example 11 illustrates the differences between the equivalent present worth of estimated life-cycle costs and annualized costs.

Figure 4.8   Tied arch bridge across Delaware canal.

Example 11. Life-Cycle Strategies at $i = 0$ and $i = 6\%$

The direct costs of two maintenance strategies A and B are compared over a period of N years at discount rates i of 0 and 6%. Numerical values of annual maintenance costs are consistent with experience and/or recommendations at the NYC DOT (Example 9). Maintenance is expressed in terms of its annual cost per unit bridge deck area. Structures are already existing; therefore, cost of initial construction is not considered.

CASE I

Strategy A: Annual preventive maintenance m_A is designed such that it can remain constant.

Strategy B: Annual maintenance starts at a minimal level m_{BI}. Demand is assumed to grow exponentially at a rate of 6%.

m_{An} = U.S. $21.50/m^2/year (U.S. $2.0/ft^2/year), without painting of steel

m_{An} = U.S. $64.50/m^2/year (U.S. $6.0/ft^2/year), with painting of steel

m_{BI} = U.S. $10.75/m^2/year (U.S. $1.0/ft^2/year), minimal maintenance

Table E11.1 shows the formulas computing the cumulative maintenance costs at discount rates $i = 0$ and $i = 6\%$.

Here, N is obtained by trial and error. The values suggest that strategy A becomes cost effective at annual cost of U.S. $21.50/m^2/year for a life-cycle longer than 21 years (at $i = 0$) or 26 years (at $i = 6\%$). By the end of this period the annual maintenance cost according to strategy B would exceed U.S. $36.55/m^2/year (presumably as a result of declining conditions). Rehabilitation would be the only option.

Table E11.1 Cumulative Annual Costs Discounted at 0 and 6%

		Maintenance Cost ($US/m^2/year$)	
	i (%)	Strategy A	Strategy B
Year n	0	$m_{An} = m_A 1.06^n$	$m_{Bn} = m_B$
	6	$m_{An} = m_A$	$m_{Bn} = m_B/1.06^n$
Years 1 through N	0	$\sum_{n=1}^{N} m_{An} = m_A \dfrac{1.06(1.06^n - 1)}{0.06}$	$\sum_{n=1}^{N} m_{Bn} = m_B N$
	6	$\sum_{n=1}^{N} m_{An} = m_A N$	$\sum_{n=1}^{N} m_{Bn} = m_B \dfrac{(1 - 1/1.06^n)}{0.06}$

	m_{An}		N at $\sum_{n=1}^{N} m_{An} = \sum_{n=1}^{N} m_{Bn}$	m_{BN} at year N	
i (%)	$US/m^2/$ year	$US/ft^2/$ year	(years)	$US/m^2/$ year	$US/ft^2/$ year
0	21.50	2.00	21	36.55	3.40
	64.50	6.00	50	198.01	18.42
6%	21.50	2.00	26	10.75	1.00
	64.50	6.00	100	10.75	1.00

At a cost of U.S. $64.50/m^2/year$, strategy A becomes cost effective only for cycles longer than *50* years (at $i = 0$) or 100 years (at $i = 6\%$). At that time strategy B would already have escalated to the unrealistic amount of U.S. $198.01/m^2/year$. That, however, is only apparent at $i = 0$. At $i = 6\%$ strategy B has constant annual present worth. Yanev (in Frangopol and Furuta, 2001, pp. 299–312) described the result as *economically admissible* but *structurally inadmissible*. While the exaggerated impracticality of the exponentially growing maintenance assumed herein for demonstration purposes is not likely to be overlooked by experts, policies similar to strategy B have been adopted over short periods. The corresponding life cycles have declined to 25 years, as in the example.

Because of their separate funding sources, expense and capital, respectively, maintenance and (re)construction, have been occasionally optimized independently. The present example should demonstrate that such "partial" optimization cannot be meaningful.

This example, as well as Example 24, illustrates difficulties arising from the high cost of preventive maintenance when the painting of steel with full containment is included. The cost of painting and the painting cycle (12–20 years) make it similar to "component rehabilitation" rather than to "routine preventive maintenance."

CASE II

Strategy A: Constant annual preventive maintenance m_A = U.S. $21.50/m^2/year$ (U.S. $2.0/ft^2/year$). Repainting and repair at 12-year cycles at a cost of U.S. $516/m^2$ (U.S. $48/ft^2$).

Strategy B: Demand annual maintenance starts at m_{B1} = U.S. $10.75/m^2/year$ (U.S. $1.0/ft^2/year$) and grows exponentially at a rate of 6%.

Table E11.2   Strategies A and B over Cycle of 48 Years (U.S. $/m²)

	i = 0	i = 6%
Strategy A	3 × 516 + 48 × 21.5 + 4840 = 7420	447.19 + 336.47 + 295.08 = 1078.74
Strategy B	2924 + 4840 = 7764	516 + 295 = 811
Strategy A/B	0.96	1.33

Strategies A and B: Reconstruction after 48 years at a cost of U.S. $4840/m² (U.S. $450/m²).

Applying the formulas of Example 10 obtains the following total direct costs for the two strategies shown in Table E11.2:

The two strategies appear comparable at $i = 0$, but at $i = 6\%$, demand maintenance and no painting are a clear winner over 48 years. There is no evidence that such practices were deliberately selected following any present worth analysis, but intuitive considerations of the presented type may have contributed to their implementation. Strategy B errs in assuming that demand maintenance measures can sustain bridges in safe operation without major maintenance for extended periods. Example EA.46 discusses the growing number of potential hazards toward the end of the projected life cycle.

CASE III

Strategy A: Constant annual preventive maintenance $m_A = $ U.S. $21.50/m²/year (U.S. $2.0/ft²/year). Repainting and repair at 12 year cycles at a cost U.S. $516/m² (U.S. $48/ft²). Reconstruction after 120 years at U.S. $4840/m² (U.S. $450/m²).

Strategy B: Demand annual maintenance starts at $m_{B1} = $ U.S. $10.75/m²/year (U.S. $1.0/ft²/year) and grows exponentially at a rate of 6%. Reconstruction every 30 years at a cost of U.S. $4840/m² (U.S. $450/m²).

Strategies *A* and *B* are compared over a 120-year cycle in Table E11.3.

Strategy A is cost effective independent of discounting; however, it relies on the assumption that maintenance $m_A = $ U.S. $21.50/m²/year coupled with painting and repairs at U.S. $516/m² every 12 years would obtain a 120-year life cycle. Maintenance expenditures are justified by the resulting delays in bridge deterioration; however, these delays are speculative. The modeling of maintenance benefits in terms of retarded deterioration is discussed in Example 24.

Table E11.3   Strategies A and B over Cycle of 120 Years (U.S. $/m²)

	i = 0	i = 6%
Strategy A	9 × 516 + 120 × 21.5 + 4840 = 12,062	508.79 + 357.97 + 4.41 = 871.17
Strategy B	4 × (901 + 4840) = 22,964	390 + 1018 = 1408
Strategy A/B	0.53	0.62

Another difficulty illustrated by the example in case III arises from an 120-year-long life cycle. Although such a life cycle is structurally and economically optimal, it is unlikely to offset the more pressing priorities of the present and the immediate future.

In certain cases neither continuous maintenance nor reconstruction satisfies the users. The West Side Highway in Manhattan (e.g., Route 9-A) (also discussed in Section 4.3.1) was scrapped over a length of 30 city blocks following the collapse of one span at West 57th Street in 1973. Attempts to replace it with a new highway were blocked by the public (Gibble, 1986). After extensive negotiations with the local community, the 64 spans north of 57th Street, known as Miller Highway (Fig. 4.9a), were rehabilitated at a cost of roughly U.S. $90 million (1990). With the bridge rehabilitated and the area underneath developed as a park (Fig. 4.9b), J. Barron reported in the *New York Times* (June 23, 2006, p. B4, column 1) that U.S. $180 million of private, City and State monies are set aside and work has begun on the "relatively simple construction" of a box that would divert the highway traffic to a tunnel under future apartment blocks. An organizer of the Committee for Environmentally Sound Development terms the project "absolutely the wrong thing to do," because "if they wanted to do something for the people in the neighborhood, they'd complete Riverside Boulevard and forget about the highway."

The 322-span Gowanus Expressway in Brooklyn (Fig. 4.10) has engendered similar debates. The main challenge of managing multispan urban bridges lies in reconciling the services they provide with the diverse interests of the many users.

Construction/Reconstruction Cost

Billington (1983) singled out Eiffel as a rare designer, famous not only for the quality of his work but also for his ability to achieve it within the available budget. Project cost and duration overruns are commonly denounced as trademarks of mismanagement, although some are easier to avoid than others. The preliminary estimates of cost and scope of work on big construction projects are highly uncertain for a number of objective reasons, which vary from case to case. New construction invariably entails unforeseen developments. A highly publicized recent example is the "Big Dig" in Boston where costs reportedly escalated from *U.S. $2 billion* to *U.S. $14 billion,* mainly as a result of soil conditions. The scope of reconstructions can be fully assessed only during their execution, when the structural condition can be examined "hands-on." Construction duration is easily underestimated, particularly when every effort must be made to minimize traffic closures.

The rehabilitation of the bridge shown in Fig. 4.11 surpassed its estimated scope considerably, because the structure was found to be in much worse condition than was first estimated. Furthermore, the bridge was considered a local landmark and repairs had to preserve the original appearance.

Constructability may necessitate the replacement of structurally sound components. The orthotropic deck shown in Fig. E3.6 required very precise vertical alignment. As a result, the supporting floor beams had to be replaced even though they had been installed less than 10 years earlier and were in good condition.

(a)

(b)

Figure 4.9 (a) West Side Highway, Manhattan, 1994. (b) West Side Highway, 2006.

Table E3.1 shows the total cost estimates for the rehabilitation of the East River bridges in New York City as they were reported between 1990 and 2004.

Cost overruns appear somewhat exaggerated, particularly for long-term projects, because monies are not adjusted for inflation over the years. The last column of Table E3.1 adjusts the estimates of 1990 for annual cost and inflation increase of *4%*.

If *U.S. $1 million* were annually committed to a project over a period of *10* years, the original estimate would amount to *U.S. $10 million*. Adjusting for annual cost and inflation increase of *4%*, the total expenditure at completion would amount to

Figure 4.10 Gowanus Expressway, Brooklyn.

$$\frac{1 \text{ million } (1.04^{10} - 1)}{1.04 - 1} = 12 \text{ million}$$

An actual 15% cost overrun could be justifiable for a U.S. 10 million project over 10 years. Adding another 20% to it would, in this case inaccurately, imply mismanagement.

Figure 4.11 Riverside Drive at West 125h Street, Manhattan, was rehabilitated in kind.

Halpin and Riggs (1992, p. 269) argue that "two of the most popular network scheduling techniques, e.g. the Critical Path Method (CPM) and the Probabilistic Evaluation and Review Technique (PERT) . . . do not provide adequate information regarding the potential for schedule overruns." The authors consider PERT "but a slight improvement [over the deterministic CPM], in that it attempts to evaluate the probability of project duration by giving the expected completion time." Underestimates by PERT are attributed to a "merge event bias," occurring when several paths converge on a single node. Monte Carlo simulation is recommended as a model avoiding the singularities.

Experience shows that total costs for large projects should be specified only within relatively broad ranges. The alternative proposals for the capital upgrading of the Tappan Zee bridge across the Hudson River (Fig. 2.2), for instance, are estimated to cost between *U.S. $3.5 billion* (for a rehabilitation of the existing structure) and *U.S. $20 billion* (for a new tunnel, carrying highway and rail traffic) (*New York Times,* May 2004).

Average and Worst Cases

Einstein and Infeld (1942) observed that statistical laws explain the behavior of crowds better than that of individuals. Averaging is common in aggregate estimates of conditions and anticipated needs of large networks over long periods. Averages grow increasingly predictable as sample spaces expand, but not so the extreme cases likely to absorb the limited available resources. According to Barlow et al. (1965) most distributions modeling the behavior of large populations differ primarily in their tail ends, where data are the most scarce (as well as critical). The scope and urgency of the worst case needs cannot be extrapolated from models of average conditions. Example 12 compares average with worst-case deterioration rates of bridge conditions obtained from visual inspections.

Example 12. Average and Lowest Condition Ratings

The comparison of bridge condition ratings and their respective ages during any year of inspections is a snapshot in time; however, it implies a rate of condition rating deterioration. The result is meaningful if the network is large, the structures are of varied ages, and the governing conditions are roughly uniform. Alternatively (and much more elaborately) the history of condition ratings, related work, and type of service must be followed over a significant period of time for each structure. The two approaches to obtaining deterioration patterns are somewhat analogous to the studies of structural response to dynamic excitations by spectral analysis and time histories.

Figs. E12.1*a* and 12.1*b* show the distributions of structural condition and sufficiency ratings of the bridges in New York City, obtained from visual inspections and computed from inventory and inspection data respectively (also compared in Example 9). The rating systems are described in Appendixes 40 and 41.

The three different patterns in Fig. E12.2 are obtained from the data of Fig. E12.1*a* as follows:

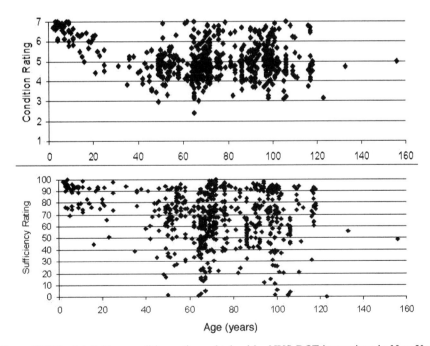

Figure E12.1   (*a*) Bridge condition ratings obtained by NYS DOT inspections in New York City, (2005). (*b*) Sufficiency ratings computed for the NYC vehicular bridges (2005).

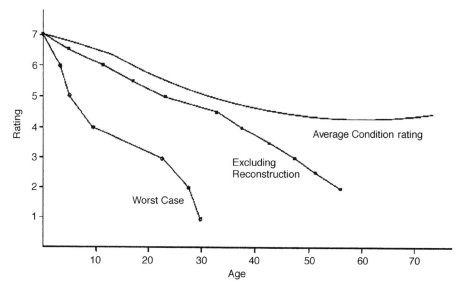

Figure E12.2   Condition rating patterns, obtained by reducing data of Fig. E12.1.

1. *Average Condition Ratings For All Bridges of All Ages.* The condition ratings decline rapidly during the early stages (and higher ratings), then tend asymptotically toward the rating of 4 on the scale from 7 to 1. This pattern includes the effects of rehabilitations and repairs to an unknown degree. After rehabilitations, the bridge construction date usually remains unchanged in the inventory. Thus the deterioration appears to have been reversed. Pont Neuf, the oldest bridge in Paris, is an extreme example. It dates back to 1603 and its present condition justifies its name, signifying "new." The apparently nonexistent deterioration of the bridge is the result of continuing rehabilitations, such as the ones shown in Fig. E12.3 and 4.77, as befits the structure's landmark status.

   The average condition rating history could become a meaningful source of information if bridge conditions at every age were associated with corresponding expenditures. In the absence of such data, construction dates could be adjusted in the inventory to the completion of the latest rehabilitation. That solution again has its limitations, since it would prescribe to Pont Neuf and to the East River bridges (Example 3) an even more misleading contemporary construction date. Alternatively, bridges showing condition rating improvements could be eliminated from the data. The result is the second pattern.

2. *Average Condition Ratings Excluding Those of Rehabilitated Bridges.* Records of capital rehabilitations exist but they must be correlated with condition ratings by a case-by-case screening. The resulting curve suggests a useful life (between rehabilitations) approaching 60 years. This finding does not agree with the capital expenditures incurred in maintaining a constant

Figure E12.3 Rehabilitation at Pont Neuf, Paris.

average bridge condition during the corresponding period (Examples 9 and 19).

Capital rehabilitation projects are scheduled according to the condition ratings generated by biennial inspections, but their scope is determined by independent in-depth inspections (Section 14.3). Final construction costs do not describe sufficiently the scope of the rehabilitation work, because they are subject to competitive bidding (Appendix 20) and include many additional site-specific expenditures (such as traffic control).

As bridges approach the end of their useful life, hazard mitigation and emergency repairs become increasingly necessary. Condition ratings may or may not be affected. Consequently the actual condition of the bridge is not uniquely defined by the inspection rating at the time of rehabilitation and cannot be ascertained at a later date. It is therefore concluded that, even after excluding documented rehabilitations, the obtained average condition rating pattern does not represent the shortest useful life cycles that determine rehabilitation needs.

3. *Lowest Ratings.* The condition rating history including only the worst condition ratings observed at any age suggested a shortest useful life of 30 years, corresponding to the expenditures described in Examples 9 and 19. For the bridge elements comprised in Eq. A41.3, Yanev (1997) reported shortest observed lives as shown in Fig. E12.4. The result is consistent with the steepest overall bridge condition deterioration rate r, discussed in Examples 23 and 24. The weighted average result, computed according to the linear assumption

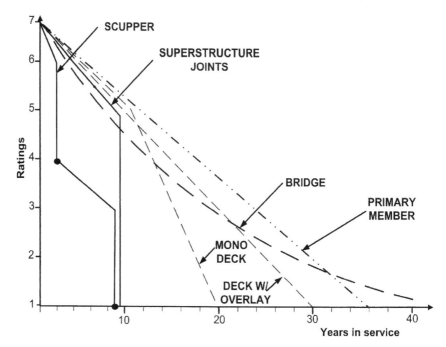

Figure E12.4 Shortest lives of bridge elements.

of Eq. A41.3 (also Eq. E24.1) is $r \approx 0.24$, as shown at the bottom of Table E23.2, column 8.

According to Eq. E19.1, the corresponding useful life is $L = (7 - 1)/0.24 \approx 25$ years. The annual rehabilitation expenditures described in Examples 9 and 19 are consistent with a 25- to 30-year life cycle.

The comparison of the three deterioration histories in Fig. E12.2 has suggested several observations. According to O'Connor and Hyman (1989) the majority of reported deterioration histories following the trajectory of case 1 appear to suggest that, with age, the *average condition ratings* stabilize at an average level. The many factors contributing to this result must be sorted out.

A bridge deterioration pattern averaging all available inspection ratings is misleading. As the bridge age advances, the need for rehabilitation and repair work increases. At the age of 100 years, numerous rehabilitations are a certainty, but their scope is unknown. The number of data points vary over the ages, with few at both ends and many in the central portion. The segments of the deterioration curve for different ages have different reliability.

Reverting the age of rehabilitated bridges to zero would be inaccurate (even if they were called new, as Pont Neuf, Fig. E12.3), because restored conditions and deterioration rates are never "as new." Excluding the rehabilitated bridges or adjusting their age significantly reduces the data for structures of advanced age. The deterioration history up to the first rehabilitation (as in case 2 herein) becomes more realistic; however, it still contains undocumented improvements and, more importantly, repairs delaying further decline in the ratings.

The "worst-case" history is the simplest to obtain and the most meaningful, since it shows conditions that automatically become the highest priority. The steepest deterioration rates of bridge elements (Fig. E12.4) are particularly significant. They offer a glimpse of the most vulnerable links in the structural system that should be targeted by intensive maintenance or early replacement.

Figure E12.4 identifies joints (Figs. 4.74 and 4.75) and scuppers (Fig. E12.5) as the foremost "weak links" in the structural chain. The NYS DOT bridge condition formula (Eq. A41.3) does not include scuppers and assigns a relatively low weight (see Table E23.2) to joints. Another structural element essential to maintenance but absent from Eq. A41.3 (presumably because of its irrelevance to direct load-bearing capacity) is paint (Fig. E12.6). Examples 23 and 24 explore the advantages of treating paint as a rehabilitation, rather than a maintenance item.

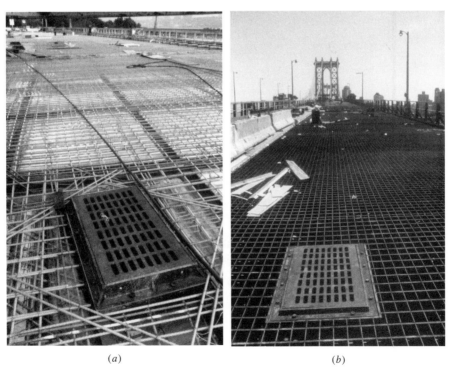

(*a*)   (*b*)

Figure E12.5   (*a, b*) New scuppers in concrete and steel grating decks.

(*c*)

Figure E12.5   (*c*) Failed scupper.

Figure E12.6   Steel girders with cover plates, diaphragms, bearings, pedestals, and stay-in-place forms in excellent condition. Failed paint.

Deferred Maintenance

Maintenance is the hardest bridge-related activity to optimize, because its tasks recur over the shortest cycles whereas its benefits are realized only over the longest cycles. Maintenance deferral is not considered critical to safety, and it affects structural performance to an unknown degree. By the time malfunctions become manifest, the structural useful life has been irreparably shortened and managers are blaming their predecessors. Recurring maintenance costs are continually minimized, even though construction has already minimized first cost, assuming future maintenance "as needed." As needed is never defined. NCHRP Report 285 (1986, p. 7) pointed out: "Deferred maintenance is a relative term and one must have a reference service level before differences can be evaluated."

So long as maintenance level is linked to the services provided by the structures, the deferral of tasks or expenditures is subject to interpretation and optimization, typically leading to minimization. The alternative of expressing maintenance level in terms of quantified tasks, in contrast, eliminates the essential role of management (e.g., to optimize expenditures) and exposes the responsible bridge owner to tort liability (e.g., negligent maintenance). NCHRP Report 285 (1986, p. 6) quoted NCHRP Digest 80 to the effect that "action taken by government is *discretionary* and, therefore, immune." However, the report added: "An individual engaged in the exercise of nondiscretionary, ministerial duties could be held liable for the consequences of his negligence." (See also Sections 4.1.1 and 13.2.)

NCHRP Synthesis 153 (1989, p. 9) summarized the early maintenance neglect as follows: "By 1976, the FHWA had recognized that the backlog of deferred maintenance was reaching crisis proportions on many Interstate and primary highways. It was also evident that routine and preventive maintenance programs were unable to meet this need and correct structurally deficient pavements."

NCHRP Synthesis 330 (2004, p. 3) reported: "The annual cost of preserving the U.S. National Highway System's pavements at existing conditions is nearly $U.S. 50 billion. Improving the system from its current condition to a 'good' level (and then, presumably letting it deteriorate back to current conditions by doing nothing more) would cost $U.S. 200 billion."

The reported "crisis proportions" were reached at a time when industrial and financial management were prospering both in theory and in application (Appendices 6–8). The differences between structural and operational management emanate to a great extent from the differences between industrial and infrastructure maintenance. The former has been extensively optimized; however, the solutions have limited application to the latter.

Industry must produce cost effectively. Mobley (1990, p. 1) estimated that the industrial sector of the United States was losing more than $60 billion each year "as a result of ineffective maintenance management." He recommended *predictive maintenance*. This is possible for mechanical and electronic components, because their remaining useful life is either directly measurable or analytically predictable.

Maintenance and inspections of bridges have been targeted for similar optimization. Infrastructure priorities are different, however. Whereas the timely replacement of parts is the only maintenance required of many operations, the maintenance of bridges must minimize replacements in order to serve the transportation function without interruption.

Neale (2001, p. 52) described the different priorities as follows:

A difference between mechanical engineering failures and civil and structural engineering failures is that the mechanical failures tend to arise from wear and fatigue, often after a long time in service. They therefore tend to be accepted as being to some extent inevitable, and part of a pattern of normal operation. A judgment is often made, actually or intuitively, on the basis of the total life cost of owning or operating a particular machine. . . . The ideal [failure] pattern is one where failure occurs not only after a long time, but where the failure time bandwidth is narrow, so that reliable performance can be guaranteed for a long time in operation.

A feature of projects in civil and structural engineering is that . . . the majority of the cost is incurred in the construction work and not so much in the subsequent operation of the facility. There is therefore a general tendency to . . . keep the initial costs down to a minimum."

Among "civil and structural engineering projects," bridges suffer the most from the reasoning exposed in the preceding paragraph, because they represent a transition between operating machines and static utilities.

NCHRP Synthesis 153 (1989, p. 31) identified the following two most common errors in the logic of maintenance cost analysis:

- Accepting labor and overhead costs as fixed because they are already budgeted and paid for and therefore concluding that only materials, equipment, and contract services will require additional cash expenditures out of budgeted dollars.
- Making judgments only on a first-cost basis, without considering the duration of the repair. This ignores annual costs, simple present worth, the time cost of money, and life costing.

Carroll et al. (in TRR 1877, 2004, pp. 10–16) identify "seven unique barriers to the development and implementation of preventive maintenance programs" for roadways, as follows:

- Lack of evidence of cost effectiveness of preventive maintenance (PM) treatments
- Alternatives to PM (such as deferred maintenance)
- Lack of top management commitment and experience
- User inconvenience and delays during work periods
- Intergovernmental politics
- Internal interest groups
- External interest groups

With examples from Montana, Michigan, Kansas, and Nebraska, the authors demonstrate the benefits of surmounting the powerful resistance against PM implementation. Bridge PM encounters similar difficulties. The management of maintenance (including inspection) suffers from the following main vulnerabilities:

- The scope of work is not as clearly defined as that of (re)construction projects. There is always a range of desirable, acceptable, and recommended levels, conveying the message that the subject is negotiable.

- The benefits are not readily quantifiable. The output is service rather than production. The objective is to minimize penalties to the users, rather than to maximize profits to the responsible owners.

- Infrastructure management is discontinuous. Future penalties and current benefits do not affect the same entities. In contrast, maintenance applies to the same facilities and requires continuity.

- Optimization algorithms must consider a "do nothing" alternative for maintenance. That option can be mistakenly interpreted to imply that nothing changes. In reality, the structure deteriorates and future needs accumulate to uncertain degree.

Because of their apparently low sensitivity to maintenance deferral (also discussed in Section 4.4), structures became a source of internal borrowing by budget managers for several decades. Examples 10 and 11 illustrate how, given appropriate discounting and other assumptions, deferred maintenance can be defended by seemingly rational arguments. Such reasoning neglects the following:

- "Maintenance-free" bridges do not exist. The recent interest in "zero-maintenance" bridges (FHWA, 2005b) can be constructive so long as it produces new structures with reduced maintenance demands. Traffic and the environment invariably require some form of maintenance. In the broadest sense maintenance includes inspections.

NCHRP Synthesis 327 (2004, p. 16) observed: "Effective bridge maintenance programs are those that are planned and systematic in their application." NCHRP Synthesis 153 (1989, p. 35) concluded: "Maintenance is forever." Thus a project-level implementation stressing discipline and accountability is at least as essential as an enlightened network-level optimization.

- The consequences of maintenance deferral are irreversible to both structure and organization. A valid analogy can be drawn between maintenance and medication, which, if discontinued, becomes detrimental to the patient. The savings from deferred maintenance are promptly absorbed elsewhere with no future benefits to the bridges. The organization in charge of maintenance grows defunct if it does not perform its function. Reorganizing is too costly and slow.

- Beyond a certain level of structural deterioration, maintenance is no longer effective and reconstruction is inevitable. Carroll et al. (in TRR 1877, 2004, p. 11) significantly note that (highway) maintenance is often effective only if applied consistently from the completion of construction.

- Hazard mitigation costs quickly surpass preventive maintenance savings. Conservative design may allow *preventive* maintenance to be deferred within limits but not *demand* maintenance. A deteriorating structural condition generates hazards to the public. Safety of traffic will have to be guaranteed, if necessary by a bridge closure.

- User costs incurred as a result of maintenance deferral accrue imperceptibly to the bridge owner (and to the users) until industry and commerce begin to stagnate or relocate.

The subjectivity of user cost estimates becomes a critical vulnerability if popular perception guides asset management policies. NCHRP Synthesis 330 (2004) concluded (p. 26): "Little evidence was found to confirm that road users and other stakeholders prefer or attribute greater value to efficient highway maintenance strategies, that is, to those with lower total lifecycle cost."

Preventive maintenance is often perceived as a luxury. It is usually funded by local taxes, whereas construction may be supported by federal and other sources. Not surprisingly, maintenance is the first target of austerity measures. Thus, while bridges are increasingly evaluated according to performance, maintenance may still have to be managed by rigid prescriptive indicators, at least until advanced monitoring techniques (Chapter 14) allow for more accurate performance evaluation.

Maintenance deferral is so persistent that measures to eliminate or reduce the need for it are constantly investigated. A recent example is the zero-maintenance policy (Section 11.4) under investigation in Switzerland and recommended for further research by FHWA (2005b). In contrast to deferred maintenance (Section 4.1.4), zero maintenance envisions the elimination of routine bridge maintenance tasks by design. The tasks ensuring traffic safety are not eliminated by this policy. The zero-maintenance policy can only apply to structures presumably designed and built for it.

4.1.6 External Causes and Sphere of Competence

Some of the critical vulnerabilities defining the outcome of bridge management efforts are beyond its competence. The limitations are both administrative and technical. The matrices of Tables 4.1, 4.4, and 4.5 are confined to bridge-related tasks and their products. For vulnerabilities external to its competence, bridge management must rely on expertise from other fields. Leading examples are the forecasts related to extreme events. Earthquakes, hurricanes, floods, collisions, and acts of aggression repeatedly demonstrate an inexhaustible potential of generating loads beyond the specified loading conditions. Experts in geotechnics, meteorology, law and traffic enforcement, and particularly statistical analysis provide estimates that may determine bridge management priorities.

The limitations of managerial competence do not appear in Fig. 1.33a. Fig. 1.33b however shows management to be multilayered and exceeding by far the domain of infrastructure operations, most significantly in the political, financial, and administrative spheres. Bridges cannot be managed independently of the larger infrastructure and overall economy. NRC (1995, p. 22) stated: "Many people assert that . . . government below the federal level is unable to deal effectively with issues of urban development and infrastructure. . . . Others argue that such problems are not new and in any case are the result of factors that extend well beyond the influence of infrastructure."

Banks (2002, Section 2.6) summarized the difficulties with transportation funding in the United States as follows: "Facility-oriented public agencies have rarely been able to find revenue sources adequate to meet the 'needs' for facilities they have identified. . . .

Although transportation-related expenditures are usually able to command public support, they rarely achieve much priority compared with other needs."

On the limitations of fuel taxes and tolls as funding sources, the author added:

> Financing of public transportation facilities has been complicated not only by the limitations of their funding sources, but also by the tendency of costs to increase more rapidly than revenues. . . . During a period of rapid inflation in the 1970s, unit construction costs, as measured by the FHWA Price Trends Composite Index, increased more rapidly than did prices in general. Also, transportation agencies have faced increased litigation costs as a result of changes in tort law.

Decentralized and privatized management of public transportation facilities are the options reconsidered after every perceptible economic or political change. Appendix 23 briefly describes attempts to model the linkages between the economy and the investments in the transportation infrastructure. Liability is discussed further in Section 13.2.

Most of the quoted analyses are conducted in purely economic terms, without attempting to quantify the consequences to the structures whose future is being decided. Budget constraints are not only among the most frequently adjusted variables in a BMS but occasionally have discontinued the very use of the BMS. Stridger (2004) listed the following typical management responses to imposed austerity:

- Better accountability
- Juggling funds
- Reducing staff
- Raising new funding
- Better planning
- Heads in the sand

All efforts to mitigate external effects have some limitations. Design and management can anticipate to a degree the vulnerabilities typical of the network, such as accidents, neglect, and extreme events. Ultimately, as stated in Section 2.4. service is a variable, safety is an imperative. (There is a parallel between this statement and Turing's aphorism, quoted in Section 1.8). The responsible owner must close any potentially unsafe bridge. The Williamsburg in New York City was temporarily closed in 1988 because of structural safety concerns. The East River bridges were temporarily closed after September 11, 2001, because the city had been attacked.

A bridge closure is a failure of one sort or another. It entails repercussions for the responsible owner and the users and is never taken lightly. Engineers are dedicated to maintaining their bridges in service as much as possible. Yet, maintaining an unsafe bridge open is a potentially fatal management failure. Many accidents could have been prevented or limited if engineers had enforced closure on safety grounds. Classical is the case of the Quebec Bridge, where the designer telegraphed his concern but construction continued until the collapse. The Tacoma bridge collapse would have been a much worse disaster if management had not undertaken to stop traffic during high winds.

Every stage and task of Table 4.5 is potentially vulnerable. If the means to identify and evaluate the vulnerabilities are lacking, they should be developed as part of the BMS system.

4.1.7 BMS/MIS

Any knowledge-based system serving as the principal source of knowledge is vulnerable to ignorance. Once the BMS begins to support critical decisions, it becomes the focal point of critical vulnerabilities.

BMSs are vulnerable internally and in their relationships with engineering, management, and general management information systems (MISs). Internally, data can be lacking (ignorance type A) or poorly correlated (ignorance type B). Databases are typically unable to correlate rates of condition deterioration with levels of maintenance (Section 4.4) and condition ratings with potential hazard reports (Section 10.3.3).

NCHRP Report 483 (2003, p. 24) cautions:

> A considerable amount of effort is required to use [BLCCA] properly and to collect project-specific data. This proficient use of software may require staff dedicated to conducting lifecycle costing as their primary job description. . . . Virtually all states consider the accuracy and availability of cost data and other information required to run the BLCCA to be inadequate. . . . Effective use of lifecycle costing requires that many state practices be modified to accommodate the concepts of whole life costing versus initial cost and to consider user and risk costs. There may be resistance to these types of changes unless the benefits are clearly identified and there is support from senior management.

Bridge management and management information systems are related in the processing of information as engineers and managers are in generating and implementing that information. Consequently many of the vulnerabilities observed in Section 4.1.1 pertain. Structurally and functionally (Appendix 18), information systems both reflect and influence the operations they are intended to support. To keep the flows of information and responsibility coherent, Mittra (1988, p. 8) recommended maintaining operation and database systems separate from the MIS. The author considered mixing the top-down and ground-up approaches during the analysis and design phases (i.e., *what* functions are provided vs. *how* the requirements are met") vulnerable to potential difficulties in backtracking. He favored the *structured* approach, i.e., designing the flow of information before any application can be undertaken. Engineering structures are similarly designed before and not concurrently with construction. The difference is that engineering structures physically impose their demands by failing whenever the latter are not satisfied. Partial failures and nonperformance are harder to spot in the MIS.

The following fundamental properties of the BMS must be designed and tested:

- Boundaries and interfacing with MIS
- Structure
- Function

Mittra (1988) identified five phases of MIS design along with their respective end products (Appendix 24).

Many BMSs are designed and implemented before a more general MIS can be considered. Under the pressure of technological innovations and economic demands, bridge managers are as likely to adapt (or "customize") a readily available BMS as they are to develop their own. The final choice is negotiated between the bridge managers, who use and maintain the BMS, the budget managers, who fund it under financial constraints, and the available expertise. Over a longer period of use the process and the support system inevitably blend; however, they should retain some independence.

BMSs/MISs are particularly sensitive to budget and personnel shortages, because they depend on the continuity and quality of data flow. Since databases must accommodate management methods and priorities evolving over time, they should be able to survive the downsizings, upgradings, priority shifts, reengineerings, and so on, typical of the political and economic environments. Public agencies should not have to retool their MIS after every election or retirement of key personnel.

On page A1 of *The New York Times* of January 14, 2005, E. Lichtblau reported that the U.S. Federal Bureau of Investigation is on the verge of scrapping a $170 million overhaul of its Virtual Case File (VCF) system, which has been "riddled with technical and planning problems." Cited are technical and financial missteps, rapid personnel turnover, employee resistance to new technology, and poor data accessibility and security. The process was compared to "changing the tires of an automobile driving at 70 mph."

Two million dollars were allocated to determine how much of the original effort could be salvaged. On March 14, 2006, *The New York Times,* (p. A24, column 1) quoted a Justice Department report: "The overhaul of the Federal Bureau of Investigation's antiquated computer system could cost another half-billion dollars to complete and runs the risk of continuing overruns. . . . The potential weakness in cost control [is] a project risk. . . . If information-sharing is not built into the system, a result may be costly and time-consuming modifications."

According to the U.S. inspector general, "the program management has been fragmented and ill equipped." Twenty-five of 76 staff positions remain vacant. "Simply put, the V.C.F. has been poorly managed."

4.2 ANALYSIS AND DESIGN

As engineering analysis matures, it either confirms or amends empirical assumptions. Boller (1885, p. 43) stated (italics are Boller's): "A general rule that will lead to satisfactory results is *to ignore any plan of bridge that can not be accurately analyzed as to the character and amount of strain occurring in all its parts*."

Contemporary analysis, data acquisition, and processing appear to offer unlimited modeling capabilities. As a result, conservatism is no longer tolerated. Modeling errors and misinterpretations therefore gain in significance.

Reference sets of minimum requirements for standard analysis, design, and construction procedures, anticipating common failures, are assembled in specifications. Specifications both define and follow established professional knowledge. Consequently they reflect its internal contradiction; they advance safe practice but cannot guarantee it. Specifications set lower (and occasionally upper) bounds of acceptable analysis complexity and design performance under prescribed demands.

4.2.1 Specifications

Scope

The guidelines for building design often amount to recommendations to be applied at the discretion of the owner. The state of Louisiana is considering the imposition of a prescriptive building code following hurricane Katrina (September 2005), but the measure is opposed, because it would increase private construction costs by an estimated 8%. As publicly owned and funded facilities, bridges must be designed and built in conformance with the governing design and construction specifications. The AASHTO bridge design and construction specifications have evolved in the United States from 1931 to their seventeenth and final edition (AASHTO, 2002). New LRFD specifications were introduced in 1994 and, by 2004 their third ed. was published. The guidelines, recommendations, and manuals published by the FHWA are voluminous. These documents have defined the professional practice and advanced its reliability and, in many cases, are legally binding.

Modern bridge design texts, such as Xanthakos (1994), Taly (1998), Barker and Puckett (1997, Section 3.2), and Chen and Duan (1999), review the governing design specifications. In contrast, that subject is at best peripheral to studies of bridge design masterpieces, such as Billington (1983) and Leonhardt (1990).

Indispensable as they are, specifications address a limited domain. The Mianus Bridge design was consistent with current specifications. After its pin and hanger failure in 1983, the use of that detail was discontinued. The Cypress Avenue Viaduct carrying Route I-880 in Oakland (Fig. 3.10) was designed in compliance with contemporaneous specifications. Updates reflect changes, precipitated by every new type of malfunction. Independently, publications, such as ASCE (1990) urge for professional quality transcending codes.

In the domain of mathematical logic, Godel's incompleteness theorem states that "any formal theory, adequate to embrace the theory of whole numbers must be incomplete" (Kline, 1980, p. 261). Similarly incomplete are any specifications codifying the practice of design and construction. As it is always possible to formulate an absurd statement in a formal theory, it is also possible to design an inadequate structure in compliance with governing specifications. AASHTO LRFD (1998) acknowledged that limitation as follows (p. 1-1): "These Specifications are not intended to supplant proper training or the exercise of judgment by the Designer, and state only the minimum requirements necessary to provide for public safety. The Owner or the Designer may require the sophistication of the design or the quality of materials and construction to be higher than the minimum requirements."

The science and art of engineering do not evolve from specifications. Rather, they inform them. The scope of specifications is limited in quantity as well as in quality. LRFD (1998, p. 3-7) discusses load calibration valid for span length up to 200 ft (60 m). Spans longer than 350 ft (100 m) have more unique than standard features. Spans exceeding 500 ft (180 m) are entirely beyond the scope of the specifications.

In the interest of safety and standardization, specifications, such as AASHTO, have simplified the modeling of structures and design loads. Structures have been reduced to equivalent strips or linear elements subjected to bending, shear, and axial loads. Loads have been represented by amplified (concentrated, uniformly or linearly distributed) static forces. Thus torsion and dynamics, although potentially critical for bridge service conditions, have been treated indirectly and prescriptively. Safety has been mostly achieved by deliberate conservatism. The same conservatism has been criticized as a weakness of the modeling, no longer justified in the presence of analytic solutions and computational power. Specifications are consequently evolving in response to several changing conditions, including the following:

- Demand: different vehicular modes and traffic conditions
- Supply:
  New materials
  Analytic advances
  Computational capabilities
  Database of structural behavior
  Cost–benefit reconsiderations

Allowable versus Ultimate and Prescriptive versus Performance Based

During the latter part of the twentieth century, design specifications evolved from the *allowable stress* to the *ultimate strength* method (see following section). The *safety factors* of the earlier allowable stress design specifications are occasionally referred to as *ignorance factors*. *Ignorance* however is an ever-present form of uncertainty that does not necessarily constitute *incompetence*. Recent specifications recognize its presence by the *load and resistance factors* based on the probabilistic considerations outlined in Appendix 7. The following shift in terminology reflects the key innovations in contemporary design specifications:

AASHTO (2002)	**LRFD**
Prescriptive	Performance based
Safety	Reliability
Loading combinations	Limit states

The limit states, consistent with the performance-based aspect of design criteria, are compared to the prescribed loading combinations in Appendix 25. Performance based is a vague term, charging design with the task to obtain desired modes of performance *and* failure. The innovations of the new specifications include the following:

Transition from allowable stress to ultimate load criteria

Reliability index, based on probability distributions, rather than safety factor, assuming Fixed margins of safety

Limit states, loading combinations, and loads (AASHTO 1998a; LRFD, Chapter 3)

Load and resistance factors (Chapter 3)

Impact factors (Chapter 3)

Live-load distribution factors (Chapter 4; also a recommendation in AASHTO 17th ed.)

No superstructure deflection requirements

"Strut and tie" method of concrete design (Chapter 5)

Prestressed concrete design (Chapter 5)

Decks and deck systems (Chapter 9)

Parallel commentary

Checklists (Chapters 5 and 6) for concrete and steel design satisfying strength requirements

Whereas the transition from allowable stress to ultimate strength design is unambiguous, the one from *prescriptive* to *performance-based* requirements is less so. The deterministic stipulations of the allowable stress method are prescriptive and relatively easy to implement but fail to recognize the complexity of actual load–structure interaction. The ultimate strength method has also been applied in a prescriptive manner. It is referred to as performance based after uncertainties are introduced into the model; however, they too can be prescribed. Descriptive might have been simpler; however, it would have implied an a posteriori position, incompatible with the purpose of specifications. AASHTO 1998 (LRFD, 1998, p. 3-1) stipulates: "Where multiple performance levels are provided, the selection of the design performance level is the responsibility of the Owner."

The quality of the product depends on the engineer of record. The owner enjoys much greater latitude of options and consequently carries a heavier responsibility. Burke (2000) cautioned: "LRFD may be hazardous to your bridge's health."

Intended to eliminate unsafe practice, codes are safe for use only by those who are familiar with the theoretical and practical foundations. Post (2005) wrote of "code anxiety." The author observed that "complexity and confusion stress structural engineers" and concluded that "designers fear future collapse from misinterpretation or miscalculation." Perhaps with reason.

4.2.2 Model

Structural Scheme

Bridge structures are analyzed and designed as discrete assemblages. The terms *member, element,* and *component* usually connote the relative size and importance of the constituents. Typical members are girder, stringer, diaphragm, brace, slab, column, bearing, pedestal, cap beam, stem, footing and pile. Elements are systems of members serving the same function, such as deck, primary member, secondary member, pier, and abutment. Components are larger subsystems comprising elements engaged in similar functions, such as superstructure, substructure, approach, foundation, utilities, and protection system. Figure 4.12 illustrates a typical vehicular bridge superstructure.

Strength

Bridge designers of the last two centuries keenly appreciated the distinctions between ultimate and effective strength, elastic limits and the respective factors of safety. Boller (1885, p. 11) argued that a factor of safety obtained as a ratio of the service and the

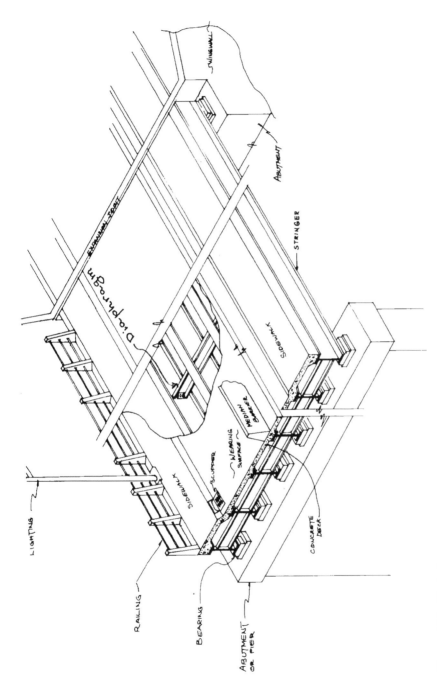

Figure 4.12 Typical span superstructure.

ultimate loads (e.g., causing a breakdown) is unreliable. The author discussed the elastic limits of the members and their connections as more indicative of actual strength. Anticipating the effects of fatigue, he recommended different safety factors for repeated loadings.

During the twentieth century design strength philosophy has evolved from allowable stress (ASD), or working stress, to ultimate strength (USD), or load factor (LFD), and, finally, to load and resistance factor (LRFD) methods. AASHTO 19998a (LRFD, Eq. 1.3.2.1-1) defines structural "strength" in supply–demand terms as

$$\sum \eta_i \gamma_i Q_i \le \varphi R_n \tag{4.1}$$

where γ_i = statistically based load factor
φ = statistically based resistance factor
η_i = load modifier
Q_i = force effect
R_n = nominal resistance

The demand of all loads (ΣQ_i) must be met by structural resistance (R_n) with acceptable reserve, ensured by the η_i, γ_i, and ϕ factors. A reversal of the inequality of Eq. 4.1 signifies failure. That can be due to an overestimate of the resistance or underestimate of the load, more likely to occur in new structural types, possibly not envisioned by design specifications.

LRFD, USD, and ASD differ primarily in the way a reliable and acceptable superiority of R_n over ΣQ_i is prescribed. According to ASD, materials are linearly elastic and isotropic. Plain sections remain plain in bending. Stresses caused by all design loads must remain within a safe margin of the elastic limit of materials (say 50%), usually obtained in uniaxial tension. Thus internal and global load redistribution and the variability of structural loads are not adequately modeled.

ASD has been criticized for its subjective conservatism. It did, however, successfully standardize highly diverse design and construction practices and foster durable structures throughout much of the twentieth century. Inevitably, the linear elastic stress–strain relationship and the uniform safety factor became outdated.

AASHTO (1998a) introduces load *and* resistance factors based on the broadest survey of empirical data, stochastic analysis, *and* expert judgment. Ang and Tang (1975) are quoted on the use of probabilistic analysis for engineering problems. Thoft-Christensen and Baker (1982) are referenced on the application of reliability theory to structures.

Equation 4.1 quantifies the supply of and demand for strength in terms of forces that can be obtained from Hooke's and Newton's laws. It has been modified to reflect the nonlinearities and uncertainties of material behavior as well as the randomness of phenomena. It remains, however, a quantitative equilibrium of supply and demand and, as such, overlooks qualitative changes representing instability.

Stability

Bažant and Cedolin (1991, p. xxi) point out that instability is a design pitfall, because, to be detected, it must be anticipated. Instability implies losing not the design strength but the form. Not only mechanical but all formalized systems can fail in that mode. Schopenhauer (1942, p. II-104) wrote: "He is a prudent man who is not only undeceived by apparent stability, but is able to forecast the lines upon which movement will take place."

The movement of an unstable system is indeed theoretically unpredictable. It is not coincidental that after a management failure operations are thoroughly reorganized, just as after a buckling failure redesign avoids the original structural scheme. The potential loss of stability is overlooked when design is limited to structural strength or when management insists on improving performance under an outdated organizational structure.

The supply of strength can be perceived as a quantity, whereas form is a quality. Strength computations are therefore quantitative, whereas the analysis of stability is qualitative. Compounding quantitative (strength) and qualitative (stability) effects is analytically elaborate. The practical manifestations of instability are also elusive. Buckling failures can be instantaneous; however, unstable structures (and organizations) can function over unpredictably long periods. Although mechanical stability has been thoroughly explored, it is not sufficiently taught. An engineering degree can be obtained without completing a course on the subject, thus entering the market with significantly vulnerable qualifications. Appendix 26 cannot amend that deficiency but highlights the subject in order to emphasize its critical importance.

Stress–Strain

If perceived as three-dimensional continua, materials are internally redundant. Engineering, however, reduces all models to the minimum significant variables. Such are the linear structural models. Structural elements are analyzed as linear or at most two dimensional. Structural materials are modeled as linearly elastic–perfectly plastic. Besides being generally rational, the assumed geometric and material linearities are considered conservative. The opposite is also possible however. Fully elastic structures absorb more energy during earthquakes than yielding ones. Three-dimensional stress can cause unexpected failures, as at the Hoan bridge (NCHRP Synthesis 354, 2005). Restrained warping can result in buckling.

On the linear assumptions paramount throughout the engineering practice AASHTO 1998a, LRFD (1988, C1.3.1) cautions: "The resistance of components and connections is determined in many cases on the basis of inelastic behavior although the force effects are determined by using elastic analysis. This inconsistency is common to most current bridge specifications as a result of incomplete knowledge of inelastic structural action."

Bending/Shear

Even when linear modeling of material behavior is justifiable, there is still a risk of ignoring some of the resulting stresses. In a discussion of compound stress Den Hartog (1949) pointed out that maximum shear is the most likely to cause material failure.

Shear V is the first derivative of the bending moment M of Eq. A26.3, as follows:

$$V = -EI \frac{d^3 y}{dx^3} \qquad (4.2)$$

Despite the direct relationship, design considers bending and shear independently. This is so because structural members are designed as a sequence of sections, normal to their centerlines and subjected to stresses along orthogonal axes. Since $V = 0$ where $|M|$ = max and vice versa, certain sections, for instance over simple supports, are designed primarily for shear, others for flexure (as at midspan). The alternative of addressing principal (normal) stresses of varying orientation is considered in the strut and tie method, recommended by AASHTO 1998a for reinforced-concrete design.

The analogy of material failure can be extended within limits to the bridge structure and to decks in particular. Decks fail primarily as "punch-outs" (Figs. 4.13 and 4.14), preceded by gridlike patterns of cracks (Fig. 4.15). NCHRP Report 495, 2003) refers to

Figure 4.13   Punch-through deck failure.

Figure 4.14 Detail of Fig. 4.13.

failures in "shear fatigue." However, longitudinal deck cracks above primary members have also been noted. They may be caused by tension as well as by shrinkage or even corrosion of the rebars. The result is usually not a catastrophic failure; however, it creates a safety hazard and precipitates early rehabilitation.

Under the combined effect of axial load, bending, and shear, reinforced-concrete column cores develop plastic hinges. The cracked section can fail gradually in a ductile mode if it is adequately confined. If the ties fail, the reinforcing bars buckle and the concrete is crushed (Fig. 4.16).

Statics and Dynamics

Civil engineering structures are vulnerable to dynamic loads because, in contrast to aeronautic, mechanical, and naval structures, they are analyzed predominantly for static loads. Dynamic effects are represented in most cases by load amplification factors and their effects are analyzed quasi-statically. Combining the effects of eccentricities and impact is up to the user. The complexity of the analysis depends on the importance of the structure and the level of the credible hazard; however, these parameters are defined by the responsible owner. In order to select the appropriate level of analysis, the owner must have an understanding of the available options, i.e., the assumptions underlying the adopted models. A brief summary is presented in Appendix 26.

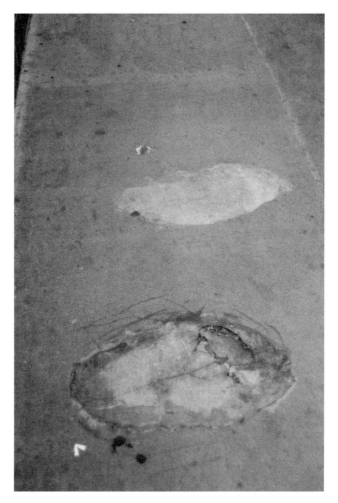

Figure 4.15   Typical bridge deck surface failure.

Primary and Secondary Effects

Effects that do not contribute to the principal load-carrying structural functions can be considered as secondary. Design specification provide for the effects of temperature, shrinkage, and support settlements. Typically secondary effects may be overlooked for the following reasons:

> The analyzed structural scheme is two dimensional, whereas the actual structure is three dimensional. Example 13 illustrates the possible consequences of analyzing separately a truss in its vertical plane and the roadway system it supports in the horizontal plane.

> Both instantaneous and long-time structural responses produce stress distributions diverging from the linear elastic models assumed by analysis. Short-time response

(a)   (b)

Figure 4.16 Poor confinement of column reinforcement: (a) Cypress Avenue Viaduct, Oakland, October 29, 1989; (b) Hanshin Expressway, Kobe, January 17, 1995.

is often rigid. Under slow-acting phenomena, such as creep and shrinkage, materials behave inelastically and nonlinearly.

The performance of "load release" devices (e.g., articulations), such as expansion joints and bearings, varies with the rate of the applied load and the amplitude of the induced displacements.

Temperature and shrinkage deformations are never monitored. Support displacements draw concern when they become manifest.

The likelihood of secondary stresses increases with the degree of structural redundancy (Section 5.2). Added articulations have been used to avoid the problem; however, the associated loss of continuity and the potential for malfunction of the details contribute more critical vulnerabilities (Example 13).

Design provides for temperature and shrinkage stresses and can similarly take into account other statically indeterminate stresses. In contrast, it is assumed that residual stresses have been minimized during fabrication (Section 4.3.1). Residual stresses typically arise because of different thermal conditions of the various parts of cross sections, particularly within metal sections rolled, galvanized, or welded under high temperatures. Their presence can be detected by X-ray diffraction (Chapter 15).

Professor A. Gjelsvik of Columbia University has pointed out that the stresses induced by daily cycles of uneven exposure to heat are taken into account for ships, but typically ignored in large steel bridges, where they can be equally significant.

Parallel wires in suspension cables are assumed to be subjected to uniaxial tension, however they carry high bending stresses as shown in Example 13, because their natural curvature is eliminated during construction.

Example 13. Primary and Secondary Stresses

A. This example was suggested by M. Bieniek, Columbia University. An idealized simply supported steel truss is considered as shown in Fig. E13.1. The real structure shown in Fig. E13.2 is similar.

The following dimensions are assumed:

$$L = 110 \text{ m } (360 \text{ ft})$$

$$E = 21,000 \text{ kg/mm}^2 \text{ (29,000 kdi)}$$

$$\sigma^{LL}_{\text{Bot. ch.}} = 7.24 \text{ kg/mm}^2 \text{ (10 ksi)}$$

(stress in bottom chord due to live load)

$$\Delta^{LL}_{\text{Bot. ch}} = \frac{\sigma^{LL}_{\text{Bot. ch}} L}{E} = 37.5 \text{ mm } (1.5 \text{ in.})$$

(extension of bottom chord due to live load)

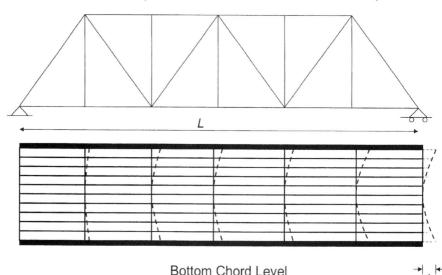

Bottom Chord Level

Figure E13.1 Simply supported truss with stringers and floor beams.

The expansion bearings and joint at one end of the truss are likely to be designed for much greater movement. If, however, the stringers and floor beams are fixed at all intermediate panel points, the floor beams would experience a progressively increasing deflection with respect to their weak axis, as shown in Fig. E13.1. The likely consequences are cracks similar to those shown in Fig. 4.22.

Figure E13.2   Simply supported truss with stringers and floor beams.

The out-of-plane deflection caused by extension of the truss is a secondary effect, unrelated to the primary function of the floor beams. They are fracture-critical elements, supporting the stringers in a nonredundant manner (e.g., without an alternate load path). The problem appears to be resolved by adding expansion joints and bearings at each panel point. However, deterioration of the stringers at the expansion bearings is accelerated, as shown in Figs. 4.54, 4.62, and 14.12.

B. The high-strength wires in suspension bridge cables typically have 5 mm (0.198 in) diameter. They are manufactured by extrusion and during cooling assume a curvature with a radius much smaller than the one of the cables. Thus they are straightened out during construction by the air-spinning method. Eliminating a curvature with radius R induces bending moment M as follows:

$$M = EI/R \tag{E13.1}$$

where E = modulus of elasticity
   $I = \pi R^4/4$ is the section moment of inertia

The corresponding maximum bending stress σ is:

Figure E13.3 Anchorage eye-bars with strands of cable wires built by the air-spinning method.

$$\sigma = M/S = E \qquad \text{(E13.2)}$$

where $S = I/R$ is the section modulus

Suspension bridge wires tend to regain some of their original curvature if they are extracted from a cable, even after many years of service. During laboratory tests, cracks invariably develop on the concave side of the wire, e.g., where the straightening would cause tension. A complex bending stress develops in the cable wires also in the anchorages where they must turn around sheaves of relatively small diameters, as in Fig. E13.3. That behavior is among the reasons leading the designers

Figure E13.4 Anchorage with prefabricated straight wire strands.

of the most recent suspension bridges, for instance in Japan, to use the prefabricated straight wire strands, originally developed under a U.S. patent (Fig. E13.4).

Bridge Superstructure

The useful life of bridges is determined mostly by the condition of the superstructure. That in turn is governed by the performance of decks and primary members. Records suggest that during the second half of the twentieth century the average useful life of superstructures in the United States has declined from 50 to 30 years. The declining trend in structural life expectancy cannot be reversed without appropriate analysis and design provisions for new construction.

Deck and Primary Members

Typical superstructures consist of decks and primary (girders) and secondary (bracing) members, as in Fig. 4.12. Modeling the response of superstructures to live loads is central to all bridge design specifications. The early objective of the AASHTO was to combine empirical evidence and analysis into a relatively simple and adequately conservative procedure. The *equivalent-strip* and *equivalent-beam* methods serve that purpose. The method discretizes the continuous superstructure into strips of repetitive stiffness, carrying fractions of the design live load in a mostly flexural mode. Two factors are essential:

- *Effective width* of the deck contributing to the strength of the equivalent beam supporting the load (zero if there is no composite action)
- *Fraction of the live load* supported by the primary member

Whereas the former parameter depends on the shear transfer in the plane of the deck, the latter depends on the action of the deck in bending and shear under the normally applied live loads.

The equivalent beam could be the primary member, such as a girder, acting alone or compositely with the deck over a specified contributing width. The contributing or effective width of a composite member depends on the capability of the concrete deck to participate in the deformation of the top fibers of the primary members under live loads. Beyond a certain width, the shear stress associated with this participation lags and the deck is not engaged significantly in the flexural deformation of the girder. Appendix 27 describes the AASHTO recommendations for the selection of appropriate effective width of bridge decks in composite superstructures.

Live loads applied between the primary members (composite or otherwise) are transmitted by the deck, spanning between them and acting in bending and shear. The *distribution factors* recommended by the AASHTO (Appendix 28 herein) determine the fraction of the truck wheel loads to be applied to the designed member, depending on the type and spacing of the members, and the deck thickness. Three- and two-dimensional systems are designed as one-dimensional elements. Factors are provided for longitudinal and transverse members (such as floor beams), for exterior and interior girders, and for two-way thick plate action of decks and for cantilevers. The formulas are calibrated for the AASHTO design trucks, briefly discussed in Section 4.2.4 herein. Tonias (1995) supplied helpful illustrations and tables explaining the AASHTO live-load distribution factors.

Decks are typically 7 in. (15.25 cm) to 9 in. (23 cm) thick. The truck wheel lines are 6 ft (180 cm) apart. The AASHTO distribution factor is designed to range roughly from 1.09 to 1.87, so that one equivalent beam would support more than one line of wheel loads. That assumption has been considered conservative. Zokaie et al. (1991) found the method conservative only for larger spacings of primary members. The alternative view holds that the superstructure redistributes live loads much more efficiently in the transverse direction. Limited destructive tests show that the equivalent strip method has produced superstructures of considerable strength reserve and load redistribution capability. Concrete decks have been observed to function safely despite suffering from extreme deterioration, as illustrated in Fig. 4.14.

The equivalent strip acting in bending does not attempt to model modern deck systems realistically. The torsional stiffness of the superstructure is ignored. Fatigue life estimates and dynamic analysis cannot be reliably performed. Even though many serviceable superstructures have been designed as equivalent strips, more realistic models are developed. AASHTO 1998a and LRFD (2004) provided more detailed calibration based on empirical data and finite-element models (Zokaie et al., 1991).

Chen et al. (NCHRP Report 543, 2005) developed a new method for computation of the effective slab width of composite steel bridge members (Appendix 27). Report 543 (2005) concluded that the current AASHTO LRFD Specifications (1998a) use conservative effective slab width compared to results from finite element three-dimensional models.

Barker and Puckett (1997, Chapter 6) defined 1.5- and 2.5-dimensional analysis for two- and three-dimensional superstructure systems. An example is the "beam-line method," which obtains distribution factors by two-dimensional analysis and applies them to equivalent one-dimensional girders as follows:

Distribution factor = critical structural response $^{2\text{-D}}$/equivalent beam response

The distribution factors in LRFD (2004) (Appendix 28) are based on Zokaie et al. (1991) and the above definition. Applicability is limited to superstructures of constant section, parallel and similar members, without curvature and overhangs.

The following definitions apply (AASHTO 1998a, ed., p. 4-3):

Equivalent beam—A single straight or curved beam resisting both flexural and torsional effects.

Equivalent strip—An artificial linear element, isolated from a deck for the purpose of analysis, in which extreme force effects calculated for a line of wheel loads, transverse or longitudinal, will approximate those actually taking place in the deck.

Finite strip method—A method of analysis in which the structure is discretized into parallel strips. The shape of the strip displacement field is assumed and partial compatibility is maintained among the element interfaces. Model displacement parameters are determined by using energy variational principles or equilibrium methods.

Deck as Primary Member

The deck is an integral part of the primary member in box girders (Fig. 4.17). Earlier versions of the design specifications envisioned open sections with a relatively small Saint-Venant torsional constant *J*, as in Fig. 4.12. Distribution factors for box sections include the torsional constant *J* explicitly, allowing for more realistic representation of superior torsional stiffness.

Schlaich and Sheef (1982, p. 3) consider the Risorgimento Bridge by Hennebique (1911) as the first box-girder three-hinged arch. In later years, box girders with higher resistance to torsion and bending in alternating directions have become "the most widespread bridge type" (Schlaich and Sheef, 1982, p. 2). Orthotropic decks (Fig. 4.18) are replacing reinforced-concrete (Fig. 4.19) and concrete-filled steel gratings (Fig. 4.20) in conventional structures requiring stiffer, lighter decks. The open gratings used on many movable bridges, because of their minimal weight (Fig. 4.21), are prone to corrosion and

Figure 4.17   Posttensioned box girders.

Figure 4.18   Orthotropic deck.

Figure 4.19 Reinforced-concrete deck.

fatigue and propagate the same defects throughout the superstructure. Their replacement with lightweight decks providing continuous surface is highly desirable.

Grillage or two-dimensional finite-element analysis can be used to model plane components, such as decks. Folded-plate or three-dimensional finite-element analysis is used for box girders. Calgaro and Virlogeux (1988), for instance, begin their text on superstructure analysis directly with torsion. Simple torsion, warping, folded-plate, and finite-element modeling results are explained and compared. Complex bridge deck behavior is also the subject of Hambly (1976). Certain authors define "folded plates" as assemblies of plates with hinged boundaries, i.e., a first approximation to a complete solution, whereas others, such as Schlaich and Sheef (1982), include the restraining moments for a consistent load–displacement solution.

New superstructure types and improved modeling of existing ones add significantly to the options bridge managers must consider and hence test their competence. Superstructure schemes must be selected taking into account the ability to construct, maintain, inspect, and replace. In a relatively rare address to the bridge *manager,* rather than the *designer,* Barker and Puckett (1987, p. 275) emphasize the long-term consequences of design choices to the bridge service: "Long-term material properties and deformations are difficult to estimate and can cause the calculated values to vary widely."

NCHRP Report 495 (2003) concluded that deck behavior is critical to serviceability and harder to evaluate than design specifications might suggest. As shown in Figs. 4.13 and 4.14, decks fail under repeated impact in a combination of flexure and shear exacerbated by fatigue (Section 4.3.1).

(*a*)

Figure 4.20   (*a*) Concrete-filled steel grating, no overfill.

NCHRP Report 495 (2003) offers estimates of the fatigue life of bridge decks, depending on the type of truck loads they carry. For a useful life estimate, that relationship must be correlated with significant deck properties, such as span, thickness, wearing surface, waterproofing, and anti-icing. NCHRP Synthesis 333 (2004) synthesizes concrete deck performance data nationwide (United States). Section 9 of LRFD (2nd ed.) is dedicated to the design and construction of bridge decks.

Superstructure Deflections

The treatment of deflections in AASHTO (2002) and AASHTO (1998a) is briefly compared in Appendix 29. While "serviceability" is a distinct limit state, represented in LRFD by three design loading combinations and one for fatigue (Appendix 25 herein),

(b)

Figure 4.20 (b) New steel grating, designed with overfill.

past deflection limits and thickness-to-depth ratios have been relaxed to *recommendations*. This is motivated by the intent of LRFD to avoid prescriptions. Rather than disregard deflections altogether, designers and managers are cautioned, as in Barker and Puckett (1987, p. 275) and in the AASHTO (1998a) commentary, to set their own limitations. Beyond the psychological effects of deflections on bridge users, acknowledged in the code, consideration must be given to the fatigue design of sign and light structures on excessively flexible bridges. Bridge owners would benefit if, limitations notwithstanding, deflection calculations were a mandatory item in design documents. The improved capabilities to monitor structural performance allow the veracity of such documents to be checked during field inspections.

Design specifications cannot address every serviceability and extreme-event requirement. Bridge owners are urged to reanalyze their structures under site-specific limit states. Figure 4.22a shows a steel floor beam cracked under global structural torsion. The sandwiching of the crack with steel plates, shown in Fig. 4.22b, is a temporary solution. Since opening to train and vehicular traffic in 1908, the Manhattan Bridge has cracked systematically (see Fig. 3.7). Figure 4.23 shows a global stiffening of the superstructure consisting of trusses, floor-beams, and stringers against torsion.

Computer Software

Computer software packages performing structural analysis are available for every set of contemporary design specifications. The risk of using a package without fully understand-

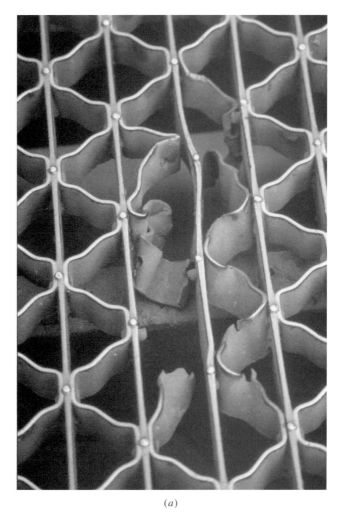

(a)

Figure 4.21 Typical breaks in steel gratings: (a) Open.

ing what it provides is always present. NCHRP Report 485 (2003) developed software validation guidelines and examples.

4.2.3 Connections

Most bridge failures either occur directly at connections or involve them. Boller (1885) wrote: "Joints [are] the most vital parts of the whole structure, the work must be judged entirely by them. [p. 47] . . . Great stress is laid upon 'the strength of joints,' since the essence of good bridge-building lies in their proper design. A joint must be as strong as the parts it serves to connect; as in a chain, wherein a defective link determines *its* strength, so in a bridge the absence of a necessary rivet would determine *its* strength [p. 8]."

(*b*)

Figure 4.21 Typical breaks in steel gratings: (*b*) With fill.

(*a*)

Figure 4.22 (*a*) Cracked floor beam web.

(*b*)

Figure 4.22 (*b*) Sandwich web plate.

Figure 4.23 Stiffening added to reduce global structural torsion.

Despite the greatly improved modeling of energy dissipation, partial releases, and geometric and material nonlinearities, the points of force transfers between discrete members remain the most sensitive both analytically and in practice.

The performance of nonarticulated connections (e.g., bolted, riveted, and welded) largely depends on design and construction. Inspections rate them as integral parts of the components they connect. Figures 4.24 and 4.25 show, however, that such connections are more susceptible to corrosion than the connected members.

Articulated connections include pins, eye-bars, hangers, linkages, clevises, and bearings, many of which are designated as fracture-critical details (Section 14.2). Expansion joints are deck articulations. Inspection records demonstrate that joints are the first to malfunction and the rest of the structure follows (Section 14.2). Figure 4.26 exemplifies the common leakage from failed expansion "compression" joints.

The supports of long-span bridges undergo frequent displacements on the order of 15 in. (375 mm). Debris accumulates freely under the traditional finger joints (Fig. 4.27) to the detriment of the highly sensitive bearings. Modular joints are a modern alternative. They must be accurately aligned, installed, and maintained in order to avoid fatigue failures of the type shown in Figs. 4.28 and 4.29.

Expansion bearings restraining the displacements of long spans invariably fail (Figs. 4.30–4.34, Example 7).

The collapse of the West Side Highway in New York City in 1973 originated at a floor beam–girder connection. The San Francisco–Oakland Bay Bridge failed during the Loma Prieta earthquake in 1989 because the two fixed bearings (Fig. 4.32) at one end

Figure 4.24 Corroded rivet heads, alkali–silica reaction in deck, failing rebars.

Figure 4.25   Failed anchor bolt.

of the collapsed span and the expansion bearings (Fig. 4.33) at the opposite end malfunctioned. One-foot (305-mm) displacement capability at the expansion bearings might have prevented the collapse. The discontinuity between the first and second levels of the "double-deck" frame at the Cypress Avenue Viaduct (Fig. 3.10) proved critical during the same earthquake (Housner, 1990), whereas the continuous two-tier roadway in San Francisco survived during the same event (Fig. 3.9).

The failure of the suspended walkways of the Hyatt Regency in Kansas City on July 17, 1981 (briefly discussed in Section 4.3, also occurred at points of load transfer.

After the Hyogo-Ken Nanbu earthquake in Kobe on January 17, 1995, an area near the failed section of the Hanshin Expressway was designated as a "bearing and restrainer cemetery" (Fig. 4.34). Postearthquake assessments (Shinozuka, 1995) found that bearings, linkages, and particularly restrainers were among the primary casualties during the event. Restrainers are mostly provisional retrofits of connections intended to absorb energy during extreme events. They have evolved from rigid or tight (Fig. 4.35) links to loose ones (Fig. 4.36), to be used with bearings accommodating larger displacements.

Connections are the most vulnerable link of the structure, which is the managed product. Discontinuity is a primary vulnerability in the management process (Section 4.1). Hence, the two share many characteristic features, including the following:

- Causes contributing to connection and management failures can be found in analysis, design, construction, maintenance, and service, i.e., throughout all stages of the process and all areas of the product.

Figure 4.26 Leakage from failed compression joint.

The eye-bar chain at the Silver Bridge and the pin and hanger assembly at I-95 over the Mianus River are examples. Both were designated as fracture-critical details by Harland et al. (in FHWA, 1986) and are no longer used by design.

- At connections, as in management, failures are often the only incontrovertible evidence of poor performance.

Up to the present, it has been difficult to monitor the transfer of forces at connections. The malfunctions described in Example 7 were noticed only because of gross geometric distortions. New data acquisition techniques are broadening the range of performance parameters that can be monitored (whereas decision support systems are enhancing BMSs). In management, similarly, there is a constant search for the truly significant performance indicators.

Figure 4.27 Finger joint at suspension bridge.

- Redundancy and overdesign can mitigate connection flaws only if adequately applied and combined. Funding can eliminate certain managerial vulnerabilities but not indiscriminate overspending.

The connection failures of Example 7 did not escalate into global failures because of successful load redistribution.

- Once identified, the failures of connections and management are eliminated by revisions of the decision-making process.

Early demolitions would have precluded the catastrophic failures of the Silver Bridge and Cypress Viaduct; however, they would have certainly caused extreme hardships to the users. No bridge manager could have taken such essential structures out of service

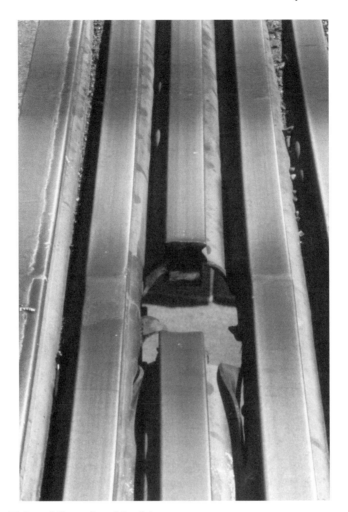

Figure 4.28   Fatigue failure of modular joint.

without precedent or clear evidence of distress. The perception of potential hazard had to be reinforced by an act of Congress after the Silver Bridge collapse and by a presidential executive order after the Loma Prieta earthquake.

The appropriate choice of connections depends (as in management) on local conditions, ranging from the economy and construction practices to traffic and the environment. After the Mianus Bridge failure, pin and hanger assemblies were eliminated from new designs. Such existing details were retrofitted in certain areas or intensively monitored in others, depending on the magnitude of the task.

For future bridges, these vulnerabilities are effectively addressed at the design phase. State and local engineering instructions disallow seismically nonperforming bearings. New bearings have evolved to meet the demand for seismic loadings, as reported in Kelly (1993), Mander et al. (1996), and Imbsen et al. (1997). Elastomeric, pendulum, and pot

(a)   (b)

Figure 4.29   Modular joint bolt fatigue failure.

bearings are replacing the rocker and steel sliding bearings of the past, as illustrated in Figs. 4.37 and 4.38. Seismic safety provisions are described in Appendix 33.

- Expertise is presented in continually updated "state-of-the art" reviews, recommendations, and amendments to current specifications. Literature is abundant on the subjects of both connections and management.

- Continuity is desirable but never entirely possible. Despite the advantages of continuous structures (physical and organizational), discontinuities and articulations are sometimes inevitable.

- Connections and management are perpetually redesigned.

In the case of connections, this can occur during "value engineering," peer review, construction, rehabilitation, and retrofitting (which happen to be key managerial challenges as well). When possible, structural discontinuities are eliminated by retrofitting measures, as in the case of expansion joints (Section 4.3.3). Seismic retrofitting is not limited to but predominantly addresses connections. Along with the possibilities to improve the structural performance come the managerial risks of failing to supervise, review, and inventory all changes in the original design.

Figure 4.30 The 4–6-in. (100–150-mm) displacement at sliding bearing of suspension bridge.

Figure 4.31 Exhausted capacity of suspension bridge sliding bearing.

(a)   *(b)*

Figure 4.32   (*a*) Shifted fixed bearing. (*b*) Failed sliding bearing, San Francisco–Oakland East Bay Bridge, October 29, 1989.

4.2.4 Loads

"Overload" and the corresponding "underdesign" are quantitative vulnerabilities. They are corrected by adjusting the design estimates and by regulating the service. Misjudging the load qualitatively is of greater interest, because it is caused by systemic oversights. Analysis divides loads qualitatively into *static* and *dynamic* as well as *primary* and *secondary*. Specific loading conditions are identified by their sources, that is, dead, live (vehicular, pedestrian, etc.), wind, earthquake, and so on (see Appendix 25). AASHTO 1998a; LRFD (2nd ed.) separates loads into *permanent* and *transient*. FHWA (2002a, p. P.1.7) provided a list of secondary loads, including those induced by material behavior (shrinkage), use (impact, breaking, and centrifugal forces), environment (ice, water, wind, earth, earthquake, temperature), and so on.

Dead Load

Absent a dynamic excitation, such as an earthquake or a transient load, the weight of the structure and the soil acts statically. Specifications define them as *dead* or *gravity loads* (DL in AASHTO, 2000b; DC in AASHTO 1998a).

Figure 4.33   Failed 6-in. (150-mm) expansion bearing, San Francisco–Oakland East Bay Bridge, October 29, 1989.

Gravity loads are most likely to cause failure as they increase during construction, in other words, while they are not "dead." Consequently, construction sequences must be explicitly designed. LRFD (1998, Section 3.4.3.1) prescribes a 1.3 (minimum) factor for jacking forces.

Equation 4.1 takes the anticipated range of dead or gravity load uncertainties into account by the load factor γ:

AASHTO, 2000b, p. 31 $\gamma_{DL} \times \beta_{D} = 1.3 \times 1.0$, except

$\gamma_{DL} \times \beta_{D} = 1.3 \times 0.75$ for minimal axial load and maximum moment or eccentricity in column design (see Eq. 3.4).

Figure 4.34 Failed bearings and bearing plates, Kobe, January 1995.

AASHTO 1998a, p. 3-11 $\gamma_P = 0.90$–1.25, except

$$\gamma_{DC} = 1.5 \text{ for load combination strength IV}$$

On long-span vehicular bridges dead loads can amount to 90% of the total loads (80% for trains). The ratio drops significantly for short spans and does not strictly apply to superstructures. The weight of various structural components is not equally predictable over the structural life cycle. Barker and Puckett (1997, p. 141) noted the larger variations observed in the weight of wearing surfaces. Cobblestone pavements assumed to have been removed may have been paved over with asphalt, as was the case shown in Fig.

Figure 4.35   Rigid restrainers.

4.39. Records of curb height suggest that asphalt deck overlays grow thicker with each resurfacing. Figure 4.40 shows inadequate and correct curb reveal after resurfacing.

Thin-bonded overlays eliminate the hazard of wearing surface accumulation (but risk an increased fragility; see Section 4.3.3).

Changes in the weight of utilities are easily overlooked, because utilities are managed by different owners.

Live Load

Live loads are dynamic and vary both randomly and systematically. Historically, design has modeled the time dependence of loads as another imperfection, by amplifica-

Figure 4.36 Tensioned and loose restrainers.

tion factors. The quasi-static approach no longer satisfies seismic, wind, and blast considerations for longer spans; however, it is still prescribed for less sensitive structures and for traffic loads. The magnitude and amplification factors for train loads, regulated by AREMA, are significantly greater than for the AASHTO vehicular loads (Appendix 30). During the early twentieth century trains were replaced by vehicular traffic lanes on a number of bridges. Figure E3.1 illustrates that process on the East River bridges in New York City. The supply of bridge serviceability was thus upgraded relative to the demand of the reduced live load, even though the structural condition (e.g., the strength) was declining. Lighter than train loads, vehicular loads are also less predictable in occurrence, geometry, and magnitude. The demand for statistical treatment has therefore increased.

Vehicles
Taly (1998, 3.2.2.2) presented the evolution of vehicular loads in a comprehensive historic context. Barker and Puckett (1997, 4.2.2) summarized concisely the recent position of AASHTO design specifications. Vehicular traffic is modeled by concentrated forces, representing wheels or combinations of axles, and uniformly distributed loads, representing fully loaded lanes. Figure 4.41 shows a typical traffic configuration for which bridges must be designed.

The forces and the distributed loads are calibrated to represent the geometry and the weight of typical trucks. AASHTO design truck loads have included HS15-44, HS20-

Figure 4.37 Rocker bearings, strengthened pier.

44, HS25, and special and military loads or overweight vehicles. AASHTO 1998a (Section 3.6.1.1) refers to HL-93 (Appendix 30).

The short-term dynamic effects of vehicular loads are modeled by impact factors I (or IM), briefly summarized in Appendix 31. Modeling the continuous flow of traffic (Fig. 4.41) must resolve the questions whether:

(a) Design lane loads should be subject to dynamic amplification

(b) Vehicles with more axles induce greater dynamic amplification

Current specifications are ambivalent on both issues. Heavier trucks are believed to exert smaller dynamic amplification. The amplification is determined assuming a linear

(a)

(b)

Figure 4.38 (a) Seismically vulnerable rocker bearing. (b) Retrofit with elastomeric bearing and pedestal.

Figure 4.39 Demolition of 60-year-old reinforced-concrete deck with cobblestone pavement.

correlation between structural deflections and loads, which may not always apply, for instance at wearing surfaces and joints, fracturing in fatigue.

Fatigue damage is modeled in terms of expected equivalent stress cycles according to the Palmgren–Miner rule (Fisher et al., 1997, p. 34). The rule offers a *probabilistic estimate* of a desired fatigue life, as shown in the Section 4.3.1.

Short- and long-term effects are inseparable. The estimated fatigue life depends on the stress ranges obtained by the live-load impact factors. The effect of live loads on bridge superstructures is central to the frequently reviewed Load Rating procedures (AASHTO, 2000; NCHRP Report 12-46, 2000). TRB Special Report 225 (1990) traces

Figure 4.40 Inadequate (left, 3 in., 75 mm) and correct curb reveal (right, 8 in., 200 mm).

Figure 4.41   Typical live load modeled by AASHTO design lane loads.

the increase of the maximum gross weight limit on interstate highways from 73,280 lb (33 tons) in 1956 to 80,000 lb (36 tons) in 1974 and the corresponding speed limit reduction to 55 mph (90 km/h).

The report encouraged the redistribution of increasing truck weight over a greater number of axles. It noted that weight limit enforcement by traffic control is ineffective, because truck companies have an incentive to exceed the live-load limits and pay the imposed fines. The BRIME 2002 report by the European Commission expressed similar views.

Bailey (1996) and Moses (in Frangopol, 1999b, pp. 1–19) discuss the probabilistic modeling of truck loads. NCHRP Report 505 (2003) reviewed the geometric characteristics of trucks. All probabilistic models must be designed and updated according to the quality and quantity of the traffic data. NCHRP Report 538 (2005) addressed traffic data collection for mechanistic pavement design.

Pedestrians
In 1830 Brown's bridge at Montrose collapsed under a pedestrian crowd rushing from one fascia to the other (Hopkins, 1970). Similar crowd behavior significantly excited a modern suspension bridge without causing failure (Irvine, 1986, p. 29). Washington Roebling issued instructions against large groups of pedestrians marching in step over the Brooklyn Bridge. The Tacoma bridge failure demonstrated the danger of closely

spaced vertical and torsional mode frequencies. Lateral motion frequencies, in turn, attracted attention after moving crowds excited them at notable pedestrian bridges. The investigations of the Millennium Bridge in London (Dallard et al., 2001; Fig. 3.1) and Passerelle de Solferino in Paris (Fig. 3.2) showed that pedestrians pace themselves according to the natural period of the bridge sway and thus exert a synchronized lateral load, rather than a random one. As in the case of wind-induced flutter, the phenomenon had already been identified (Fujino et al., 1993).

Pedestrian loads can reach hazardous magnitudes statically as well. Pedestrians overloaded the Golden Gate Bridge (Figs. 4.42 and 4.43) on its 50th anniversary, causing spectacularly photogenic deflections midspan (along with some inelastic deformations).

Figure 4.42 Golden Gate Bridge.

Figure 4.43   Seismic retrofit at Golden Gate Bridge.

The AASHTO lane load is 64 psf (3.1 kN/m^2) over 10-ft (3-m) *design* lanes, subject to 35% multiple presence reduction. Pedestrians can surpass that by exerting a uniform load of 34.7 psf (1.7 kN/m^2) over all 12-ft (3.7-m) *traffic* lanes, not an unlikely prospect. Taly (1998, p. 181) attributed the first live-load recommendation in the United States to Whipple in 1846. That consisted of pedestrians and was estimated at 100 psf (4.8 $\times$ 10^{-3} MPa). Design live loads for sidewalks according to AASHTO 1998a; LRFD (2nd ed.) range from 75 psf (3.6 $\times$ 10^{-3} MPa) for sidewalks to 85 psf (4.1 $\times$ 10^{-3} MPa) for purely pedestrian bridges.

Particularly hazardous is the crowd reaction to a *perceived* structural distress. On May 30, 1883, a week after the opening of the Brooklyn Bridge, a crowd estimated at about 20,000 panicked and rushed to the bridge exits for no particular reason. Twelve people were trampled to death.

Extreme Events

All measures preparing for extreme events are fundamentally vulnerable, because the proactive approach assumes a hypothetical (and therefore debatable) hazard, whereas the reactive approach follows after the disaster has caused losses. (Emergency response is discussed in Section 4.1.3.) Thus cost–benefit analyses recommending preventive measures against extreme events are perceived as either too late or alarmist. NCHRP Report 489 (2003) by Ghosn, Moses, and Wang developed risk analysis for highway bridges under extreme events, including earthquakes, wind, collisions, and scour. ASCE summarized current practices for the evaluation of the risk associated with natural hazards in Taylor and Van Marcke (2002).

Floods and Scour

Floods are the leading cause of bridge failures worldwide. A noteworthy example is the New York State Thruway bridge over Schoharie Creek. Its failure during torrential flooding on April 3, 1987, confirmed an anticipated vulnerability to scour. A preliminary design had recommended a longer span with piers outside the creek channel. That alternative had been rejected because of its higher construction cost. Similar considerations have been quoted after the failure of the levees at New Orleans during hurricane Katrina in September 2005.

The evaluation of scour-related hazards and some methods of their mitigation is discussed in TRR 1351 (1992). FHWA (1998) describes a demonstration project on scour monitoring and instrumentation. The New York State Department of Transportation (NYS DOT) has developed a hydraulic vulnerability manual.

Floods can result from earthquake–induced ocean waves (tsunami). The destructive capacity of the floods on the west coast of North America in 1964, southeast Asia on December 27, 2004, and Louisiana in September 2005 surpassed the design provisions. Warning systems have great value, if rapid response, and particularly evacuation can follow.

Earthquakes

The Seismic Safety Committee to the U.S. Department of Transportation expressed the following concern [Parsons Brinkerhoof Quade & Douglas (PBQ&D), 1993, p. vii]: "Very few of the individuals responsible for the operation of various transportation enterprises, in either the public or private sector, in all the modes have an adequate appreciation of the seismic vulnerabilities they face."

That awareness is abruptly adjusted after every major earthquake, such as the San Fernando (1971), Loma Prieta (1989), Northridge (1994), and Kobe (1995), all exposing bridge vulnerabilities.

The collapse of the elevated Route I-880 over nearly two miles along Cypress Boulevard in Oakland, California, during the Loma Prieta earthquake (Fig. 3.10) claimed 41 victims. One motorist was killed during the same event at the collapsed 50-ft (16-m) span of the San Francisco–Oakland East Bay Bridge (Fig. 4.33). These were the only fatalities throughout the otherwise massively affected Bay area.

The seismic retrofit and strengthening of existing structures is a bridge management task performed with limited public funds, often under uninterrupted traffic. The hazard level is uncertain, public scrutiny is vigorous, and the potential losses can be high. Remedial measures bring incalculable benefits.

In response to the San Fernando earthquake in 1971, the Cypress Viaduct in Oakland was strengthened with longitudinal restrainers in 1977. During the Loma Prieta earthquake in 1989, 48 bents of the two-tier viaduct failed under the mostly transverse ground excitation. Housner et al. (1990) and Nims et al. (1989) found the longitudinal restrainers of little consequence to the failure mode. Nims et al. (1989) concluded that without the restrainers the structure might have failed longitudinally as well. Transverse strengthening had also been anticipated, pending funding availability.

Thus the two most perilous twentieth-century bridge failures in the United States, the Cypress Viaduct (41 fatalities) and the Silver Bridge (46 fatalities), had very different but similarly wide-ranging causes. The Cypress Viaduct failure would qualify as class B according to Thoft-Christensen and Baker (1982) because it was designed according to specifications that did not address seismic loading. The overall structural scheme was inappropriately discontinuous, both longitudinally and transversely. The concrete bents lacked internal ductility, and column reinforcement had no confinement. Footings were located in "bay mud" and stood on piles of average depth 50 ft (15.25 m). Limited funding only allowed for incremental retrofitting. As at the Silver Bridge and at Scoharie Creek, the absence of any one of the critical deficiencies could have prevented the global structural failure. Although the comparison is not rigorous, a two-tier structure in the Embarcadero area of San Francisco with continuous columns (Fig. 3.9, Section 3.2) survived the earthquake with considerable damage localized at the girder-to-column connections and was later demolished.

The ensuing investigations, including Housner et al. (1990), resulted in several modifications adopted by bridge design. Longitudinal continuity was introduced across expansion joints (Figs. 4.35 and 4.36). The confinement requirements for longitudinal reinforcement in columns were strengthened. Concrete two-tier viaducts were to be avoided. Some, including the Embarcadero (Fig. 3.9), were dismantled; others were strengthened. The design displacement of sliding bearings was increased.

The Northridge earthquake confirmed the vulnerability, already known since San Fernando (1971), of rigid pier columns to shear failure (Goltz, 1994). Continuity and ductility of frames became a high design priority.

The failure of rigid restrainers between primary members at expansion joints was particularly widespread during the Hyogo-Ken Nanbu earthquake in Kobe, Japan (January 17, 1995) (Fig. 4.34). During that event 18 statically determinate hammerhead piers failed at the Hanshin Expressway, reinforcing bars failed at welded connections, tubular steel columns buckled, and some cable stays ruptured.

Since 1989 the essential San Francisco–Oakland East Bay crossing (Fig. 1.43) has carried traffic with temporary strengthening and enlarged expansion bearings. The cost of a retrofitting is estimated to surpass that of a new bridge. The choice of a new structure satisfying all the complex requirements of the region (Fig. 1.44) is under review (Section 1.5).

The failure modes and loads at the Cypress Viaduct and the Silver Bridge were different, but they triggered comparable developments in seismic engineering and in bridge management. "Emergency management" permeated routine operations (Section 4.1.1). "Extreme events" became a governing condition state for design in general.

In 1990 Presidential Executive Order No. 12699 directed all federal agencies to produce an implementation plan assuring that seismic design considerations would be incorporated into new buildings constructed with federal monies or leased for federal functions.

The FHWA has sponsored numerous studies of the seismic vulnerability of new and existing highways at the three National Earthquake Engineering Research Centers (see

Appendix 33). Seismic bridge design, retrofit, condition evaluation, and risk assessment manuals have been published and repeatedly updated. The Applied Technology Council (ATC, 1983) was followed by FHWA (1995a). Rojahn et al. (1997) summarized seismic design criteria worldwide (Appendix 32). Guidelines for complex bridge structures and substructures are part of FHWA project DTFH61-98-C-00094. Response spectra and time–history analyses evaluate the elastic and inelastic behavior of sensitive and important structures (Fig. 4.43).

Software for postevent transportation network management was developed by Werner et al. (2000) for specific geographic areas and is currently generalized. Cooperation with Japan and Europe is particularly active in the seismic analysis and design field.

The concepts of earthquake-resistant design formulated by Newmark and Rosenblueth (1971, Chapter 14) influenced design philosophy in general. The essential text on structural dynamics by Clough and Penzien (1975) evolved concurrently with their work at the Earthquake Engineering Research Center at the University of California at Berkeley. The seismic design and retrofit recommendations by Priestly et al. (1996) similarly have a general application. NCHRP Report 489 (2003) applied the reliability concepts recommended in AASHTO LRFD (1998) to the design of highway bridges for extreme events.

The probabilistic risk analysis of random seismic hazards eventually propagated throughout design procedures (Chapter 5). The algorithms for prioritizing the mitigation of seismic hazards (Appendix 33, Example 14) can be modified to prioritize general improvements according to different vulnerability criteria.

Example 14. Prioritizing Seismic Vulnerability of Bridges According to Inventory Data

Appendix 33 refers to several formalized procedures for evaluating the level of seismic hazard and the potential losses from credible earthquakes. The scope can be limited to individual structures or can include transportation and mixed-use networks. Yanev and Tran (1997) identified potentially vulnerable bridges in a densely populated urban area by scanning the bridge inventory. The following conditions governed:

Moderate seismic hazard, AASHTO zone 2, peak ground acceleration of 0.19g

Seven hundred and fifty bridges with 4500 spans

Average condition 4.5 on a scale of 1–7, including considerable numbers of deteriorated structures, as shown in Example 12

Average daily traffic exceeding 100,000 vehicles on many of the structures

The following items are scanned:

- Number of spans
- Alignment
- Maximum span length
- Main span material

- Main span structural type
- Maximum pier height
- Pier material
- Number of pier columns
- Pile length
- Bearings (fixed/expansion)
- Annual average daily traffic (ADT)

Expansion joints were assumed at every second span. The condition of pier columns, primary members, bearings, and pedestals was taken into account to sort out structures with otherwise comparable features.

Essential information was lacking on the following subjects:

- Pedestal width
- Pile length
- Soil type
- Footing condition

The following observations emerged from the screening:

- Steel is the main structural material in more than 60% of the bridges.
- Steel stringer and girder bridges with concrete decks comprise 53% of the stock.
- Bridges with more than five spans comprise 15% of the stock but account for 65% of the spans.
- Bridges with ADT greater than 50,000 are 22% of the total number.
- Spans longer than 60 ft (18.3 m) are found on 34% of the bridges.
- Skew angle piers are found at 32% of the bridges.
- Soil type S3 supports 12% of the bridges (carrying the same routes).

The structures identified as most vulnerable had been already included in rehabilitation plans, confirming the consistency of independent prioritizations in moderate seismic zones. Governing design specifications (when local requirements are more stringent, they supersede general ones) stipulate acceptable detailing and recommend appropriate structural performance or elements, such as bearings, footings, and connections.

A preliminary screening of this type must be followed by specific studies of essential and vulnerable structures subjected to credible (500- and 2500-year-return-period) earthquakes. Traffic corridors and alternate routes must be considered. The software developed under FHWA-MCEER Project 094 (Appendix 33) allows users to analyze their asset inventory under prescribed events, taking into account the interaction of structural conditions, transportation demands, and lifeline performance.

Despite the concentration of effort and the resulting improvement in seismic resistance, G. Housner's caution that "earthquakes repeatedly prove us stupid" remains valid. Earthquakes are a hazard even in moderate seismic zones (Section 9.2, Appendix 33). AASHTO 1998a (Section 4.7.4) reminds bridge owners that design specifications provide only minimum requirements for seismic analysis. Management similarly remains vulnerable, even if resources are adequate.

Soil Liquefaction

Soil liquefaction is a seismic consequence strongly amplifying structural displacements. Design in liquefiable soils is highly constrained. The discovery of liquefiable soils under existing foundations, particularly in seismically active zones, calls for emergency planning. Vick (2002) suggests detailed methodologies of risk analysis for this type of hazard.

Wind

From their inception, suspension bridges were vulnerable to dynamic and particularly wind loads, as was the case with the Menai Straits Bridge in 1826 and 1836 and the Wheeling Bridge in 1854. J. Roebling brilliantly secured his record-breaking suspension spans against wind by adding stay systems. His solution, however, was an empirical one and the profession still had to comprehend dynamic stability in analytic terms.

Von Karman and Biot (1940) defined *flutter* as the loss of stability in an airplane wing with nearly equal natural frequencies in the torsional and flexural modes. The relevance of flutter to suspension bridges was fully appreciated after the wind-induced failure of the Tacoma Narrows Bridge on November 7, 1940. That failure remains a subject of discussion because different factors contributed to it. The natural frequencies of fundamental dynamic modes were too close to each other. The structure was exceptionally light, with a narrow roadway and little damping. The stiffening girders trapped wind vortexes. The absence of any one of these factors might have saved the bridge. The similar but wider Whitestone Bridge in New York City (1937) is sensitive to high winds to a manageable degree. When its designer Othmar Ammann cited luck as his best asset (Talese, 1970), he might have been referring to the similarities between the Tacoma and Whitestone Bridges. Engineers traditionally aspire to independence from fortune. Managers, however, know that this is never entirely possible.

The aerodynamic stability of long spans remains a highly active research topic, as discussed, for instance, in Larsen and Esdahl (1998). A concise review of structural instability due to nonconservative loads can be found in Chapter 2 of Bǎzant and Cedolin (1991). Paradoxically, the advances in bridge dynamics only underscore the need for empirical knowledge obtained by in situ structural monitoring and by model tests in wind tunnels. Even though AASHTO design specifications do not address long-span bridges, AASHTO 1998a (Section 4.8) recommends wind tunnel tests. An historic overview of the subject is provided by Scott (2001).

The active and passive control of wind-induced bridge response has been investigated by Fujino (in Juhn et al. 2005) and other researchers, based on data from the Hakucho, the Akakshi-Kaikyo, and other long-span bridges in Japan.

Collision

Vessel collisions are exceptionally destructive, as in the case of the Sunshine Skyway in Florida. Figure 4.44 shows a pier column broken by barge impact. Traffic was immediately shifted away from the weakened strip of the deck. A continuous fascia girder was installed over the lost support. The bridge will be replaced by a longer span at higher elevation. Ice and floating timber pressure is estimated in detail by the AASHTO. A vessel can push ice against piers without exiting the navigation channel.

Figure 4.45 illustrates train impact on a steel column. Whenever possible, bridge columns must be protected from train collision by crash barriers. Steel columns are encased in concrete for added resistance to impact.

Truck collisions with piers and superstructures are common. Providing crash protection by the same measures as in railroad bridges may be impractical because of geometric constraints or because of concern for the motorists involved. The structure shown in Fig. 4.46 is a landmark and cannot be easily modified. It was strengthened to absorb the impact without significant damage. After sustaining numerous hits, the structure in Fig. 4.47 was removed. On pedestrian bridges (Figs. 4.48a and 4.48b), vehicular impact is likely to be the governing loading condition.

Given a good record of past incidents and their consequences, risk analysis (Chapter 5) can evaluate the costs and benefits of the remedial options, including change in the geometry, warning signals, and traffic enforcement.

Figure 4.44 Concrete pier column and cap beam, broken by impact.

Figure 4.45 Train impact on steel column.

Figure 4.46   Collision of truck with vehicular bridge.

Figure 4.47   Impact damage to vehicular bridge superstructure.

AASHTO 1998a (Section 3.14) evaluates the hazard of vessel collision in particular detail. The owner must classify the bridge according to its importance. The consequences of collisions to the structure, the protective systems (e.g., dolphins, fenders), and the vessels must be evaluated jointly. The penalties must be minimized with the concurrence of the agencies managing navigation (such as the U.S. Coast Guard).

Technology enhancements have provided more effective warning systems; however, on most structures there is not enough site-specific data to allow statistical estimates of the reduction of the risk level and hence a life-cycle cost–benefit analysis.

Fire

Fresh in everyone's memory is the World Trade Center North Tower collapse (Fig. 4.3) on September 11, 2001, where sustained high temperatures were decisive [Federal Emer-

(*a*)

(*b*)

Figure 4.48   Truck impact damage to pedestrian bridges.

Figure 4.49 Trusses damaged by fire.

Figure 4.50 Rocker bearing and pedestal failing under thermal expansion caused by fire.

gency Management Agency (FEMA) 403, 2002] to the integrity of the simply supported floors. Even more to the point, World Trade Center building 7 collapsed seven hours later entirely as a result of fire.

Fire is not unusual on and under bridges, while fireproofing is considerably less likely than in buildings. Figures 4.49–4.51 show typical fire damage.

The modulus of elasticity decreases as temperature mounts. Under sustained fire girders acting in bending develop large permanent deflections. Expansion bearings cannot accommodate the extreme displacements and fail, as in Fig. 4.50.

Concrete spalls, often explosively, when heated. The exposed reinforcement buckles. Steel members easily buckle under fire (Fig. 4.51). The effect is demonstrated in Example 15.

Figure 4.51 Steel column, buckled under thermal expansion.

Example 15. Critical Strain Induced by High Temperature

The coefficients of thermal expansion are

$$\Delta\varepsilon_t^S = 11.7 \times 10^{-6} \text{ mm/mm°C } (6.5 \times 10^{-6} \text{ in./in./°F}) \quad \text{for steel}$$

$$\Delta\varepsilon_t^c = 10.8 \times 10^{-6}/°C \ (6.0 \times 10^{-6}/°F) \quad \text{for normal-weight concrete}$$

$$E_{\text{steel}} = 200,000 \text{ MPa } (29,000 \text{ ksi})$$

$$E_{\text{concrete}} = 0.043 \ y_c^{3/2} \ f_c'^{\ 1/2} \text{ (SI)} = 33,000w_c^{3/2} \ f_c'^{\ 1/2} \text{ (U.S.)}$$

where y_c = density of concrete (kg/m^3)
　　　f_c' = specified strength of concrete (MPa or ksi)
　　　w_c = unit weight of concrete (kcf)

Assume an unbraced simply supported straight member of length $L = 15$ ft (4572 mm), cross section W36 × 256, $A = 75.4$ in.^{2} (48645 mm^2), and $r_y = 2.65$ in. (67.31 mm).

The critical strain ε_{cr} in a member with the above properties is (as in Eq. 4.3a)

$$\varepsilon_{\text{cr}} = \left(\frac{\pi \ r_y}{L}\right)^2 = \left(\frac{3.14 \times 2.65}{15 \times 12}\right)^2 = \left(\frac{3.14 \times 67.31}{4572}\right)^2 = 2.13 \times 10^{-3}$$

The temperature difference producing ε_{cr} in the member is

$$\Delta°t_{\text{cr}} = \frac{\varepsilon_{\text{cr}}}{\varepsilon_t} = \frac{2130}{11.7} = 182°C \quad \text{or} \quad \frac{2130}{6.5} = 328°F$$

Here $\varepsilon_{\text{cr}}/2$ could easily have been induced by the applied loads. Hence $\Delta°t_{\text{cr}}/2 = 164°F$ may induce buckling in a restrained member of the assumed properties. According to SFPE (2000) the elastic modulus of steel begins a near-linear decline beyond 200°C (392°F). Buckling is typical in steel members under fire, as shown in Fig. 4.51.

The elastic modulus of concrete (Kodur and Harmathy, 2002) declines much faster, beginning at ambient temperatures and reaching 70% of its original value at 200°C (392°F).

During a recent thunderstorm, a stay of the multispan cable stayed bridge at Rion-Antirion (Fig. 4.52) was struck by lightning and caught fire. M. Virlogeux described the event as a "scenario which has never been imagined" (ENR, February 7, 2005, p. 12). It demonstrated that lightning is a fire hazard for bridges, particularly if the material used for corrosion protection (in this case of the stays) is flammable.

Sabotage
The following features distinguish sabotage from other extreme events:

- Sabotage is not purely random. The evolution of security measures over the period from the bombing of the Oklahoma FBI building to the attacks on the World Trade

Figure 4.52   Rion-Antirion Bridge, Greece. Courtesy of G. Hovanessian, Advitam.

Center in New York City and the Pentagon in Washington, D.C., demonstrates this point. Even if based on a probabilistic analysis, the recently adopted security alert levels are assigned deterministically.

- Structures are subjected to highly improbable loading combinations. The towers of the World Trade Center collapsed under the cumulative action of impact, blast, and sustained high temperature, added to all dead and live loads (Fig. 4.53). The result was a progressive collapse unlikely under the anticipated design loads. To describe the "vagaries of blast damage" (p. 13) during World War II, Lord Baker (1978) similarly coined the term "spreading collapse."

- The design and maintenance measures protecting bridges against deliberate destruction are directly opposed to the fundamental function of maximizing transportation at minimum cost. Consequently strengthening and deterrence are mutually dependent.

The estimated likelihood of sabotage of the infrastructure grew radically after the attacks perpetrated on the United States on September 11, 2001. Transportation agencies in the United States are adopting a National Incident Management System (NIMS), described in NCHRP Report 525 (2005).

Figure 4.53 South Tower, World Trade Center, September 11, 2001.

Loading Combinations

Most failures result from unforeseen combinations of independently noncritical loadings. Loading combinations are therefore particularly challenged to anticipate failures, as in the example of the World Trade Center towers in New York City, quoted in the preceding section (Figs. 4.3 and 4.53).

The history of past failures is essential but insufficient in determining the likelihood of combined load effects, particularly those due to extreme events. That likelihood is determined by a variety of probabilistic methods, relying on the "degree of belief," discussed recently by Vick (2002) and (to the extent available) statistical data. Thoft-Christensen and Baker (1982) rigorously formulate the method of combining loads treated as stochastic variables.

Barker and Puckett (1997, Chapter 4) and Taly (1998, Chapter 3) have reviewed the loading combination philosophies in the AASHTO design specifications during their transition from ASD to LFD and to LRFD. In many cases a bridge is economically feasible only if extreme-event expectations are relaxed. While the expectation of individual extreme events, such as earthquakes, collisions, and floods, is occasionally raised, specifications consider their coincidence with more than 50% live loads and with each other unlikely. Since databases constantly grow, loading combinations are periodically updated. The load-bearing capacity of the bridges also may change over time. NCHRP Report 454 (2001) by Moses describes the method for calibrating load factors for load rating of existing bridges. Ghosn et al. NCHRP Report 489 (2003) develop loading combinations for extreme events.

Table 3.22.1A in AASAHTO (2002) defines 10 loading groups in either service load (e.g., ASD) or load factor design. LRFD (1998, Table 3.4.1-1) lists 11 *loading combinations* grouped within four *limit states*. The general descriptions of the loading combinations are quoted in Appendix 25. The comparison should demonstrate a more comprehensive effort to envision structural behavior under a wider variety of demands in AASHTO 1998a. The full extent of the analysis, however, is up to the frequently mentioned *owner*. The generally helpful checklists for superstructure design, provided as Appendices A-5 and B-6 to the AASHTO 1998a Chapters 5 and 6 (on concrete and steel design, respectively), emphasize the strength limit state. Serviceability, fatigue, and extreme events remain at the discretion of the bridge manager.

4.3   DESIGN AND CONSTRUCTION

4.3.1   Materials

Material nonperformance can be traced to both design and construction. Navier attributed, at least in part, the 1826 demise of his Pont des Invalides in Paris to construction and material deficiencies (Hopkins, 1970, p. 203). Karnakis (1997), however, pointed out flaws in the analysis and the design of the anchorages.

Design selects the appropriate construction materials in part according to their known mode of failure. Boresi and Sidebottom (1985, p. 100) distinguish the following three types of material failures:

- Large deflections
- Yield
- Fracture

Large deflections are more of a global symptom than a material failure. They can result from changes in the material properties, particularly those affecting the elastic modulus, the elastic limit, the consistency, and cohesion (in the case of concrete).

In the manner proposed by Thoft-Christensen and Baker (1982) for structures (Section 3.4 herein), material failures can be divided into classes A (in a mode expected of the material) and B (in a mode not expected of the material). Overstress would be a class A failure. A class B failure would be the brittle fracture of a structure designed for elastic–ductile behavior or the burning of a structure not designed for elevated temperatures.

During fabrication and construction, steel is typically vulnerable to residual stresses, microscopic cracks, embrittling chemical content, surface defects in galvanization, and epoxy coatings. Cast-in-place concrete deviates from specifications mostly because of poor admixtures, alkali–silica reactivity, inadequate water–cement ratio, and curing. Waterproofing (e.g., of bridge decks, prestressing tendons, suspension cables) should receive special attention, because of its critical importance and sensitivity to minor defects.

In Table 4.6 typical failures of the most frequently used materials are classified according to the phenomenon (e.g., the cause) and the mode (e.g., the effect). Not all phenomena and failure modes are independent, once again exhibiting a coincidental nature. As in Tables 4.1, 4.4, and 4.5, the purpose of the matrix of Table 4.6 is to identify vulnerabilities in either the product (in this case, the failing material) or the process (e.g., the destructive phenomenon). Material vulnerabilities are likely to be examined along the rows of Table 4.6 during analysis, design, and construction. Inspection is more likely to observe the effects along the columns and seek to deduce the causes. Each material and structural type must be investigated (e.g., steel, reinforced concrete, prestressed concrete, deck, primary member, soil) and subsequently considered as part of the structure. It is evident that the mechanical response of the structure depends on the ensemble of its elements. It is less apparent that the chemical behavior of the structural materials is interdependent. For instance, the presence of a more noble metal, such as bronze, can accelerate the corrosion of steel. Inappropriate choice of aggregate can cause alkali–silica reaction in concrete.

Farrar, Lieven, and Bement (in Inman et al., 2005, pp. 1–12) distinguish between physics- and data-based predictive modeling of damage states. Statistical treatment of data governs in the absence of a definitive physical model and vice versa. The two approaches are intended to converge. The data adjust the model, while the model determines the type of monitoring that can provide pertinent data. The result is a damage prognosis system (also discussed in Section 10.6).

Fatigue

Fatigue-induced failures are typical examples of combined effects (Section 3.5). The number *and* amplitudes of loading variations jointly exhaust the material "toughness." Minor geometric irregularities, corrosion, poor detailing, and residual stresses further reduce that toughness by causing stress concentration (hence the name "stress raisers"). Whereas instability (Section 4.2.2) changes the structural shape, fatigue affects the material structure.

Fatigue fractures lead to catastrophic failures only when they occur in elements critical to the structural performance. Inspections inevitably come under scrutiny and even blame after failures attributed to fatigue (as management is scrutinized after any failure), yet they are the least likely to prevent such occurrences or to observe them with a satisfactory margin of safety. Assigning critical functions to fatigue-prone elements is a design error, as it is a managerial error to address fatigue hazards primarily by inspections. FHWA (1986) identified fracture-critical structural elements (Section 14.2) as a

Table 4.6 Type of Failure and Susceptible Materials

Type	Phenomenon	Type of Failure					
		Fracture	Yield/Rupture	Deformation/Distortion	Discontinuity	(Pre)- and Stress Loss	Other
Mechanical	Fatigue	S, HS			S, RC	PRC	
	Stress concentration	S, RC					
	Residual stress	S					
	Overstress (tension, flexure, shear)	C, S	S, HS, RC, PRC	S, RC, PRC		S, HS, PRC	
	Relaxation			PRC	PRC	PRC	
	Freeze–thaw	S, RC, SP			RC, PRC		
	Temperature	S, HS	S, HS	S, HS, RC	RC, PRC	PRC	
	Blast		S, HS, RC, PRC	S, HS, RC, PRC	S, HS, RC, PRC		
	Other						
Chemical	Corrosion		RC, PRC, S		S, RC, SP	PRC	
	Shrinkage				RC, PRC	PRC	
	Embrittlement	HS					
	Sun light				SP		
	Alkali–silica reaction				RC, PRC		
	Chemical attack				RC, PRC		
	Other						

Note: Stress/strain and material (e.g., mechanical and chemical) changes are not independent (neither are loads and the environment). Under temperature changes materials experience volumetric changes and hence loads, but their elastic moduli also change. Fatigue causes stress concentration in materials prone to become less homogeneous. Abbreviations: RC, reinforced concrete; S, steel; HS, high-strength steel; SP, steel paint; PRC, prestressed concrete.

network-level vulnerability, to be mitigated, most effectively, by redundant design. NCHRP Synthesis 354 (2005) updates the subject to the present.

Steel

Fisher et al. (1997, p. 1) state: "Fatigue fractures in metals occur when cracks are initiated and grow under the action of repetitive loads."

The likelihood of a fatigue failure is estimated according to the Palmgren (1924)–Miner (1945) rule as follows:

$$\sum n_i/N_i = 1 \tag{4.3}$$

where n_i = number of stress cycles occurring at range i

N_i = number of stress cycles occurring at range i that would cause failure

The damage at any stress range is a linear function of the number of stress cycles at that range. The total damage is the sum of all cycles. The rule is simple and obtains realistic results; however, Fisher et al. (1997, p. 34) point out that it ignores sequence effects and average stresses in the cycles. Most importantly, the rule is a *probabilistic estimate* of the fatigue life that does not exclude the possibility of a premature fracture.

The cumulative effect of local "stress concentrators" and increased load cycles (in either amplitude or frequency) is critical but hard to evaluate. The combination of fatigue and corrosion is even more complex. Corrosion creates stress "concentrators" or "raisers" by disturbing the normally smooth surfaces of structural members

Corrosion in microscopic cracks accelerates their propagation. Hence the term *stress corrosion* has been coined. This phenomenon is considered instrumental in cracking and breaking of high-strength steel wires of suspension bridge cables (Figs. E3.4) and pre-stressing strands.

A number of failures have occurred due to brittle fractures where ductile behavior was expected from the material (Fig. 4.54). Such are the Ashtabula creek cast-iron rail-road bridge in 1876 in the United States and the Viaduct d'Evaux over the Tardes in 1881.

Fatigue fractures are further exacerbated by residual stresses. Rolfe and Barsom (1987, p. 4) quote the 1962 brittle fracture at the Kings Bridge in Melbourne at ambient temperature of 40°F as follows: "Poor details and fabrication resulted in cracks which were nearly through the flange *prior to* any service loading."

Most common "stress raisers" are the terminations of tension flange cover plates. The detailing of such points offers considerable possibilities (Fig. 4.55) to reduce the fatigue hazard.

Low temperatures are, once again, doubly conducive to fatigue fractures: They embrittle metals, while contractions raise stresses. Uneven exposure to heat can cause significant daily stress cycles in steel members, including fracture-critical ones.

Aluminum is occasionally used in bridges, most frequently in sign structures, but also in some decks, for instance, at the historic Smithfield Bridge in Pittsburgh (Fig. 1.14). Aluminum does not have the toughness limit of steel and requires additional caution.

The Silver Bridge (Example 6) and other examples contributed to coining the term *fracture critical* (FHWA, 1986, p. 20), designating structural members critical to the

Figure 4.54 Brittle fracture of steel bracing.

structural integrity *and* prone to fractures. The many orthotropic decks on long-span bridges often suffer from fatigue cracking, particularly at the intersection points of their longitudinal ribs and floor beams (Fig. 4.56). The overall redundancy of the structure, however, precludes the possibility of a catastrophic fracture failure. Nonetheless, levels of stress concentration and load cycles must be investigated if the owner expects a reliable forecast of the deck useful life (Section 10.6.1).

Concrete fatigue
Reinforced concrete is a nonhomogenous material that cracks under normal operating conditions. Consequently it can be considered a priori fatigued. Yet concrete structures have failed in what can best be described as fatigue, namely under cyclic loading and in a brittle fracture mode. Figure 4.57 illustrates such a failure in a concrete bridge pier. The pier continued to function without appreciable sag after the column-to-floor beam connection failed. The likeliest cause of damage was the cyclic nature, rather than the magnitude, of the live loads. The concrete of the overreinforced detail was also affected by alkali–silica reaction, making it even more susceptible to fatigue.

Although concrete decks delaminate (bottom and top, see Figs. 4.14 and 4.15) when reinforcement corrodes, the impact of repeated traffic loads significantly exacerbates the effect. Spalling can occur when reinforcement is in relatively good condition, as in Fig. 4.58. Wearing surface and entire deck patches of higher density and strength, such as the latex-modified concrete (LMC) overlays, fracture in a fatiguelike manner.

The internal redundancy of reinforced-concrete sections usually prevents local fatigue-type failures from causing structural collapse; however, they can significantly reduce the useful life of a bridge.

(*a*)

(*b*)

Figure 4.55 Bottom flange cover plates.

(*c*)

Figure 4.55   Bottom flange cover plates.

Environment

All environment is aggressive to some degree. Structural materials are selected because of their resistance to ambient effects. Design specifications recommend provisions for temperature changes, the effect of water and soil. Of greatest consequence yet frequently neglected during construction is the minimum concrete cover of steel-reinforcing bars. AASHTO (2002) stipulates concrete cover ranging from 1 to 3 in. (25 to 75 mm). AASHTO 1998a increases the cover to 4 in. (100 mm) under exposure to salt water. Inadequate concrete cover is most common in bridge deck bottom reinforcement. Example 16 illustrates the spalling of deck undersides where cover appears to have been relatively thin. Excessively thick cover may also become a hazard due to its weight.

Example 16. Spalling of Deck Underside

On June 1, 1989, a concrete fragment with an estimated weight of 300 lb (135 kg) spalled from the underside of the F.D.R. Drive in New York City at East 19th Street. A motorist was killed by the impact. During the following emergency inspection a large portion of the deck underside was found to be on the verge of spalling and was immediately removed (Figs. E16.1 and E16.2).

(*continued on page 252*)

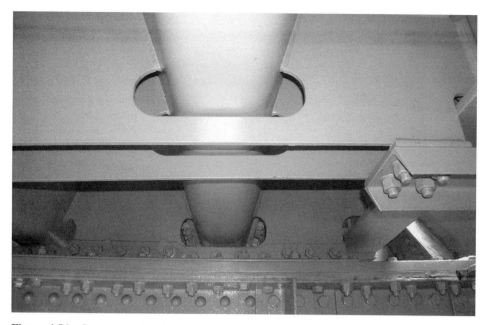

Figure 4.56 Cut-outs designed to reduce stress concentration in orthotropic deck ribs and floor beams.

Figure 4.57 Fatigue-type failure in concrete bridge pier.

Figure 4.58   Exposed deck top reinforcement showing no corrosion.

Significant consequences of the incident included the following.

INVENTORY

The city bridge inventory was immediately scanned for concrete decks rated less than 5 located above any form of traffic and without stay-in-place forms. A list of approximately 400 bridges was compiled and prioritized for emergency inspections according to the ratings and the traffic. Knowing the magnitude of the task, city management was able to promptly allocate funds and retain professional expertise under emergency contracts. Inspections and temporary repair work continued for roughly two years at a cost exceeding U.S. $50 million. Remedial measures included removal of the concrete (Fig. E16.2), installation of timber shielding (Fig. E16.3), netting, and expanded metal mesh (Fig. E16.4)

INSPECTION

Concrete deck undersides were designated as "100% hands-on details." Since the entire bridge underside areas could not be sounded with hammers during biennial or monitoring inspections, guidelines had to be developed for the recommended sample sizes. Considerations included the deck appearance, such as evidence of al-

Figure E16.1 Concrete deck underside, ready to spall.

Figure E16.2 Spalled deck underside.

kali–silica reaction (Fig. 4.59), cracks, hollow sound, already existing spalls (Figs. E16.1 and E16.2), thickness of the concrete cover, when exposed (excessive or too thin), rebar corrosion, and type of traffic overhead and underneath. Each indication of a deficiency increased the need for further sounding. Sounding was criticized as a labor-intensive and uncertain method of inspection; however, no substitute has been found.

The widespread areas of exposed reinforcement at pier girders necessitated more elaborate analysis of the structural integrity of such members. Figure E16.5 shows strain gauges attached to the exposed reinforcement of a cap beam (also discussed in Chapter 15) for the purpose of determining if the structure is still capable of carrying live loads.

Notification

Evidence of potential spalling was designated as a "safety flag" requiring prompt interim action (see Appendix 46). The potentially hazardous area had to be kept off limits until remedial action.

Forecasting

Example EA46 points out that, in the case of spalling deck undersides, the most sensitive ratings are 4 and 3 (on the 7-to-1 scale used in New York State and described in Appendix 40). By the time ratings have declined to 2 and 1, much of the loose material has already fallen or has been removed.

Figure E16.3 Timber shielding, impact damage.

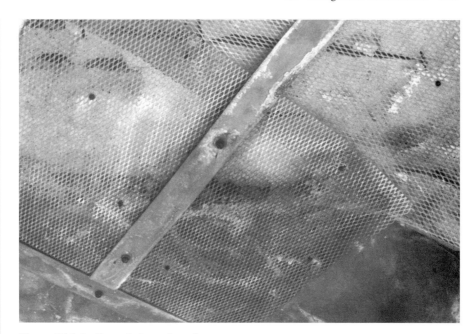

Figure E16.4 Expanded metal mesh.

Figure E16.5 Strain gauges attached to exposed reinforcement of cap beam.

DESIGN

The mechanism of spalling is roughly illustrated in Fig. E16.6.
Several design measures were recommended for consideration, including:

- Stay-in-place forms (Fig. 4.67)
- Deck waterproofing
- Rebar waterproofing, for instance, by epoxy coating (Fig. 4.67)
- Noncorrosive rebars (carbon fiber, stainless, galvanized steel)
- Cathodic protection, passive or active
- Corrosion and alkali–silica reaction inhibiting admixtures
- Concrete cover designed according to climate and aggressive environment

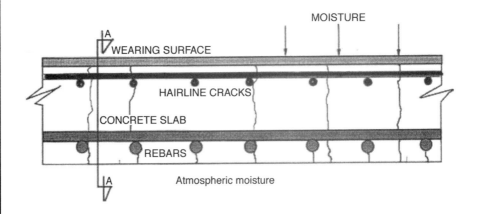

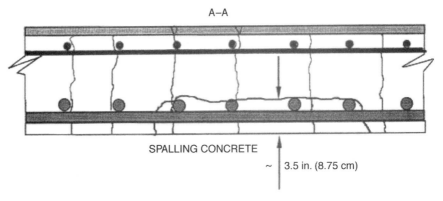

Figure E16.6   Spalling of concrete deck.

Figure E16.7 Failed granite cladding of wingwall.

CONSTRUCTION

The following features were targeted for particularly vigorous quality control when applicable:

- Bottom reinforcement concrete cover, for example, 1.5 in. (375 mm)
- Aggregate alkali–silica reactivity
- Concrete water–cement ratio
- Humidity during curing
- Waterproofing integrity
- Rebar coating integrity

NONSTRUCTURAL ELEMENTS

Nonstructural elements such as decorative stone cladding and brick veneer were found to present a similar hazard. Such elements are destabilized when their anchoring devices corrode. Sounding is ineffective, because it invariably suggests a concealed cavity. The deck emergency make-safe measures do not apply. Hence suspect stone and brick masonry cladding is removed (Figs. E16.7 and E16.8). Design is directed to seek other decorative means.

Occasionally, elements designed for decorative purposes assume structural functions without adequate verification of their capacity. Figure 16.9 shows a decorative

Figure E16.8 Removal of unstable brick veneer from wingwall.

Figure E16.9   Decorative stone parapet.

stone parapet with no anchorage for resisting the vehicular impact specified by AASHTO.

Breysse (in Cremona, 2003, Chapter 10) introduced the subject of uncertainties and reliability in the modeling of materials with an account of the following incident occurring in France in October of 1999. A student attempted to hang from a concrete canopy which broke and killed him. An order was immediately issued to make safe 172 schools with balconies, terraces, and/or canopies.

Ambient Temperature

Although ambient temperature variations are much milder than the high temperatures developing during fires, their repetitive nature and extreme values can be hazardous. Freezing temperatures contribute to brittle fracture of steel. The temperature at the Point Pleasant Bridge on December 15, 1967, is estimated at about $-23°C$, thus adding to the factors contributing to the collapse (Section 3.4). The collapse of the suspension bridge at Sully-sur-Loire on January 16, 1985, is attributed to temperature-induced brittle fracture (Persy and Raharinaivo, 1987; Gourmelon, 1988).

Freeze–thaw cycles are a primary cause of concrete degradation. Their effect is exacerbated by high levels of humidity (as in poor waterproofing and ponding) and poor concrete air entrainment. Cramer and Carpenter (in TRR 1668, 1999, pp. 1–17) discuss the effect of aggregate gradation on freeze–thaw durability.

Inadequate provisions for thermal expansion have caused numerous failures of structural members, particularly in long-span bridges. Figure 4.31 shows a sliding bearing supporting railroad tracks on a suspension bridge. Under low temperature, support was almost completely lost due to misalignment of the rail and beam expansion joints. Service had to be interrupted for an extensive rehabilitation.

Chemical Reactivity

Atmosphere, soil, and water invariably produce chemical changes in structural materials over time. Analysis considers shrinkage and creep, corrosion, and alkali–silica reactivity (ASR). Design selects materials according to their performance in a given environment.

The reaction between the concrete aggregate and the cement forms the highly hygroscopic alkali–silica gel. Deicing salts may act as a catalyst. There is no apparent strength loss; however, once water permeates the structure, the gel expands and splits the concrete (Fig. 4.59a). ASR has often occurred contrary to expectations. Tests must demonstrate that the materials to be used in concrete mixes are not alkali–silica reactive.

NRC (1993a and b) investigated the mechanisms of ASR, proposed rapid methods of detection, and developed methods of mitigation. Fournier, Bérubé, and Rogers (in TRR 1668, 1999, pp. 48–53) proposed guidelines for the prevention of alkali–silica reaction in new concrete structures. Figure 4.59b shows a bridge deck where cracks have been sealed wih epoxy in order to arrest ASR.

Corrosion

Metals

Corrosion is the electrochemical oxidizing of metals. The subject is addressed comprehensively in texts such as Jones (1992). The author (p. 3) quoted estimates of the economic costs of corrosion to the United States ranging between U.S. $8 billion and $126 billion per year. According to FHWA sources, corrosion causes annual losses of U.S. $29.7 billion to the U.S. transportation network. Roughly 80% of that amount is sustained by automobiles, while vessels suffer about 10% of it. The quoted assessment illustrates the difficulty in estimating the losses caused by corrosion. It is not clear, for instance, how much damage is sustained by the infrastructure. The corrosion of the water supply (Fig. 4.1), the power, and communication networks is evaluated indirectly. The corrosion of exposed steel and reinforced-concrete structures, such as bridges, although visible, is not easy to measure or relate to damage and cost.

The collapse of a span at the West Side Highway in New York City (1973) was attributed to unchecked corrosion. A corroded diagonal stay at the Brooklyn Bridge ruptured, killing a pedestrian in 1981. On June 1, 1990, the concrete deck spalled at the F.D.R. Drive in New York City and killed a motorist underneath (Example 16). Rebar

(a)

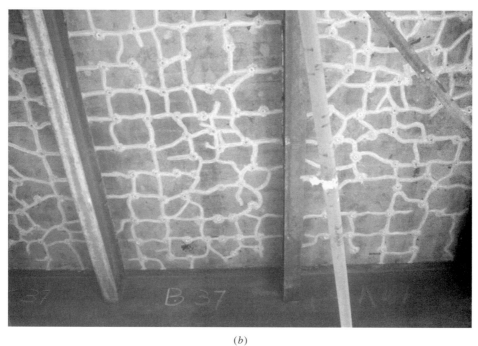

(b)

Figure 4.59   (a) Efflorescence, indicating alkali–silica reaction. (b) Cracks sealed to arrest ASR.

corrosion was among the causes. The 1988 biennial inspection of the Williamsburg Bridge in New York City closed the bridge until the consequences of the observed corrosion could be sufficiently assessed (Example 3 herein).

Corrosion steadily gained importance during the twentieth century for the following reasons:

• The usage of deicing salts increases with the volume of vehicular traffic.

The application of deicing salts can be considered as an environmental hazard to the extent that it is a function of the climate. Sanitation departments in large cities are committed to a "black-top" roadway surface policy. During an average winter, 300,000 tons of salt are applied on the streets and bridges of New York City (Fig. 4.60).

The effect of deicing salts stands out in a comparison between railroad and vehicular bridges of the same age and structural type. Figure 4.61 shows a railroad structure suffering from fatigue but with no significant corrosion. At the portion of the same structure

Figure 4.60 Stockpile of rock salt for winter deicing.

Figure 4.61 Cracked bearing plate in stringer supporting railroad tracks at bridge aged approximately 85 years.

carrying vehicular traffic (Fig. 4.62) fatigue is a secondary effect, precipitated by severe corrosion.

- New types of steel, higher stress levels, and more demanding performance requirements (as in prestressed bridges) increase the structural sensitivity to corrosion.

Phenomena such as stress corrosion (see preceding section on fatigue), pitting corrosion, and embrittlement are becoming gradually understood and addressed by design specifications.

The use of high-strength steel in suspension bridge cables provided important lessons in the effect of corrosion on critical and vulnerable structural elements.

The original cables at Pont de Tancarville were neither galvanized nor wrapped (Virlogeux, 1999). They consisted of helical lock-coil strands, one of which ruptured due to corrosion. The new cables are not wrapped either—again on the argument that wrapping would only trap moisture that penetrates inevitably. Strands are galvanized and individually protected.

When painting and cleaning have been neglected, corrosion is a maintenance failure. Section 11.4 argues that painting is better managed as a rehabilitation project than as a continuing maintenance task. The FHWA concurs: Bridge painting has become eligible for federal funding as a capital expenditure.

Cleaning is among the costliest but most effective maintenance tasks. In the absence of statistically useful data on cleaning and washing of bridge decks, drainage underdeck

Figure 4.62 Corrosion and fatigue contributing to crack in stringer supporting vehicular traffic in bridge as in Fig. 4.61.

surfaces, tops of piers, and so on, management cannot conduct a rigorous cost–benefit analysis and justify maintenance expenditures.

The task is further complicated by environmental restrictions. The requirement for full containment of lead paint, for instance, radically changed the economics of painting. The cost of painting of steel bridges in New York City escalated from U.S. $5/m² of steel in 1990 to $200/m² in 2000. The Clean Water Act prohibiting any work over water without containment has a similar effect on the cost of bridge repair and reconstruction. Such management constraints eventually change not only maintenance but also design.

The direct effect of salt water is extremely powerful, particularly in tidal "wet" zones. If possible, noncorroding materials should be considered as alternatives to concrete and steel in such areas. Timber is a popular alternative to steel and concrete piles; however, it is vulnerable to rotting and marine borer attacks (Fig. 4.63). Plastic and carbon fiber piles or wrappings have been used.

On both federal and local levels numerous salt substitutes and technologies are under development and testing. Seasonal washing and regular cleaning are moderately effective albeit relatively expensive (Section 11.4).

Corrosion is occasionally exacerbated by the galvanic action between different metals. The Manhattan Bridge anchorage saddle (Fig. 4.64) was encased in bronze that induced corrosion in the cable wires.

Corrosion-resistant weathering steel (Fig. 4.65) should not be considered corrosion free, particularly if it remains continually wet.

NCHRP Report 333 (1990) by Kulicki et al. provided guidelines for the evaluation of corrosion in steel bridges. Mitigating the corrosion of reinforcement in concrete bridges is discussed, for instance, by Mallet (1994). Vesikari (1988) and Clifton (1991) develop models for estimating the remaining useful life of concrete as a function of various deterioration agents.

Concrete

Even when the original cracking and resulting permeability of steel-reinforced concrete is caused by freeze–thaw effects, corrosion of the rebars causes the ultimate structural failure. Oxidizing steel expands up to several times (some reports cite 8 times) its original

Figure 4.63   Timber piles in wet zone.

Figure 4.64   Encasement of anchorage saddle.

Figure 4.65   Weathering steel bridge.

volume. The surrounding concrete splits, as in Fig. 4.66. The particularly hazardous spalling of bridge deck undersides is illustrated in Example 16.

Design must create maintainable concrete by providing adequate protection against water and other corrosive agents (Section 4.3.3). Considerations about the cost-effectiveness of waterproofing are discussed in Section 4.3.3. In order to impede corrosion, waterproofing must be essentially impregnable. Figure 4.66 shows that corrosion causes spalling even when concrete cover is abundant and stresses are negligible if water can penetrate.

Stay-in-place deck forms have been used against underdeck spalling (Fig. 4.67). They raise the objection that moisture may be retained and early deterioration can remain concealed. Limited inspection records in New York City do not show an accelerated deterioration of such decks.

To mitigate spalling, bridge decks have been designed without top reinforcement. High-performance concrete and corrosion-inhibiting admixtures are used in sensitive structures. Increasingly carbon fiber reinforcement is considered as a viable alternative. The extra cost is justified by the obtained life extension.

NRC (1993a) discusses the protection, repair, and rehabilitation of concrete bridges against corrosion.

Prestressed Concrete

Prestressed (pre- and post-) tensioned bridges are a twentieth-century advance in design and construction. With their potential and limitations still being explored, they must conform to higher construction, material, and performance criteria. Smaller deflections and lower corrosion levels can be significant. Menn (1986) provides a comprehensive

Figure 4.66 Splitting of concrete caused by corrosion with possible contribution of freeze–thaw effects.

Figure 4.67 Stay-in-place deck forms and epoxy-coated rebars.

reference on the subject. Prestressed structures are particularly demanding to the manager, because they can fail suddenly and with little if any warning. Prestressing tendons are both fracture critical and inaccessible. The construction of prestressed bridges was temporarily stayed in the United Kingdom during the early 1990s until a reliable means of tendon inspection could be demonstrated. The ban has since been lifted in part as a result of advances in nondestructive testing and evaluation (Chapter 13). The number of prestressed bridges is growing and their spans are getting longer (NCHRP Report 517, 2004).

The primary concerns are prestress losses caused by combinations of geometric inaccuracies, corrosion, shrinkage, creep, relaxation, and anchorage failures. At the viaduct shown in Fig. 4.68, the curing of the segments is thought to have produced an unintended curvature, leading to indications of prestressing loss (such as crumbling at the cold joints between the segments). The bridge was retensioned with external tendons (Fig. 4.69). Those tendons were subsequently found to suffer from corrosion.

Figure 4.68 Posttensioned segmental box-girder viaduct.

Figure 4.69 External prestressing tendons of viaduct in Fig. 4.68.

Tendon corrosion was cited at the collapse of the pedestrian bridge in Concord, North Carolina, over Route US 29 on May 21, 2000.

The reports by the U.K. Highway Agency/Labortoire Central dews Ponts et Chaussée (LCPC, 1999), and NCHRP Report 496 (2003) provide an overview of typical problems and some solutions. The burden of "vulnerability awareness" is increased for all engineers involved with prestressed structures, including designers, constructors, inspectors, and owners.

4.3.2   Construction

AASHTO (2002) provides separate design and construction specifications. AASHTO 1998a prescribe construction load allowances but refer to the LRFD construction specifications.

The Occupational Safety and Health Administration (OSHA), U.S. Department of Labor, regulates construction safety.

Accidents during or directly upon completion of construction are often caused by departures from design specifications or from the approved construction sequence. Concrete arch and slab formworks have collapsed because of delays in concrete pours.

Demolitions are highly sensitive, particularly when conducted above or under traffic. Figure 4.70 shows a cantilever floor beam which collapsed during the demolition, because the rivets were removed from the tension zone of the connection to the pier. Figure 4.71 shows the collapse of concrete bridge deck segments during demolition. The original structure had been a rigid frame; however, the deck was cut into simply supported beams in order to maintain traffic both underneath and on top of the partially removed structure.

The collapse of the suspended walkways of the Hyatt Regency in Kansas City on July 17, 1981 (Levy and Salvadori, 1992; Wearne, 1999), was primarily caused by a construction error: Suspension rods were interrupted at each floor rather than run continuously. Design, however, was already susceptible: Floor beams consisting of channel sections transferred loads to the rods through unstiffened flanges rather than through the webs.

Construction determines the life of structures and yet it is not systematically taught. Some knowledge can be derived from errors if they gain notoriety, but most of the information is presented in a popular media version. ENR (May 24, 2004, p. 12) reports the collapse of a 153-ft (47-m) girder during construction at Interstate 70 west near Denver on May 15, 2004. Inadequate temporary bracing appears to have caused instability; however, all assessments are speculative. The lack of temporary lateral bracing appears to have caused the collapse of a pedestrian bridge shown in Fig. 4.72.

The collapse of the 30-m-long covered pedestrian passageway at the Charles de Gaulle International Airport in Paris (Fig. 3.11) occurred 30 months after completion (ENR, Sept. 27, 2004, p. 16). *The New York Times* (May 25, 2004, p. A10) commented: "Architects and engineers asked whether the fashion for increasingly innovative buildings has strained the limits of what a group of specialized construction companies can safely build together."

Clough (1986) reported "a disproportionate number of business failures" among construction contractors. The losses in terms of direct and user costs incurred during

Figure 4.70   Steel cantilever floor beam, collapsed during demolition.

construction delays are considerable. Clough (1986, p. 28) quotes Dun & Bradstreet on the following causes for such failures (listed in descending order of importance):

1. Incompetence
2. Lack of company expertise in sales, financing, and purchasing
3. Lack of managerial experience

Figure 4.71   Concrete bridge deck, collapsed during demolition by precutting.

Figure 4.72   Buckling failure during construction of pedestrian bridge. Courtesy of S. Summerville, Weidlinger Assoc.

4. Lack of experience in the firm's line of work

5. Fraud, neglect, disaster, or unknown

The first four causes represent incompetence of one type or another, and the fifth includes poor control (e.g., incompetence on the part of the owner) and random events. Clough (1986, Chapter 5) describes pre- and postqualifying for contractors, intended to ensure their competence. The rigorous implementation of these procedures can be potentially onerous and litigious, requiring thorough familiarity and expert supervision on the part of the owner.

4.3.3   Maintainability, Repairability, and Inspectability

The terms *maintability, repairability,* and *inspectability* are awkward, at least in part, because the concepts they represent are still struggling for recognition. The more conventional *durability* is perceived as the ability of structures to resist deterioration. Maintainability, inspectability, and repairability are qualities either enhancing structural durability or compensating for its deficiencies. In contrast to the infrastructure, industry perceives maintainability as the ability to replace before failure and without interruption of the process. Inspection, for instance in electronics, may consist of the statistical analysis of failure data. Thus, definitions intended for industrial production and equipment have significant yet limited applications to bridges. MIL-Hdbk-472 (1984) defined maintainability as "the ability of an item to be retained in or restored to specific conditions by appropriate maintenance action at prescribed intervals."

Raheja (1991, p. 94) expanded the definition to "the science of minimizing the need for maintenance and minimizing the downtime if maintenance action is necessary." He stated (p. 93): "Maintainability is mainly a design function."

Drawing on experience in aviation, Kraus (in Ireson and Coombs, 1988, Chapter 15) traced the origins of maintainability to the reliability research of the 1960s. The author stressed the following important distinction:

> Maintainability is a characteristic of design which, when achieved, contributes to fast, easy maintenance at the lowest lifecycle cost. The word "maintenance" is often incorrectly interchanged with maintainability. Maintainability is design-related and must be incorporated into the design of a system or equipment during the conceptual design, definition, and full-scale development phases of the lifecycle. Maintainability expenditures are heaviest during these periods and taper off to sustaining engineering once the system or equipment is in the field. Maintenance, on the other hand, is operation-related since it refers to those activities undertaken after a system is in the field to keep it operational or to restore it to operational condition after a failure.

Design achieves maintainability when it does not rely on future maintenance and inspections. Boller (1885) recommended adding 0.25 in. (6.25 mm) to the thickness of steel sections to compensate for poor painting. More generally, Waddell (1916, p. 1546) concluded: "It is exceedingly bad practice to skin the life out of a bridge in order to save metal."

The maintenance of sensitive bridge elements can consist of routine tasks, such as cleaning, oiling, bolt tightening, and so on, and/or timely replacement (as in the case of wearing surfaces, joints, scuppers, and paint). Maintenance-intensive bridge details are often designed independently of the structures and eventually selected according to availability. Section 4.2.3 draws a parallel between the structural discontinuities created by articulated connections and devices and the discontinuous nature of their existence, beginning with analysis and design and moving to construction and inspection. Waddell (1916) cautioned against the latter discontinuity: "A consulting engineer should never trust the detailing of a bridge to the manufacturing company, but should prepare complete plans therefore in his own office."

Moreover, installation should be inspected and the required maintenance should be specified.

The demands for maintainability and inspectability increase as the levels of structural redundancy and strength reserve are reduced. The failure of the Silver Bridge in 1967 (Section 3.5) emphasized the vulnerability of fracture-critical details (two-eye-bar suspension chains) as well as the inability to inspect them.

The Mianus River Bridge in Connecticut collapsed again due to the failure of a fracture-critical detail (FHWA, 1986). That detail was a pin-and-hanger assembly simply supporting a skewed suspended span. It did not fracture, however. Rather, rust and debris accumulated behind the pin cap and pushed it out. Unrestrained, the hanger slipped off the pin. The ensuing forensic investigation and litigation revealed deficient maintenance and inspection (see Section 14.2). The nonredundant design, standard at the time of construction, although exculpated, was discredited.

Figure 4.73 shows a steel column corroded to the point of failure. Maintenance and inspection have been clearly deficient; however, redundant design prevented the collapse of the structure.

When in 1973 a span of the West Side Highway in New York City collapsed at West 57th Street, the bridge had served since 1929 without systematic maintenance. Corrosion attacked the connections of floor beams to girders and stringers to floor beams. The ensuing investigation, however, revealed that of all the bridge spans, covering a distance of 30 city blocks, only the failed one had a particularly vulnerable bearing detail.

The maintainability of bridges can be viewed as inversely proportional to the number of their expansion joints (Section 4.2.3). As a result, joints are the subject of numerous studies, such as NCHRP Report 141 (1989) by Burke and NCHRP Synthesis 319 (2003) by Purvis. NCHRP Report 319 identifies the "cushion" joint (Fig. 4.74) as the least performing of the widely used expansion joints. Modular joints (Figs. 4.28 and 4.29) are discussed in NCHRP Report 467 (2002). A survey of joint performance (Fig. 4.75) can suggest that "the only good joint is no joint." Accordingly, jointless bridges are recommended whenever possible. Miller et al. (2004) provide guidance in eliminating the discontinuities between simple-span precast concrete girders. For shorter spans, integral and semi-integral abutment bridges (Burke, 1993, 1994) offer superior seismic performance and maintainability. For spans up to 400 m (1200 ft) continuity can be maintained with appropriate bearings. For shorter spans with maximum expansion–contraction of up to 2 in. (50 mm) "plug" joints may provide a continuous elastic surface easier to repair.

Decks literally and figuratively interface with the "customers," (e.g., the vehicles). Decks, and particularly their wearing surfaces, are also the common elements between bridge and roadway networks. The vague maintenance becomes quantifiable when it determines the effective methods of cleaning, deicing, and repairing decks, their joints,

Figure 4.73   Failed 45-ft (14-m) steel column.

scuppers, waterproofing, and wearing surface. The importance of these tasks can be deduced by deck and superstructure inspection records (Example 12) as well as by the performance of structures where they are absent. The open steel gratings (Fig. 4.21) appear to eliminate the roadway wearing surface along with all of its maintenance demands; however, they have a short life, are difficult to replace, and amplify the deterioration of the primary structure.

Wearing Surface and Waterproofing

The corrosion of bridge deck reinforcement is mitigated by diverging methods as is that in suspension bridge cables. Sources in Western Europe (and the state of Florida) argue that protective coatings of the rebars cannot compensate for adequate deck waterproofing.

Figure 4.74 Failed cushion joint and plug joint replacement.

By the same logic, the designers of the Williamsburg Bridge cables (preceding paragraph) did not use galvanized wires. In contrast, many of the U.S. bridges use epoxy-coated rebars in combination with thin-bonded deck overlays without separate waterproofing.

Nongalvanized high-strength wires are no longer used. Similarly, unwrapped cable strands are increasingly rare. The performance of bridge decks using different corrosion protection is much harder to compare because of the different environmental, construction, and service conditions. Furthermore, the significant data accumulate over decades and are not always rigorously documented according to a unified standard. The maintainability and longevity of wearing surfaces of both roadways and vehicular bridges determine the life of these assets. The two widespread wearing surface options are thin bonded overlays and asphalt overlay with waterproofing.

Inspection records in New York City indicate that the average useful life of bridge decks decreased from 50 to 25 years, while waterproofing was eliminated. Waterproofing membranes have been used on older bridges in the United States and are still in use elsewhere along with asphalt wearing surface. Resurfacing usually uncovers water under the top asphalt course. The asphalt can be porous ("open grid"), allowing water to penetrate to the membrane and then evaporate again rather than remain trapped under the surface.

Waterproofing is effective only if it is continuous over the entire bridge and sidewalk. Alternatively, recent monodecks have thin bonded overlays with waterproofing additives. Such decks are reinforced with epoxy-coated, galvanized, and even stainless rebars. Cathodic protection is an effective but rarely applied alternative. Any defects in the coatings must be avoided, because they become the focal points of accelerated corrosion. Recent

Figure 4.75 Failed armored compression joint.

monodecks fail primarily in a fatigue-type crushing mode at the wearing surface, as in Fig. 4.76.

The benefits of waterproofing have been under discussion since the advent of roadway bridges. Chapter XLIII of Waddell (1921), contributed by J. B. W. Gardiner of New York City, is entitled "The Economics of Waterproofing." It makes no categorical pronouncement either for or against waterproofing but stresses its importance and the high sensitivity to good quality.

The FHWA has devoted considerable effort to the development and use of thin bonded overlays. The basic advantages and disadvantages of the two types of surfaces can be summarized as follows:

Bituminous wearing surface with waterproofing

+	−
Relatively easy to apply, remove, and replace.	Asphalt accumulates during each resurfacing, the dead load increases, and the curb vanishes (Figs. 4.40 and 4.41).
With good waterproofing extends the life of concrete decks.	Water accumulates under the waterproofing.
Asphalt surface may act as a damper against live-load impact.	Traction declines with use.
Traction can be improved by concrete admixtures.	

Open-graded friction course (OGFC) mixes provide porous surfaces, allowing water to penetrate to the top of the waterproofing but also to evaporate freely (NCHRP Synthesis 284, 2000).

Thin bonded overlay (monodeck)

+	−
Good rideability.	Fatigue type failures (Figs. 4.14 and 4.15).
Reduced deck weight.	Hard to repair.
Requires no resurfacing.	May require more frequent deck replacement.

Wide temperature variations and the associated use of deicing salts are particularly damaging to thin bonded overlays. Figure 4.76 illustrates fatigue-type failures in concrete overlays. Figure 4.77a shows waterproofing for an asphalt overlay. Figure 4.77b shows an 'open grid' asphalt overlay during application. Figure 4.77c shows a similar wearing surface ready for opening to traffic.

Microsurfacing combines some features of thin bonded overlays and asphalt surfaces. It is a thin (3.125-mm) bituminous course applied on top of the structural deck.

FHWA (2002b) provides guidelines for detection, analysis, and treatment of material-related distress in concrete pavements. The extensive information available from research on roadways has a limited validity on bridges, because neither the physical characteristics nor the cost–benefit considerations apply directly.

4.4 MAINTENANCE (SECTION 11.4)

AASHTO (1999a, p. 15) argued for *preventive* maintenance, represented by the adage "a stitch in time saves nine." in preference to the *reactive* maintenance, expressed by the familiar "if it ain't broke, don't fix it." The stated objective of preventive maintenance (PM) to prolong the duration of existing conditions renders it inferior to the tasks seeking tangible improvements, such as management and design. Management tends to underestimate the value of maintenance, whereas design overly relies on it. Both *maintenance deferral* (Section 4.1.5) and poor *maintainability* (Section 4.3.3) encumber the "assets" (Fig. 4.4b) with a deficit, which remains undefined until it becomes insurmountable.

(a)

(b)

Figure 4.76   Fatigue-type wearing surface failures: (a) in thin bonded overlay; (b) in steel grating overlay.

(c)

Figure 4.76   Fatigue-type wearing surface failures: (c) in substrate.

Within their domain, maintenance tasks suffer from the following typical vulnerabilities:

Performing a maintenance task without achieving the intended effect

Selecting ineffective maintenance tasks

Failing to determine and document the quality and quantity of maintenance

Failing (jointly with inspection) to model the correlation between maintenance and structural condition

There is no optimal or codified rate of deterioration and hence no definitive maintenance. The network-specific maintenance manuals (Example 23) in Bieniek et al. (1989) and Vaicaitis et al. (1999) have surveyed existing practices and records of past performance. The latter, however, are highly uncertain.

NCHRP Report 285 (1986) made the following observation (p. 28): "Highway agency bridge expenditures are only 3% to 5% of those recorded on pavements. Consequently, maintenance optimizations consolidate bridge treatments into broad activity categories. No record is ever made of specific treatments to specific elements and the required cause and effect data base is never created."

Although maintenance is considered decisive in prolonging bridge useful life, it has yet to become quantified in bridge inventories as NCHRP Report 285 recommended. Even when annual maintenance expenditures are documented, their benefits in retarding bridge deterioration remain unknown. Thus maintenance has no track record.

(a)

(b)

Figure 4.77   (a) Waterproofing for asphalt overlay, Pont Neuf. (b) Open-grid asphalt overlay on orthotropic deck, Tsing Ma Bridge.

(*c*)

Figure 4.77 (*c*) Wearing surface on Great Belt Bridge.

Most new structures are automatically vulnerable to maintenance neglect, because it is often assumed that they are built to resist deterioration better than their predecessors. That assumption may indeed pertain under the still experimental zero-maintenance policy (FHWA, 2005b) but is not the general case. In contrast, rehabilitation plans for old structures usually include recommendations for future scheduled maintenance.

4.5 INSPECTION (SECTION 14.5)

The discussion on error and experience in Section 3.1 summarizes the fundamental vulnerability of field investigations. The findings are difficult to obtain and have no uniquely defined interpretation. In the absence of actual failures, alarmist assessments are ignored;

following disasters, they are found lacking. The all-inclusive skepticism of Murphy's law (Section 3.3) aptly expresses the resulting frustration.

As a process, inspections fail by misdiagnosing or overlooking departures of real conditions from a state called "as built" although it is actually ideal. The failure may be caused by the inspection design (e.g., unrealistic expectations), the execution, or (typically) both. The safety of the inspection personnel is a vulnerability of all field operations (Section 4.6). The inspection product (e.g., the reported information) can be improperly presented or processed. All of these vulnerabilities are usually present to a degree and contribute to the outcome. The condition discussed in Example 16 had been under inspection at the time of the accident. The subsequent investigation found that inspection equipment had been inadequate, the knowledge of the potential hazard had been insufficient, and the processing of the information had been delayed.

Inspections are generously funded after structural failures. Once the level of risk is perceived as acceptable, management develops an interest in minimizing inspection costs. Biennial inspections were a national mandate in 1980, but in 2003 an improved and better studied bridge network might allow the relaxing of that rule to a recommendation. An alternative view holds that inspections as well as maintenance should be regular, as are the seasonal climate changes and traffic. Both positions can be satisfied by regular inspections with a scope, varied according to the structural condition, always including the task to identify potential hazards. Nonetheless, inspections cannot guarantee safety in the absence of maintenance, just as maintenance cannot improve design and construction.

While bridge management cannot be optimized in isolation from other social activities, the frequencies and scope of inspections can adapt to the needs of the costlier bridge-related tasks, for example, design, construction, maintenance, operation, and ultimately management. The numerous inspection types described in Chapter 14 are designed to match the diverse condition evaluations of Chapter 10. The process can be reversed, however. If structural assessment cannot meet the demands of bridge management, revisions of maintenance, construction, and design practices might be appropriate. Consequently a "shoot the messenger" reception is among the inspector's occupational hazards.

Inspectors can be "rotated" at inspection sites, adding a touch of randomness to the mix of assessment uncertainties. Unfortunately, bias is not excluded, as the old reports increasingly influence the new. (Withholding previous reports is considered cost ineffective.)

Quantification of the findings is repeatedly stressed but hard to enforce. Consistency with measurements promotes objectivity but does not reduce the demand for subjective judgment. The wrong experience is also possible. Meaningless measurements are as likely as is their misinterpretation. Loss of steel cross section to corrosion is typically exaggerated during visual inspections unless exactly measured. Even if accurately quantified, however, it does not reveal the risk of cracking due to stress concentration (see Fig. 4.62). The correlation between hollow-sounding concrete, the corrosion rate in reinforcement, and deck spalling is hard to quantify. It is safe to assume that quantification will remain incomplete to a degree, allowing sufficient need for subjective (qualitative) judgment.

The deduction of a reliable database from inherently unreliable inspections of uncertain conditions shares the classical difficulties encountered by reasoning since antiquity

(Section 1.8). Seneca's admonition that "knowledge is not justified unless shared with others" must be reconciled with Descartes' mistrust of diverse opinions. Inspections invariably report *deterioration*. As the phenomenon is universal, the term is ubiquitous. The present text uses it 186 times. The adjectives *minor, serious,* and *advanced* introduce a qualitative gradation but no quantifiable distinction and are often interchanged. The actions expected to remedy the observed conditions are the equally vague "repair as necessary" and "restore to as new."

FHWA (2001b) investigated whether the opinions obtained by visual inspections contain any knowledge worth sharing. Ten inspection tasks were conducted at seven "precalibrated" bridges by 49 inspectors from 25 states. The results of the field tests are presented along with a literature review and a survey of existing practices. The findings include the following:

> Ninety-five percent of the condition ratings (Section 10.4, Appendix 34) assigned during *routine* inspections spread over five contiguous condition levels with 68% varying within one level.

> *In-depth* inspections were found unlikely to correctly identify significant specific defects, such as welding cracks in steel primary members. The use of the terms suggests that one owner's *routine* inspection is another's in-depth inspection (Section 14.3).

> The inability to recognize structurally significant features, such as support condition, bridge skew, fracture-critical members, and fatigue-sensitive details, was common.

> Factors contributing to the deficiencies included:

> • Fear of traffic
> • (Lack of) formal bridge inspection training
> • Reported structural maintenance level
> • Accessibility
> • Visibility
> • Time constraints
> • Wind

Significantly, licensed professional engineers were less likely to misdiagnose a bridge condition. The findings indicated a shortage of both expert systems and experts. Recommendations included better quantification of the findings, possibly by an expanded use of nondestructive testing and evaluation (NDT&E Section 15.1) methods.

4.5.1 Structural Diagnostics and Health Monitoring (Sections 10.6 and 15.1)

For the first time in engineering practice, NDT&E methods (Section 15.1) are measuring structural response with high precision in real time. A new branch of condition assessment has resulted. Structural health monitoring (SHM) is a luxury as well as a challenge to the bridge management process (e.g., the optimization of costs and benefits) and to the product (e.g., the performance of the structures) for the following reasons:

- Structural inspectability does not necessarily improve with the acquisition of new data. (Data can be unreliable or meaningless.)
- An inspectable structure does not guarantee a superior service. (A structure should not be selected only because its performance can be monitored.)
- Data acquisition incurs continuing maintenance and expertise costs. (An abandoned data acquisition system is a conspicuous management failure.)

With the data provided by NDT&E, condition assessment has evolved to fairly advanced diagnostics. Diagnostics perform system identification; i.e., the parameters of the structural model must be determined from the measured behavior. Both the model and the measurements are vulnerable. Bucher and Pham (2005) wrote:

> System identification requires the solution of inverse problems, which unfortunately leads to rather ill-conditioned mathematical formulations, e.g. linear equations. This tends to become more pronounced as the number of parameters to be identified increases. The consequences of the ill-conditionedness is that any small errors in the measurements will be considerably amplified, so that large errors in the identified parameter values will occur. When using optimization methods, eventually several local minima may be found which do not represent the real system parameters.

The authors proposed effective methods of overcoming the quantitative numerical difficulties in parameter identification.

Models may not be truly representative in qualitative ways as well. Linear elastic models do not represent yielding, strength models do not capture instability, and dynamic solutions do not adjust the structural stiffness as it changes (Appendix 26). It is particularly difficult to enable knowledge-based and expert systems to adjust not only the model parameters but also the models.

4.6 OPERATION

The term *operation* covers activities conducted at the site of the structures (e.g., field work). So defined, operations are vulnerable to poor quality and unsafe practices.

4.6.1 Movable Bridges

Movable bridges as well as any structures relying on human operations (such as active collision protection systems or traffic-monitoring centers) are susceptible to performance errors. Figure 4.78 shows the ruptured cables of the auxiliary counterweight of the vertical-lift bridge in Fig. 4.79. The incident occurred during manual operation.

The main vulnerability of operations is the management of personnel (Section 4.1.4). Movable bridges must be staffed with operators for regular and emergency openings at the request of the authorities in charge of navigation (such as the U.S. Coast Guard). Emergency communication provisions must be maintained between the personnel and between the responsible agencies.

Beyond their structural complexity (Fig. 4.80, Section 14.3) movable bridges require highly specialized mechanical and electrical maintenance. The complex needs add to the life-cycle costs. In order to eliminate the dependence on sea-faring navigation, the mov-

Figure 4.78 Ruptured auxiliary counterweight cable of vertical-lift bridge.

able bridge shown in Fig. 4.80*b* will be replaced by a longer fixed span at a higher elevation.

4.6.2 Safety of Field Operations

All field operations, including maintenance and inspection, are associated with a risk to the personnel. Despite OSHA regulations, construction accidents occur. Maintenance and inspections are similarly regulated. The responsible owner may initiate more stringent standard operating procedures (SOPs). The hazards most significant to personnel with routine presence on bridges are related to traffic; however, structural conditions are also

Figure 4.79 Vertical-lift bridge.

a source (Section 14.4). After one of the prestressing tendons shown in Fig. 4.69 broke (as a result of corrosion), inspections required special protection from further fractures.

4.7 ANTICIPATION OF VULNERABILITIES

A comprehensive checklist of possible vulnerabilities in the bridge-related tasks and products would have been a useful tool; however, it cannot be compiled. An incomplete one entails all the dangers of ignorance. Knowledge is by definition incomplete and that is why engineering remains partly empirical. Experience produces incomplete lists of "usual suspects," specifications undergo revisions, and forecasts are updated. Analysis

(a)

Figure 4.80 (a) Bascule bridge.

Figure 4.80 (b) Bascule bridge to be replaced by fixed bridge at higher elevation.

289

assumes, design prescribes, and construction approximates. Maintenance, inspections, and operations are even less definite.

In his notebooks, Leonardo da Vinci speculated that "the unexpected always happens." The message of Murphy's law (Section 3.3) is similarly inconclusive.

Vulnerabilities of disparate uncertainty interact over the structural life cycles. Their cumulative effects produce *critically coincidental* rather than *incidentally critical* causes of failures. In order to anticipate their likelihood and their consequences, engineering employs the diverse tools of probabilistic and deterministic modeling.

Chapter 5

Probability of Failure

"The optimist and the pessimist dispute only that which is not."

P. Valéry (1941, p. 105)

Failures of civil engineering structures are relatively rare, opening a wide field for the speculations of optimists and pessimists. This is primarily due to the rigorous application of the following two principles:

- Structures (particularly those designated for public use) are designed and built to meet demands *much* greater than the aniticipated ones (Chapter 4.2.2).
- Actual and potential causes of failures (e.g., the vulnerabilities discussed in Chapter 4) are systematically identified and eliminated from practice.

An outstanding example is the collapse of the World Trade Center (WTC I and WTC II) in New York City on September 11, 2001 (Figs. 4.3 and 4.53). The Twin Towers were clearly assaulted far beyond the design specifications considered conservative at the time of their construction (1968–1974). The engineering community immediately began investigating the failure modes in order to introduce measures for precluding failures under such extreme conditions.

As a result of theoretical, physical, and operational uncertainties (Sections 1.8 and 1.9), the estimates of the safety of engineering structures are ambiguous. In the deterministic universe where engineering originated, ambiguity was a failure in itself. One hundred and fifty years later it is classified as fuzziness (e.g., a form of uncertainty).

The uncertainties to be resolved by managers in the operational sphere and by designers in the physical domain are both different and differently perceived. It has been suggested that the differences may have become institutionalized. The estimates Feynman (1999) obtained for a failure likelihood of the Space Shuttle *Challenger* ranged from 0.01 from designers to 0.00001 from managers. Vick (2002) attributed this 1000-fold opinion gap to the different attitudes of designers and managers. Since both of the surveyed groups consisted of highly competent NASA engineers, Descartes (Section 1.8) might have concluded that they were not considering the same failure by the same method.

Feynman (1999, p. 157) found that the mathematical model used to calculate the erosion of the *Challenger*'s solid rocket booster O-rings was "based not on physical understanding but on empirical curve fitting." He further criticized NASA's notion of safety factors by drawing an analogy to bridges (p. 156). He argued that if a beam cracks "one third of the way" under a load it was designed to carry, the "safety factor is not 2/3, it is none at all." Management could interpret this design failure as a "close call."

The possibility of a failure that can be neither eliminated nor fully predicted constitutes a risk. The uncertainty of risk assessment necessitates that its professional treatment be codified.

5.1   RISK ASSESSMENT

Risk assessment is expected to maintain relatively well-defined present actions and hypothetical future consequences in a desired equilibrium. It is therefore urgently demanded and relentlessly disputed.

Taylor and Van Marcke (2002) evaluated acceptable risk for lifelines and natural hazards. Dowrick (1977) defined seismic risk by the following product:

$$\text{Risk} = \text{hazard} \times (\text{vulnerability}) \, (\text{value}) \tag{5.1}$$

Vick (2002, p. 123) used the more general current definition:

$$\text{Risk} = (\text{probability of failure}) \times (\text{consequences of failure}) \tag{5.2}$$

The author described risk analysis as the following sequence (p. 126):

$$\text{Initiator} \rightarrow \text{response} \rightarrow \text{consequences}$$

Consequences, effects, or penalties can be quantified as costs; however, social constraints and imperatives can influence the prioritization qualitatively. Seismologists, meteorologists, statisticians, and others determine the types and likelihood of initiators or extreme events. Engineers estimate the structural consequences of such events. That task poses a dilemma: Failure of any likelihood is neither fully acceptable nor entirely avoidable.

FHWA (2005d, p. 13) reported a method of calculating the likelihood L of a risk event adopted by transportation asset managers in England as follows:

$$L(\text{risk event}) = L(\text{cause}) \times L(\text{defect}) \times L(\text{exposure}) \times L(\text{effect}) \tag{5.2a}$$

Not all failures result from underestimated likelihood. A structure may have been designed to sustain damage during the maximum credible earthquake with a return period of 2000 years without loss of life. After an event of such magnitude, if designed to specifications, that structure may have to be demolished or substantially repaired. The community and the media question any level of damage, just as they question the excessive costs of construction. At best, engineering management can provide a range of probable outcomes and credible scenarios.

The constitutive likelihoods of Eq. 5.2a can be rated as certain (1), high (0.7–0.99), medium (0.3–0.69), and low (0.0–0.29). Thus, the essentially probabilistic risk indicator is the product of four expert opinions. Numerical values are assigned from look-up tables in a highway agency manual, for example, as follows:

$$L(\text{risk event}) = 0.85 \times 0.50 \times 0.15 \times 0.85 = 0.054 \tag{5.3}$$

Hence, unconditional belief in "objective" probability is discouraged. De Finetti (1974, p. x) opened his fundamental text on the subject with the following assertion:

"PROBABILITY DOES NOT EXIST" (De Finetti's capitalization).

Fittingly for a nonexistent entity, probability has numerous definitions. Vick (2002, p. 2) lists five: classical, subjective, logical, personalistic, and frequency. They emanate from the objective and subjective interpretations of uncertainty, discussed in Sections 1.8 and 1.9 and Appendices 2, 4, and 9. The author stresses the role of the *degree of belief* in any adopted, assumed, or calculated probability. Subjective and objective probabilities are combined in fuzzy-set theory.

In some cases *error* appears to overlap with *uncertainty*. For economic analysis, DeGramo et al. (1979, p. 146) equate *risk* and *uncertainty.*

5.2 STRUCTURAL RELIABILITY

According to Ang and Tang (1975, p. 12), "the most desirable solution is one that is optimal in the sense of minimum cost and/or maximum benefits."

Chapter 2 discusses the quest for desirable transportation solutions through maximizing benefits, minimizing penalties, and setting priorities under negotiable and rigid constraints. The relative value of any outcome resulting from such an informal process can be perceived as a compromise, which is the opposite of the objective or rigorous optimization as it was originally conceived. This type of optimization can evaluate options assuming a formalized procedure, and sets of scenarios, or vulnerabilities (Chapter 4).

The structural property to be optimized is no longer its strength but a more generally described performance, for instance, a level of reliability (Section 4.2.1). It can be quantified by probabilistic methods and even negotiated, based on combinations of statistics, determinism, and "degrees of belief." The level of confidence in each type of assessment (e.g., based on physics or statistics) must also be evaluated. In probabilistic terms failure is the loss of reliability. Hence, reliability is the likelihood of survival. Uniform acceptable level of structural reliability (Appendix 7) is a "utility" or objective to be attained under the constraint of minimized life-cycle cost (Section 1.5, Appendix 34).

Cornell (in Freudenthal, 1972) described the need of a method "capable of carrying out the engineering balance of initial investment versus potential future benefits and losses, with appropriate respect for all the uncertainties in the situation." The author elaborated (p. 48):

> The benefits and losses may not be solely economic, but also professional and social (for example, the loss of prestige and lives in a serious collapse). The uncertainties may not be solely probabilistic (in the relative frequency sense), but also statistical (i.e. due to practical limitations upon the data available to estimate parameters) and professional (i.e. incomplete information or knowledge about the underlying model). At the same time the method should provide a mechanism for incorporating the information contained in professional judgment.

Bayesian statistical decision theory (Appendix 2) is applied to reliability-based design as follows (p. 48):

> If one takes this [Bayesian] position consistently, then it is *acceptable* to treat the unknown parameters of the distributions of these random variables as being, themselves, random variables. To treat distribution parameters as random variables is, of course, not

permitted within the framework of classical statistics. In the same vein, if useful to design, it may not be unreasonable to treat all uncertain factors as random, including, for example, the errors implicit in approximate stress analyses, the errors in professional simplifications in load distributions, or quality of an as yet unknown building contractor.

For systems typical in electronics, Barlow et al. (1965) defined up to five types of reliability. Hudson et al. (1997, p. 240) refer to the six generic patterns of life-cycle reliability for mechanical and electrical equipment (Appendix 7).

Different conditions govern the transportation infrastructure, however (see Section 4.1). For civil structures, Thoft-Christensen and Baker (1982) deemed extreme events more critical than deterioration. The authors developed structural reliability theory (Appendix 7) based on random models for physical, statistical, and model uncertainties. In order to reflect deterioration at different maintenance levels, Thoft-Christensen (in Das et al., 1999, pp. 15–25) estimated bridge reliability distributions by Monte Carlo simulation. The author defined structural reliability as follows:

General. The ability of a structure to fulfill its design purpose for some specified time.

Mathematical. The probability that a structure will not attain each specified limit state (ultimate or serviceability) during a specified reference period.

Methods of structural reliability analysis according to the Joint Committee on Structural Safety [sponsored by IABSE (International Association for Bridge and Structural Engineering), RILEM (Réunion Internationale des Laboratoires et Experts des Materiaux, Systémes de Constructions et Ouvrages), CIB (International Council for Research and Innovation in Building and Construction), and others] are grouped in three levels as follows:

Level 1. Safety Checking. Design methods in which appropriate degrees of structural reliability are provided on a structural element basis (occasionally on a structural basis) by the use of a number of partial safety factors, or partial coefficients, related to predefined characteristics or nominal values of the major structural and loading variables.

Level 2. Methods involving certain approximate iterative calculation procedures to obtain an approximation of the failure probability of a structure or structural system, generally requiring an idealization of a failure domain and often associated with a simplified representation of the joint probability distribution of the variables.

Level 3. Methods in which calculations are made to determine the "exact" probability of failure for a structure or structural component, making use of a full probabilistic description of the joint occurrence of the various quantities which affect the response of the structure and taking into account the true nature of the failure domain.

Specifications provide level 1 or level 2 design, depending on their methods. New structures are designed to perform under a range of probable loads. Structural response to actual and anticipated loads is a function of presumed strength, condition, and mode of operation. The probability of structural failure therefore depends on knowledge of the failure modes of the materials, the individual members, the representative system, and the loads.

Materials have been empirically shown to fail in the brittle or ductile modes (Section 4.3.1). Members fail according to their mode of operation, which is under combinations of axial loads, flexure, shear, and torsion. Thoft-Christensen and Baker (1982, p. 242) distinguish between mode types A (consistent) and B (not consistent) with design, for example, expected and unexpected modes of failure. Loss of strength, for instance, may be expected, whereas loss of stability may have been overlooked (Section 4.2 herein).

Independence and Correlation

As in structural stability, only the probability of failures by anticipated causes is estimated and their combinations are easily overlooked. The Bayes theorem (Appendix 4) assumes that the various failure causes are mutually exclusive and some prior knowledge is available. Formulas such as Eq. 5.2a reflect these assumptions, even though neither condition can be entirely satisfied. Melchers (1987, p. 52) defined his subject as "a reliability analysis for *imaginable* events." Thoft-Christensen and Baker (1982, Chapter 8.3) and Ghosn and Frangopol (in Frangopol, 1999b, Chapter 4) model the probability of failure in series systems by Ditlevsen's bounds. That method assumes knowledge of the safety margins of individual structural members *and* their correlation.

Not all loads and their combinations can be anticipated. The combination of mechanical damage and sustained high temperature proved particularly destructive at the World Trade Center Towers in 2001 (FEMA, 2002). Design had considered both of these hazards but independently. The stochastic approach to the correlation problem is presented by Thoft-Christensen and Baker (1982, Chapter 10). The Ferry Borges–Castanheta model is recommended for repeated independent loadings. The approach applies to known conditions. Design for unexpected and unprecedented conditions is deterministic; it relies on expert opinion.

Barlow et al. (1965, p. 9) caution that "the differences among the gamma, Weibull and log normal distribution functions (see Appendix 9) become significant only in the tails of the distribution, but actual observations are sparse in the tails because of limited sample sizes." The authors conclude: "Unfortunately, the choice of a failure distribution on the basis of these physical considerations is still largely an art."

More recently, Z. Bažant has proposed to graft a Weibull distribution tail on a Gaussian distribution core for the purpose of modeling the failure probability of concrete and other quasi-brittle materials. The heuristic aspect of reliability analysis implies gradually replacing intuitive selections with statistically based ones; however, data on the extreme cases (e.g., the tails) should remain rare.

Example 17 illustrates the joint contribution of deterministic and random factors to the pattern of structural condition ratings.

> #### Example 17. Rate of Change in Condition Ratings
>
> Examples 9 and 12 described the unquantifiable but significant contributions of phenomenological and subjective factors to the normal distribution of the bridges in large networks with respect to age and condition ratings. Deterioration histories were shown to reflect that distribution. The results are illustrated with the following simplified examples.
>
> Table E17.1 and Fig. E17.1 describe an idealized condition rating distribution of 750 bridges as follows.

Table E17.1   Idealized Condition Rating Distribution of 750 Bridges on Scale of 7 (New) to 1 (Failed) (NYS DOT)

Bridge Condition, $7 \geq R \geq 1$	Age (years)	Number of Bridges
$7 \rightarrow 6$	0–3	50
$6 \rightarrow 5$	4–9	100
$5 \rightarrow 4$	10–21	200
$4 \rightarrow 3$	22–33	200
$3 \rightarrow 2$	34–39	100
$2 \rightarrow 1$	40–42	50
$1 \rightarrow 7$	43–45	50

Note: Average condition $R_{av} = 4$ (Fig. E17.1).

If the seven rating groups of Table E17.1 deteriorate (piecewise) linearly over time and if bridges are distributed uniformly with respect to age, then 16.7 bridges decline to a lower rating annually. A steady state can be maintained if rehabilitation restores bridges at reaching the end of their useful life to a rating of 7 over three years. The cumulative distribution of the bridge population along the ordinate is the concave–convex, or flat-S, rating history shown in Fig. E17.1.

The delay of deterioration in the middle of the rating scale can be attributed to subjective and objective causes. Limited funds are usually allocated to maintenance by "triage" (also discussed in Sections 4.1 and 14.3 with respect to postdisaster

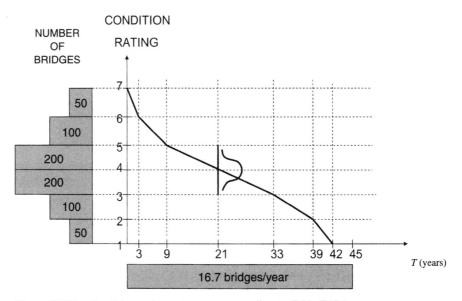

Figure E17.1   Condition rating pattern corresponding to Table E17.1.

inspections). Under extreme demands, resources are dedicated to the group of the population closer to the mean, assuming that the lower end is beyond repair and the upper end has no urgent needs. Such a strategy would account for the concave–convex rating history shown in Fig. E17.2.

Management must resolve whether a useful life of 42 years, the corresponding demand for a continual rehabilitation of 50 bridges, and the completion of 17 projects every year are optimal. The average deck area of the approximately 750 bridges managed by New York City is 15×10^6 ft^2 (1.4×10^6 m^2). The "average" bridge would have an area of 20,000 ft^2 (1860 m^2). At a cost of U.S. \$450/ft^2 (\$4840/m^2), direct annual rehabilitation costs under this regimen would amount to U.S. \$150 million. Example 9 illustrated the low reliability of such general estimates.

The pattern of Fig. E17.1 does not refer to the size of the stock, because not only the rating scale but also the life expectancy within each rating level is fixed. Two critical assumptions of the hypothetical equilibrium are overly simplistic:

- Rehabilitations do not restore bridges to as-built condition or to a rating of 7.

- The assumed deterioration history is an average. If the ratings at every age fit a normal distribution, then the worst, rather than the average, conditions would determine the rehabilitation needs.

The bridges in Table E17.2 are distributed uniformly along the rating scale and normally with respect to age. The convex–concave cumulative distribution is shown in Fig. E17.2.

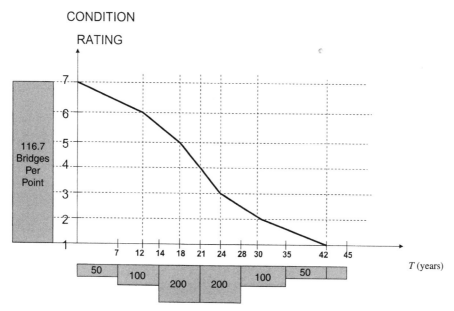

Figure E17.2 Condition rating pattern corresponding to Table E17.2.

Table E17.2 Idealized Condition Rating Distribution of 750 Bridges on Scale of 7 (New) to 1 (Failed) (NYS DOT)

Bridge Condition, $7 \geq R \geq 1$	Age (years)	Number of Bridges
7 → 6	0–12	116.7
6 → 5	13–18	116.7
5 → 4	19–21	116.7
4 → 3	22–24	116.7
3 → 2	24–30	116.7
2 → 1	30–42	116.7
1 → 7	43–45	50

Note: Average condition $R_{av} = 4$ (Fig. E17.2).

Combining the rating distribution of Fig. E17.1 and the age distribution of Fig. E17.2 inevitably produces the straight-line pattern of Fig. E17.3.

Closer to reality, the average deterioration pattern of the more than 750 bridges managed by New York City (Fig. E17.4) is almost linear, as in Fig. E17.3 (see also the second of the three cases in Fig. E12.2).

The findings were borne out independently. Example 9 points out that New York City had to close a number of bridges in the late 1980s. By committing annually up to U.S. $500 million to rehabilitation over 15 years, the city eliminated bridges rated less than 3 from its stock by 2005. The average bridge condition moved from 4.5 in 1992 to 5.0 in 2004. The potential of maintenance to affect the slope of deteri-

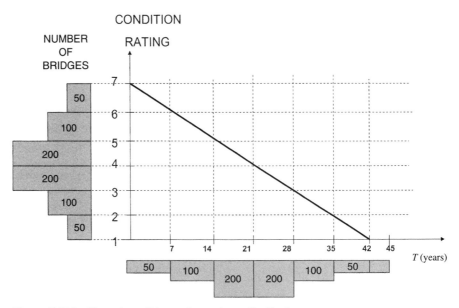

Figure E17.3 Normal condition rating and age distributions.

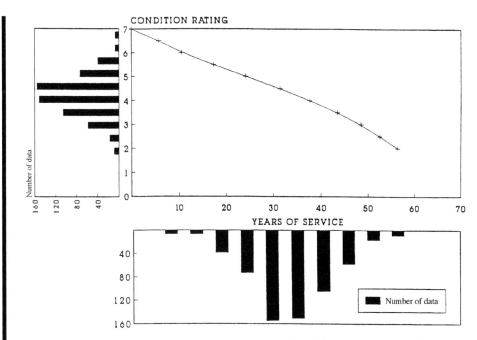

Figure E17.4 Average bridge condition and distribution with respect to age and condition ratings (Yanev and Chen, in TRR 1389, 1993).

oration, particularly in the high- and middle-condition-rating ranges, is explored in Examples 23 and 24.

The modeling of systems and networks can also use a touch of art (as in the preceding quote from Barlow et al., 1965). Cremona (2003) presents its current state comprehensively. Structures are modeled as combinations of parallel and series subsystems and members. The members of a parallel system function concurrently. If one fails, the load is redistributed among the others until the entire capacity is exhausted. The series system is a chain that fails at the weakest link (which can be any one). Statically determinate structures are series systems, whereas indeterminate ones are parallel systems or combinations. Connections must be modeled as members. The totality of a system's (structure or process) failure modes is a series system. Ghosn and Frangopol (in Frangopol, 1999b, Chapter 4) point out that the *summation* of many random variables produces a normal distribution, whereas their *product* results in a lognormal distribution. Appendices 7 and 17 summarize some design provisions for estimating the reliability of systems with parallel and series links.

Ghosn and Frangopol (in Frangopol, 1999b, Chapter 4) state that estimating "the reliability of structural systems whether in series, parallel or combined is possible only if all the failure modes can be identified." Hence, before it qualifies as *reliable*, design must be *predictable*, or, in the words of De Finetti (1974), *previsible*. The author emphasizes the difference between indeterminacies, uncertainties, and the probability one deterministically assigns to an outcome. Estimates of probability of structural failure and remaining service life can be improved by stochastic analysis when data are abundant. Limited knowledge leads to deterministic choices based on the best assessment of the available information and the constraints. Optimality holds only with respect to a particular set of benefits and constraints. What then should be the target reliability of design specifications?

The "Deacon's Masterpiece" by Oliver Wendell Holmes (1895) is an example of uniform strength design (Appendix 1). Its unique mode of failure consists of instant disintegration. In the absence of such perfection in design and construction, specifications (Section 4.2.1) must recommend the structural "reserve." The amount of reserve has been alternatively quantified in terms of strength and, more recently, in terms of a reliability index β.

Frangopol (1999a,b) reviews and expands on the significant contributions of reliability theory to the practice of bridge design and management in the United States and Japan. Ghosn and Frangopol (in Frangopol, 1999b, Chapter 4) describe methods for determining bridge reliability compatible with current specifications. The authors model bridges as combinations of parallel and series systems of components. A convergence to combinations of loads producing collapse mechanisms (Appendix 10) is sought by incremental techniques, combining probabilistic and deterministic analyses.

Moses (in Frangopol, 1999b, Chapter 5) summarized the probabilistic procedures underlining the selection of live-load magnitudes for design and for load rating (Section 10.5). He anticipated (p. 133) that data correlating actual live loads with structural responses would be a major future contribution of BMSs to structural reliability assessments.

Partial and gradual failures (Section 3.3) are clearly preferable to sudden and global ones. Many sensitive structures employ tell-tale signs and, more recently, health-monitoring techniques as precursors of potential failures. The fatigue fuses used in aircraft, for instance, are nonstructural components designed to fail first under service conditions.

Risk analysis is a priori limited so long as it considers essentially anticipated hazards. This is more realistic for structures with known failure modes and, consequently, predictable failure load combinations. AASHTO (1998a, C4.5.2.3) encouraged the anticipation of failure modes by design as follows: "Where technically possible, the preferred failure mechanism should be based on a response that has generally been observed to provide for large deformations as a means of warning of structural distress. . . . Unintended overstrength of a component may result in an adverse formation of a plastic hinge at an undesirable location."

The optimization of failure modes relies on load redistribution, which in turn requires ductility and redundancy.

Indeterminacy and Redundancy

Structural Analysis

The redundancy of structures is synonymous with static indeterminacy (also known as hyperstaticity); however, the two usually imply different connotations and attitudes. *Redundancy* is taken to mean reserve (associated with safety in structures, waste in management). *Indeterminacy* is blamed for complicating static analysis and for generating thermal and secondary stresses.

Shrinkage, creep, temperature variation, and support settlements are considered secondary effects, producing secondary stresses in statically indeterminate structures (prestressed members are internally indeterminate). The terminology is somewhat vaguer than in the case of the second-order analysis discussed in Section 4.2.1. Trusses, for instance, were traditionally analyzed by assuming hinges at panel points and subsequently adding "secondary" moments to account for the neglected fixity. With improved computational power and software, the actual truss redundancy is modeled directly. The geometric stiffness can be determined according to the axial loads and bending moments are obtained as primary member forces.

Actual stresses and deformations in complex redundant structures according to Heyman (Appendix 35) are best approximated by plastic analysis if (no small caveat) adequate *ductility* and *stability* are provided. Structural analysis is shown to be meaningless if it ignores the physical limitations imposed on design by construction.

Several analysis and design procedures recommended by AASHTO (1998a) reflect Heyman's view. One is the *strut and tie* (or *stress field*) analysis for modeling of two-dimensional stress distribution in reinforced concrete. Another is the ductile design and plastic hinge analysis for seismic response. The support imperfections (or uncertainties) discussed by Heyman (1998) are taken into account by a load factor γ_{SE}, to be established on a "project-specific basis" (AASHTO, 1998a, p. 3–10). The innovations are consistent with the general trend from *prescriptive* (e.g., uniquely but hypothetically defined) to *performance-based* (e.g., vaguer but more realistic) design specifications.

Structural Design

The first manual on the inspection of fracture-critical elements by Harland et al. (in FHWA, 1986) was strongly influenced by the failures at the Silver Bridge and at Mianus River (1983). The identification and retrofit of fracture-critical members became a top management priority. Design and assessment focused on evaluating the effect of redundancy on structural performance and, ultimately, on reliability. For further quantification of load rating calculations, the *Manual for Condition Evaluation of Highway Bridges* (AASHTO, 2000b) refers to Ghosn and Moses (NCHRP Report 406, 1998) (Appendices 7 and 17). Liu et al. (NCHRP Report 458, 2001) define three types of redundancy as follows:

Internal: The failure of an element remains limited to that element.

Structural: Providing both static indeterminacy and adequate ductility.

Load-path: More than two load paths.

In terms of serviceability, structures are considered redundant if they provide an *alternate load path*. A simply supported span, for instance, is statically determinate and hence nonredundant. Although detailed inventories take that into account, another level of redundancy is derived from the number of primary members in that span. A system of two primary members is nonredundant even if they are continuous. According to certain bridge inventories, systems of three primary members are redundant and others require four primary members for redundancy. Still, two-girder bridges can redistribute loads through a variety of lateral bracing systems and their level of redundancy should be assessed accordingly (Daniels et al., 1989).

AASHTO 2004, 1998a accounts for redundancy by the η factors shown in Eq. 3.1 and in Appendix 17 herein. The commentary (LRFD, p. 1–3) acknowledges that the η_i factors are subjective and may undergo revisions, particularly η_R. The numerical values of the factors determining η_i have been selected by observing their effect on the reliability index β (Eq. A7.7 herein).

It is inevitable that η_D should modify η_R. The value of redundancy lies in allowing load redistribution. That is only possible if structures are ductile at yield. Thus, in order to improve reliability, redundancy must be present globally and internally, i.e., on a system and on a member level.

Global and Internal Redundancy

A structure is redundant *globally* (or as a system, in the functional sense) if load redistribution is possible after its members or elements reach their ultimate loads. Static indeterminacy is therefore necessary but not sufficient. Hambly's four-legged chair (Appendix 35) would qualify as globally redundant only if it could carry the ultimate design load on any three of its legs. If not, the chair could be susceptible to progressive collapse. From the standpoint of reliability, globally redundant structures should meet the following requirements:

Be parallel rather than series systems

Be able to redistribute their functions after losing some of their elements

Overall reliability therefore implies a combination of global and internal redundancies. Structural elements or members can be considered *internally* redundant if stress redistribution is possible within their cross section. That distinction is illustrated by a comparison between eye-bar chains and parallel wire cables. Globally both are nonredundant, since structures are not typically designed to survive without them. Internally, however, a crack in an eye-bar is likely to propagate throughout the entire member. In the case of an internally nonredundant two-bar chain (as in the Silver Bridge, Example 6) the structure would be compromised. Some internal redundancy may be gained if the eye-bars are numerous, as in Fig. 5.1, rather than one and two, as in Fig. 5.2.

In contrast, a suspension cable consists of thousands of parallel wires (7696 in the case of Fig. E3.5). Unless it is symptomatic of a general malfunction, the loss of one or several (out of several thousand) wires is relatively insignificant at the location of occurrence and becomes moot at a distance along the cable due to friction with healthy wires. Wire breaks cannot be ignored, but they can be managed, their causes identified and eliminated (see Example 3). A broken strand at one of the cables at Pont de Tancarville (Virlogeux, 1999) was treated as redundant in the short term and the bridge remained in service while a cable replacement was in preparation. The lesson of this partial failure (Section 3.3) was applied uneventfully at the Pont d'Aquitaine (Fig. 5.3), where the similar cables were also replaced but by different methods.

The concrete deck shown in Fig. 4.14a has failed in a brittle mode; however, its steel reinforcement has remained ductile. More importantly, the deck is a two-way continuous thick plate with considerable redistribution capacity.

Figure 5.1 Internally redundant multi-eye-bar chain.

In the case shown in Fig. 4.71, a deteriorated reinforced-concrete deck was able to redistribute its weight, working as a thick plate. When the plate was reduced to parallel beams by precutting, 2 out of 20 such beams failed in simple bending.

Ductility

Redundancy is the structural ability to redistribute loads and ductility takes that ability to the level of the critical structural sections. Whereas loads redistribute globally (e.g., between structural elements), stresses do so internally, within the element cross section. Hence, ductility can be seen as a form of internal redundancy. In contrast, brittle behavior is internally nonredundant. Under extreme loads, ductile sections yield (in the case of steel), cracks open (as in concrete), and inelastic deformations develop. Brittle materials fracture. Figures 3.7, 3.10, 3.11, 4.1, 4.16, 4.22, 4.28, 4.29, and 4.44 illustrate brittle

Figure 5.2 Internally nonredundant two-eye-bar chain.

failures in steel and reinforced concrete, whereas the behavior shown in Figs. 3.3, 4.14, and 4.47 is ductile.

The ultimate strength method of reinforced-concrete design assumes nonlinear load redistribution under normal working conditions. Under extreme loadings, a reinforced-concrete section should be able to deform inelastically and sustain permanent damage without failing. The Cypress Avenue Viaduct in Oakland (Fig. 3.10) lacked adequate continuity at the base of the second tier and failed during the Loma-Prieta earthquake. The continuous one in San Francisco (Fig. 3.9) survived without collapse or injury; however, its ductility was minimal and it was irreparably damaged.

A demand for ductility at failure clarifies the role of deck reinforcement, discussed in the discussion of stress–strain in Section 3.1.1. Since, as argued, decks function primarily in shear, it is in shear that they also fail. Even if reinforcement is not essential to the deck resistance under typical live loads, it becomes indispensable at failure. The local deck failure in Fig. 4.14 would have been catastrophic without the steel reinforcement.

Figure 5.3 Pont d'Aquitaine, Bordeaux.

Also significantly, the exposed steel is not particularly corroded, suggesting that the failure was caused by shear and fatigue, rather than by expansion of the rebars.

AASHTO (1998a, Eq. C1.3.3-1) recommends to the owner the following minimum ductility factor μ "as an assurance that ductility failure modes will be obtained":

$$\mu = \frac{\Delta_u}{\Delta_y} \tag{5.4}$$

where Δ_u = deformation at ultimate limit
  Δ_y = deformation at elastic limit
  Δ_y = limit of elastic analysis

and μ is not the same as in Appendix 8.

Structures must be checked for stability under deformations equal to Δ_u.

The method, proposed by Ghosn and Moses (1998), for evaluating structural reliability as a function of its redundancy is briefly summarized in Appendix 17.

Nonredundant structures and processes where every local failure can become catastrophic are particularly targeted by reliability methods, as are cases where redundancy depends on a few critical details. Upper and lower bound approaches do not necessarily converge to the same failure mode. "Upper bound" reasoning, i.e., the search for failure modes, can consider causes other than overstress, such as loss of stability, fatigue fracture, or even extreme events.

Database Management System (DBMS) Redundancy

FHWA (2001e, p. 15) defined DBMS redundancy as follows: "The storage of multiple copies of identical data. The process of limiting excessive copying, update, and transmission costs associated with redundant data is called "redundancy control." *Database replication* is a strategy for redundancy control intended to improve system performance. *Redundancy* may also refer to backup systems that take over data processing and transmitting functions when a primary system fails."

As defined in the preceding paragraph, redundancy can be perceived as a source of both reliability and overspending.

5.3 NETWORK RELIABILITY

The design of network reliability and the estimates of the reliability of existing networks advance in the following two directions:

Models of the networks and the potential hazards become more detailed and realistic.

The networks under consideration grow more inclusive and integrated.

Cremona (2003; Appendix 17 herein) defined networks as parallel and series systems. Bridges are parallel and series systems of structural elements. Vehicular transportation networks are parallel and series systems of roadways with nodal points at intersections and at structures. The relative complexities of the two are roughly comparable to those of calculus (of functions) and calculus of variations (e.g., of functionals). The reliability analysis of networks, for instance, in the work of Von Neumann and Morgenstern (1964), has preceded and stimulated that of engineering structures largely because certain systems (e.g., in electronics) are sufficiently defined by the remaining useful life of their components. Pass–fail are the only performance criteria and timely replacement is the only need. The reliability of such networks is analyzed by Barlow et al. (1965) and Gertsbakh (2000).

Reliability and management developments were applied to roadways before bridges because of the relative uniformity and simplicity of the former networks. It was always understood, however, that disconnected models of roadways as approaches to bridges and bridges as nodal points in the road network could not be optimized separately. Within a short-range planning horizon, transportation demands would focus on road surface quality while neglecting structural life-cycle needs. The less likely opposite alternative of improving structural conditions while neglecting traffic demands would be equally wasteful. Functioning transportation networks include roadways, bridges, lifelines, and other infrastructure assets. An application of reliability theory to structural systems is offered by Thoft-Christensen and Murotsu (1986).

The reliability of the transportation networks under seismic loads has received considerable attention. Thoft-Christensen and Baker (1982) analyzed the reliability of structures subjected to random occurrences, such as quality lapses and extreme events (Section 5.3). Werner et al. (2000) developed a risk-based methodology for assessing the seismic performance of highway systems. Deterministic and probabilistic results are obtained for the level of damage to a transportation network of defined inventory under a prescribed earthquake. Basöz and Kiremidjian (1996) assess the risks for highway transportation systems under more generally defined natural hazards, including tornadoes, hurricanes, floods, and earthquakes. The risk assessment combines vulnerability and importance rankings of the individual assets. Two types of network analysis are performed: connectivity and serviceability. Both studies estimate losses and recommend plans of pre- and post-event action. Chang et al. (1996) estimate losses caused by extreme-event-induced failures of various utility networks, including light, gas, and water supply.

Mohamed et al. (in TRR 1490, 1995, pp. 1–8) recognize the following four types of network prioritization or optimization models in use:

Sufficiency ratings

Level-of-service deficiency ranking

Incremental benefit–cost analysis

Mathematical programming (used, e.g., by Pontis and North Carolina DOT).

The authors distinguish between optimization in the time domain, to be performed by a dynamic programming model, and in the network domain, for which they develop an artificial neural network (ANN) model.

5.4 PROCESS RELIABILITY

Structural reliability is optimized under cost and other constraints with respect to assumed failure modes. Processes do not have distinctly defined failure modes. Rigorous optimization (Appendices 36 and 37) quantifies "utility." The only truly quantifiable managerial resource is money, and its use cannot be optimized without qualitative choices. Budgets are the only exclusively managerial products and management fails directly only on budgetary terms. A wider range of managerial deficiencies is deduced from the failures of the products and/or services, i.e., the physical and logistical structures (Chapter 4). Hence, reliability is the objective (or target) of design, but it can be viewed as a managerial constraint. Bridge management became the practice on qualitative grounds, after spectacular lapses in public safety, rather than for the purposes of economy (Chapter 3).

Processes can be modeled as combinations of parallel and series systems and optimized in terms of simplified constraints, such as time and money. The critical path method (CPM) minimizes the time of construction projects by identifying the nonredundant sequence of key operations in the process and ensuring the smooth transition between them. The same approach would be nonredundant and hence vulnerable to discontinuities in safety-related transportation activities (Section 4.1.3). Organizational redundancy, on the other hand, amounts to duplication of effort. A process can be designed for a targeted reliability under the constraint of economy if unacceptable modes of performance can be defined. The levels of process redundancy could be expressed in terms of organizational reliability factors equivalent to the structural reliability indices β (Appendix 17), just as the loss of safety is quantified, by assigning appropriate monetary values to expected losses (Appendix 38).

In the general terms of Table 4.3, functional redundancies make a difference at the points of transition from one task to the next, for example, from design to construction, and then to maintenance and operation. The technical staff responsible for infrastructure networks is typically organized in units dedicated to the tasks of analysis, design, construction supervision, maintenance, repairs, operation, and inspection. The result is a globally nonredundant chain. If each of the units develops expertise in the domain of the others (e.g., construction checks design, maintenance reviews design and checks construction), a level of redundancy is achieved (along with some duplication of effort). Alternatively, staff can be dedicated to groups of facilities sharing common features, such as bridges of a certain type (e.g., over water, over land, movable, suspension, steel, concrete). The respective teams would manage the structures over all stages of the life cycles shown in Table 4.2, ensuring a functional continuity. The overall organizational structure would be redundant. Whereas nonredundant processes would be likely to fail in as many discontinuous modes as there are transitions between tasks, redundant ones would tend to a unique "incremental collapse," for instance, if the staff lacks manpower, expertise, or other capabilities. Table 4.5 suggests a variety of QC and supervision measures ensuring adequate structural and personnel performance.

5.5   REANALYSIS

Reliability analysis evaluates two sets of uncertain conditions, namely, those related to the occurrence of certain events and those related to the consequences (or penalties) that the engineering structures and services might suffer. The range of uncertainty is somewhat narrowed down by focusing on the known vulnerabilities discussed in Chapter 4. The latter in turn were suggested by the catastrophic or partial failures enumerated in Chapter 3. Such reasoning is speculative, incomplete, and vulnerable to ignorance. Applied to thinking, rather than to things, Murphy's law (also discussed in Section 3.3) might state that "anything that might be overlooked will be."

Santayana (1928, p. 45) believed that history is always "written wrong, and perpetually rewritten." That view applies to the continuing reanalysis of past failures. Acknowledging the growing numbers of rewrites and revisions, current engineering documents have adopted the legalistic catch-all escape clause "including but not limited to." The present text similarly precedes any enumeration or list with the term *partial,* as in Tables 4.2 and 4.3. This fuzziness has some value in planning and analysis but is unacceptable in the implementation of procedures and the execution of tasks. The list shown in Table E26.1, for instance, cannot be "partial."

To reduce the effects of incompleteness, the probabilistic (degree of belief) and statistical (frequency) approaches complement each other, as do upper and lower bound iterations or form and content considerations. Reliability analysis is therefore an iterative heuristic reanalysis, continually refined within its own scope and expanded beyond that scope.

PART II

ASSESSMENTS: BRIDGE MANAGEMENT SUPPORT SYSTEMS

Chapter 6

System and Structure

While vulnerabilities of the product (e.g., the structures; Section 4) were increasingly suggesting the need for systematic management, the production process was organizing into structured systems. In *Webster's New College Dictionary* there are 10 definitions of system and 5 of structure, including the following:

System

1. A group of interrelated, interacting, or independent constituents forming a complex whole
2. A functionally related group of elements
    d. A network of structures and channels, as for communication, travel, or distribution (a rail system)
9. A method: procedure

Structure

3. Interrelation of parts in a complex entity

Philosophy can sound ambivalent on the distinction between *structure* and *system*. Boudon (1968, p. 17) pointed out that "the definitions of the notion of structure proposed by economists, sociologists, psychologists can teach about these definitions, but cannot produce . . . a definition of the notion of structure." The author quoted Flamant (p. 14) as follows: "A structure is an ensemble of elements related so that any modification of one element or one relationship results in modifications of other elements or relationships."

Civil engineering and Webster's definition (5) avoid the semantic stalemate by defining structures as "built to perform a function," such as to carry loads. Heyman (1996) begins with the statement: "A structure (from the Latin *struere*) is anything built."

Thus, *systems* define the organization or function of *structures*, for example, two or three dimensional, truss, girder, suspension. Whereas structures have evolved since the origin of the universe, they have been designed according to known systems for centuries. Whether spontaneous, intuitive, or analytic, however, structures exist because they comply with the governing constraints, such as gravity. Hence, Salvadori ("Why Buildings Stand Up," 1980) and Gordon ("Why Things Don't Fall Down," 1978) explain what demonstrably works. Their investigations identify the underlying systems which should be refined and reused.

By the same definitions, an organizational structure would be "anything that works" for the users. Since organizations are constrained by society rather than by physics, the extent to which their structures work is limited and debated. They are typically investigated for the purpose of change. Hence, despite the permanence of their fundamental principles (Section 1.1), management systems obsolesce much faster than structural ones. The notions of competence in the two fields also have different shelf lives (Section 1.11).

When, in the mid-nineteenth century professional engineering abolished "building as usual," "thinking as usual" had already been questioned repeatedly (Section 1.2). Descartes had engineered a proto-information system for constructing reliable knowledge in the Aristotelian tradition. The Cartesian method divides information into its smallest units, proceeds from the simpler to the more complex, accepts only ascertainable truths, and stores them readily retrievable for reference.

By the twentieth century, the quality and quantity of mathematical methods and tools approached the Cartesian specifications. Computerized operational algorithms and data processing could finally raise decision making above the "common sense," distrusted by Descartes as "the most widespread human quality" (Section 1.8). Information management systems (IMS) could be demonstrably rational. Systematic management was identified with management by computerized systems. J. Bailey (1996) argued that parallel rather than sequential reasoning would have been even better suited for automation, but it was not pursued early on because of the reigning commitment to the Cartesian deductive model. The latter model allowed Hudson et al. (NCHRP Report 300, 1987) and OECD (1992) to declare the end of "business as usual" in bridge management. Hawk (in NCHRP Report 483, 2003, p. 63) defined bridge management systems (BMS) as follows: "A set of rules, guidelines, and procedures used to identify a *management strategy;* the term has been usually used to refer to computer programs such as PONTIS and BRIDGIT that organize and automate these rules, guidelines, and procedures; often include storage and organization of inventory and inspection information, maintenance scheduling, and work-program organization."

For engineering systems, Mittra (1988) adopted one of the many definitions proposed by Alexander (1974) as follows: "A *system* is a group of elements, either physical or non-physical in nature, that exhibit a set of interrelations among themselves and interact toward one or more goals, objectives, or ends. The elements that comprise a system may be of several different types. . . . The nesting of smaller subsystems within larger ones forms a hierarchy that is characteristic of any system. A system and all its subsystems accept input and produce output."

Bridges are physical subsystems of the transportation network, carrying dead and live loads. Managing engineers are a system of experts charged to operate them safely and cost effectively. IMS must store and transmit the structurally and socially significant information between the two. Lucas (1985, p. 4) has defined information systems as follows: "Information system is a set of organized procedures that, when executed, provide information for decision making and/or control of the organization."

Any systematic procedure prescribing and tracking bridge-related actions based on knowledge of conditions and constraints is a bridge IMS or a BMS. All responsible bridge owners have had one of sorts. In the past it may have been limited to as-built drawings, annual capital and expense budgets, and contract and accounting procedures.

In 1776 Peronnet reported 2090 road plans and 757 bridge drawings, corresponding to 3135 previously known highway locations in France (Picon, 1992, p. 47). In 1903 the New York City Bridge Commission reported the condition of 45 bridges. In 2005 their number had grown to 796. In the meanwhile the NYS DOT and the state, throughway, metropolitan transit, and port authorities have emerged as major managers of bridge networks in the same geographic area (Example 1). Information systems evolved along with the growing structural networks.

OECD (1992, p. 75) reported first attempts to automate the storing and processing of bridge-related information in Denmark, Belgium, and by the French National Railway System (SNCF) during the early 1970s. The evolution of the U.S. National Bridge Inventory (NBI) is summarized in Appendices 11 and 14.

Construction evolved from trial and error during various applications to the design of complex structural systems. Information systems, in contrast, adapt their general structures to specific applications. Mittra (1988, Section 1.8) described this approach as *top down,* a global view, proceeding concurrently with analysis of the functions and design of the means "from the more general top-level requirements to the more detailed lower-level requirements." As a quantitative improvement ("in degree only, not kind"), the author recommends a *structured* method as follows:

> The structured system design starts with the logical system prepared as the end product of the system analysis phase and converts it into a physical system. It changes the data flow diagrams into implementation-bound system flowcharts and develops design specifications first at system level and then at the program level. Thus the development of the complete logical system is a prerequisite for the development of the physical system. There is a total separation of the logical system from the physical system.

This definition parallels the engineering requirement that design precede construction. It was pointed out in Section 4.1.7 that systems emphasizing this approach can underestimate the theoretically fuzzy but practically decisive contribution of empiricism. The implementation phase of a system must compensate for such vulnerability by extensive tests of "β-versions" prior to final system approval. Five general phases of system development are described in Appendix 24. The development process is cyclic, with an average life of the system defined by the design, once again mirroring the life-cycle process typical of bridge management in general.

Consistent with the thinking expressed in Section 2.2, FHWA (2005a, p. 6) defines BMS as "a system designed to optimize the use of available resources for the inspection, maintenance, rehabilitation, and replacement of bridges." However, a BMS can guarantee neither the optimality of the decisions nor their implementation. In contrast, a BMS can strive not only to optimize but also to maximize the use of bridge-related information. Postdisaster investigations typically conclude that decisions have been based on "the full extent (or state of the art) [of the] knowledge available to management."

Although such findings avoid liability, they imply ignorance (Section 4.1.1). Once a BMS is in place and provides decision support, either information must be available or its lack must be treated as a vulnerability. There is a broad correlation between the structural competence implied in a license to practice civil engineering and the managerial competence of the BMS managers. Table 6.1 illustrates the correspondence between the

Table 6.1   Bridge Structures and Bridge Management Systems

Requirement	Bridge Structures	Bridge Management Systems
Reliability (Appendices 5–9, 17)	*Process:* Current specifications express the necessary and acceptable standards of bridge design in terms of reliability. In reliability analysis, a system is a set of elements with an estimated likelihood of failure (Appendix 17). *Product:* A structure is designed to serve with an acceptable margin of reliability, e.g., with an acceptably low likelihood of failure.	*Process:* If a BMS is adopted for production line use, the reliable operation of the assets depends on it. *Product:* Data management systems are tools for estimating the following: • System reliability, e.g., its likelihood of failure under design conditions • Design conditions • Outcomes of alternative life-cycle strategies
Cost effectiveness (Chapter 2 and 10)	Bridges must provide service at a cost acceptable to the community (Section 1.5). Management decisions are increasingly based on life cycle, rather than on first-cost analyses.	BMS were developed concurrently with life-cycle cost analysis. Mandated by the Intermodal Surface Transportation Efficiency Act (1991), BMSs were only recommended by the National Highway System Act in 1995 (see Appendix 11). The implication was that BMSs must prove their cost effectiveness.
Quality of service	Reliability and economy cannot capture fully the demand for bridge service. The quantity of services provided by alternative solutions can be compared (Section 10.1); however, the satisfaction of the users must also be considered directly. In the developing interaction, users select the structures and the structures influence the users' choices.	BMSs must adapt to the specifics of the structures and the technical staff who manage them. Eventually BMSs influence the quality of the structures and the qualifications of the personnel.

basic requirements to be satisfied by both BMS logistical structures and the physical structures they manage.

Both physical and logical systems process input (loads are the inputs of engineering structures) and deliver services. Consequently, the several elementary system configurations summarized in Appendix 18 have widespread application in both domains.

The BMS flowcharts, shown in Appendix 16, combine, to various degrees, the database management options described in Appendix 18. Some, such as OECD (1992), separate the input from the output modules. Others (e.g., NCHRP Report 300, 1987) assign modules to activities that may store input and generate output, such as maintenance.

In order to be compatible with more general MISs, BMSs must take into account global management requirements, for instance, as described in TRB Special Report 234 (1992) and NCHRP Synthesis 238 (1997) (Appendix 39 herein).

A functioning BMS is a cycle of information related in some manner to the life cycle of the managed assets. In one form or another, BMS modules correspond to the

Table 6.2   Bridge Management Cycle

External	→	→	Input	→	Database	→	Output	→
Sources	←	↑						
		↑	*Product*					↓
		↑	• Inventory		*Process*			↓
		↑	• Conditions		• Contracting			↓
		↑	• Calculations, drawings		• Consultant selection			↓
		↑	• Costs		• Analysis			↓
		↑	• Forecasts/estimates		• Design			↓
		↑	• Standards/priorities		• Construction			↓
		↑	• Work records		• Maintenance/operation			↓
		↑	• Options/alternatives		• Inspection			↓
		↑	• Network data:		• Management:			↓
		↑	maps, traffic counts		personnel, QC&QA, audits			↓
		↑	• Budget		recommendations, prioritization,			↓
		↑	• Operation procedures		emergency response,			↓
		↑	• Legal requirements		BMS operation, management			↓
		↑						↓
		← ← ← ← ← ← Actions ← ← ← ← ← ← Decisions ← ← ← ← ← ←						

products and activities listed in Tables 4.2 and 4.3, forming a cycle of the type shown in Table 6.2.

As a process, the bridge life cycle is defined by its activities (Table 4.3). Management reduces these activities to prioritized tasks. The BMS database is in a crucial but supporting position in Fig. 4.4*b*. The flowcharts of OECD and DANBRO (Figs. A16.3 and A16.4) emphasize the process by tracing activities between data blocks. Management should be able to shift priorities and reconfigure the database according to changing needs, vulnerabilities, and perceptions.

The product (or output) of a BMS consists of past, present, and projected conditions and needs, as in Table 4.2. In that context and in the flowcharts of Fig. 4.4*a* and PONTIS (Fig. A16.7), the database is dominant. On the role of the database in society, Santayana (1928, p. 44) stated: "There can be no serious history until there are archives and preserved records."

De Tocqueville (2000; see Section 1.2 herein) found that democratic societies "mistrust systems and like to stick very close to the facts." He warned, however, that "he who tries to uncover facts illustrating the real influence of the laws on the fate of humanity is liable to great mistakes, for nothing is so hard to appreciate as facts" (p. 214). "The taste for analysis comes to nations only when they are growing old, and when at last they do turn their thoughts to their cradle, the mists of time have closed around it, ignorance and pride have woven fables around it, and behind all that, the truth is hidden."

Shaped by the supply of information and the demand for (as well as a resistance to) analytic evaluation, the database serves the BMS as memory serves thinking.

Chapter 7

Data Management

The FHWA (2001e, p. 10) defined the *electronic databases* and their *management systems* as follows:

> A repository for information or data organized in such a way that a computer can quickly select desired pieces of data. . . . A *database management system* (*DBMS*) is needed to access information from a database.

> *DBMS* is a collection of programs that enables information to be stored in, modified, and extracted from a database.

Hudson et al. (1997) distinguished between *data base* as a collection of computerized data and *database* as a specific group of data within the structure of a software management system. A database system was considered to consist primarily of an operating system, database management software, and application programs. DBMS structures are summarized in Appendix 18.

Software design (Date, 1973, 1983) considers databases as computerized systems carrying out "transactions." Each transaction is a unit of work. In this view the BMS is a DBMS applied to bridges. Database structures (Appendix 18) often grow to reflect the properties of the physical networks of structures they represent (Table 6.1). Bridges conform to structural specifications. Data structures must meet specifications for definition, storage, access method, and integrity.

The terminologies of the BMS users, managing bridges, and BMS designers, managing data, correspond without being identical. The physical inventory of the former is represented by data objects in the inventory of the latter, whereas the physical inventory of the latter is office hardware to the former. A bridge designer may consider the BMS as a modernized archive describing physical structures and transportation networks. A database designer sees it as a network of substructures transporting information.

Database and structural management collaborate continually, despite their distinctly different pace. Computer software and hardware capabilities are upgraded, even in mature databases, on the average every 3–6 years. Multimillion infrastructure capital budgets are continually adjusted for 10 and 15 years and for longer planning horizons in life-cycle cost analysis. After more than 30 years of development, the U.S. federal bridge management database is integrating into *asset* management (Section 12.1). The latter term has always been standard in financial management. The NBIS specifications (FHWA, 2005a) propose the following definitions:

Land Management Highway System (*LMHS*): Adjoining state and local public roads that provide major public access to Bureau of Land Management administered public lands, resources, and facilities.

Linear Referencing Systems (*LRSs*): A set of procedures for determining and retaining a record of specific points along the highway. Typical methods used are kilometer point (mile point), kilometer post (milepost), reference point, and link node. LRS data are required for the annual Highway Performance Monitoring System (HPMS) data submittal from states to the FHWA.

To facilitate maintenance, bridge databases must be structured into coherent independent modules, compatible with different environments and suitable for integration into larger information systems. Quantitatively, database design must strike a balance between brevity and detail. Qualitatively, the equivalent conflict arises between simplicity and complexity. Each of these universal trends has eloquent champions and faithful followers. Gertsbakh (2000, p. 17) recounted Einstein's advice: "Everything should be made as simple as possible, but not simpler."

Barrow (1991, p. 1) quoted the science fiction writer and engineer Poul Anderson as follows: "I have yet to see any problem, however complicated, which when you looked at it in the right way, did not become still more complicated."

When the classics diverge as extremely as in this selective review, the path of least resistance follows all of them, so far as resources allow. Reflecting the conflicted trends, the flowcharts describing bridge management functions in Appendix 16 run the range from the basic to the elaborate. Since the complexity of natural phenomena cannot be fully modeled, it is practical to maintain simplicity. Nonetheless, overlooked information occasionally proves significant at a later date. Even after bridge conditions became the subject of regular inspections, data on the factors influencing conditions (e.g., maintenance, environment) remain lacking or uncorrelated.

Database design choices become harder to optimize as the limitations on data acquisition and storage (both quantitative and qualitative) relax. Text and images are readily processed. Digital photography provides more visual information than assessment modules can fully exploit. Online health-monitoring systems record direct structural response parameters (in "real time").

The newly gained data abundance shows that unstructured information is wasted.

"It is tough maintaining astronomical sums of bridge maintenance data," said H. Fujikawa, president of the Honshu-Shikoku Bridge Authority (*Roads & Bridges,* Aug. 2003). He continued: "Since we have a limited amount of staff members efficiently maintaining the bridges, the maintenance performance data method helps."

The developers of DANBRO (Appendices 16 and 40) similarly resist unnecessary data accumulation. To qualify for acquisition and storage, data must satisfy the following requirements:

Be necessary and sufficient for the envisioned type of assessment

Be reliably acquired, stored, and retrieved

In order to serve their purpose, data archives must be accessible. Valuable information is wasted when "as-built" drawings of old bridges are lost, damaged, or discarded in various archaic storage rooms (Fig. 7.1).

Figure 7.1 Storage of hard-copy bridge-related data.

As data become more readily obtained and processed, the contained information grows harder to ignore. Decisions are eventually judged according to the sources they could have used. Drucker (1973, p. 538) described information as "bi-valent," because it is directly connected to top executive management and to primary sources. He concluded: "Information work should be kept separate from other kinds of work."

Existing decision-making stereotypes (see Chapter 12) underscore the distinction between the two different function a BMS can assume, namely:

Bridge information management system

Bridge management support system

The former designation is limited to data management. The latter implies optimization. In the general framework of a BMS, data management is one of the modules and decision support in the form of optimization is another (Appendix 16). Hudson et al. (1997, p. 330) concluded: "Modularity of a system requires that the analysis tools be independent of the database."

Patterson and Scullion (1990) have identified five functional levels of infrastructure management data needs: *sectoral, network, project, operational, and research and de-*

velopment. Data must be adequately referenced to the relevant conditions at the time of input. Cross-referencing (e.g., "linkages") between standards, costs, conditions, constraints, and options must be "user-friendly." Inputs must identify their sources, in order to allow for estimates of their reliability. Costs are only meaningful in the context of concurrent economic indicators. Structural conditions must be linked to the type of evaluation, the level of maintenance, traffic volumes, and environmental factors. Hudson et al. (1997, p. 80) recommended: "Element-specific data detail is best defined for groups of similar information, since these groups have features in common for more than one application and form a natural basis for collection and storage in a database. . . . The amount of detail required for these various applications [increases] progressively from the overall summary statistics through planning, programming, design and research."

The authors suggested four levels of information quality and quantity with corresponding standards of reliability.

The following are typical bridge-related data blocks:

Inventory of assets
- As built: structure, material, service
- As is: conditions

Standards (past, present)
- Design
- Construction
- Operation
- Maintenance
- Inspection
- Environment
- Legal

Costs
- Analysis
- Design
- Construction
- Reconstruction
- Maintenance
- Inspection
- Operation
- Emergency/extreme event
- Users (individual, community)

Service (traffic)
- Type
- Volume
- Accidents
- Enforcement

Operation
- Work records
- Procedures
- Design/repair options
- Maintenance schedules
- Contracts
- Community outreach

Network
- Global positioning system (GPS)
- Road map
- Lifelines
- Essential services, importance
- Emergency management
- Perceived risks

BMS Management

Most data modules are designed while the data are only anticipated and require refinements at later stages. A module dedicated to database management, updating, and upgrading becomes increasingly valuable with prolonged use.

Seeking the "perfect" structural condition data set is theoretically futile, financially wasteful, and professionally paralyzing. Maintaining a thorough inventory of the assets, on the other hand, is attainable and constitutes a sound policy, regardless of the imminent needs. The element specific inventories of Alabama, Indiana, North Carolina, Pennsylvania, and New York State exceeded the NBI detail and provided relatively advanced condition and needs assessments.

Chapter 8

Inventory

The inventory is a natural first block or module in the database. The idealized model of the as-built (or as-designed) network is relatively stationary and "objective." The performance of the as-built assets (according to governing specifications) is implied in the inventory and provides a "serviceability" assessment of a network "in a state of good repair." The latter term is vague, whereas the responsibility of the owner for the inventoried items is binding. Legal disputes over the liability for malfunctioning items in the bridge inventory repeatedly remind of the need for clear and presentable records.

The evolution of the National Bridge Inventory (NBI) in the United States (Appendix 11) shows that some conclusions could be drawn early on, others evolved over 25 years of data collection. By the late 1990s, with the number of inventoried bridges surpassing 600,000 (e.g., more than twice the original number), the FHWA was considering new inventory guidelines. FHWA (2004a) documents the discussion over the Notice of Proposed Rulemaking (NPRM) for a coding guide amendment. The proposed NBIS specifications (FHWA, 2005a) are summarized in Appendix 14.

Other countries, for instance, the members of the European Union (BRIME 2002), are taking stock of their assets in similar steps. BRIME (2002) reported the following bridges on the national highway systems of some of the member countries (p. 51):

	France	Germany	Norway	Slovenia	Spain	England
Bridges	21,549	34,824	9,163	1,761	13,600	9,515
Area (1000 m²)	7,878	24,349	2,300	660	5,526	5,708

For bridges longer than 2 m, OECD (1992) reported a total number of 209,300 in France and 651,869 in Japan.

Itoh and Liu (in Frangopol (1999a, p. 137) reported that between 1956 and 1975 about 61,000 highway bridges longer than 15 m were built in Japan. By 1996 the overall number in that category had reached 130,196, with a total length of 7481 km.

The level of integration varies widely between different bridge owners, localities, and countries. In certain countries, where as many as 75% of the regional bridges have been constructed prior to the twentieth century, the first bridge management efforts are limited to the modern national highway systems. Two further developments inevitably follow:

All bridges (and culverts) are included in the inventory.

The bridge and culvert inventory is integrated into multimodal asset inventories, including roads, railways, and lifeline utilities.

NCHRP Report 437 (2000) discussed the use of GPSs for the collection and presentation of roadway (including bridges) inventory data for the NBI.

8.1   ESSENTIAL PARAMETERS

Essential structural parameters fall into the following categories:

Structural type

Main superstructure material
Type of primary structure
Type of deck
Number of spans
Joints, bearings
Abutments
Approaches
Pier type
Pier material
Footings/piles
Sidewalks

Geometry

Total length
Longest/shortest span
Tallest/shortest pier
Longest/shortest piles
Skew
Horizontal/vertical curvature
Slope
Width out to out and curb to curb
   (min, max)
Width of sidewalks
Parapets, railings
Clearance above (min, max)
Clearance below (min, max)
Lateral clearance
Channel depth, high, low

Service/condition

Feature carried
Feature crossed
Number of lanes (width)
Number of tracks
Average daily traffic (ADT)
Utilities
Load rating
Condition rating
Vulnerability rating
Serviceability rating
Potential hazards

History

Date of completion
Dates of reconstruction/major
   rehabilitation
Maintenance record
Inspection record

Network data

GPS reference
Alternate routes
Closest intersections
Importance

The preceding entries must be readily available for data searches. Digitized drawings, photographs, and cross references to GPSs are standard.

8.2   TYPES OF BRIDGES

The NBI (FHWA, 1995b) classifies bridges according to material (see Table A14.1), main structure type (see Table 14.2), feature carried (see Table A14.3) and feature crossed (see Table A14.4). Most bridge texts, such as Barker and Puckett (1997), Taly (1998), and Xanthakos (1994), provide examples of all types of bridge structures, listed in Table A14.2.

Figure 8.1 High Bridge, 1848, 1928, New York City.

Railroad bridges that also carry vehicular traffic (no. 4 of Table A14.3) are managed by the FHWA. Uniquely railroad bridges (no. 2 in Table A14.3) are part of the NBI but under the management of AREMA.

Number 0 (other) in Table 14.3 can be an aqueduct or utility-carrying structure. The oldest bridge in the New York City inventory is a dismantled aqueduct, modeled after the Roman Pont du Gard near Nimes, France, and refitted with a steel three-hinge arch to facilitate navigation. The High Bridge over the Harlem River, dating from 1848 and modified in 1928 (Fig. 8.1), has also served as a pedestrian crossing and will be rehabilitated in order to provide this service again.

Of the features crossed in Table 14.4, no. 5, waterway, has the greatest significance for bridge design and operation. Also significant are railroad operations under a bridge, because the two modes of transportation must coordinate their activities.

NBIS definitions are not always conclusive about the level of detail of the inventory (e.g., bridge, span, component, and element specific). Adequate decision support requires data on the components, elements, and members in all spans of the bridges on the network. Management decides to what level that information should be centralized.

8.3 COMPONENTS, ELEMENTS, AND MEMBERS

In most texts, elements are subsets of the sets of components. A bridge element in turn is a set of members. A span primary member may consist of floor beams and steel or concrete girders, as shown in Fig. 4.12 (Section 4.2.2). Figures 8.2 and 8.3 illustrate

Figure 8.2   Typical steel superstructure: stringers, diaphragms, bracings, floor beams, and girders.

Figure 8.3   Typical concrete superstructure: prefabricated prestressed girders, composite with deck, and diaphragms.

Table 8.1   Bridge Components and Elements According to NYS DOT (1997)

Components	Elements
Abutments	Joint with deck, bearings, anchor bolts, pads, bridge seat and pedestals, backwall, stem, erosion or scour, footings, piles
Wingwalls	Walls, footings, erosion or scour, piles
Approaches	Drainage, embankment, erosion, pavement, guide railing
Stream channel	Stream alignment, erosion and scour, bank protection
Deck elements	Wearing surface, curbs, sidewalks and fascia, railings and parapets, scuppers, gratings, median, monodeck surface
Superstructure pier	Deck structural, primary members, secondary members, paint, joints bearings, anchor bolts, pads, pedestals, top of pier cap or beam, stem solid pier, cap beam, pier columns, footings, erosion or scour, piles
Utilities	Lighting standards and fixtures, sign structure, utilities and support

typical steel and concrete spans. Table 8.1 describes the bridge components and elements according to the New York State inventory.

The FHWA *Coding Guide* records the number of spans and identifies the main structural types of the components, as shown in Table A14.2, but does not discretize into spans and elements for evaluation purposes. That practice was augmented in 1997. PONTIS refers to the commonly recognized (CoRe) structural elements introduced by AASHTO (1998b, Interim 2002). The CoRe guide considers certain components in greater detail, still not on a span level. The NYS DOT bridge inventory records the elements in every span and their condition is rated. Appendix 14 enumerates typical bridge components and elements according to NBIS and NYS DOT. "Erosion and scour" are listed as bridge elements, in order to ensure the regular inspections and assessments of these vulnerabilities (Section 10.2).

The inventory contains an appraisal of the as-built state of the assets. The as-is condition must be evaluated or assessed and rated or prioritized periodically. The concurrent use of synonymous terms underscores the great variety of the used methods, their output, and their relative significance. Appendix 40 describes some of the widely used condition rating systems. Condition ratings are assigned according to the inspection criteria adopted by the owner (Section 10.4). Load ratings reflect the capacity of the weakest *member* (Section 10.5) according to the governing design specifications.

Chapter 9

Assessments: Conditions, Needs, and Resources

Descartes concluded that opinions vary because people "think along different routes and do not consider the same things" (Section 1.8). Management must assign appropriate weight to the different but significant opinions in order to take all of them into account. The diversity of the evaluations relevant to bridge management stems from the broad range of interests, most importantly the following:

Community needs (e.g., transportation)

Structural needs (e.g., maintenance, repair, rehabilitation, replacement)

Assessment and response capabilities (e.g., expertise)

Resources (e.g., funding)

Thus evaluations are a priori subjective in different ways. Structural condition assessments may appear more objective than those of needs, because they pertain to actual structures; however, they evaluate uncertain phenomena subjectively. The equilibrium between conditions, needs, and resources cannot be rigorously optimized or even quantified, because these three essential variables are not always sufficient and sufficiently known.

"Every scientific law, every scientific principle, every statement of the results of an observation," stated Feynman (1999), "is some kind of a summary which leaves out details, because nothing can be stated precisely."

Since all formalized assessments focus on specific aspects of reality, any one of them is limited. Despite their inconsistencies and contradictions, however, in a broader context they can serve as a redundant system of improved cumulative reliability.

In the BMS flowchart of Fig. A16.6 (BRIME, 2002), project-level management communicates to the network level the assessments of costs and structural conditions. No communication is shown in the reverse direction. Small et al. (in TRC 498, 2000, A-1) locate the determination of costs on the project level and that of policies on the network level.

In the bilateral model of Fig. 1.33a, network-level managers maximize service (e.g., traffic), whereas project-level managers maximize structural performance. Decisions are reached by the various methods discussed in Chapter 2. Ultimately they are the product of expediency, precedent, negotiations, personal and popular preferences, and rigorous cost–benefit optimization. Management expresses costs and benefits in monetary terms.

Engineering qualifies and quantifies structural conditions. Infrastructure asset management must establish a language able to correlate the two types of assessments. When all involved are "bilingual," this critically important two-way transfer of information between the domains of physics and finance is fully redundant. Continuity (Section 4.1.1) is thus achieved at the expense of some duplication of effort (or expertise). In the trivial case when engineer and manager are the same, the conflicting demands for needs and condition assessments are integrated. The process becomes more efficient but also nonredundant and unpredictable. For practical purposes, indicators more nuanced than mere costs, for instance, describing the levels of service (Chapter 11 herein), are developed.

The following assessments typically span between economics and mechanics:

Estimates of the "remaining useful life" of the assets

Investment required for restoring the assets to as-built or as-new condition

Public asset management agencies are required to report annually the estimated differences between the existing and the as-new states of their assets in monetary terms. Such assessments are fraught with the following chronic inaccuracies (see also Section 4.1.5):

Structural condition assessments are vague in engineering terms and become even less precise when translated into monetary terms.

Condition assessments do not define the scope of rehabilitation work. Actual projects include structures and components in good condition as well as many costs dictated by logistics and constructability. Structures may be rehabilitated before they have reached the end of their useful life if they are located along an already closed traffic corridor.

The needs of the structures and the users are estimated separately and, often, independently. The effort to maintain a continuous traffic flow during reconstructions adds significantly to the construction costs. Table E3.2 does not identify explicitly either the user costs or the hardship assumed by the owner in order to reduce them.

Tentative as they are, the estimated costs of the assets' replacement and restoration alert the community of the potential deficit carried by a deficient infrastructure, as do the statistics of trade deficits and national debt.

Adey and Hajdin (2005) treated structural needs/conditions as an inventory deficit. Five levels of structural deterioration are expressed in terms of a debit from an inventory, which would be fully supplied if bridges were as new. Bridge conditions can then be modeled as in an inventory stock problem, for instance, by Markov chains, particularly if the network is relatively homogeneous and the supply of condition improvement is assured on demand.

Qualitative ratings (as in Appendices 33 and 40) can prioritize limited resources (Section 2.3) and provide approximate needs estimates. Hearn (in Frangopol, 1999b) called them "priority rankings." The author distinguished between needs assessments based on priority ranking functions (Appendix 41) and those resulting from cost optimization. The NBIS *sufficiency rating,* the NYS DOT *bridge condition rating,* and Pennsylvania's *total deficiency rating* (Appendix 44) are typical priority ranking functions. The complementary alternative is an assessment based on optimized cost–effect of preventive and corrective interventions. Such an approach reflects the management view of bridges as a set of annual expenditure needs. Inevitably, the two interact. Specific actions such as closures, load postings, and retrofits can be correlated to certain ratings provided that each decision is confirmed by a site-specific evaluation. Example 18 illustrates how bridge condition ratings must correspond to the needs for capital expenditures of a bridge owner. Caltrans (Appendix 41) uses a *health index,* translating inspection ratings into costs.

Example 18. Bridge Management Equilibrium in Terms of Condition Ratings

Yanev (2003) defined a "bridge management steady state" in terms of the following notation:

A	deck area of bridge stock
A_{Rec}	deck area under reconstruction
A_{Rep}	deck area under repair
R	average condition rating of bridges with area A
ΔR_{Rec}	average annual change of R of A_{Rec}
ΔR_{Rep}	average annual change of R of A_{Rep}
r	annual rate of negative change of R (e.g., deterioration) of $A - A_{Reh} - A_{Rep}$
C_C	bridge reconstruction cost (U.S. \$/unit of bridge deck area/year)
C_R	bridge repair cost (U.S. \$/unit of bridge deck area/year)
C_M	bridge maintenance cost (U.S. \$/unit of bridge deck area/year)
L	bridge useful life (years)
L_0	bridge useful life at no maintenance (years)
L_M	bridge useful life at full maintenance (years)

If the average condition rating of the stock remains constant, the following condition must be satisfied:

$$(A - A_{Rec} - A_{Rep})\, r = A_{Rec}\, \Delta R_{Rec} + A_{Rep}\, \Delta R_{Rep} \qquad \text{(E18.1)}$$

Figure E18.1 illustrates the bridge management equilibrium under the described assumptions. Figure E18.2 is the flowchart of actual activities maintaining this equilibrium.

For simplicity, repairs and reconstructions are merged into a general rehabilitation such that

$$A_{Reh} \approx A_{Rec} + A_{Rep}$$

where C_{Reh} = average bridge rehabilitation cost (U.S. \$/unit of bridge deck area/ year

ΔR_{Reh} = average annual change of R of A_{Reh}

Equation E18.1 reduces to

$$(A - A_{Reh})\, r = A_{Reh}\, \Delta R_{Reh} \qquad \text{(E18.1a)}$$

or

$$\frac{A_{Reh}}{A} = \frac{1}{\Delta R_{Reh}/r + 1} \qquad \text{(E18.1b)}$$

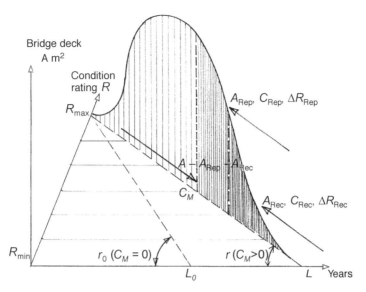

Figure E18.1   Bridge management equilibrium.

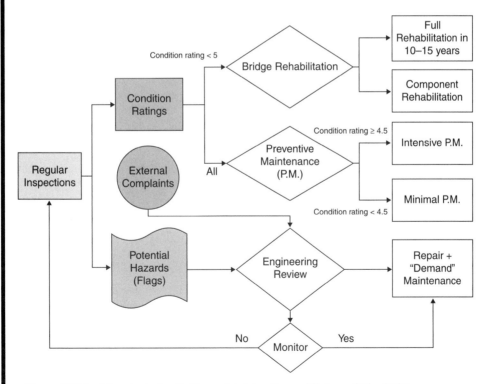

Figure E18.2   Flowchart of activities maintaining the equilibrium of Fig. E18.1.

The ratio $\Delta R_{\text{Reh}}/r$ represents condition improvement versus deterioration:

$$\frac{\Delta R_{\text{Reh}}}{r} \to 0 \text{ if either } \Delta R_{\text{Reh}} \to 0 \text{ (e.g., rehabilitation is ineffective) or}$$

$$r \to \infty \text{ (e.g., deterioration is instant)}$$

In either case, it follows that $A_{\text{Reh}} = A$, i.e., the entire stock must be constantly rehabilitated.

Equations E18.1, E18.1a, and E18.1b describe a steady state such that the average bridge condition remains constant. Rehabilitation (representing reconstruction and repairs) counteracts the effects of deterioration, as in the simplified illustrations of Example 17. The direct annual costs C_{DA} associated with maintaining this equilibrium are

$$C_{\text{DA}} = (A - A_{\text{Reh}})C_M + A_{\text{Reh}}\,C_{\text{Reh}} \qquad (E18.2)$$

Substituting A_{Reh} from Eq. E18.1b into Eq. E18.2 yields

$$C_{\text{DA}} = \frac{A\,(C_M\,\Delta R_{\text{Reh}} + r\,C_{\text{Reh}})}{\Delta R_{\text{Reh}} + r} \qquad (E18.3)$$

The C_{DA} of Eqs. E18.2 and E18.3 can be reduced by setting $C_M = 0$ as follows:

$$C_{\text{DA}} = A_{\text{Reh}}\,C_{\text{Reh}} \qquad (E18.2a)$$

$$C_{\text{DA}} = \frac{A\,C_{\text{Reh}}}{\Delta R_{\text{Reh}}/r + 1} \qquad (E18.3a)$$

Equations E18.2a and E18.2b state the "zero-maintenance" policy. The simplicity of the equations and of that policy stems from ignoring the effect of C_M on the rate of bridge deterioration r.

In practical terms, C_{DA} can be minimized by the following means:

Reduce C_{Reh} and Increase ΔR_{Reh}. Rehabilitation cost and quality are well-recognized targets of bridge management optimization.

Reduce r. In terms of Eq. E18.3a, the deterioration rate r is highly uncertain and subject to elaborate modeling (Examples 9 and 17). It is believed that r depends on the level of maintenance. In Fig. E18.1, that dependence is represented as a change of the slope from r_0 to $r\,(C_M)$. The implication is broadly stated as

$$r = f(C_M) \qquad (E18.4)$$

Substituting r from Eq. E18.4 and minimizing C_{DA} with respect to C_M yields

$$(C_{\text{Reh}} - C_M)\,\frac{\partial r}{\partial C_M} + f(C_M) + \Delta R_{\text{Reh}} = 0 \qquad (E18.5)$$

Equation E18.5 would allow the optimization of C_M and C_{Reh} if $f(C_M)$, the effect of maintenance cost on deterioration, were known. To the contrary, however, C_M and

r are related through many uncertain correspondences, most prominently the following:

Maintenance *cost* does not model maintenance *effectiveness* with sufficient accuracy.

The effects of maintenance on the longevity of structural elements and on their *condition ratings* are not known and are not identical.

The *bridge condition rating R* is obtained by prescribed formulas (Appendix 41) from the condition ratings of components and elements, which in turn reflect subjective visual assessments.

Examples 23 and 24 illustrate how a variety of maintenance strategies can be conducted at the same annual cost C_M.

The equilibrium model assumes knowledge of the relationship between the deterioration rate $r(C_M)$, the maintenance cost C_M, and other variables, such as traffic and time.

As in all modeling exercises, $f(C_M)$ can be represented by purely formalistic considerations, by phenomenological ones, or by combinations of the two. Examples 20–24 illustrate some of the possibilities.

9.1 SUPPLY OF AND DEMAND FOR EXPERTISE

The structural assessments summarized in Fig. 4.4a and Tables 4.5, and 9.1 have evolved through the demand for information and the supply of expertise. They fall into the groups

Table 9.1 Bridge Conditions, Sources, Types of Evaluation and Response

Condition	*Source*	*Evaluation*	*Typical Response* *(Except Do Nothing)*
Serviceability (Section 10.1)	Inspection, analysis, inventory	Appraisal	Repair, rehabilitation, replacement, reconfiguring of traffic
Vulnerability (Section 10.2)		Rating	Review, strengthening, retrofit, rehabilitation, replacement, closure
Potential hazards (Section 10.3)	Inspection, analysis, NDT&E, diagnostics	Flags	Closure, review, repair, monitoring
Condition assessment (Section 10.4)		Condition rating, priority ranking	Review, repair, rehabilitation, replacement, emergency repairs, load rating, maintenance
Load rating (Section 10.5)		Inventory and operating rating	Load posting, strengthening, closure

of in situ inspections, inventory review (e.g., data search), and analysis. The findings may reflect average conditions (for general network-level assessments) or worst cases (for immediate action). They may emphasize professional opinion (or "recommendation") or seek to minimize it by close adherence to formulas and instructions. Determinations calling for prompt remedial actions are critical. The different assessments call for different types and levels of expertise. The demand for advanced expertise implies that the required assessment is critical and therefore that management is vulnerable.

In the absence of life-cycle management, conditions are rated primarily according to their level of risk. Newly appointed management of existing facilities has no choice but to address potential hazards first (Section 10.3). Once hazards have been mitigated, structural conditions are evaluated according to a broader (e.g., life-cycle) range of anticipated needs.

9.2 NEEDS/RESPONSE OPTIONS

"Proactive" assessments must correspond to available options. The choice of options in turn depends on the findings. In his recommendation of Bayesian statistical decision theory for structural reliability analysis, Cornell (in Freudenthal, 1972, p. 47) pointed out: "An immediate but often overlooked conclusion is that, as with stress analysis, only as accurate and detailed a reliability analysis as will have a significant economic influence on the final design is justified."

This principle is fundamental to operational research. Beale (1988) singled out the need to maintain models only as realistic as necessary (Appendix 8). System identification is guided by *frugality*. An iterative (or heuristic) process of updated "least-cost" (e.g., least number of significant detail) models emerges, as shown in the following elementary flowchart.

$$\text{Condition assessments} \rightarrow \text{Needs estimates}$$
$$\uparrow \qquad\qquad\qquad\qquad \downarrow$$
$$\text{Actions taken} \leftarrow \text{Forecasts} \leftarrow \text{Action alternatives}$$

Bridges can be assessed according to a selected evaluation system until the accumulated data suggest modifications. The spread of conditions along the adopted rating scale must correspond to the capability of management to address them (Examples 17–20). A "bipolar" scale (e.g., safe/unsafe) would be adequate for rating potentially hazardous conditions (Section 10.3) if identification were unerring and prompt remedial action were assured. Table 9.1 assembles a set of typical condition assessments, each satisfying specific management needs and consistent with engineering capabilities.

The summary of structural assessments and responses in Table 9.1 is typical. It is conducive to the "responsive" mode which management seeks to outgrow. Alternatively, bridge-related needs can be summarized as in Table 9.2.

Table 9.2 is less specific but more adaptable to long range proactive planning. It reminds us that qualitative needs (such as maintenance and inspection for structures or safety for the users) are essential and should not be satisfied merely in response to specific quantified demands (e.g., conditions) but as a routine management practice.

Certain aspects of qualitative needs can be quantified in terms of costs; however, their overall impact remains unquantifiable. In a democratic context, qualitative choices

Table 9.2 Bridge-Related Needs

Needs	Users	Structure	Engineering Management
Quantitative (measurable, objective)	Speed, capacity, geometry, ridability, pollution control	Repair, replacement, rehabilitation: strengthening, widening, clearance, etc.	Funding, personnel, equipment, time and material, database, support (legal, administrative)
Qualitative (perceived, subjective)	Comfort, accessibility, safety, aesthetics, environmental compatibility	Maintenance, repair, upgrading, hazard mitigation	Support, expertise (technical, decision—BMS, popular, political)
Cost ($)			

are made with broad public participation, as in the examples of Figs. 1.43, 1.44, and 1.45. Needs are discussed in Chapter 11.

9.3 QUANTITY/QUALITY AND DETERMINISM/UNCERTAINTY

Structural condition assessments seek to establish the following uncertain correlations:

$$\text{Real structure} \leftrightarrow \text{Structural model}$$

$$\text{Structural model} \leftrightarrow \text{Rating system}$$

$$\text{Parameters modeling structural behavior} \leftrightarrow \text{Methods of assessment}$$

$$\text{Experts performing assessment} \leftrightarrow \text{Assessment reports}$$

Socioeconomic, environmental, and transportation conditions contribute their own uncertainties to the assessment process.

The types of uncertainties inherent in engineering and management evaluations and the methods appropriate for their treatment are discussed in Appendices 5–9 and 36–39. Because they are significant to varying degrees in condition assessment, statistical and deterministic methods supplement each other in the process. As in structures and operations, redundancy can improve reliability—hence the numerous (redundant) forms of assessment described in Chapter 10. Cornell (Freudenthal, 1972) is quoted on the application of statistical decision theory to structural design in Appendix 2.

Quantities may enhance the reliability of engineering assessments, but they do not readily convert to quality. Sections 4.1 and 5.3 caution that duplication of tasks and redundancy is not automatically conservative, particularly if one assessment is used to override the restrictions of another. Management must set procedures that preclude such a course. For instance, the NYS DOT has established deterministically a set of structural vulnerabilities (Section 10.2) according to the inventory, potential hazards according to field inspections, and load ratings according to analysis. Any of these can require prompt remedial action.

The number of rating levels discussed in the preceding sections and illustrated in Appendix 40 must correspond to the range of anticipated conditions, the ability to identify them, and the possible actions. Relatively few levels (4–5 in the AASHTO CoRe and DANBRO scales) reflect a higher confidence in the evaluation, response, and forecasting. Added rating nuances (7 levels in NYS DOT, 10 in NBIS) facilitates statistical treatment of the data but reduces the decisiveness of the output.

9.4 CHANGE OVER TIME

The service life of a properly managed bridge does not end with a collapse but with a project for its replacement or, in the worst case, with a closure. At such a time, assessments must still rate the bridge fit for a minimal service, such as the adequacy for dead load. (The subject is discussed further in Section 11.3.)

In order to estimate future *conditions,* management relies on forecasts of *condition ratings* likely to result from future inspections. Condition ratings are therefore useful only so long as they change at a predictable rate over time.

Time is the argument in all forecasting (Section 10.4.5), even though it is somewhat coincidental to bridge deterioration. The Roman aqueducts seem to have risen above it largely because they have been relieved of all service loads. Golabi et al. (1992) weigh the pros and cons of modeling deterioration as a function of structural "age" and, despite reservations, find it justifiable. Ultimately, the significant data points on the time scale are those marking changes in the type of service, corrective interventions, and extreme events (each distinguished by their respective properties).

Most network-level BMS algorithms model the probability of future changes in structural conditions by Markov chains (Appendix 44). The Markov chain model has been both praised and criticized for ignoring the structure's past. It is more applicable under invariable service and environmental conditions.

Changes in the structural conditions are assumed to be either continuous or stepped. By stipulating biennial inspections, the NBIS imply that conditions would not decline significantly over less than two years. Annual interim inspections were subsequently recommended for bridges in advanced states of deterioration (e.g., by the NYS DOT).

The scheduling of inspections can be optimized as a function of structural condition but should be adjusted for random events such as traffic accidents. For larger networks with many uncertainties and scattered conditions, inspections at regular intervals can balance cost with reliability. Examples 9 and 17–20 describe elementary relationships between NYS DOT condition ratings, their change over time, the present condition of the assets, and the estimates of future needs.

Condition ratings support long-term bridge management decisions involving large portions of national and local budgets. Consequently, once defined, they should remain as consistent as possible. Nonetheless, an iterative adjustment of the rating system appears inevitable. The "heuristic" nature of the process reflects the accumulated knowledge of the prevalent conditions and the evolving bridge management priorities.

9.5 SIZE, COMPLEXITY, AND IMPORTANCE

Size, complexity, and importance are purely relative and hence network-specific. A transportation network is likely to include, for example, simple span decks and girders, multispan continuous prestressed beams, and suspension, cable-stayed, and movable bridges, that is, both routine and unique features. The importance and the complexity of the structures are usually related, so that unique bridges also provide critical services. The same management procedures do not fit all the structures. Such is the case of New York City (Examples 1–3) where four organizations manage 2200 bridges. One of them, the NYC DOT, manages 800 bridges. The average number of spans is 5; however, included are the highly complex and unique East River bridges, 25 movable bridges and bridges with 431 spans, all of which require specific management. In order to satisfy both network and project management needs, condition evaluations must achieve the following:

> The condition of unique and complex structures must be quantified in sufficient detail. At unique structures, the regular inspections can be supplemented by special ones, designed to address critical details (Chapter 14) and the performance of essential maintenance tasks.

> Ratings should distribute the existing conditions of large and diverse inventories into subsets of manageable size. Priority ranking formulas (Appendices 39–41) are indispensable in sorting out numerous candidates. On the level of a specific structure they merely point to a need for more specific evaluations, such as load ratings (Section 10.5) and diagnostics (Section 10.6).

BRIME (2002, Chapter 4) illustrated (rather than prescribed) a possible functional relationship between five (or six) condition assessment levels and the need for load ratings (Appendix 45). Itoh and Liu (in Frangopol, 1999a, pp. 136–151) described a similar five-grade rating system adopted by the Japanese Ministry of Construction for assessing the concrete decks of 130,192 bridges with a length of 7481 km. An importance factor can be assigned to structures or routes, depending on the volume of traffic, the level of network redundancy, safety, environmental, and other considerations. That factor can be used as a modifier of the independently obtained structural ratings, reflecting management priorities. Because of their subjectivity and their origin at the network level, indicators of importance and complexity (as well as risk exposure) are perceived differently at the various project levels. Decisions based on such indicators are therefore the subject of vigorous disputes in the spheres of popular opinion, professional, and political management.

Chapter 10

Structural Conditions

Engineering evaluation cannot be entirely independent of needs, particularly the more urgent ones. Nonetheless, structural conditions are not negotiable and, as much as possible, should be assessed by objective engineering methods. Condition evaluations therefore must serve the common purpose of engineering and management by translating primarily objective condition assessments into essentially subjective needs estimates. In order to satisfy these possibly conflicted demands, conditions assessments, as all languages, are "fuzzy." Engineering could not deliver nor could management use an uniquely defined condition assessment system.

The assessments described herein are designed to support decisions under the constraints of minimizing hazards and expenditures, while maximizing service and satisfaction.

10.1 SERVICEABILITY

Distinguishing between as-built and as-is structural performance allows managers to measure disrepair and obsolescence of their stock independently. According to the NBIS (FHWA, 1995b) structural conditions are *rated,* whereas serviceability is *appraised.* The numerical scales are similar, as shown in Appendix 40. *Appraisal* appears more appropriate for the assessment of the inevitably subjective service, whereas *rating* is reserved for the more readily quantifiable strength.

The serviceability of the as-built structure is quantified to a degree in terms of traffic type, capacity, and average traveling speed. Quality of service can be estimated according to design parameters, such as number and width of lanes, curve radii, accessibility, and congestion. If not directly available, traveling speed can be deduced from traffic volumes.

The serviceability appraisal can decline due to upgraded traffic demands, while the structural condition remains unchanged. Poor structural condition reduces the ADT and increases the likelihood of safety hazards. That effect is taken into account by a composite rating (such as the NBIS *sufficiency rating,* discussed in Section 11.1 and Appendix 41). The service under the bridge is also affected, but to a different degree.

With nine grades, including "somewhat better than minimum adequacy," the NBIS rating scale is refined to the point of becoming "chromatic." The main advantages of such narrow distinctions are the following:

The number of worst cases requiring immediate action appears manageable.

The changes in the appraisals are likely to be more apparent and the trends more pronounced.

The NBIS serviceability appraisal of structures supports network assessments. The serviceability of the network as a whole requires complex assessments, for instance, as discussed in Appendix 15. Environmental requirements, such as the presence of toxic materials, can be linked to serviceability.

10.1.1   Serviceability Forecasts

Both the serviceability with respect to the standards adopted by the network and the standards themselves evolve. Consequently, separate forecasts must anticipate the physical degradation and the transportation demand. Socioeconomic assessments (Section 10.3) are essential for serviceability forecasting. NCHRP Report 538 (2005) discusses traffic data collection for mechanistic forecasts. The *deficient* and *obsolescent* bridges in the NBI are equally significant (Appendix 14).

Bridges are constructed in response to anticipated traffic demands, which, to a degree, they are also expected to generate. The Brooklyn Bridge attracted not only its own traffic but that of the three later East River crossings (Example 3) and the entire New York metropolis. In the mid-twentieth century escalating demands for automobile routes supported predominantly expansionist traffic forecasts. The George Washington Bridge (Example 2) opened with four traffic lanes in 1931, but another four were ready for opening in 1943. The anticipated lower deck added six more lanes in 1962. Had the demand developed otherwise, the lower deck was designed to carry train tracks as well.

Recent megastructures, such as the Akashi-Kaikyo Bridge in Japan (Fig. 1.40a), the Great Belt (Fig. 1.41), the Öresund between Denmark and Sweden, and the proposed Messina Straits Bridge, must take into account ferry services and charge comparable tolls, thus essentially aiming at a balanced rather than a competitive number of passengers.

Towards the end of the twentieth century, demands for environmental quality began to outweigh those for traffic quantity. Urban communities sought to limit vehicular traffic volumes by restricting the expansion of the bridge and roadway infrastructure. Tunnels were constructed to steer traffic away from the densest areas. Rail mass transit has become popular again.

Figure 10.1 illustrates the different behavior of serviceability (represented by load rating) and structural condition over time. The former must remain constant over the life of the structure and must meet certain limited requirements under extreme (ultimate) conditions. The latter may follow various deterioration trajectories. For roadway surfaces, the present serviceability index (PSI) combined several quantitative pavement characteristics according to the AASHTO Road Test of 1962 (NCHRP Synthesis 330, 2004, p. 10). The PSI was correlated with user judgment, resulting in a standard model of serviceability, declining over time. The result is similar to the convex curve shown in Fig. 10.1. A concave (or piecewise concave) extension at advanced age and low ratings is assumed possible with corrective interventions.

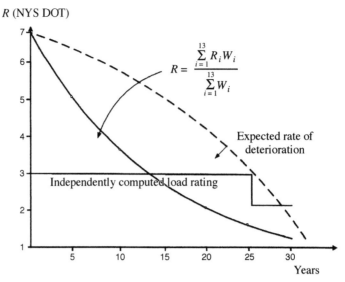

R (NYS DOT)

$$R = \frac{\sum\limits_{i=1}^{13} R_i W_i}{\sum\limits_{i=1}^{13} W_i}$$

Expected rate of deterioration

Independently computed load rating

Years

Figure 10.1 Convex and concave models of condition deterioration and load rating estimates.

Figure E9.3 shows that the relationship between the ratings of serviceability (as defined by the FHWA sufficiency rating) and structural condition (for instance as defined by NYS DOT) can be represented by either a convex or a concave curve, depending on the orientation of the coordinate system. Structural conditions decline asymptotically with respect to serviceability (of which they are the cause rather than the effect), as well as with respect to time (Fig. 10.1). As structural conditions deteriorate, the decline of serviceability accelerates and is the first to reach unacceptable levels.

Sections 1–3B of the *Recording and Coding Guide* (FHWA, 1988) link the serviceability appraisal ratings to levels of ADT and bridge geometry. That information must therefore be regularly updated (ADT inventory data are always worth verifying).

10.2 STRUCTURAL VULNERABILITY

BMSs treat the identification and prioritization of structural vulnerabilities as a condition assessment (e.g., a form of system identification). The inventory should reveal potentially vulnerable combinations of structural and operational features. Data mining can use generic algorithms or develop criteria as required by local conditions.

The as-built stock is evaluated according to the inventory (such as the NBI in the United States), load estimates, extreme-event forecasts, and a variety of other sources, some of them outside the immediate scope of bridge management. In contrast, imminent and potential hazards can only be observed by direct inspection. The NBI is updated annually according to the NBIS. For certain bridge types, computer programs (such as Vitris in the United States) obtain load ratings, taking into account section loss and local damage of the structural elements.

Among the structural vulnerabilities discussed in Chapter 4, most frequently identified as critical are excessive live loads, nonredundancy, fracture-critical or poorly performing details, inadequate clearance, seismic and high-wind zoning, foundations in liquefiable soils, and submerged pier footings. The seismic vulnerability governs in California and in the New Madrid area; the hydraulic one (scour) is critical in New York State. Vessel collisions (discussed at length, e.g., in Gluver and Olsen, 1998) are critical in Florida and Maine. Floods and hurricanes govern in Louisiana, Florida, and Texas. Security rose abruptly among priorities after September 11, 2001 (still to a different degree depending on the location).

The NBI specifications (FHWA, 2005a) recommend the coding of seismic (Appendix 14) and scour vulnerabilities on a scale of 8 (not vulnerable) to 1 (critical or failed).

The New York State bridge safety assurance program recognizes the following seven vulnerabilities (also listed in Chapter 4):

Hydraulic (NYS DOT, 1991)

Seismic (NYS DOT, 1995)

Overload (NYS DOT, 1993)

Collision (NYS DOT, 1995)

Steel details (NYS DOT, 1993)

Concrete details

Security

The list combines familiar structural flaws and extreme events. They persist as vulnerabilities in part because of changing structural conditions, traffic, and extreme-event expectations. Regional seismicity forecasts are often revised. Hurricanes, prestressed concrete, suspension, and cable-stayed details are not included but could be foremost in different geographic settings.

10.2.1 Vulnerability Forecasts

The consequences of vulnerabilities, i.e., the "likelihood of occurrence" portion of the risk analysis defined by Eq. 5.1, are the subject of forecasting. "Penalties" are discussed in Section 10.4. Extreme events such as earthquakes, floods, and collisions are treated as random. The probable (or credible) magnitudes and return periods of such events are usually selected by a combination of probabilistic and deterministic reasoning. Some types of hazards (such as the seismic one) provide a larger statistical databank than others (e.g., acts of terrorism). For example, credible earthquake magnitudes are selected in a "two-tiered" approach for 500- and 2500-year return periods.

Babaei and Hawkins (1993) prioritized seismic retrofitting needs according to the number of structural deficiencies and the estimated remaining life. Appendix 33 describes the deterministic prioritization for seismic vulnerability adopted by NYS DOT (1995), and the probabilistic model developed by Basöz and Kiremidjian (1996).

When a relatively low seismic hazard is coupled with a relatively poor average bridge condition, as in New York State during the 1980s and 1990s, the seismic vulnerability program replaces all substandard details during regularly scheduled bridge rehabilitations. Figures 4.38, 10.2, and 10.3 are examples.

The FHWA is funding a comprehensive pre- and postearthquake bridge safety program through its three national centers. A number of publications on seismic bridge design and retrofit are available on federal (Multidisciplinary Center for Earthquake Engineering Research; FHWA, 1995, 2002a; AASHTO, 1999b) and state levels. The current FHWA project DTFH61-98-C-00094 will update the seismic bridge retrofit manual and issue new vulnerability guidelines for the existing highway infrastructure. The LRFD AASHTO (1998a) specifications recognize vulnerabilities in a number of ways, for instance, by specifying extreme-event limit states and by penalizing nonredundancy.

10.3 POTENTIAL HAZARDS

Whereas vulnerabilities (Section 10.2) should be discernible from the inventory, potential hazards are identified during field inspections. The two are not independent and the database should allow cross referencing between them.

So long as they exist, potential hazards override other optimization constraints (Section 4.1.2). The NYS DOT has designated potential hazards as FLAGS (NYS DOT, 1997). Appendix 46 describes the flagging procedure. Chapter 4 discusses numerous typical hazards. Example EA46 describes the history of flag incidence in New York City.

Figure 10.2 Retrofitting of bridge bearings under traffic.

Figure 10.3 Retrofitted fracture-critical pin-and-hanger assembly.

NBIS 23 CFR 650 Subpart C specifies the *follow-up on critical findings* as follows:

§650.313 Inspection procedures

(h)Establish a statewide or Federal agency-wide procedure to assure that critical findings are addressed in a timely manner. Periodically notify the FHWA of the actions taken to resolve or monitor critical findings.

10.3.1 Potential-Hazard Forecasts

If traffic-related hazards are well documented, they can be modeled stochastically, so long as the governing conditions remain constant. The events are never entirely independent of structural and operational conditions. Therefore the samples must be as homogeneous as possible (e.g., similar structures, surface conditions, traffic type and count, weather).

The hazards resulting from structural conditions are neither random nor purely deterministic. Hazards can be related to condition *ratings* if the bridge population offers adequate samples in each level. The model can be probabilistic (if data are abundant), deterministic (if little input is available), or mixed (in order to take advantage of both approaches).

The "flag" forecasting method described in Example EA46 is deterministic. It was used successfully at the NYC DOT during the early 1990s when the reports of potential hazards had reached 3000 per year. At U.S. $10,000 to $15,000 per repair and given the potential risks to the traveling public, hazard mitigation dominated management priorities. The described method relies on the relatively consistent relationship between the condition ratings of structural elements and the flags these elements have generated. Based on the forecasts, the expected conditions were subjected to a triage (Sections 4.1.2 and 14.3). The options were to repair permanently or temporarily or to monitor until the next regular inspection.

The correlation of "actual" hazards and reports of potential hazards must be addressed separately. A number of factors (accidents, work conditions, qualifications, etc.) influence the highly subjective process of identifying and reporting potential hazards.

Damage Assessment

Damage is assigned two essentially opposite meanings as follows:

The consequence of a concrete occurrence or event. In this view *deterioration* is a continuous process punctuated by instances of discrete *damage.*

Any departure of the managed asset from the as-built condition and hence all deterioration.

Yao and Furuta (1986) recognized three types of damage assessment: numerical, monetary, and verbal. Adeli (1988, p. 163) divided damage assessment into qualitative damage descriptors and quantitative damage levels. The latter are determined according to the following three criteria:

1. Structural integrity
2. Functionality
3. Repairability

Depending on the definition, damage may require immediate corrective measures or may be the subject of more elaborate qualitative and quantitative condition evaluations.

In the most general terms, *damage* also signifies financial loss. Thus *minimizing damage* can imply lowest costs over a given planning horizon. That minimization problem, however, requires a two-way conversion between structural qualities and monetary quantities. That is the domain of structural condition evaluation.

10.4   STRUCTURAL CONDITION EVALUATION

Condition evaluations are perceived differences between the as-designed, as-built, and as-is states of structures. They are the most complex and the vaguest engineering assessments. The subject can be a bridge element, a group of similar elements within a span, or in all spans (see Table 8.1), components, and ultimately the entire bridge. The predominant method is visual inspection. Condition evaluations determine present and future needs, the adequacy of maintenance, and the effects of traffic and the environment. They are used both qualitatively and quantitatively (see Fig. 4.4a).

Project management is familiar with field conditions (hands on) and tends to assess them qualitatively. Network management compensates for the lack of hands-on experience by quantification. In order to accommodate both top-down and ground-up (e.g., centralized and decentralized) needs, condition evaluations must be descriptive and prescriptive; they must employ analogy as well as analysis. Essential prerequisites include the following:

Qualified staff (e.g., licensed professional engineers)

Reliable inventory

Established inspection and rating procedure

Response capability

Inspections perform the following tasks:

Verify the inventory (Chapter 8)

Identify potential hazards and schedule remedial action (Section 10.3)

Rate the conditions of structural elements or components in terms of departures from the presumed as-built state and describe them in details corresponding to their gravity

Rate the load-bearing structural capacity

Recommend remedial actions

The condition rating systems presented in Appendix 40 attempt to incorporate these diverse tasks according to the demands of the networks they serve. They fall into two general groups that can be roughly described as *rating/descriptive* and *defect/action*.

10.4.1   Rating/Descriptive

The process consists of the following task sequence:

Establish condition rating scales → inspect → evaluate/rate → describe → recommend prompt action → analyze → recommend short- and long-term action → update database

A high level of expertise on site is critical for this essentially ground-up approach. The resulting independent expert assessments of the conditions of structures and their elements become the reference until the next scheduled inspection (see Table 4.5). Quantified data are included for independent analysis, such as load rating and posting or for the design of emergency repairs. Appropriate follow-up on a network level is assumed.

The rating scales adopted by FHWA (1995b) and NYS DOT (1997) are of this type. The maintenance rating scale introduced by FHWA (2002c, Section 4.2.4) appears designed to conform to the FHWA condition rating scale (Appendix 40).

The four grades of corrosion in high-strength steel wires are described in visual terms; however, the resulting loss of strength and imminent failure can be inferred. Consequently the findings of this visual inspection amount to a diagnostic evaluation (Section 10.6).

10.4.2 Defect/Action

Tasks are organized in the following sequence:

> Establish repair options → inspect → compare to catalogued condition → recommend immediate actions → analyze → recommend long-term action → update database

Inspections localize and quantify precatalogued bridge malfunctions. Corrective action is recommended (in some cases also selected from a database), simplifying the qualitative evaluation. The available options are usually three or four, ranging from "do nothing" to issuing work tickets (as in DANBRO).

Whereas the *rating/descriptive* method determines needs based on observed conditions, the *defect/action* method rates the findings according to a predetermined set of response options. The merger of the assessment and decision stages is efficient. Limiting the findings to predefined options may create a bias.

Significantly, the defect/action method does not emphasize the recommendation to close the bridge immediately as a result of the findings, whereas the rating/descriptive method does so categorically. The defect/action rating system is suited for relatively predictable and serviceable conditions. Such is the choice of DANBRO, Denmark, as well as the Honshu-Shikoku and the Hanshin Expressway Authorities in Japan and AREMA (Appendix 40). The *condition states* used in PONTIS (AASHTO, 1998b) and the ones proposed by Hearn (in Frangopol, 1999b, Chapter 8) are of this type (Appendix 40). The CoRe manual rates every element in a span from 1 to 4 or 5, quantifies the defect, and prescribes the appropriate action. Hearn proposed states (e.g., protected, exposed, vulnerable, attacked, damaged) which imply the recommended actions.

In practice, the rating/descriptive and defect/action assessments are combined to various degrees. Hazards (Section 10.3) were originally reported strictly by the defect/action method. They were identified for immediate remediation. As the potential presumably imminent hazards proliferated (Example EA46), they had to be prioritized and consequently were rated according to increasingly elaborate scales.

For a large bridge network (and hence a broader range of conditions), the rating/descriptive method offers certain advantages. A refined rating/descriptive scale makes prioritization more flexible. If misused, it can degenerate into "procrastination by inspection." For hazard mitigation (Section 10.3) during emergencies (Section 4.1.1), the rating/descriptive and defect/action methods are combined according to the expertise of the emergency response teams.

Computerized expert systems seek to fill the resulting gap, once again relying on experts for appropriate updates and application. Bridge management and expert systems as well as other developments in artificial intelligence (Appendices 42, 43, and 47) follow a similar path. The rating/descriptive evaluation system uses the *declarative* knowledge representation, whereas the defect/action system is an example of *procedural* representation. A heuristic capability allowing assessments to learn from past experience is a prerequisite for modern condition rating and forecasting models (Appendix 47).

10.4.3 Elements

Bridge managers and database designers must select the level of detail of condition evaluation. With a stock approaching 650,000 bridges, the NBIS opted originally for a *bridge-specific* evaluation of significant elements.

NYS DOT inspections evaluate *span-specific* data for 12,000 bridges. Example 17 is intended to show that span-specific information is indispensable for estimating the needs of a large network. On long spans, such as those of suspension bridges, the distances between panel points of the stiffening trusses or between suspenders can be inventoried and evaluated as equivalent spans.

Appendix 41 describes the rating/descriptive span condition ratings developed by the NYS DOT. Bridge elements are rated from 1 to 7 in every span. The ratings reflect the *average condition of primary members* if the latter are *more than three* (e.g., if they are redundant). The *worst of the three or less* is rated. The *worst bearing* in a span is rated. The resulting qualitative condition database supports network-level estimates of rehabilitation needs but is too broad for project-level estimates. As described in Example EA46, hazards and the condition ratings of elements can be correlated.

Example 19 shows an expansion joint inspection form developed by the NYC DOT. It applies the defect/action approach to an element already shown to be the weakest link in the structural network by rating/descriptive evaluations.

Example 19. Joint Inspection Form

The joint inspection form in Fig. E19.1 serves a specialized response to general preliminary findings (e.g., condition ratings). It was introduced by the NYC DOT in the mid-1990s after expansion joints were shown (by more than 10 years of inspections) to be a primary weak link in the bridge structural system (Figs. 4.26–4.29, 4.75, and 4.76). Once joints were identified as targets for replacement, the report form could become highly specific.

NYC DOT DIVISION OF BRIDGES
BRIDGES INSPECTION / RESEARCH & DEVELOPMENT UNIT
JOINT INVENTORY AND COMPONENT CONDITION REPORT

SHEET _____ OF _____

BIN: _____

INSPECTED BY: _____

FEATURE CARRIED: _____

TITLE: _____

FEATURE CROSSED: _____

AMBIENT TEMPERATURE: _____

WEATHER CONDITION:
(CLOUDY / SUNNY / RAIN)

DATE: _____

		CONDITION OF JOINT ABOVE DECK												CONDITION OF JOINT BELOW DECK														
																			BEARING					DRAINAGE				
																		BEFORE JT.		AFTER JT.							
LOCATION: ABUTMENT / SPAN No.	TYPE OF JOINT	TYPE OF W.S.	COND. OF W.S./DECK (Y/N)	JT. OPENING DIM. (INCH)	VERTICAL DISPLACEMENT (INCH)	COND. OF SEAL (Y/N)	LOOSE PLATE OR MISSING PLATE (Y/N)	MISSING BOLTS OR LOOSE BOLTS (Y/N)	BROKEN WELD (Y/N)	DEBRIS (Y/N)	POOR FINGER JOINT ALIGNMENT (Y/N)	JOINT PAVED OVER (Y/N)	CONDITION OF DECK AROUND JOINT (Y/N)	CONDITION OF PRIMARY MEMBER / DIAPHRAM (Y/N)	COND. OF SEAL (Y/N)	CONDITION OF JT. ANGLE, JT. SUPPT. (Y/N)	STAINING OF SUB STRUCTURE (Y/N)	BEARING TYPE	COND. OF BEARING / A.B. (Y/N)	CLEARANCE (INCH)	BEARING TYPE	COND. OF BEARING / A.B. (Y/N)	TYPE OF DRAINAGE	CONDITION OF DRAINAGE (Y/N)	PHOTOS (YES / NO)	FLAG / NON-FLAG ISSUED (YES / NO)	RECOMMENDATION (YES / NO)
1	2	3	4	5	6	7	8	9	10	11	12	13	14	15	16	17	18	19	20	21	22	23	24	25	26	27	28

N = NO COMMENT (IN GOOD CONDITION)
Y = COMMENT WRITTEN
N / A = NOT APPLICABLE
9 = NOT ACCESSIBLE DESCRIBE

JOINTFORM.XLS

Duong Tran 9/97

Figure E19.1 Joint inspection form.

10.4.4 From Element to Bridge Condition Rating

Bridges are complex combinations of parallel and series systems. Their overall condition cannot be uniquely defined. The level of specificity varies (bridge, span, element) along with local priorities, but all BMS use the evaluations of members or components as input to a priority ranking function of the type shown in Eqs. A40.1 and A41.1.

All bridge condition ratings combine deterministic and probabilistic methods as well as top-down and ground-up approaches. Despite efforts to quantify the findings (see Section 14.5), element condition ratings are mostly subjective and qualitative. The entire structure is rated both subjectively and as a chain of (subjectively rated) elements by weighted procedures which can be designed (and modified) after the trends have been stochastically analyzed. Semideterministic priority rankings computing a bridge condition from the conditions of the structural spans and components are shown in Appendix 41. The most commonly cited NBI *sufficiency rating* evaluates bridges in their entirety.

The NYS DOT computes a *bridge condition rating* (Eq. A41.3, Table E23.2) but requires the inspecting engineer to assign independently a *general recommendation* (an integer from 1 to 7) as an expert opinion formed during the inspection. The two con-current ratings achieve, if not the full benefits of both the deterministic and probabilistic aspects of structural condition evaluation, at least those of two deterministic ones.

Yanev and Chen (in TRR 1389, 1993, pp. 17–24) noted that the NYS DOT general recommendation and condition ratings are well correlated. The two ratings, however, are not truly independent. Equation A41.3 can be perceived as overly conservative; hence Eq. A41.4 was introduced for more realistic assessments. Ultimately, the two did not produce significantly different prioritizations.

Frangopol and Furuta (2001) proposed *reliability-based condition ratings* from 1 (unacceptable) to 5 (excellent), corresponding to values of the reliability index $4.6 < \beta < 9$ (Appendix 41). As all reliability-based methods, this one depends on calibration and therefore on prior assessments. Similarly to other bridge reliability β factors, this condition rating estimates the failure probability of structural members based on their estimated loss of strength. Estes and Frangopol (in Ratay, 2005, p. 42) consider the idea "still in its infancy." The method might benefit from considering connections as independent elements, since their reliability appears to be critical to the structure (Section 4.2.3). Equation 5.2a quotes a "likelihood of risk" rating, obtained as a product of the tabulated likelihoods of cause, defect, exposure, and effect.

The actual scope of rehabilitation work (Section 11.3) must be determined, rather than estimated, by site-specific in-depth evaluations, which in turn influence the priority assessments, as anticipated in Section 9.1. Service d'Etudes Techniques des Routes et Autoroutes (SETRA) originally attempted to evaluate all rehabilitation needs from available inspection reports. Once the available data proved insufficient, a sample group of bridges was inspected and rated on the described numeric scale developed for that purpose. The alphanumeric intermediate and extreme ratings were added in order to reconcile the findings with feasible options (Appendix 40). The federal coding guide similarly needed the ratings poor, critical, and imminent failure between fair and failed in 1988 and in 1995 but sought to reduce their number in 2005.

As the rating levels of numeric or other priority ranking systems increase in number, so does the difficulty in separating the subjective and objective influences informing the results. The group of urgent cases appears smaller, but a larger group appears poised to join in. The distinctions between groups fade. The comparison of ratings assigned according to scales with even and odd rating levels can provide some clues.

The AASHTO (1998b) CoRe guide uses 4 to 5 quantified "condition states" (compared to the NBI's 10 "rating levels") and recommends corresponding actions (Appendix 41). Hearn and Frangopol (in TRC 423, 1994, pp. 122–129) translated the CoRe response-oriented quantified ratings into NBI qualitative priority rankings. Since CoRe ratings imply a response, the objective is prioritization. The two proposed methods (Appendix 41) do not correlate PONTIS *condition states* and NBI *condition ratings* uniquely.

Another approach would be to adjust the NBI ratings. The original highway bridge condition formulas were designed to estimate urgent repair and replacement needs of a declining and relatively uncharted network. As conditions and the knowledge of their state improve, inspections increasingly focus on the identification of long-term maintenance needs. The NBIS specifications (FHWA, 2005a, Appendix 40) reduce the condition rating levels from 10 to 8. The question may arise whether evaluations will be adjusted or the extreme ones will be eliminated.

10.4.5 Structural Condition Forecasts

"Knowledge," R. Feynman lectured in 1963 (1998, p. 25), "is of no real value if all you can tell is what happened yesterday. It is necessary to tell what will happen tomorrow. . . . Only you must be willing to stick your neck out."

Condition evaluations, as all engineering output, are designed to produce results within an anticipated range. Assessments and their forecasts focus on specific aspects of the overall condition, such as the degradation of materials, the deterioration of structural elements, and the entire structures. Structural condition forecasting consists essentially of selecting and updating a model reflecting some change in properties. All bridge condition evaluations (Section 10.4.3) imply a deterioration rate forecast. When condition ratings recommend repairs within a prescribed time frame, as they do in the bridge evaluation systems of SETRA/LCPC (IABSE, 1995, pp. 407–412), DANBRO (Appendix 40), and the Chicago rail (Walther and Koob, 2002), a forecast of the deterioration rate is implied. The NBIS biennial inspection interval similarly implies that serviceable bridges will not deteriorate significantly in less than two years. When the inspection interval is reduced to one year for bridges considered in "poor condition," that condition is essentially prioritized.

The improvements produced by the various forms of structural upgrading, discussed in Chapter 11, are no less uncertain than the rates of deterioration; however, they receive less attention. Once the BMS database grows to a significant size, forecasting increasingly predicts the future behavior of the database rather than physical conditions. The database in turn is updated predominantly by inspection reports of deterioration rather than by construction, maintenance, and repair assessments.

Deterioration Models

Patterns commonly used to represent the change in conditions and their ratings with respect to time are discussed herein and in Appendix 44. Example 17 approaches the problem empirically, whereas Example 22 seeks formal solutions. Acknowledging its dependence on deterioration models, bridge management has developed a number of them based on statistics, material behavior, and experience. The inconclusive data prompted Veshosky (1992) to assert that "life-cycle analysis does not work for bridges," at least at the time.

Statistical and Physical Considerations

Models can be based on statistical (stochastic) or physical (phenomenological) interpretation of the data. Statistical and physical considerations are often combined deterministically, as by Miyamoto (in Frangopol and Furuta, 2001) and Ellingwood (in Frangopol, 1998). Ng and Moses (in Frangopol, 1998) refer to deterministic deterioration models as *static decision models,* as proposed by Kulkarni (1984) for pavement management.

Frangopol (1998, 1999a), Frangopol and Furuta (2001), Miyamoto and Frangopol (2001), and Frangopol and Liu (Miyamoto et al., 2005) (Appendix 44) discuss the probabilistic modeling of the uncertainties inherent to both bridge deterioration and the forecasting methods.

So long as the database consists of engineering judgment alone, forecasting consists of expert opinion or statistical surveys of various opinions. Thompson (in TRC 423, 1994, p. 39) reflected on the need for expert judgment in the early stages of implementing PONTIS and BRIDGIT. Both packages are updated as data accumulate.

The Markov chain model is chosen as the forecasting tool of BMS packages deterministically. Physical properties are no longer assigned fixed values deterministically but are modeled as probability distributions. These in turn are assigned more or less deterministically according to the best fit of available data and convenience (Appendix 9).

Ng and Moses (in Frangopol, 1998) and Hearn (in TRC 498, 2000, C-1) propose a semi-Markov model introducing a "holding time" parameter that can be calibrated by tests to reflect actual conditions (Appendix 44).

A detailed treatment of Markov chains and Monte Carlo simulation is presented in Bremauld (1998). Bruhwiler et al. (in Miyamoto and Frangopol, 2001) found the method more appropriate for large populations than for specific structures. Frangopol and Das (in Das et al., 1999, pp. 46–58) identify the following limitations of the Markovian approach to forecasting:

Future condition state predictions are independent of the element history.

Deterioration is assumed to be a single-step function.

Transition rates are not time dependent.

Transition probabilities do not vary from year to year.

To a bridge owner, annual budgets consist precisely of individual projects. Elaborate probabilistic models of network deterioration easily yield to simple deterministic ones on the project level.

Deterministic and Probabilistic Modeling

The BRIME (2002, p. 29) report defined the scope of the task as follows: "Deterioration models describe the slow degradation and change in strength of a material and are used to predict the change in structural as well as functional parameters due to the accumulated levels of structural loading, environmental conditions, and maintenance practices."

Morcous et al. (2002) classify models into three categories that are not necessarily mutually exclusive, namely:

Deterministic

Stochastic

Artificial intelligence

Adams and Sianipar (1995) observed that "most project-level models are based on deterministic approaches, while most network-level deterioration models use a stochastic approach" and deal with optimizing the allocation of the scarce resources to maintenance, repair, and rehabilitation (MR&R) of the network.

The artificial intelligence models (Appendices 42 and 43) apply stochastic methods to deterministic (knowledge-based) and statistically obtained information. Miyamoto et al. (in Miyamoto and Frangopol, 2001, pp. 143–170) obtain comparable results by a hierarchical decision support algorithm and by intuitive deterministic reasoning.

Shape

Since all models are shaped by a mix of determinism and statistics, it is essential to keep track of the constitutive ingredients. The various combinations of deterministic and probabilistic considerations have produced the full range of deterioration patterns, including straight lines, polygons, and stepped, convex, concave, convex–concave and concave–convex (flat-S) shapes (Examples 9, 17 and 20–22).

Example 20. Average Linear Bridge Deterioration

For a first approximation of the supply–demand equilibrium of the bridge network, it is assumed that:

- Bridge deterioration is linear
- $R_{max} = 7$ is the highest condition rating (as in NYS DOT (1997), Appendix 40)
- $R_{min} = 1$ is the lowest condition rating (NYS DOT)
- $\Delta R_{Reh} = 0.833 \ (R_{max} - R_{min})/3 = 5/3$, i.e., rehabilitations are completed in three years and improve the condition rating by five rating points
- $\Delta R_{Rep} = 1$, i.e., repairs are completed in less than one year and improve the rating by one point
- $R = $ const., i.e., the average bridge condition rating remains constant from year to year:

$$r = \frac{R_{max} - R_{min})}{L} = \frac{6}{L} \qquad (E20.1)$$

If L were terminated at a rating of $R_{min} + 1$ or if rehabilitations only attain $R_{max} - 1$, then

$$r = \frac{R_{max} - R_{min} - 1}{L} = \frac{5}{L} \tag{E20.1a}$$

Combining Eq. E20.1 with Eq. E18.1b and assuming that $\Delta R_{Reh} = 5/3$ allow the average useful life L of the stock to be expressed as a function of the ratio A/A_{Reh} as follows:

$$L = 3 \left(\frac{A}{A_{Reh}} - 1 \right) \tag{E20.2}$$

or

$$A_{Reh} = \frac{A}{L/3 + 1} \tag{E20.2a}$$

As in many bridge management simplifications, the only condition evaluation relevant to Eq. E20.2a is the average useful life L, which implies the time to bridge failure. The AASHTO has recommended a default value of $L = 75$ years. During the 1970s and 1980s the useful life of bridges in New York City (Yanev, 1997) declined to $L \approx 30$ years, corresponding to a linear deterioration rate of $r \approx 0.2/$ year.

Chase et al. (in TRC No. 498, 2000) reported average bridge life of 42 years according to the NBI. These life cycles correspond to the annual rehabilitation needs A_{Reh} shown in Table E20.1.

The NYC DOT has been responsible for 750–850 bridges with a deck area of roughly 15×10^6 ft^2 (1.5×10^6 m^2). Nine percent of that stock amounts to approximately 70 bridges, or 1.35×10^6 ft^2 (0.13×10^6 m^2) of deck area.

The following costs were established or recommended for New York City bridges during the 1990s.

- $C_{Reh} = $ U.S. \$450/ft$^2 = $ U.S. \$4840/m^2 (over 3 years)
- $C_M = $ U.S. \$6/ft$^2 = $ U.S. \$65/m^2 (annually)
- $C_{Rep} = $ U.S. \$150/ft$^2 = $ U.S. \$1612/m^2 (annually)

At a rehabilitation cost of U.S. \$450/ft^2 (U.S. \$4840/m^2), the required annual expenditures would amount to U.S. \$608 million. In the mid–1990s capital reconstruction contracts at the NYC DOT exceeded U.S. \$500 million. In the meanwhile Example 9 shows that a number bridges had to be fully or partially closed.

Table E20.1   Annually Rehabilitated Deck Area as Fraction of Total Deck Area

Useful Life	L	A_{Reh}
Recommended (AASHTO)	75	0.04A
Average (NBI)	42	0.07A
New York City (1990)	30	0.09A

The estimates of annual rehabilitation needs depend not only on L but also on the condition rating range, $R_{max} - R_{min}$. Bridges rated $R_{min} + 1$ often must be partially closed or rehabilitated before they reach a rating of R_{min}. If A were distributed uniformly along the rating scale, the fraction of the stock in each rating group would be equal to $A/(R_{max} - R_{min})$. The average condition rating of any group would decline by one point over the following period:

$$\frac{1}{r} = \frac{L}{R_{max} - R_{min}} \quad \text{years} \quad \text{(E20.3)}$$

The engineering tasks are facilitated by smaller quantities, $A/(R_{max} - R_{min})$, or a wider range, $R_{max} - R_{min}$. Planning is more comfortable with longer periods, $L/(R_{max} - R_{min})$, favoring smaller ranges, $R_{max} - R_{min}$. For the multibillion-dollar rehabilitations of the recent decades (Examples 3 and EA46), the NYC DOT found both types of ratings useful. Up to 1989 the city maintained a four-level system (as in Table A40.4). In 1989 that rating was correlated with the seven-grade inspection system of the NYS DOT as follows:

New York City	New York State
Poor	1.0–3.0
Fair	3.01–4.5 (changed to 5.0 in 1996)
Good	4.51–6.0
Very Good	6.01–7.0

The rating of 3.0 is a logical poor–fair boundary from an engineering standpoint, because it implies "not functioning as designed." Management had reported in 1989 that poor bridges would be eliminated by the year 2000. By 2004 that commitment was met.

The good–fair boundary is more debatable (as is the role of maintenance, compared to that of rehabilitation). Purely technical considerations would place it at 5.0. In 1990, however, the number of fair bridges falling in the interval 3.01–5.0 would have been unmanageable; therefore, 4.5 was selected. By 1996, with conditions brought under control, the limit was raised from 4.5 to 5. By 2004 the average bridge condition rating had risen from 4.5 to 5.

Example 21 compares two simple life-cycle strategies based on linear models.

Frangopol (1999b) developed uncertainty propagation graphs for the performance level of bridges. They are bilinear, as is the deterministic model of the simple life cycle in Example 21. Each point of the performance-level graphs, however, has an associated probability density function. The resulting *whole-life reliability profiles* have a flat-S shape. In contrast, the linearized flat-S shape of Fig. E17.2 is obtained by a regression treatment of actual condition ratings. Each point of the latter curve represents a numerical reduction of data scattered similarly to the assumed probability density function. Example 22 shows simple expressions modeling convex, concave, and double-curvature patterns on purely formalistic considerations.

Example 21. Life-Cycle Strategy Comparison Based on Linear Deterioration Models

The condition deterioration rate r is central to the modeling in Examples 17, 18, and 20. As shown in Example 20, linear deterioration rates and average condition distributions can have a limited usefulness for general estimates of needs in large-asset networks. The rough comparison of essentially opposite life-cycle strategies illustrated in Fig. E21.1 is based on such a simplified model. Yanev (in Vincentsen and Jensen, 1998, pp. 11–22) presented a similar example. The two illustrative strategies A (e.g., minimum + demand maintenance) and B (e.g., preventive + demand maintenance) are illustrated in Fig. E21.1 and summarized in Table E21.1.

The comparison of cases A and B suggests the following observations: Over a 60-year cycle strategy B consistently appears cost effective (e.g., $A/B > 1$); however, the degree varies depending on discount rates and inclusion of original construction costs. If first costs are not considered, discounting emphasizes the benefits of increased maintenance costs (strategy B). Considering first costs *and* discount rates substantially reduces the perceived advantages of strategy B on direct costs. Dis-

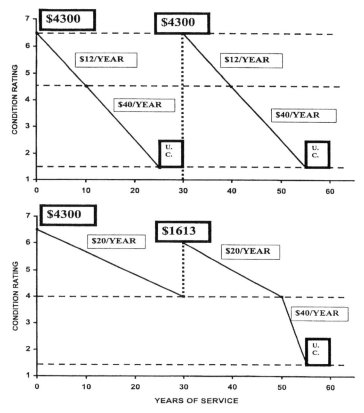

Figure E21.1   Two life-cycle strategies.

Table E21.1 Comparison of Life-Cycle Strategies *A* and *B* at Different Discount Rates *i*

Tasks/Expenditures	Unit Cost (U.S. $/m²/year)	Duration (years)	Total Cost (U.S. $/m²) at Discount Rate i		
			0	4%	8%
Case A					
Minimum maintenance	12	12	144	117	98
Demand maintenance	40	15	600	290	147
Full rehabilitation	4840/3	3	4840	1493	481
User costs	UC		3UC	1.04UC	0.38UC
Total first cycle		30	3UC+ 5584	1.04UC+1900	0.38UC+726
Second cycle		30	3UC 5584	586	72
Including first costs	4840	60	6UC+11,168	1.36UC+2486	0.41UC+ 798
			6UC+16,008	1.36UC+7326	0.41UC+5638
Case B					
Preventive maintenance	20	50	1000	447	265
Demand maintenance	40	7	280	35	5
Component rehabilitation	1612		1612	497	160
Rehabilitation	4840/3	3	4840	460	48
User costs	UC		3UC	0.3UC	0.03UC
Total		60	3UC+ 7732	0.3UC+1439	0.03UC+ 478
Including first costs	4840		3UC+12,572	0.3UC+6279	0.03UC+5318
A/B					
Direct total cost *A/B*			1.44	1.73	1.67
Excluding first cost					
Including first cost			1.27	1.17	1.06
User cost UC$_A$/UC$_B$			2.00	4.50	13.70

Note: UC = user costs.

354

counting has the opposite effect on user costs; i.e., the benefits from maintenance increase with it.

Example 22. Typical Deterioration Paths

The linear history of a condition rating R over time t (the straight line in Fig. E22.1) is modeled by

$$R = R_{max} \left(1 - \frac{t}{T} \right) = R_{max} (1 - a) \qquad (E22.1)$$

where R_{max} = highest rating at $T = 0$
$\quad\quad\quad T = t$ at which $R = 0$, e.g., structural life span
$0 \leq t \leq T$ = time
$0 \leq a \leq 1 = t/T$

The rate of deterioration is constant:

$$\frac{\partial R}{\partial a} = -R_{max} \qquad (E22.1a)$$

A convex history can be represented by

$$R = R_{max} (1 - a^n) \qquad (E22.2)$$

In the simplest case of $n = 2$ (Fig. E22.1) the negative rate of deterioration increases linearly from 0 to $-2R_{max}$ according to

$$\frac{\partial R}{\partial a} = -2R_{max}a \qquad (E22.2a)$$

The corresponding concave trajectory of Fig. E22.1 is symmetric with respect to the straight line defined by Eq. E22.1 and has the equation

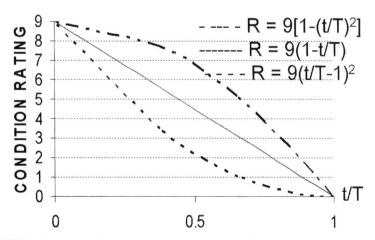

Figure E22.1   Examples of convex and concave deterioration patterns.

$$R = R_{\text{max}} (a - 1)^2 \qquad \text{(E22.3)}$$

As expected, the slope at $a = 0$ $(t = 0)$ is $-2R_1$ and increases to 0 at $a = 1$ $(t = T)$:

$$\frac{\partial R}{\partial a} = 2R_1(a - 1) \qquad \text{(E22.3a)}$$

A trajectory undergoing an inflexion from a convex to a concave curvature can be modeled by the equation

$$R = R_{\text{max}} (1 - a^{\beta(1-a)}) \qquad \text{(E22.4)}$$

where β is a shape factor.

The negative slope increases from zero at $a = 0$, then decreases to zero at $a = 1$, as shown in Eq. E22.4a) and Fig. E22.2 $(\beta = 2)$:

$$\frac{\partial R}{\partial a} = -R_{\text{max}}\beta\, a^{\beta(1-a)} (a^{-1} - 1 + \ln a) \qquad \text{(E22.4a)}$$

Convex–concave curvatures can also be modeled by

$$R = \frac{R_{\text{max}}}{1 + \beta a^2} \qquad \text{(E22.5)}$$

The slope of the curve, known as the "witch of Agnesi," is zero at $a = 0$ $(t = 0)$ and tends towards zero again as $a \to \infty$, passing through an inflexion point according to

$$\frac{\partial R}{\partial a} = \frac{-2R_{\text{max}}\,\beta a}{(1 + \beta a^2)^2} \qquad \text{(E22.5a)}$$

The cumulative normal distribution along the abscissa has a similar convex–concave pattern (Fig. E17.2). This deterioration trajectory was reported for the pavement life cycle by NCHRP Synthesis 153 (1989) and Report 285 (1986). It is consistent with the expectation that the condition ratings of new and old structural

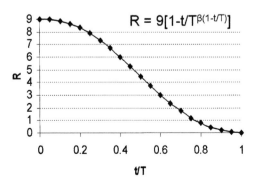

Figure E22.2   Example of convex–concave deterioration pattern.

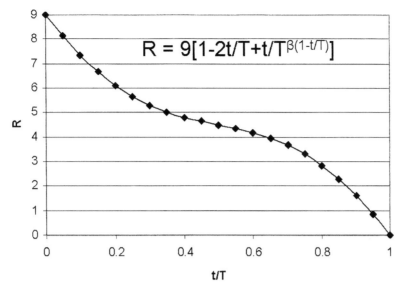

Figure E22.3 Example of concave–convex deterioration pattern (flat S).

elements decline slowly, the former because good conditions are durable, the latter because of repairs and subjective considerations.

To the contrary, the steepest decline of bridge condition ratings in New York City (Fig. E17.1) was shown (Yanev, 1997) to follow a curve symmetric to the one in Eq. 22.4 as shown in Fig. E22.3 and the equation

$$R = R_{max} \left(1 - 2a + a^{\beta(1-a)}\right) \qquad \text{(E22.6)}$$

The slope of the curve is

$$\frac{\partial R}{\partial a} = R_{max} \left[\beta a^{\beta(1-a)} \left(a^{-1} - 1 - \ln a\right) - 2\right] \qquad \text{(E22.6a)}$$

The slope is $-2R_{max}$ at $a = 0$ and can be adjusted to a suitable positive value at $a = 1$ by the shape factor β. Abed-Al-Rahim and Johnston (TRR 1490, 1995, pp. 9–18) obtain a similar flat-S deterioration model by a semiprobabilistic procedure. Example 17 shows how this shape can be deterministically related to a normal distribution of bridges along the rating scale. The latter is statistically expected of larger networks and has been observed in New York City.

A slow deterioration is intuitively expected of a new structure accelerating toward failure, as in the convex curve of Fig. 10.1 and the PennDOT model (FHWA, 1987). Instead, histories of average condition ratings show the concave trajectory of Fig. 10.1 (Yanev and Chen, in TRR 1389, 1993; Yanev, 1997; Example 12 herein). O'Connor and Hyman (1989) concluded: "All the studies on bridge deterioration to date imply that the rate of deterioration tends to slow down markedly after 15 years or so. In fact data from many studies—when taken at face value—suggest that the average bridge condition actually improves or heals with age at some point."

Two recognized factors contribute to the concave shape of condition rating histories:

Inspectors readily downgrade a new bridge but are reluctant to entirely condemn an already low-rated one (Veshosky et al., 1994).

Undocumented repairs, such as hazard mitigation work, do not improve conditions but delay their further decline toward the end of a bridge life. Figure E12.1 shows that not only repairs but also entire rehabilitations are clearly contributing to the general pattern (explaining the excellent condition of an 80-year-old bridge).

The former factor influences condition ratings, the latter applies to actual conditions. New bridges are rarely rated "perfect," leading to a steep initial deterioration pattern. Moreover, while the top rating on the federal scale is excellent, New York State defines it as new, rarely the same. Once a condition rating system is established, forecasts predict ratings or needs for action, not actual conditions.

Emergency repairs typically mitigate potential hazards (Section 10.3) prior to a bridge closure. The useful life of a technically failed structure can be thus extended, contributing to the impression of a slowing rate of deterioration. Since condition ratings and actual conditions may reflect partially independent influences, the two should be forecast separately.

Condition Rating Forecasts

Forecasting of condition ratings determines the expected shape of the deterioration model. Deterioration decelerates in concave models and accelerates in convex ones. Thus, in "midlife" the structure would be rated above average according to a convex curve, whereas it would be rated below average if the pattern were concave. Fatigue failures would fit the convex model. Concrete decks (dominant in the life cycles of highway bridges) with intermediate patching and resurfacing are likely to deteriorate in a concave mode; however, the "standard model" of pavement performance (NCHRP, Synthesis 333, 2004) has a convex shape.

The concave shape is a property of condition rating histories obtained by weighted average formulas. Let the bridge deterioration rate be defined according to the New York State bridge condition formula of Eq. A41.3 as follows:

$$r = \frac{\partial R}{\partial t} = \frac{\sum\limits_{i=1}^{n}(\partial R_i/\partial t)W_i}{\sum\limits_{i=1}^{n}W_i} = \sum\limits_{i=1}^{n}\left(\frac{\partial R_i}{\partial t}\right)k_i \qquad (10.1)$$

where R_i = condition rating of element i
$\quad W_i$ = weight of element i
$\quad r$ = bridge deterioration rate (points/year)
$\quad t$ = time (years)

Let the condition rating histories R_i contributing to the overall bridge rating be as in Fig. E12.4. According to Eq. 10.1, as the bridge elements with shortest useful life, such as expansion joints, fail, their contribution to r vanishes and the deterioration rate is reduced. If all elements i follow the steepest linear deterioration path, r_i (as in Table E23.1), the resulting bridge deterioration history will assume the concave curvature of

Fig. 10.1. The bridge rating will reach its lowest value at the deterioration rate of the elements with the longest life, i.e., the primary member and the deck. A "failed bridge" (e.g., $R = 1$ or $R = 0$, depending on the rating scale) does not deteriorate ($r = 0$). It follows that the improvement of joint condition ratings cannot reduce significantly the rate of overall bridge condition rating deterioration. That effect will become apparent when the useful life of decks and primary members has been extended.

In one of the earliest BMSs, PennDOT (FHWA, 1987) modeled the bridge useful life in terms of NBIS bridge condition ratings (Appendix 40) as follows:

$$CNR = \left(1 - \frac{EQA}{ESI} \right)^{0.7} \tag{10.2}$$

where EQA = equivalent age of bridge element (years)
      ESL = equivalent life of bridge element (years)
      CNR = condition rating at equivalent age

The model assumes knowledge of the bridge element expected life. A link to load bearing capacity is provided (see following section).

Double-curvature models are generally considered more realistic than the single-curvature ones. The deterioration of roadway surfaces, sealants (NCHRP Report 285, 1986), and some bridge components has been modeled by the shapes shown in Examples 17 and 22, i.e., with a convex beginning, a concave ending, and an inflexion point. Thus the fastest deterioration occurs in the middle of the useful life cycle.

Abed-Al Rahim and Johnston (in TRR 1490, 1995) attributed this impression to undocumented repairs at the end of the useful life of structures. Excluding such effects, the authors constructed models following a symmetrically opposite (e.g., flat-S) trajectory. Concrete deck deterioration was modeled by a flat-S shape by OECD (1992, p. 63) and attributed to FHWA Demonstration Project 71 (1989). Yanev (1997) reported similar "worst-case" bridge and component condition deterioration rates, i.e., concave at the high end and convex at the bottom of the rating scale (Fig. E12.2).

The flat-S and the convex–concave deterioration patterns resemble cumulative normal distributions along the ordinate and the abscissa, in this case the age and the condition rating scale, respectively. Figure 10.4 and Examples 12 and 17 show that this is not purely coincidental. If the distribution is normal along both axes, the corresponding pattern is the product of the two (e.g., a straight line). Since (near-) normal distributions are likely to represent the age and the condition of large bridge populations, modeling the deterioration of dense networks by straight lines might prove not only simple but also realistic.

Rates of condition deterioration become meaningful when their dependence on governing factors is known. Although structural conditions are modeled with respect to age, they depend on traffic and maintenance. Deterioration models do not reflect these variables. Measuring the service life of vehicles in mileage alone similarly ignores riding surface and maintenance conditions. A deterioration rate of a unit bridge deck area per unit maintenance cost for a given structural type and traffic would require knowledge well beyond the ratings obtained from visual inspections.

Examples 12 and 17–20 illustrate the relationship between condition ratings and estimated needs in New York City for 1990 and 2004. The poor–fair boundary (Example 18) determines rehabilitation needs (e.g., capital reconstruction budgets). The fair–good

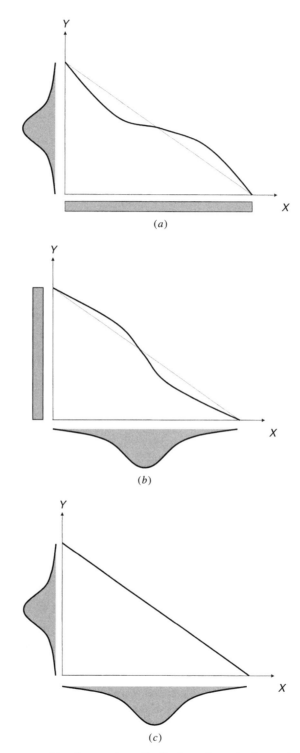

Figure 10.4   Patterns resulting from various distributions along abscissa and ordinate.

boundary is essential for maintenance (e.g., expense) needs. The conditions in 1990 placed an emphasis on rehabilitation and hazard mitigation. The distribution was close to normal, with a mean of 4 on the seven-grade scale established by the NYS DOT. A four-grade rating scale proved too crude for the prioritization needs. By 2004, after 15 years of rehabilitation, repair, and "demand" maintenance, the mean of the condition distribution had shifted to 5 on the seven-level scale.

As rehabilitation improves conditions, the preoccupation with hazard mitigation is superseded by interest in cost-effective life extension. Early condition ratings refined at the lower end of the scale may have to be superseded by a later generation more nuanced in the center and the top end of the range. The forecasting of failures grows vaguer in the absence of actual data, but the quality of structures gains.

Given the inevitable ignorance and vagueness implied in condition ratings superimposed on the actual randomness of the conditions they represent, management should not invest exorbitant resources and confidence in the precision of the modeling procedures. Yanev (1997) correlated the capital expenditures of a large bridge network (e.g., New York City) with a simple deterministic model based on worst-case condition ratings (Example 20).

At the lower end of the rating scale, straight-line models imply abrupt bridge closures which cannot be put into effect on the strength of any general model. Ang and Tang (1975, p. 12) cautioned that assuming "consistently worse conditions . . . could be too costly as a consequence of *compounded conservatism.*" In general, Kierkegaard (1849) found "fatalists and determinists" prone to despair (Section 1.8). Managers take note. They extend the service life of their assets by detailed inspections and emergency measures. Thus the asymptotic decline of the structural condition is objectively enforced, but its cost remains unknown.

Condition State Forecasts

Condition ratings reflect the uncertain combination of perception and physical change. Consequently they change both continually *and* in a stepwise manner. Even though condition ratings and states both emanate from visual inspections, the ratings can be descriptive (e.g., no action is recommended), whereas the states are prescriptive (e.g., action is recommended). Condition states envision the use of Markov chain models and their updating capability is clearly specified, for instance, in PONTIS (Thompson, in TRC 423, 1994, p. 39). Frangopol and Das (in Das et al., 1999, pp. 46–58) propose five reliability states linked to the reliability index β (Appendix 41). Approximately linear or bilinear deterioration profiles are considered.

Whatever the model (convex/concave/bilinear/linear), condition ratings, states, and their forecasts must be more conservative than load ratings, as shown on Fig. 10.1, because inspections lead the cyclic reassessment process, which ultimately reappraises serviceability (Section 10.1.1). The responsible owner may initiate the rehabilitation process while the bridge is rated "functioning as designed" (e.g., 5 on the NYS DOT scale; see Fig. E18.2) on the premise (valid in the early 1990s) that, by the beginning of construction, the rating will have reached 3 ("not functioning as designed"). Schedules can be adjusted over the elapsing 10–15 years as priorities change or conditions deviate from the forecast.

Forecasts of condition ratings or states are indispensable for large networks where statistical and physical considerations interact. Diagnostics and load ratings are more appropriate for individual structures than either deterministic estimates of condition ratings or probabilistic forecasts of condition states.

Condition ratings are more specific than *serviceability* but fuzzier than *load* ratings or diagnostic assessments. Compared to the static (e.g., fixed over time) load ratings, they are a priori dynamic (both because they change over time and because they can be represented by a margin with respect to a fixed point in time). Figure 10.1 shows condition ratings declining by more than 50% of their range before the load rating is reduced.

10.5 LOAD RATINGS

In contrast to the preceding ratings based on visually obtained information, load ratings result from analysis. They establish the capacity of bridge elements in as-built or as-is condition to resist loads according to the governing design specifications. Load ratings are legally binding; i.e., the bridge owner is responsible for the service of the structure as rated. Consequently, the terminology must be used consistently. Appendix 45 describes the load rating procedures for vehicular bridges in the United States according to NCHRP Report 12-46 (2000) and the NBI Specifications (FHWA, 2005a). The NBIS require load ratings to be performed by engineers licensed to practice in the United States.

For railway bridges, AREMA (2001) defines *normal* and *maximum* load ratings. BRIME (2002) shows recommended procedures for the determination of load ratings, defined as *assessments,* for the European Union (Appendix 40).

10.5.1 Load Posting

Load posting is an extreme and binding interpretation of a load rating to be imposed by the owner. It specifies the maximum allowable traffic load. The allowable load categories are specified in Appendix 45. A bridge deemed incapable of supporting a minimum of 3 tons gross live load should be closed. The owner may close a bridge at a higher capacity level, for instance, 5 tons, if a more restrictive load posting cannot be reliably enforced. Load posting should be based on load rating calculations for AASHTO load types 3, 3S2, and 3-3. A bridge need not be posted if the legal load rating factor RF > 1. For RF < 1, the bridge evaluation manual (NCHRP Report 12-46, 2000) proposed the following linear guideline:

$$\text{Safe posting load} = \frac{(\text{RF} - 0.3)\ W}{0.7} \tag{10.3}$$

where W is the weight of the rating vehicle in tons, $W = 40$ tons when the load rating factor is governed by a specified lane load with superimposed axle loads. Equation 10.3 implies that, if RF ≤ 0.3 for all W, closure is appropriate.

10.5.2 Load Rating Forecasts

At their best, condition ratings, combined with inventory data ought to provide estimates of current and future load ratings. In practice, they usually indicate when load ratings should be reviewed. The links between load and condition ratings are continually redefined, reflecting the dynamic interaction between inspection and analysis. A condition rating designated as "not functioning as designed" does not necessarily imply a reduced load rating, because

it is assigned without calculations and

the original structure may have been designed for higher than the current live and dead loads and safety factors.

A number of bridges have been converted from rail to vehicular use, concrete decks have been replaced with orthotropic ones, and so on.

The indirect correspondence between condition and load ratings is illustrated in Fig. 10-1.

Condition ratings influence the load ratings (Appendix 45) by the terms φ and φ_C. The *Guide Manual for Condition Evaluation* (AASHTO, 2003) proposes the following values for φ_C, shown in Table 10.1.

The PennDOT BMS (FHWA, 1987) assumed the following relationship between load capacity LC, design load capacity LC_D, and EQA and ESL of (Eq. 10.2), resulting in a convex normalized curve:

$$\frac{LC}{LC_D} = 1 - \left(\frac{EQA}{1.1\ ESL}\right)^5 \tag{10.4}$$

where LC = load capacity
$\quad$ LC_D = design load capacity

The equivalent age of the bridge element EQA is defined in terms of the condition rating, while its estimated life ESL is assumed. An alternative table linking LC/LC_D directly to condition rating thresholds is also available, yielding a stepwise convex relationship.

A bridge member load rating is not automatically equal to zero if the respective element condition rating is zero. PennDOT makes that provision in a table, setting LC = 0 for condition ratings equal to zero. A load rating can be zero even if the condition rating is not, and vice versa.

For the bridges in North Carolina, Johnston (in TRC 423, 1994) developed the relationship between NBI condition ratings and load ratings, shown in Table 10.2.

If properly implemented, the two ratings should form a conservative system. A bridge can be found unable to sustain the design live and dead loads by load rating calculations or it can be independently termed unsafe by inspection. As in every redundancy, overlapping assessments do not automatically guarantee safety. Bridge closures can be forestalled by load rating calculations when condition ratings are unacceptable and by

Table 10.1 Effect of NBI Condition Ratings on Load Ratings (AASHTO, 2003)

Condition Rating of Member	Item 59 (FHWA, 1995a)	φ_C
Good or Satisfactory	≥ 6	1.00
Fair	5	0.95
Poor	≤ 4	0.85

Table 10.2 Estimated Load Capacity Deterioration Rates (Johnston, in TRC 423, 1994, p. 144)

Lower Rating of Superstructure and Substructure (NBI)	Load Capacity Deterioration Rate (tons/year)		
	Timber	Concrete	Steel
6–9	0.00	0.00	0.00
5	0.30	0.20	0.20
4	0.60	0.30	0.30
≤ 3	1.00	0.50	0.50

frequent inspections when the load ratings fall short. Inspections and analysis are equally responsible for preventing the latter course by exercising rigorous quality control. Management has the ultimate responsibility of deciding on the side of safety.

Present and future load ratings are central to the estimates of network serviceability, and consequently there is considerable incentive to improve their reliability. Structure-specific load rating forecasts are particularly meaningful when they fully utilize the available diagnostic methods, as recommended in Section 8 of AASHTO (2003).

10.6 DIAGNOSTICS

Diagnostics has come to signify direct measurement of the physical properties of structural elements and site-specific modeling of their behavior. Diagnostic investigations typically determine with greater reliability the capacity of a structure that is not satisfactory according to the conventional condition and load ratings. The range and reliability of such work is expanding rapidly with the application of NDT&E techniques (Chapter 15).

Investigations of reinforced-concrete decks, members, and pavements have been reported by NCHRP Reports 292 (1987), 304 (1988), and 312 (1988) and Vesikari (1988), Clifton (1991), and Wadia-Fascetti et al. (2002). Advanced Technology and Large Structural Systems (ATLSS, 1995) investigated the orthotropic deck of the Williamsburg Bridge (Example 3). The results were compared to field stress measurements over a relatively brief period and extrapolated to obtain an expected useful life.

Lichtenstein (in NCHRP, Research Digest, 1993) distinguished between diagnostic and proof load tests as alternative sources for load rating (Appendix 45).

Section 10.4.1 cited the four stages of corrosion in high-strength wires as an implicit diagnostic assessment (Appendix 40). This is possible because the scope of the inspection is limited to corrosion and the function of the wire is limited to tension. A diagnosis of a suspension bridge cable, however, cannot be easily derived from the diagnosis of the individual wires. Betti and Yanev (in TRR 1654, 1999, pp. 105–112) refer to a wire damage index (WDI), proposed by M. Bieniek of Columbia University, for prioritizing the conditions of suspension cables based on the number of corroded wires. The WDI is defined as

$$\text{WDI} = \% \text{ of broken wires} + 0.5 \times \% \text{ of grade 4 wires} +$$
$$+ 0.2 \times \% \text{ of grades 3 and 2 wires}$$

The manager of a single suspension bridge does not "prioritize" the needs of the cables. Table A14.2 in 2004, however, lists 97 suspension structures in the United States in 2004, down from 110 in 1992. Of those listed, 33 are structurally deficient and 39 obsolete. On the level of that network, prioritization is inevitable.

The WDI does not evaluate strength. For the latter purpose NCHRP Report 534 (2004) developed a number of diagnostic models based on the wire ductility and other properties obtained from field inspections and laboratory tests. The report recommended the development of diagnostic methods for the inspection of suspension bridge parallel wire cables. This is the objective of a current FHWA project.

10.6.1 From Diagnosis to Prognosis

Condition rating forecasts serve to prioritize needs. Load ratings (Section 9.5) quantify structural strength; however, they should remain constant so long as a bridge is in full service. Therefore, they are exact but static (e.g., valid only at a fixed time). Future serviceability (Section 9.1) depends on physical conditions and transportation demands. Physical conditions can deteriorate considerably before they begin to affect the service. On the other hand, the quality of service can be declared inadmissible due to changes in the perception of hazards and service demands without significant physical change. Consequently, the forecasting of conditions combines stochastic and phenomenological models.

Phenomenological prognoses model material and hence structural degradation as a function of loads and environmental effects. Vesikari (1988) modeled concrete deterioration caused by corrosion in the reinforcing steel. The effects of alkali–silica reaction, chloride attack, corrosion, and repeated impact have also been modeled, for example, in Frangopol (1998) by Ellingwood (pp. 88–97) and Thoft-Christensen (pp. 181–193) and by Roelfstra et al. (in TRC 498, 2000, pp. C2/1–13).

Fisher et al. (1977), Fisher (1984), and Rolfe and Barsom (1987) estimate the fatigue life of steel elements as a function of stress cycles and geometry. ATLSS (1995) estimated the expected life of a new orthotropic deck on the Williamsburg Bridge (Example 3) at 68 years.

Walther and Koob (2002) combined field inspection findings and laboratory tests of material to forecast the fatigue life of 100-year-old riveted bridge details for the Chicago rail system. A five-tier condition rating system is developed for the purpose, including the levels critical, poor, marginal, fair, and good. Consistent with the AREMA approach to inspection, each condition level is defined by the urgency of the repairs needed, namely immediate, within one year, within three years, depending on further investigation, and none. The assessment determines the scope of required rehabilitation. Repairs are designed following field and laboratory tests.

Drawing a correlation between the element condition states defined in PONTIS and nondestructive evaluation (NDE) capabilities, Hearn (in TRC 498, 2000, C-1) concluded that "data from tests can be used as condition ratings." NDEs (Chapter 14) are also recommended by FHWA (2001b) as a means of improving the reliability of condition ratings obtained by visual inspections (Chapter 13).

Phenomenological models can treat data stochastically. Farrar, Lieven, and Bement (in Inman et al., 2005, pp. 1–12) outline a cyclic damage–prognosis solution distinguish-

Table 10.3 Structural Models and Data Sources

Life-Cycle Stage	Model	As-Built Inventory	Material Properties (Design Specs, Tests)	Dead and Live Loads (Specs, Analysis)	Construction QA&QC	Live Loads (Weigh in Motion, ADT)	Inspection/ Diagnostics	Extreme Events	Other
					Data Source				
Proposal/project	Theoretical (analysis and design)		Y	Y				Y	
Asset/liability	Serviceability	Y		Y		Y		Y	Y
	Vulnerability		Y[a]	Y	Y[a]	Y		Y[b]	Y
	Hazards						Y	Y	Y
	Condition rating						Y	Y	Y
	Load rating	Y		Y		Y	Y		
	Sufficiency	Y	Y	Y	Y	Y	Y	Y	

[a] QC&QA may determine that the materials or processes used in construction have not met the design specifications.
[b] Extreme-event forecasts change over the structural life cycle.

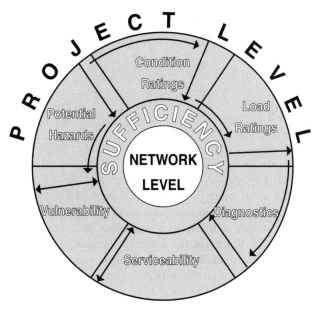

Figure 10.5   Flow of information between condition assessments on project and network levels.

ing between physics and database activities. The process departs from an original model, collects and analyzes data, updates the physical model, and determines future data needs. Table 10.3 cites typical structural models along with the contributing data sources. The models are developed during the bridge life cycle for the purposes of the BMS. The modeling process requires a decision support system of its own, i.e., a data acquisition system (DAS).

10.7   SUMMARY

The assessments described in sections 10.1 through 10.6 are parts of a system. Each assessment contributes to or emanates from the others. Inspection reports must be correlated with load ratings. Hazard reports must be linked to condition ratings. The results obtained by the various forms of assessment circulate between the network level database and the actual structures roughly as shown in Fig. 10.5. The resulting knowledge of the structures and the network must exceed the sum of the individual assessments.

The described data flow is suitable for incorporating into data management systems (Section 14.5) and *expert systems* (Section 15.3).

Chapter 11

Needs

OECD (1992, p. 58) observed that "the functional life of bridges is less than the structural life," e.g., 25 to 50 years (in countries of high traffic growth), compared to 50 to 100 years (barring disasters). In this case, serviceability (Section 10.1), although more ambiguous than structural strength, is a leading indicator of transportation infrastructure needs. Example 9 illustrates the different impressions conveyed by assessments based solely on structural conditions (e.g., condition ratings, Section 10.4) and those emphasizing serviceability (as in the sufficiency rating, described in Appendix 41). Service-oriented management may overrely on performance indicators and general perceptions, while losing touch with engineering realities. The "if it ain't broke, don't fix it" attitude results. Structural condition assessments show earlier signs of decline, and can therefore be perceived as alarmist. The two extreme attitudes correspond to the two funding boundaries, as follows:

1. *Minimal funding.* A triage is performed and unsafe structures that cannot be effectively repaired are closed. A "worst first" policy is implemented among the rest.

2. *Full funding.* No visible deterioration is tolerated.

According to Das (Frangopol, 1999a, p. 29):

Current BMS are based on the following two principles:

(1) Maintenance needs are directly related to the condition states of the structures.

(2) Proposed work is justified now if it will cost more later.

However, contrary to the first assumption, the extent of bridge rehabilitation, replacement and even preventive work largely depend on the load carrying capacity (or structural adequacy) of the structures rather than on their condition alone.

In this discussion, maintenance appears to imply corrective measures. Repairs are more readily associated with threshold conditions than maintenance tasks, because the former produce positive change, whereas the latter prevent negative change.

Furuta et al. (2003) treat the development of performance measures for bridge management as a multiobjective optimization problem. The following objective functions are considered:

LCC is minimized.

Service life is maximized.

Target safety level is maximized.

The authors consider five damage states: very severe, severe, moderate, slight, and very slight as fuzzy sets. The structure is modeled as a combination of series and parallel systems. Dominant failure mechanisms are developed. As in the preceding view, maintenance consists of improvements such as painting, section restoring, and desalting.

Cho (in Juhn et al., 2005, pp. 95–113) summarized: "Most traditional optimization algorithms, including mathematical programming are problem-dependent. They usually use gradient information to guide the search process and often assume continuous-valued design variables. This may lead to significant difficulties when design variables can only take discrete values, such as those in the present maintenance optimization problems."

Cho recommended genetic algorithms (GAs; see Appendix 43) as "effective tools for present maintenance planning problems where different maintenance actions have to be scheduled at discrete years."

Under budget shortages, structural deterioration may have to be tolerated so long as it does not affect service. Figure E16.6 shows the underside of a concrete deck and girder with exposed reinforcement monitored by strain gauges. The respective condition ratings of the shown structure are justifiably low; however, the governing load-bearing capacity in shear was found to be serviceable. Over 15 years, the structure carried traffic restricted to 5 tons. The media were eager to question the safety of the public but not to discuss the maintenance costs, which would have precluded this expedient.

The critical importance of the subjective correlation between conditions and needs was noted in Sections 9.1 and 9.2. Whereas conditions are real but modeled subjectively, needs are subjective but produce real consequences. Needs are quantified in terms of resources and qualified in terms of urgency and priority (e.g., absolutely and relatively). Table 9.2 describes bridge-related needs and the objectives they satisfy within the bilateral relationship of Fig. 1.33a. Table 11.1 divides the assessments of these needs among the interested groups shown in Fig. 1.33b. A comparison of Tables 9.2 and 11.1 shows that neither the rating/descriptive nor defect/action condition evaluations (Sections 10.4.1 and 10.4.2) can capture fully the needs of all parties interested in the structural performance.

The network or asset management level view (Section 12.1) allows for a better correlation of structural and user needs. "Infrastructure management" is the subject of Hudson et al. (1997).

Whereas structural needs are unavoidably cyclic, network needs should be more uniformly distributed over time. Over longer periods, the conditions and needs of numerous assets scatter along assessment scales in patterns, fitting the normal or Weibull distribution (Appendix 9). Das (in Frangopol and Furuta, 2001, p. 3) wrote: "The nature of (condition) distribution is such that, if the 'unacceptable' elements are the only ones repaired or replaced at year 0, after a period of time, say at year N, the number of elements to be repaired or strengthened will be much greater. . . . For this reason, it is not sufficient to repair, strengthen or replace only those elements which are found to be inadequate."

Figures E9-3 and 10-1 show that "serviceability" remains deceptively independent of above average "structural" conditions. If structural deterioration is allowed to advance, the sufficiency and the load ratings decline precipitously and eventually become the causes for bridge closures.

All assumed distributions, however, model subjective findings rather than quantifiable variables. The needs for maintenance, repair, rehabilitation, and replacement are deduced

Table 11.1 Domains and Criteria of Participants in the Bridge Management Process

	Structural Engineering	Engineering Management	Asset Management	Asset Users
Domain	Structures	Transportation network	Transportation/ economy	Economy/transportation
Criterion	Condition/performance	Performance/service	Service/utility	Utility/satisfaction
Quantitative (measurable, objective)	Geometry, materials, load-bearing capacity, redundancy, life span, hazard mitigation, costs	Resources (budget, personnel, equipment, time and material, database); services (loads, traffic counts, performance indicators; support (legal, administrative, legislative)	Economic growth, monetary costs, sustainability	Speed, capacity, geometry, ridability
Qualitative (perceived, subjective)	Condition ratings, risk, reliability, benefits	Support (popular, political, technical, decision—BMS), safety	Environmental compatibility, availability, aesthetics	Aesthetics, comfort, perceived benefits

from the condition evaluations (Table 4.2), expressed in terms of engineering tasks (Table 4.3), and management options. Risk assessments (Chapter 5) prioritize the needs.

In order to prevent, rather than respond, management needs economic as well as structural forecasts. The linkages between investments in transportation and economic performance are continually examined. Recent conclusions are quoted in Appendix 23. The modeling contains considerable uncertainties and interpretations vary. The findings suggest that the economy heavily influences transportation development and a smaller reverse influence also exists.

Table 10.4 and Fig. 10.4 show that conditions are most frequently assessed on the project level (deterministically), whereas needs are estimated and prioritized on the network level using larger data banks and stochastic analysis. Thus, although the conditions discussed in Chapter 10 and the needs reviewed in Chapter 11 are related, there is no direct correspondence.

11.1   QUALITY AND QUANTITY OF SERVICES

The services bridges provide in terms of transportation and those they receive in the form of maintenance, rehabilitation, repair, replacement, and operation must be measured in qualities and quantities. It is highly significant for bridge management that the needs for and the benefits from the provided services (Section 10.1) exceed its scope.

The NBI serviceability rating (Section 10.1) appraises structures in the as-built system. The derivative sufficiency rating supports network evaluations, such as: "More than 40% of the nation's bridges are structurally deficient or functionally obsolete" (Chase et al., in TRC 498, 2000, p. C-6).

The annual needs for upgrading the U.S. national bridge network are estimated to exceed U.S. \$7 billion (in direct costs).

The NBI sufficiency rating (Appendix 41) reflects service needs as a combination of structural adequacy and safety, functional (e.g., geometric) obsolescence, and "essentiality" for public use. The prescribed ratio of their relative importance (55:30:15%) is not intended to assess individual structures, particularly the unique and essential ones. On a national level, current and future bridge infrastructure needs can be estimated from the sufficiency rating as described by Small et al. (in TRC 498, 2000, A-1):

Sufficiency Rating (SR)	Federal Funding Eligibility
$SR \leqslant 50$	Replacement
$50 < SR\ 80 \leqslant 80$	Rehabilitation

NCHRP Synthesis 238 (1997) reviewed the performance measurements in state highway agencies (SHAs). FHWA (2001b) investigated the reliability of the inspections producing structural condition evaluations (Section 14.5). The application of NBI standards to the evaluations submitted by local bridge owners for prioritization of funding on the federal level was audited by FHWA in 2003 (Section 13.2).

11.1.1 Options and Definitions

Examples 17 and 18 describe the equilibrium between the activities at the bridges of New York City (ca. 1990). Optimizing that equilibrium is the main bridge management task. In a continuous bridge management operation, the options of maintenance, repairs, and rehabilitation may have to be evaluated not only according to their respective costs and benefits but also according to their funding sources (e.g., capital or expense) and the preferred method of execution (e.g., in-house or under contract). The difficulties in selecting and implementing management strategies or policies are reflected in the vagueness of the associated terminology. According to NCHRP Synthesis 330 (2004): "There is little consistency among practitioners and researchers regarding definitions and distinctions among various maintenance activities. Also, various analyses of benefits are not necessarily comparable."

NCHRP Synthesis 327 (2004) refers to the AASHTO *Maintenance Manual* (1999a) and lists the following definitions:

Maintenance—the technical aspect of the upkeep of the bridges; it is preventive in nature. Maintenance is the work required to keep a bridge in its present condition and to control potential future deterioration.

Rehabilitation—the process of restoring the bridge to its original service level.

Repair—the technical aspect of rehabilitation; action taken to correct damage or deterioration on a structure or element to restore it to its original condition.

The definitions have the following significant implications:

Maintenance does not improve condition ratings.

Rehabilitation and repair do not match identically the "original condition."

The definitions so far represent management categories. Specific tasks are further defined in terms of their effect on the structure, as follows:

Stiffening. Any technique that improves the in-service performance of an existing structure and thereby reduces inadequacies in serviceability (such as excessive deflection, excessive cracking, or unacceptable vibrations). Figure 4.23 illustrates a bracing installed for that purpose.

Strengthening. The increase of the load-carrying capacity of an existing structure by providing the structure with a service level higher than the structure originally had (sometimes referred to as *upgrading*). Figure 11.1*a* shows an historic timber bridge (combination of truss and laminated arch) in Switzerland strengthened with carbon fiber laminates.

Maintenance. NCHRP Synthesis 330 (2004, p. 8) quoted a classification of maintenance into *routine, corrective, preventive, proactive,* and *reactive* (Hudson et al., 1997). The synthesis adopted the following definitions proposed by AASHTO (1999a, p. 5):

Preventive maintenance (PM)—planned strategy of cost-effective treatments . . . that preserves the system, retards future deterioration, and maintains or improves the functional condition of the system (without substantially increasing structural capacity).

Preservation—planned strategy of cost-effective . . . treatments . . . to extend the life or improve the serviceability; a program strategy intended to maintain the functional or

(*a*)

Figure 11.1   (*a*) Historic timber bridge reinforced with laminated carbon fiber.

structural condition. Preventive maintenance (PM) is commonly assumed to forestall deterioration without improving the structure.

For pavements, NCHRP Synthesis 153 (1989, p. 7) defines *routine maintenance* as follows: "A program to keep pavements, structures, drainage, safety facilities, and traffic control devices in good condition by repairing defects as they occur. . . . Routine maintenance is generally reactive."

Preventive maintenance is a cyclic planned activity defined (NCHRP Synthesis 153, 1989) as "a program strategy intended to arrest light deterioration, retard progressive failures, and reduce the need for routine maintenance and service activities."

Corrective maintenance is synonymous with *repair,* although managers may assign different scope to each of the two terms.

Demand maintenance consists of safety-related emergency repairs, as in the elimination of potential hazards (Section 9.3.) and is purely reactive. Das (in Frangopol, 1998) refers to such maintenance as *essential.* AASHTO (1999a, p. 1-5) refers to reactive maintenance in a similar sense.

Predictive maintenance was recommended by Mobley (1990) for industrial production. It seeks to reduce maintenance waste by improved estimates of machine useful life.

(*b*)

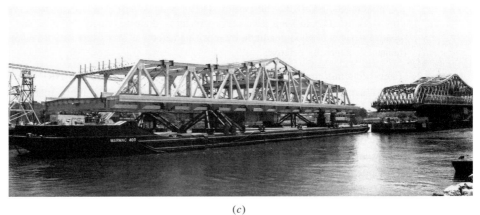

(*c*)

Figure 11.1 (*b*) Bridge rehabilitation: replacement and strengthening of corroded steel superstructure, new deck. (*c*) New 3rd Avenue swing bridge installed on top of new bearing on existing foundation, Willis Avenue bridge in background.

Reliability-centered maintenance is defined by Hudson et al. (1997, p. 240) in the same sense. Sections 4.1 and 4.5 argued that this strategy is more applicable to mechanical components where maintenance consists of replacement of components with highly predictable useful life spans. So far bridges offer more examples of under- rather than overmaintenance. Nonetheless, predictive maintenance can be costeffective at complex structures, for instance, in combination with the tools of structural health monitoring (SHM, Section 15.1).

The preceding definitions anticipate limited funding by dividing maintenance into *optional* or *inevitable* (e.g., *demand*). That distinction can encourage reductions of expenses perceived as optional.

Repair and *rehabilitation* differ in scope, depending on interpretation. Repairs are usually considered expense budget items, whereas rehabilitations are capital expenditures, eligible for federal funding. Structural improvements qualify as capital work if they exceed cost and expected useful life thresholds (e.g., more than U.S. $5 million and 5 years of useful life). In the early 1990s New York City bridge management designated such repairs as *component rehabilitation*. The Office of Management and Budget (OMB) approved their capital funding.

Rehabilitations of bridges on the NBI usually include deck replacement (Fig. 11.1*b*). *Reconstruction* is essentially synonymous.

Retrofit, as well as *stiffening* and *strengthening,* improves the structural response, most commonly to extreme events. Retrofit implies the replacement of details that, for a variety of reasons, are no longer considered adequate. Targeted are fracture-critical details such as pin-and-hanger assemblies (Fig. 10.3), sliding steel and rocker bearings, restrainers (Figs. 4.30–4.38), and fragile columns (Figs. 3.9 and 3.10). From a management standpoint, retrofit might imply a higher priority dictated by vulnerability estimates, contrasted with the rehabilitations, scheduled according to the normal course of deterioration.

Replacements can involve existing foundations or entirely new alignments. The new swing bridge shown in Fig. 11.1*c* reused the original foundation and parts of the fender system from the replaced structure.

The NBI classification described in the preceding section correlates the terms replacement and rehabilitation to sufficiency ratings below 50 and from 50 to 80, respectively.

11.2 HAZARD MITIGATION

Hazards result from combinations of vulnerabilities in the as-built (Section 10.2) system and subsequent "damage" (Section 10.3). Most life-cycle assessments consider hazards as random. Ang et al. (in Frangopol, 1998) estimate and minimize the cost of losses and repairs caused by catastrophic failures over the life cycle of structures (Appendix 38). The solution is limited to concrete buildings under earthquakes but can be adapted to other hazards. The example stresses the need to estimate the likelihood and cost associated with the loss of life. Such estimates must distinguish between bridges that cannot be allowed to fail catastrophically and bridges that could fail functionally, without endangering the life of the users. The scope and priority of remedial work are determined accordingly and expressed in terms of importance factors.

Bridge life-cycle cost analysis (BLCCA; NCHRP Report 483, 2003) estimates *vulnerability costs* incurred by natural hazards, overload, cyclic loads, collisions, and accidents. All estimates assume the likelihood of occurrence and resulting penalties as in Eq. 5.2. Average daily traffic is a relatively well quantified significant parameter. Structural conditions and costs of repairs are less specific. A variety of semiprobabilistic methods are proposed (Section 9.2.1). No prioritizing scheme can be selected on a cost basis alone.

The stable (presumably causal) relationship between structural conditions and potential hazards demonstrates that, as conditions decline, hazards proliferate and closures become necessary before the load-bearing capacity has been exhausted. Management must eventually draw a line at the condition ratings below which it cannot function cost effectively and safely. This finding confirms the trend noted by OECD (1992) and quoted in the opening paragraph of Chapter 10 to the effect that service life ends most frequently when serviceability rather than strength is exhausted. Rather than specialize in the forecasting of hazards caused by structural deterioration, management should eliminate them. However, even a network in relatively good condition is subject to extreme events and systematic deterioration. Thus regular structural inspections and hazard mitigation provisions remain indispensable.

11.3   REHABILITATION AND REPLACEMENT

Replacements and rehabilitations mark the end of the service life of structures. As that end approaches, condition evaluations become increasingly unreliable, because (a) they rely on more limited experience and (b) their uncertainty incurs greater penalty. In the absence of sufficient data for adequate probabilistic analysis, the relatively rare structural closures supply data for deterministic estimates of the life-cycle duration.

OECD (1992) referred to service life predictions as "a posteriori estimations," because they ultimately rely on known closures. Yanev and Chen (in TRR 1389, 1993, pp. 17–30) estimated the bridge useful life at 30 years, given none or negligible maintenance. According to OECD (1992), 40% of all reasons for bridge closures are structural, including the following:

Design/construction

Impossible repair

Many elements to replace

Environment (salt, frost)

Catastrophic failure (collapse)

River upstream

Traffic carried

Impact below

Maintenance failure

Other

The remaining 60% of the reasons are functional and include new standards, increased or lapsing traffic demands (geometric and weight), road and canal widenings.

Itoh and Liu (in Frangopol, 1999a) report the following principal reasons for bridge demolition in Japan:

Improvements: 45%

Functional deficiency: 21%

Damaged elements: 19%

Load capacity: 7%

Seismic strengthening: 2%

Disaster: 2%

Other: 4%

Because of their magnitude and impact, capital improvements are rigidly constrained in terms of time and funding. By the time a project is launched, design has been selected to meet the schedule and the budget, which is to satisfy first-cost and immediate requirements (Section 4.1.3). Thus construction, rather than maintenance, considerations often determine the life cycle of the assets. NCHRP Synthesis 153 (1989) recommended that the FHWA report TS-78-216, "Integration of Maintenance Needs into Preconstruction Procedures" (pertaining to roadways), should be reissued. A single comprehensive guideline is not possible for bridges.

The cost-effective practices of local bridge owners include *maintainability* measures planned by design and introduced during construction. NCHRP Synthesis 327 (2004) concluded: "It is evident from the survey responses that bridge engineers have learned from the problems of existing structures regarding the consideration of maintenance needs when constructing new bridges. It is understood that the design of new bridges should emphasize durability. The responses indicated both material choices and design philosophies to promote maintenance-free structures."

Typical maintainability or durability measures identified by NCHRP Synthesis 327 (2004) are quoted in Appendix 48. Numerous reports recommend cost-effective construction practices, including NCHRP Syntheses 327 (2004), 345 (2005), and 346 (2005).

New Materials

Metallurgy is said to have initiated modern engineering and management (Section 1.2). The professions remain committed to designing and using new materials. Metal bridges evolved from iron to steel. High-strength, high-performance (Günther, 2005), weathering, and other special steels made new structural forms feasible. Aluminum found applications. Concrete advanced from unreinforced to reinforced with steel (variously protected against corrosion), high strength, prestressed (pre- and posttensioned), and fiber reinforced (carbon, glass, etc.). Hybrid combinations of steel, concrete, wood, and plastics have emerged.

The introduction of new materials and new applications of familiar ones must overcome the conservatism fueled by inexperience. Bridge owners understandably prefer to invest in structures with a demonstrated life-cycle performance. National network management, such as the FHWA, and the industries foster innovation by funding pilot projects using new materials and applications and monitor and publicize their performance.

Innovative designs typically present first-cost advantages convincingly. Realistic designs of maintenance, repairs, and replacements can make such proposals more attractive. Eventually the cost-competitive options assert themselves.

11.4 MAINTENANCE AND REPAIR

Estimates of maintenance needs are both essential and complex because they must be consistent with a number of fairly rigid and possibly conflicted constraints, including the following:

Available funding (e.g., expense budget)

Long-term funding (e.g., capital budget)

Life-cycle performance

Operational capability of owner

Community needs

Safety

Emergency response

The many vulnerabilities of maintenance discussed in Sections 4.1.5, 4.3.3, and 4.4 appear to recommend less maintenance-intensive bridges. The more dynamic automobile industry, which generates the demand for vehicular bridges, has repeatedly reduced car maintenance to scheduled replacements of parts or blocks. Vehicular longevity may not have always increased, but cost–benefit analysis must have identified other advantages. A similar trend is already apparent in bridges. The existing stock, however, has been designed and constructed with different needs, which should persist over the coming decades.

Preventive maintenance, that is, a process that does not improve a condition but extends its duration (see also Section 4.1.4), is vulnerable by definition, because it yields benefits that do not constitute improvements. More precisely, the improvements are not reflected in regular structural inspection reports. Just as bridge safety considerations cultivated structural inspections capable of identifying hazards, maintenance management needs regular task performance assessments. Such assessments would reflect both the product and the process.

Maintenance appears to have preceded design in adopting the performance-based approach. Maintenance manuals, recommendations, and conference proceedings, such as the AASHTO *Manual for Bridge Maintenance* (1987) and *The Maintenance and Management of Roadways and Bridges* (1999a) and TRB's *Maintenance Management* (2001), are distinguished by the following features:

Maintenance is more often described than defined.

Maintenance tasks are not discussed in great detail. Rather, the emphasis is on structural elements and their common deficiencies and needs.

Repair and rehabilitation are considered as maintenance alternatives.

These distinctly performance-oriented features reflect a designer's interest in the product. A more prescriptive view of maintenance is obtained when the process is considered from an economic standpoint. TRR 1877 (2004) identified prevalent (internal and external) vulnerabilities in maintenance management (Sections 4.1). Llanos (1992) presented an economic approach to the management of bridges on the French national highway network. Performance-based and prescriptive approaches both have limitations (Section 4.2.1). The latter is inflexible and does not optimize with respect to variable conditions and constraints. The former depends on highly sensitive optimization, which easily lapses into minimization. Results strongly depend on the quality of input data, which, thus far, is lacking.

The early metal bridges were expected to last forever in the tradition of the stone arches. The maintenance implicit in this goal was duly noted, for instance, by Boller (1885, pp. 79, 81):

> Because a bridge is an iron one, it does not imply that it requires no further care after it is once finished. . . . It is with a view to permanence that it is recommended in *no case* to allow plates or parts to be used *less* than one quarter of an inch in thickness, and perhaps even five sixteenths of an inch would be still more desirable as a minimum thickness. . . . It would be good practice for the authorities of every county to examine their bridges systematically every spring for signs of rust, which, if discovered, should be attended to as soon as possible. In this way their bridges (if originally good ones) can be made to last forever.

Waddell (1921) was quoted to the same effect in Section 2.1. If funding is available, the problem is reduced to the design of the maintenance that maximizes structural condition and/or service. The management of the Shinkansen "bullet" train on the Tokyo–Osaka line, for instance, reported average annual delays of 6 seconds in 2004. Forty-year-old structures are maintained in as-built condition by repairing all visible defects and replacing elements approaching the end of their estimated fatigue life. The Sydney Harbour mixed-use bridge (Fig. 11.2a) has been spared the replacement of a single rivet since its opening in 1931. Repainting of the Golden Gate Bridge (Figs. 4.42 and 4.43) commences at one end of the structure as soon as it is completed on the other end. Suspension cables must be periodically rewrapped (Fig. 11.2b) according to diagnostic findings and maintenance inspections (Section 14.3).

In most cases, however, "lasting forever" has yielded to the search for optimal maintenance levels and costs within constrained budgets. NCHRP Synthesis 153 (1989, p. 31) expressed that position as follows:

> Only through LCC (life cycle cost analysis) can trade-offs among PM (preventive maintenance), corrective maintenance, rehabilitation, and reconstruction be evaluated. Factors to be considered in a lifecycle cost analysis are:
>
>> The engineering and economic issues that can be examined with respect to performing or deferring PM.
>>
>> Present costs for replacement or rehabilitation.
>>
>> Annual maintenance costs.
>>
>> Future increases to maintenance costs as a result of deterioration.

(*a*)

Figure 11.2 (*a*) Sydney Harbor Bridge, continually repainted.

Future rehabilitation costs.

Analysis period.

Interest rate.

The value assigned to the interest rate can be a major factor in the outcome.

The effect of interest rates is discussed in Appendix 23 and Example 10. NCHRP Synthesis 153 (1989, p. 33) acknowledged:

Benefits are much harder to evaluate in dollar terms. Some standard types of benefits are:

Safety;

Travel time savings;

Reduced tort liability claims;

Reduced vehicle operating and maintenance costs;

Reduced disruption of adjacent business activities;

Reduced discomfort;

Preservation of the investment by deferring or reducing the high future reconstruction costs associated with a "do nothing" PM policy.

(*b*)

(*c*)

Figure 11.2 (*b*) Rewrapping of Golden Gate Bridge cable. (*c*) Repainting of major urban bridge without traffic interruption.

Fifteen years later, NCHRP Synthesis 330 (2004) concluded (p. 26): "Although the theoretical benefits of maintenance are embedded in the pavement and bridge management software used by most state highway agencies, those management tools are used for the most part only as required by federal regulation. . . . Consequently only one third of these agencies reported the use of life-cycle costing or other benefit-cost methods to assess maintenance priorities; two thirds of these agencies use such methods only for major projects." In addition to a similar list of potential benefits from maintenance, NCHRP Synthesis 330 highly recommended the use of analytic tools for quantifying these benefits and the use of advanced media tools for selling maintenance needs to the public.

Level of Service

NCHRP Report 285 (1986, p. 6) recommended measuring infrastructure maintenance by *level-of-service* guidelines, with the following qualification: "Since a maintenance service level is a changing condition over time, it is not definable. What is *definable* is the condition of the maintenance element at different points in time."

Section 10.4 argued to the contrary, that element conditions are not uniquely definable. The "definables" involved are clearly relative.

According to Das (in Frangopol, 1998), "estimates of maintenance needs should be based on structural adequacy or safety rather than on condition states of the structures." The author considers that this can be best accomplished on a project level by direct structural assessment. Frangopol (1999b, Chapter 9) argued that "bridge reliability has to be explicitly quantified from the condition states of (structural) elements."

Condition states and *reliability* must therefore be derived from measurable quantities of materials with calculable resistance capacities. Hudson et al. (1997, p. 240) refer to such maintenance as *reliability centered*. For electronic and industrial equipment, *predictive maintenance* (Mobley, 1990) optimizes the utility of interventions by anticipating the expected useful life. Thus far, neither the concepts nor the underlying properties can be directly linked to the maintenance of bridges.

In the Markov chain terminology adopted by PONTIS and BRIDGIT, current PM reduces the likelihood of a structure to transit to a lower condition state during the period defined by its frequency of application. Markov chain models do not consider past performance and hence cannot evaluate the contribution of past maintenance.

Two alternatives to *level-of-service maintenance* can be formulated as *zero maintenance* and *prescribed maintenance*.

Zero Maintenance

FHWA (2005c) discussed a study by the federal DOT of Switzerland into the cost–benefit of zero maintenance and recommended the subject to the attention of the FHWA for further research. In contrast to maintenance *deferral* (Section 4.1.5), zero maintenance envisions new structures designed to minimize maintenance needs by improved durability and/or facilitated component replacement, as in the preceding quote from NCHRP Synthesis 327 (2004). The added design and construction costs should be offset by the maintenance savings.

This method does not eliminate traffic maintenance. Rather, it seems to assume that the level of service is ensured by design at no structural maintenance (until repair).

Prescribed Maintenance

Maintenance-related data are considered insufficient for adequate management. Hudson et al. (1997, p. 243), however, have examined data accumulated by (roadway) maintenance management systems over 10 years and never used. This and similar experiences indicate that attempts to optimize maintenance merely minimize the physical damage after the financial damage has been minimized, logically culminating with the zero maintenance. The alternative would be a "zero-optimization" policy. Maintenance would be prescribed by managerial decision as a set of regularly scheduled tasks until there is cause for revision (e.g., in response to changes in the assets, the usage, the funding, the information, and the management).

Examples 23 and 24 show that PM tasks are cyclic and seasonal. The exception is paint, discussed in the following section. Prescribed maintenance avoids the pitfalls of the performance-based predictive maintenance by eschewing the favorite managerial prerogative to optimize costs. Instead, it is a form of structural hygiene, conducive to good structural health. The structural soundness of this strategy is obvious; however, it relies on a financial soundness beyond most owners' means.

A broad but ultimately telling prescriptive indicator of maintenance needs is the ratio of annual maintenance expenditures to the estimated bridge replacement costs, shown in Table 11.2.

Bieniek et al. (1989), Vaicaitis et al. (1999), and BRIME (2002) have estimated annual maintenance costs ranging from 0.5 to 1.5% of bridge replacement costs. Yanev (in TRC 423, 1994, pp. 130–138) reported that the Port Authority of New York and New

Table 11.2 Maintenance as Fraction of Estimated Replacement Costs of Bridge Stock (%)

	OECD (1992), *Bieniek et al. (1989)*	*BRIME (2002)*
Belgium	—	0.3
Finland	—	1.0
France	0.3	0.5/0.6
Italy	1.5	—
Germany	1.5	1.0
Great Britain	0.5	1.0
Ireland	—	0.6
Norway	—	0.6
Spain	—	0.3
Sweden	—	1.7
Japan	2.5	—
New York City		
Pre-1989	0.05	—
Recommended (1989)	0.50	—

Jersey spends approximately 1% of the estimated replacement cost of the George Washington Bridge (Example 2) on annual maintenance (distinct from operation). Empirical evidence therefore suggests that annual maintenance level amounting to roughly 1% of the replacement cost is a threshold below which deterioration accelerates. Since the output is dimensionless, the reported ratios are relatively independent of cost adjustments. The results invite comparisons between the maintenance policies of different owners and reveal trends that can be independently confirmed. Such comparisons, however, are invariably incomplete for the following main reasons:

Maintenance definitions vary. Many owners include repairs in the maintenance costs.

Maintenance needs depend on the type and age of the structures, the traffic, and the environment.

Consequently maintenance programs must be site specific. Example 23 describes a scheduled PM program. Example 24 estimates the effect of the maintenance tasks, recommended in Example 23, on condition ratings by a deterministic knowledge-based model. Since these tasks have not been rigorously optimized or funded, they are largely hypothetical. Typical maintenance tasks (Table E23.1) such as sweeping, spot painting, drain cleaning, washing, oiling, and deicing can easily be mismanaged or inadequately performed. The current NCHRP Project 14-15 is compiling an inventory of bridge maintenance activities and their costs.

Example 23. Recommended Preventive Maintenance (PM), NYC DOT

A preventive maintenance program for the bridges managed by the NYC DOT was designed by a university consortium centered at Columbia University (Bieniek et al., 1989; Vaicaitis et. al., 1999). The researchers investigated comparable practices worldwide and recommended a "full" (100%) maintenance consisting of 15 tasks. The prescribed annual frequencies and respective costs are listed in Table E23.1, columns 1–4. The 1999 update extended the painting cycle from the 8 years (0.125/year) recommended in 1989 to 12 years (0.083/year).

The vector of maintenance task frequencies f_j in column 3 of Table E23.1 is deterministic. It is recommended as optimal and should result in a useful life L_{il} as shown in column 4 of Table E23.2. Since the statement cannot be fully confirmed by hard evidence, budget managers consider it as maximal and fund only parts of it. As discussed in Chapter 2, optimization of the inevitable budget cuts translates into minimizing the deterioration and the risk. The maintenance of mechanical and electrical equipment is essential to movable bridges and should be excluded. Partial funding could be distributed proportionally to all tasks. The resulting "partially funded" maintenance risks are ineffective, because bridge deterioration may not depend linearly on the frequency of application of all tasks. An attempt to model the "cost effectiveness" of maintenance tasks in terms of bridge deterioration is described in Example 24.

Table E23.1 Parameters of Annual Maintenance Activities and Costs

Maintenance Task (1)	Unit Cost c_j (USD/m^2) (2)	Recommended Annual Frequency f_j (3)	Annual Cost (USD, 1999) (4)	k_j (5)	$k_j/c_jf_j \times 10^{-2}$ (6)	Cost-effective Frequency f_j^{ce} (7)	Annual Cost (USD 1999) (8)	$k_j/c_jf_j^{ce} \times 10^{-2}$ (9)
Debris removal	0.13	12(52^a)	2,319,653	0.068	4.4	34.708	6,709,153	15
Sweeping	0.02	26	613,071	0.060	11.5	248.194	5,852,319	12
Clean drain	0.33	2	863,804	0.118	17.9	24.852	10,733,443	14
Clean abutment, piers	1.94	1	2,776,013	0.089	4.6	3.139	8,712,665	15
Clean grating	0.40	1	55,490	0.078	19.5	13.437	745,600	14
Clean joints	0.75	3(26^a)	3,262,730	0.101	4.5	9.191	9,995,427	15
Wash deck	1.01	1	1,455,198	0.057	5.64	3.878	5,643,897	14
Paint	301.45	0.083	36,041,997	0.050	0.2	0.011	4,982,679	15
Spot paint	66.44	0.25	23,743,128	0.044	0.26	0.045	4,275,512	15
Sidewalk/curb repair	3.72	0.25	1,328,182	0.029	3.12	0.528	2,806,598	15
Pavement/curb seal	3.22	0.5	2,334,466	0.110	6.83	2.356	11,000,178	14
Electric maintenance	0.03	12	1,107,143	0	—	12	1,107,143	—
Mechanical maintenance	0.03	12	1,010,502	0.073	20.3	80.273	6,759,670	3
Wearing surface	4.85	0.2	1,390,305	0.040	4.12	0.568	3,949,428	14
Wash underside	9.24	1	13,189,518	0.084	0.91	0.623	8,217,488	14
Total	—	—	91,491,200	1	—	—	91,491,200	—

Note: In U.S. dollars (USD) per unit deck area, unit treated area or number of applications.

^{a} East River bridges.

Table E23.2 Element and Bridge Condition Ratings (NYC DOT after NYS DOT, Eq. A41.3)

		Useful Life		Rating		Weight		Deterioration	
i	Element	$L_{i0}{}^a$	$L_{il}{}^b$	New	Failed	W_i	k_i	r_{i0}	r_{il}
(1)	(2)	(3)	(4)	(5)	(6)	(7)	(8)	(9)	(10)
1	Bearings, anchor bolts, pads	20	120	7	1	6	0.083	0.30	0.05
2	Backwalls	35	120	7	1	5	0.069	0.17	0.05
3	Abutments	35	120	7	2	8	0.111	0.17	0.05
4	Wingwalls	50	120	7	1	5	0.069	0.12	0.05
5	Bridge seats	20	120	7	1	6	0.083	0.30	0.05
6	Primary member	$30/35^c$	120	7	2	10	0.139	0.2/0.17	0.05
7	Secondary member	35	120	7	1	5	0.069	0.17	0.05
8	Curbs	15	60	7	1	1	0.014	0.4	0.10
9	Sidewalks	15	60	7	1	2	0.028	0.4	0.10
10	Deck	$20/35^c$	60	7	2	8	0.111	0.3/0.17	0.10
11	Wearing surface								
	Separate course	$10/15^c$	20	7	1	4	0.056	0.6/0.4	0.3
	Bonded monodeck	$10/15^c$	30	7	1				
12	Piers	30	120	7	2	8	0.111	0.2	0.05
13	Joints	10	30	7	1	4	0.056	0.6	0.2
						$\Sigma = 72$	1.00	$r_0 = 0.24$	$r_1 = 0.075$

Note: Footings are not included primarily because they are inaccessible for visual inspection. The relevance of paint (also missing) on steel members is discussed in Example 24.

[a] Observed at no maintenance.

[b] Assumed at full maintenance (Table E23.1).

[c] With/without joints.

Example 24. Bridge Deterioration Rate as Function of Maintenance

"Full maintenance" is defined by the vector of 15 maintenance task frequencies f_j (Table E23.1, column 3) and discussed in Example 23.

Bridge condition R is defined by the NYS DOT formula

$$R = \sum_{i=1}^{13} k_i R_i \qquad (E24.1)$$

where k_i = normalized values of weight factors W_i, shown in Table E23.2, column 7

R_i = lowest ratings of 13 components in Table E23.2, observed on rated bridge, not necessarily in same span

According to Eq. E24.1, the deterioration rate r can be expressed as

$$r = \sum_{i=1}^{13} k_i r_i \qquad (E24.2)$$

where $r_i = \partial R_i/\partial t \approx \Delta R_i/\text{year}$ = annual deterioration rate of bridge component i

ΔR_i = annual change in condition rating of component i

It is assumed that the "steepest" deterioration rates r_{i0} (column 9, Table E23.2) of the 13 components are known from inspection records and have occurred at a maintenance level $m_j = 0$.

At full maintenance (e.g., $m_j = 1$) the component deterioration rates r_{i1} (column 10, Table E23.2) must be assumed. Columns 9 and 10 of Table E23.2 allow computing r_0 and r_1 according to Eq. E24.2.

In the range between the two extreme maintenance strategies, component deterioration rates r_i depend on the 15 maintenance tasks j of Table E24.1 as shown in Eq. E24.3

$$r_i = r_{i0} - (r_{i0} - r_{i1}) \sum_{i=1}^{15} k_{ji} \, m_j \qquad (E24.3)$$

where $0 \le m_j \le 1$ = maintenance level of performance of jth task, expressed as fraction of recommended full maintenance level

k_{ji} = normalized values of importance factors $0 < I_{ji} < 100\%$, denoting importance of jth maintenance task on ith component

The procedure therefore depends on the matrix of importance factors I_{ji}. The values assigned to I_{ji} in Table E24.1 are subjectively selected first approximations. Extreme cases are relatively obvious. The importance of painting is set at 100% to steel but at 0 to concrete. It is harder to correlate the importance of sweeping and washing. The sensitivity of the factors must be explored and inspection and maintenance records must be cross referenced.

Substituting Eq. E24.3 into Eq. E24.2 obtains the sought relationship between the bridge deterioration rate r and the levels of performance of maintenance tasks m_j:

$$r = \sum_{i=1}^{13} k_i \, r_{i0} - \sum_{i=1}^{13} k_i \, (r_{i0} - r_{i1}) \sum_{i=1}^{13} k_{ji} m_j \qquad (E24.4)$$

Differentiating Eq. E24.4 with respect to m_j obtains the sensitivity k_j of the deterioration rate r to the performance level m_j of task j (columns 5 and 6, Table E23.1):

$$k_j = \frac{\partial r}{\partial m_j} = \sum_{i=1}^{13} k_i \, (r_{i0} - r_{i1}) k_{ji} \qquad (E24.5)$$

The performance level can be assumed proportional to the frequency f_j or to the annual cost. The cost effectiveness CE_j of task j can therefore be expressed as

$$CE_j = \frac{k_j}{c_j f_j} \qquad (E24.6)$$

where c_j = unit cost of task j

f_j = annual recommended frequency of task j at $m_j = 1$ (Table E23.1, column 3)

Up to this point, the vector of recommended maintenance frequencies f_j and the matrix of importance factors I_{ji} are uncorrelated expert opinions. If they were indeed optimal, the values of f_j listed in Table E23.1, column 3, should have obtained equal

Table E24.1 Importance Factors of Maintenance Activities j on Bridge Components i; Noncalibrated (100%)

Bridge Element i	1	2	3	4	5	6	7	8	9	10	11	12	13
Maintenance Task j	Bearings	Backwall	Abutment	Wingwall	Seat	Primary Member	Secondary Member	Curb	Sidewalk	Deck	Wearing Surface	Pier	Joint
1 Debris removal	70	50	20	20	80	50	50	80	80	80	90	10	80
2 Sweep	20	10	10	0	50	50	50	100	80	90	100	10	100
3 Clean drains	90	90	90	80	100	100	100	100	100	100	100	50	100
4 Clean abutments, piers	100	100	100	90	100	80	80	0	0	50	50	100	50
5 Clean gratings	100	50	70	100	100	100	100	10	10	80	100	100	90
6 Clean joints	100	80	100	50	100	100	80	50	50	90	90	90	100
7 Wash deck	50	30	20	0	60	40	40	100	0	100	100	40	100
8 Paint[a]	100/0	50	0	0	100/0	100/0	100/0	0	0	40	0	100/10	50
9 Spot paint[a]	100/0	50	0	0	100/0	100/0	100/0	0	0	0	0	100/10	0
10 Patch walks	0	0	0	0	0	0	0	100	100	10	10	0	50
11 Payment and curb	100	100	100	50	100	100	100	100	100	100	100	50	50
12 Electrical maintenance	0	0	0	0	0	0	0	0	0	0	0	0	0
13 Mechanical oiling	100	50	50	20	100	100	100	100	0	50	0	100	100
14 Resurface	0	10	0	0	10	10	10	50	50	100	100	10	100
15 Wash underside	100	100	100	50	100	100	100	0	0	80	0	100	90

[a] Alternative values apply to steel and concrete structures, respectively.

or comparable cost effectiveness CE_j. To the contrary, the values of CE_j listed in column 6 range from 0.2 for paint to 19.5 for cleaning of gratings (disregarding the mandatory mechanical maintenance). The implication is that, if paint, cleaning of gratings, and other cleaning operations are judged of comparable importance (e.g., $I_{ji} = 100\%$), the less costly cleanings must be more cost effective. That result can be explored in the following two ways.

A. I_{ji} (TABLE E24.1) GOVERNS

If I_{ji} governs the maintenance optimization, then CE_j must be set at a constant level, resulting in the following "cost-effective" maintenance vector f_j^{ce} (Table E23.1, column 7):

$$f_j^{ce} = \frac{\text{const. } k_j}{c_j} \tag{E24.7}$$

The annual maintenance cost C_M (also referred to in Examples 19 and 23) can be expressed as

$$C_M = \sum_{i=1}^{15} c_j f_j^{ce} = \text{const.} \sum_{i=1}^{15} k_j \tag{E24.8}$$

The total annual maintenance budget C_M should therefore be reallocated among the 15 maintenance tasks according to Eq. E24.8. The resulting frequencies f_j^{ce} are shown in Table E23.1, column 7. As intended their cost effectiveness (Table E23.1, column 9) is relatively uniform.

B. f_j (TABLE E23.1, COLUMN 3) GOVERNS

If f_j governs, I_{ji} must be adjusted to agree with it. A uniquely defined solution can be obtained by assuming that the rows of the original matrix I_{ji} represent correct relative importance but the columns are subject to adjustment. The assumption implies that the importance of each task to the 13 significant elements has been rated correctly but the tasks are incorrectly compared to each other. Equation E24.7 must be modified to obtain the calibrated sensitivity k_j^C as follows

$$f_j = \frac{k_j^C}{\text{const. } c_j} \tag{E24.7a}$$

The calibrated importance factor matrix I_{ji}^C can then be obtained from the condition

$$I_{ji}^C = \frac{I_{ji} \, k_j^C}{k_j} = \frac{\text{const. } I_{ji} f_j c_j}{\sum_{i=1}^{13} k_i \, (r_{i0} - r_{i1}) k_{ji}} \tag{E24.9}$$

The resulting I_{ji}^C matrix is shown in Table E24.2.

The results of A and B can be summarized as follows: The cost-effective maintenance strategy of Table E23.1, column 7, was not constrained by practical considerations and ignores them accordingly. Sweeping is recommended at an almost daily schedule (248.194 times a year), whereas painting is virtually canceled (0.011 an-

Table E24.2 Calibrated Importance Factors I_{ji}^c (100%)

Bridge Element i	1	2	3	4	5	6	7	8	9	10	11	12	13
Maintenance Task j	Bearings	Backwall	Abutment	Wingwall	Seat	Primary Member	Secondary Member	Curb	Sidewalk	Deck	Wearing Surface	Pier	Joint
1 Debris removal	4	3	1	0.6	5	3	3	5	5	5	5	0.6	5
2 Sweep	0.4	0.2	0.2	0	1	1	1	2	2	2	2	0.2	2
3 Clean drains	1.5	1.5	1.5	1	2	2	2	2	2	2	2	0.8	2
4 Clean abutment, piers	6	6	6	6	6	5	5	0	0	3	3	6	3
5 Clean gratings	1	0.5	0.8	0.1	1	1	1	0.1	0.1	0.9	1	1	1
6 Clean joints	6	5	6	3	6	6	6	3	3	5	5	5	6
7 Wash deck	2	1	0.8	0	2	1.5	1.5	4	0	4	4	1.5	4
8 Paint	100	50	2.4	1.8	100	100	100	0	0	40	0	100	50
9 Spot paint	80	40	0	0	80	80	80	0	0	1.5	0	80	0
10 Patch walks	0	0	0	0	0	0	0	15	15	1.5	1.5	0	7
11 Pavement and curb	4	4	4	2	4	4	4	4	4	4	4	2	2
12 Electrical maintenance	0	0	0	0	0	0	0	0	0	0	0	0	0
13 Oiling	2	1	1	0.5	2	2	2	2	0	1	0	2	2
14 Wearing surface	0	0.5	0	0	0.5	0.5	0.5	3	3	5	5	0.5	5
15 Wash underside	28	28	28	14	28	30	30	0	0	20	0	30	25

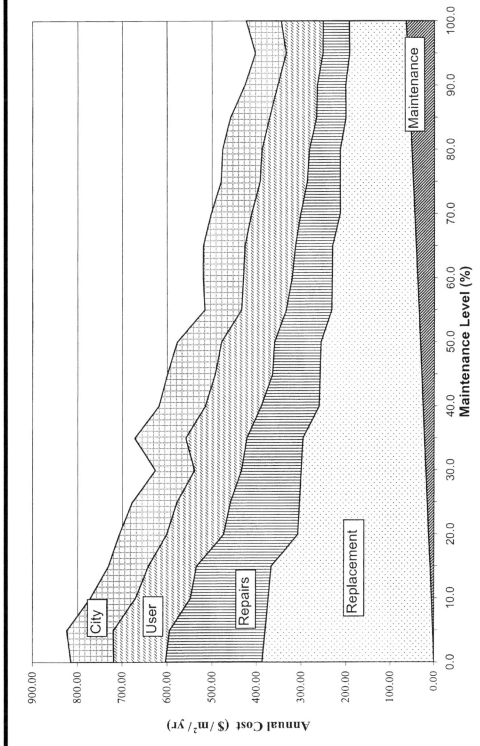

Figure E24.1 Annualized costs (U.S. $/m^2/year) of replacement, repair, users, and owner as function of maintenance expenditures.

nually, e.g., once every 90.90 years). The consistent attempt to minimize deterioration within the prescribed budget reveals the most significant vulnerability of managing maintenance. The 15 maintenance tasks are vastly disparate in magnitude, cost, and physical significance. The recommended painting comprises 40% of the annual maintenance cost (Table E23.1, column 4). Spot painting represents another 26% of C_M. Together they amount to 66% of the recommended annual maintenance cost. From a cost-effectiveness standpoint such expenditures can be justified if painting is 100 times more important than cleaning of gratings and 250 times more important than sweeping as reflected in I_{ji}^C (Table E24.2).

Yanev and Testa (2001) developed an algorithm combining the effects of maintenance (as in case B) with prescribed frequencies and costs of repairs, component replacements, and ultimately full replacement. User costs and costs incurred by the owner (e.g., business losses sustained as a result of poor transportation) were included. A sample of the output is presented in Fig. E24.1.

NCHRP Synthesis 153 (1989) refers to NCHRP Report 273 (1984) for a comprehensive study of optimal (highway) maintenance strategies.

Ultimately, prescriptive and performance-based maintenance management methods form a dynamic equilibrium subject to periodic updating. Maintenance tasks must be reviewed and adjusted to the changing needs of the community, the conditions of the structures, and the transportation demands. Recent design and construction proposals include maintenance plans typically for 30-year periods. Rehabilitation is thus implied at the age of 30 years, even though AASHTO recommends a nominal life span of 75 years. New bridge designs are beginning to consider methods of component replacement.

Paint and Painting

The cycles and costs of painting are closer to those of component replacements and rehabilitations than to those of seasonal maintenance tasks. As a maintenance task, the painting of steel (and earlier, iron) has always been vulnerable to delays and poor quality. Boller (1885, pp. 79–80) observed: "Too often insufficient painting is allowed to remain as the only protection for years . . . When iron is *neglected* [author's italic], it is only a question of time as to its final destruction. . . . It is advisable to have iron bridges so designed that all parts of the work should be open to inspection, and within reach of the paint-brush."

Example 24 shows that the cost of painting (removal, abatement, and application) and the expected useful life of bridge paint far exceed those of regular maintenance tasks. At a recommended cycle of 12 years, the costs of full and spot painting amount to 66% of the total annual maintenance cost in Table E23.1. The associated cost effectiveness has sharply declined, however, because of the increase in metal surface preparation and containment costs, including lead abatement (from U.S. \$5 to \$301/m^2 of steel surface between 1990 and 2000 in New York City). As an example, the bridge shown in Fig. 11.2*c* is to be repainted over a period of up to 5 years without traffic interruption at a contract cost of U.S. \$167 million.

The FHWA now considers painting eligible for federal funding, thus lending key support to the view that the task is much better managed, funded, and scheduled as a form of rehabilitation. Structural repairs can be included.

The needs for painting would be better reflected in the bridge inventory if paint were considered an essential element of steel structures and its rating contributed to the overall bridge condition rating.

Repair

Most texts (e.g., AASHTO, 1999a) discuss repairs as one of the maintenance options (or penalties). The design and appropriate application of steel and concrete repairs is addressed by specialized publications, such as Fisher (1984), Pritchard (1992), Emmons, (1993), and Mallett (1994). NCHRP Reports 222 (1980) and 243 (1981) discuss repairs as alternatives to replacements.

To supplement the expense-funded Preventive Maintenance Program (Example 23), NYC DOT developed a capitally funded Corrective Repair Program (Yanev, in Vincentsen and Jensen, 1998, pp. 11–22). Corrective repair work had to meet the following criteria:

Useful life >5 years

Cost > U.S. $5 million

Condition rating improvement >1 rating point (NYS DOT rating system)

Duration of repair work <1 year

The program was approved as a form of capital improvement (and hence funding) under the designation *component rehabilitation.*

Typical repairs are more labor intensive than maintenance and smaller in scope than rehabilitation. They include joint (Fig. 11.3), scupper and individual bearing replacements, roadway resurfacing, and waterproofing. Splash zone spot painting can qualify in this category. Such repairs can be perceived as alternatives to more intensive washing and surface cleaning.

Cost–benefit comparisons would have to take into account the particular specifics of the structure under consideration and the implications to traffic and to the environment (for example, if water must be contained during washing).

Repairs can be grouped into projects of larger scope, amounting to rehabilitations. The many advantages include a combined bidding process, clearer funding, and improved traffic coordination. The planned retrofitting of structural components, for example, the seismic retrofitting of bearings, is conducted along traffic corridors with minimized traffic closures (Figs. 4.38, 10.2, and 10.3).

Repairs must compete with the alternative of replacements. Waddell (1921, p. 408) pointed that out early in the twentieth century: "Quite often bridges are repaired, which, from the standpoint of true economy, should be relegated to discard."

That position was held by many designers who proposed replacements for the Williamsburg Bridge in New York City in 1988 (Example 3).

Expansion Joints

In Sections 4.2.3 and 4.3.3 expansion joints were identified as vulnerable at every stage of the bridge life-cycle process. The most common compression joints (Figs. 11.3a,b) are intended to be regularly cleaned and their seals periodically replaced.

The replacements need not be "in kind." Figure 11.4 illustrates a replacement of a "cushion" joint (Fig. 4.74a) with a "plug" joint (Fig. 4.74b). Although plug joints are not problem free, their repair may consist of a relatively easier resurfacing.

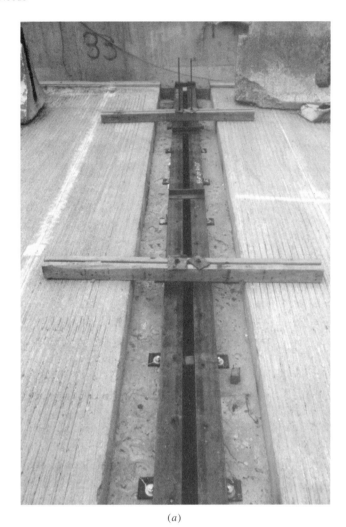

(a)

Figure 11.3 (a) Armor compression joint.

NCHRP Report 467 (2002) describes the performance of modular bridge joint systems (MBJSs) (Fig. 11.5). Their functions are much more complex and usually involve highly traveled roadways on long-span structures.

Failure modes of modular joints are illustrated in Section 4.2.3 (Figs. 4.28 and 4.29). Indications of malfunctions, such as excessive noise and vertical movement of the "center beams" or rupture of the filler material, must be observed and reported as they occur, because a fracture may be imminent.

The modular joint in Fig. 11.5 has smaller and more closely spaced center beams than the one shown in Fig. 4.28. The smaller and more numerous bars are designed to extend the fatigue life of the assembly.

(b)

Figure 11.3   (b) Installation of compression joint seal.

Figure 11.6 illustrates a model of a joint used at the suspension Rainbow Bridge in Tokyo. The joint accommodates large displacements typical of suspension bridges while providing a continuous wearing surface.

Troughs under all large-displacement joints must be regularly inspected and cleaned. Moving parts must be greased or oiled according to their specific maintenance plan.

Emergency/Temporary Repairs

Emergency repairs mitigate the potential hazards of Section 10.3. They must be preformed with minimal traffic interruptions and are often designed to last only until the forthcoming rehabilitation. The NBIS explicitly directs load rating calculations to ignore any contributions of temporary supports. Condition ratings must similarly ignore temporary repairs, even if the service has been restored. Typical temporary repairs are deck steel plates (Fig. 11.7) and shorings. The height of the supported structure and the expected service life of the temporary repair determine whether the shoring can consist of a timber post (Fig. 11.8), a steel column (Fig. 11.9a), or a tower (Fig. 11.9b). "Strongbacks" (Figs. 11.10a,b) are used when support from underneath is impractical, as over a channel or active traffic. They redistribute the load among adjacent primary members but close the lane above the affected area.

Figure 4.44 shows impact damage to a concrete pier. The spacing of the primary members is 8 ft (2.4 m). Loads from the bridge superstructure were redistributed and the

Figure 11.4 Replacement of cushion with plug joint with traffic lane closures.

damage was limited to the impacted members. After the traffic lane above the damaged column was closed, a new girder (Fig. 11.11) was installed spanning over the destroyed column. Longer spans and wider primary member spacing would have required greater pier protection against impact.

Maintainability/Repairability

Section 4.3.3 identified maintainability and repairability as vulnerabilities of design in general and the design of nonredundant structures in particular. This is recognized in the design of new structures and major rehabilitations. Fracture-critical, seismically nonperforming, and uninspectable details are avoided. The examples of the Williamsburg Bridge and the Pont de Tancarville have served to discourage the use of nongalvanized high-strength wires. Contemporary suspension bridge cables are protected redundantly.

Figure 11.5 Modular joint at suspension bridge.

Whereas the original cables at the Pont d'Aquitaine (Fig. 5.3) consisted of nongalvanized, unwrapped helical strands, the strands of the replaced ones are galvanized, wrapped, and air-dried (Kretz et al., 2006).

The Honshu-Shikoku Bridge Authority developed a system introducing dry air under the wrapping at the Akashi-Kaikyo, Kurushima, and other suspension bridges, as shown in Figs. 11.12 and 11.13 (NYSBA/HSBA, 2002).

Figure 1.40*a* shows a traveler under the main span of the Akashi-Kaikyo Bridge. The bridge also has a service roadway under the traffic lanes (Fig. 11.14).

In contrast, older bridges must be upgraded to conform to contemporary maintainability and inspectability standards. The travelers shown in Fig. 11.15 were added long after the completion of the East River bridges and have been repeatedly upgraded. Figure 11.16 shows an experimental anti-icing spray system installed on the Brooklyn Bridge.

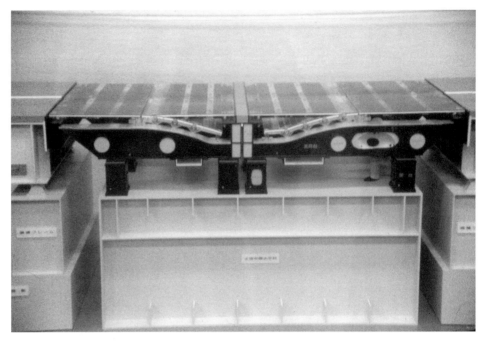

Figure 11.6 Model of large-displacement joint at Rainbow Suspension Bridge, Tokyo.

Inspectability is a related requirement, discussed in Chapter 14.

11.5 ECONOMIC ASSESSMENTS

In contrast with accounting, engineering economy recognizes nonmonetary considerations. Nonetheless, money is the only language that can quantify immediate costs exactly, future benefits vaguely, and needs and conditions uncertainly and partially. Translating past and current structural and traffic condition assessments into immediate and future costs is vulnerable to the discrepancies between engineering and accounting models (Chapter 4) and particularly suited for probabilistic analysis (Chapter 5). Economic analysis for civil engineering structures is presented, for instance, by De Gramo et al. (1973), Park and Jackson (1984), Hudson et al. (1997), and Chang (2005). Example 10 highlights some of the fundamental concepts presented therein. NCHRP Reports 285 (1986) and 483 (2003) summarize economic analysis methods and terminology typically used in bridge management.

Asset management, for instance, as defined by GASB-34 (FHWA, 2000), requires annual assessments of needs for restoring the infrastructure to as-new condition. The latter description is rather vague for the purposes of mechanics but apparently meaningful in finance. Chapter 9 pointed out that the estimates of the remaining bridge useful life and the cost of restoring bridges to as-new condition or a state of good repair are too broad for engineering purposes.

Figure 11.7   Deck steel plate with loose bolt.

Section 4.1 discusses some of the seemingly acceptable procedures through which first and life-cycle needs of a structure can be underestimated. An early twentieth-century view of life-cycle cost comparison methods (Wadell, 1916) is quoted in Appendix 12.

The capability of the bridge database to support meaningful life-cycle cost decisions has been both questioned (Veshosky, 1992) and encouraged (Veshosky et al., 1994; Frangopol and Furuta, 2001; Miyamoto and Frangopol, 2001). Veshosky (1992) argued that "lifecycle cost analysis does not work for bridges" (subsequently adding a significant "yet"), because their life expectancy cannot be predicted with sufficient accuracy. Leeming (in Harding et al., 1993, pp. 574–583) considered the following criticisms of whole-life costing to be "true to varying degrees" but more appropriately described as "problems to overcome":

Figure 11.8 Timber shoring.

For civil engineering structures with long lives the problems are more significant. There are doubts about forecasting physical life expectancy particularly when economic, functional, technological or social obsolescence is reached first. Costs and economic conditions are difficult to predict over long periods and capital costs tend to predominate. It is said that results can mislead the unwary or be adjusted to provide the desired answer. Future maintenance costs are regarded as visionary while capital costs are real.

NCHRP Report 483 (2003, part 11, p. 4) states:

Life cycle cost analysis (LCCA) is a process for evaluating the total economic worth of a usable project segment by analyzing initial costs and discounted future costs, such as maintenance, reconstruction, rehabilitation, restoring, and resurfacing costs, over the life of the project segment.

(a)

Figure 11.9 (a) Steel shoring, failed bearing stool.

The option of investing in an alternative other than the facility (or segment) in question is implied. This potential vulnerability can be avoided by stipulating that, in one form or another, the service under consideration must be provided in perpetuity (Section 4.1.3).

NCHRP Report 285 (1986) (Appendix 49) developed life-cycle cost analysis for bridges along with highways. The recommendation was reiterated by FHWA (1994). Hawk (in NCHRP Report 483, 2003) developed a BLCCA software package compatible with the BMS packages PONTIS and BRIDGIT. LCCA is presented as "a technique for evaluating the economic efficiency of expenditures." The report extends beyond cash

(*b*)

Figure 11.9 (*b*) The 45-ft shoring tower for column in Fig. 4.73.

flow analysis by considering all resources involved, including those of the public in a benefit–cost analysis (BCA) module. Appendix 34 herein contains a brief description.

11.5.1 Benefit–Cost Analysis

De Gramo et al. (1973, p. 375) (also discussed in Section 4.1.3 herein) define the conventional benefit–cost (*B*/*C*) ratio as follows:

$$B/C = \frac{\text{AW (net benefits to the user)}}{\text{AW (total net costs)}} = \frac{B}{\text{CR} + (\text{O \& M})} \qquad (11.1)$$

(*a*)

Figure 11.10 Strong-back: (*a*) top of deck.

where AW = annual worth
  B = annual worth of net benefits (gross benefits minus costs) to user
  CR = capital recovery cost or equivalent annual cost of initial investment, in-
    cluding any salvage value
  O & M = uniform annual net operating and maintenance disbursements to supplier

A modified *B/C* considers only the net benefits reduced by the operating and maintenance costs:

$$\frac{B - (O \& M)}{CR} \qquad (11.1a)$$

(*b*)

Figure 11.10 (*b*) below deck.

Benefit and Cost Value

Yao and Furuta (1986) distinguish between numerical, monetary, and verbal damage assessments (Section 10.3). Of these, the monetary alone is capable of expressing in (approximate) numerical terms both costs and benefits. The method expressed in Eq. 11.1 has limited validity relative to bridges, primarily because, as illustrated in Fig. 1.33*b*, the users and the managers of publicly owned infrastructure facilities do not have the same perception of the benefits. Immediate reductions in maintenance costs can translate into tax cuts but also imply an increase in user costs and capital expenditures over time. Table 11.1 attempts to summarize the different priorities to be reconciled by the various levels of bridge-related competence.

Extratransportation benefits, such as advertising opportunities, are explored by some bridge owners (see following paragraph).

Figure 11.11   Emergency repair of impact damage.

Alternatively, Hawk (NCHRP Report 483, 2003) recommends the "net present value," which adds all costs, positive and negative. That method determines present value by discounting (see following paragraphs).

Salvage Value

The salvage value of a structure that cannot be used in a rehabilitation is most likely to be negative (see also Section 1.5). The environmentally acceptable removal of a bridge,

Figure 11.12 Cable band with equipment introducing hot air under wrapping of cable, Kurushima Bridge, Japan.

particularly in urban settings, is technically demanding and costly. The few exceptions mostly demonstrate the generality of this practice. The lift bridge shown in Fig. 4.79 replaced a swing bridge, which in turn was floated to a nearby location and served for several decades. Chan reported in *The New York Times* (January 14, 2006, p. B3, column 1) that the Willis Avenue Bridge between Manhattan and the Bronx (Fig. 11.1c) has been offered for U.S. $1.00 to any buyer who would preserve it. The owner (NYC DOT) will assume the cost of shipping the structure within a 15-mile (22-km) radius.

Originally built for U.S. $2.4 million and opened in 1901, the Willis Avenue swing bridge, in contrast to the Macombs Dam Bridge (Fig. 1.10), is not a landmark. Because of its poor condition (caused primarily by its open steel roadway grating; Fig. 4.21), the bridge is reported to absorb approximately U.S. $1.1 million/year (2005) in demand maintenance expenditures. Its replacement cost is estimated at U.S. $300 million (ca. 2012). The offer to preserve the bridge is motivated by its perceived historic value.

Following the collapse of a span in 1973, the West Side Highway in Manhattan south of West 57th Street never carried traffic but remained standing until 1990. Eventually, it was demolished by the owner (New York City) and dumped in the Atlantic Ocean as potential habitat for marine life.

Since the 1970s, the railroad bridge known as the Conrail Highline (Fig. 11.17) has carried only dead loads, wind loads, and billboards. Following intense public debate, in 2005 the structure was transferred to New York City at no cost, to be rehabilitated as a public park.

Figure 11.13   Cable inspection and maintenance platform, Kurushima Bridge.

Figure 11.14   Service roadway, Akashi-Kaikyo Bridge, Japan.

Figure 11.15 Travelers on Brooklyn and Manhattan Bridges.

Brühwiler and Adey (2005) list the following cost–benefit methods of comparing management strategies:

Present Worth. The present worth method (PWM) applies discount rates to life-cycle costs, benefits, or both. Limitations of discounting are discussed in Section 4.1.3 and Example 10.

Equivalent Uniform Annual Cost (Hudson et al., 1997, p. 296). The method combines all initial capital costs and all recurring future expenditures into equal annual payments over the analysis period. The authors caution that the method does not include benefits in the evaluation.

Incremental Rate of Return (Hudson et al., 1997, p. 300). Costs, benefits, and rate of return are calculated with respect to a base strategy.

Incremental Benefit–Cost Ratio (Hudson et al., 1997, p. 301). The method compares the present worth of benefit–cost ratios for alternative strategies.

Cost-Effectiveness Method (Hudson et al., 1997, p. 301). The cost effectiveness of this method is the reciprocal value of the traditional one; i.e., it is the ratio of effectiveness and the present worth of all life-cycle costs. The method is recommended for the evaluation of pavements, where "effectiveness of an alternative is calculated as the area under the performance curve (i.e. serviceability versus age), multiplied by traffic volume and length of road section."

Figure 11.16 Anti-icing spray system, Brooklyn Bridge. Courtesy B. Ward.

Figure 11.17 Conrail highline, unused for 35 years.

User Costs

User costs (see also Section 4.1.3) are indirect yet critical for management decisions. No other penalty combines in the same organic manner the quantifiable and the subjective aspects of serviceability.

BLCCA (NCHRP Report 483, 2003) models the user costs caused by traffic delays and accidents. Thompson et al. (in TRR 1697, 2000, pp. 6–13) adjust the model used in PONTIS and BRIDGIT to reflect more accurately the effects of road widening. NCHRP Synthesis 330 (2004) correlates the International Roadway Roughness Index (IRI) and vehicle operating costs. On the determination of user costs, Johnston (in TRC 423, 1994, pp. 139–149) issued the following reminder: "It is important to remember that, although we engineers can calculate stresses to many insignificant figures, we only know the real loads, and thus the stresses, to one or sometimes two significant figures. Achieving the same level of accuracy in estimating costs may still be a goal, but it is probably attainable."

At present, user costs are estimated within tens of millions of dollars. Some assumptions typical of user cost models are summarized in Appendix 49.

Traffic delays and vehicle operating costs are quantifiable, but they do not capture the losses caused by overall traffic reduction. Bruhwiler and Adey (2005) treat user costs as *unattained benefits* in a supply–demand system (SDS). The supply side consists of the structural and network performance. The demand side comprises all consequences to the users in terms of costs. Loss of life can be modeled separately from other penalties. A distinction is made between costs incurred when traffic is constrained and when it is not possible, i.e., when it is reduced to zero. In the latter case, the costs are established as a fraction of the gross domestic product (GDP), thus (inevitably) linking bridge management to the economic development on a national level. National economy is beyond the scope of bridge management; however, the dependence between the two, if ignored, quickly becomes a critical vulnerability (Section 4.1.6).

Public, Private, and Privatized Services and Facilities

The linkages between the transportation infrastructure and economic development are discussed in Appendix 23. The modeled relationships are highly sensitive to local conditions, data quality, and assumptions. Of particular interest are the comparisons between the performance and management of public, private, and privatized transportation facilities. Privatization has been very popular with high-level management during the last decade (see Appendix 19 and Example 4).

Tolled facilities, in addition to generating dedicated funding, create a coherent interaction with the users (Yanev, in TRC 423, 1994, pp. 130–138). The toll revenues quantify the value of the provided service and address the structural needs. This type of equilibrium is lacking at publicly owned facilities, funded by taxes. Nonetheless, the introduction of tolls on publicly owned structures is not a popular solution. Tolls were lifted from the East River bridges (Examples 1 and 3) by order of the New York City mayor in August 1911. Since then, their reintroduction has been repeatedly considered as a source of funding (for the bridges as well as for numerous public activities). Llanos (1992, p. 72) quoted economic studies suggesting that tolls benefit the users when the demand surpasses the supply, causing traffic congestion. That finding appears to assume the availability of alternative traffic routes. Ultimately, tolls on major facilities are related to the entire multimodal transportation network. The history of tolls on the George Washington Bridge is briefly discussed in Example 2.

Physical isolation (usually on the periphery of urban areas) and administrative autonomy (as in the case of public authorities) have aided the management of tolled bridges in achieving genuine life-cycle optimization (see Example 2). Once tolled facilities are integrated into larger transportation networks (such as the Metropolitan Transit Authority in New York City), their needs are prioritized within a broader context, rather than optimized within their direct revenues.

Indirect Effects of Transportation Projects

NCHRP Report 466 (2002) by the Louis Berger Group followed up the earlier NCHRP Report 403 (1998) in providing a desk reference for estimating the indirect effects of proposed transportation projects. The main objective is to achieve compliance with the National Environmental Policy Act (NEPA). NCHRP Report 466 (2002, p. 12) pointed out that the Transportation Equity Act for the Twenty-First Century (TEA-21) of 1988, in contrast to ISTEA (1991), eliminated the requirement for a separate major investment study (MIS) to precede any substantial transportation project (Appendix 11). Indirect effects are studied as part of the general transportation planning process of state agencies and metropolitan planning organizations.

The Council of Environmental Quality (CEQ) has defined the effects of project actions as follows:

Direct Effects. Caused by the action and occurring at the same time and place. Examples include increases in traffic volume and speed, improved access, displacement of local business, and alterations of water drainage patterns.

Indirect Effects. Caused by the action and occurring later in time or farther removed in distance but still acceptably foreseeable. The three broad categories of indirect effects according to the CEQ are:

1. Alteration of the behavior and functioning of the affected environment caused by project encroachment (physical, chemical, biological) on the environment

2. Project-influenced development effects (i.e., the land use effect)

3. Effects related to project-influenced development effects (i.e., effects of the change in land use on the human and natural environment)

Cumulative Impacts. The impact on the environment which results from the incremental impact of the action when added to other past, present, and reasonably foreseeable actions. In practice, analysis of cumulative effects has been incorporated with the assessment of indirect effects because many indirect effects, including induced development effects, fall within the definition of cumulative impacts.

Secondary Effects. Some authors differentiate *secondary* from *indirect* effects by associating the former with *induced development*.

Induced Growth. Changes in the intensity of the use to which land is put that are caused by the action/project. For transportation projects, induced growth is attributed to changes in accessibility caused by the project. Such changes may influence the location of future development.

Chapter 12

Decision Making

"A BMS is a decision support tool, not a manager!"

Shirole et al. (in TRC 423, 1994, p. 34)

The ultimate objective is effective management, not reliable support. Management is a process of taking and implementing decisions. It can be formalized only within limits. In the modern version of the ancient conflict between rigorous deduction and common sense (Sections 1.7, 1.8), contemporary management is challenged to formalize its decision as much as possible and then take uniquely innovative decisions. Kline (1953, p. 232) observed that "incompleteness in the thinking of the creators of major ideas is almost to be expected." Leibnitz credited analytic geometry to the genius of Descartes, rather than to his method. Thus, managers, who, even though endowed with genius, might be incomplete thinkers, are offered decision support tools which fall short of both genius and completeness, particularly in the following areas:

The managed assets cannot be fully described or assessed.

Decisions can be formally optimized only within simplified constraints.

Just as some actions are based on reflection whereas others are reflexive, decision making can take the available information into account to a varying degree (anticipated by I. Asimov, Appendix 3). BMS decision making reflects the general patterns observed in thinking and in professional development (Section 1.2). The chart in Fig. 12.1 illustrates the flow of information expected to develop in a functioning BMS along the lines suggested in Fig. A16.3.

The objective in Fig. 12.1 is to determine the short- and long-term capital and expense plans. The capital plans in turn lead to the selection of projects and the allocation of funding. The expense budget funds continuing tasks, such as maintenance. The activities can be structured as in the chart of Fig. E18.2. The eligibility for funding from expense and capital budgets has profound implications for the management policy (Chapter 11).

Hudson et al. (1997, p. 320) distinguish the following methods of setting infrastructure expenditure budgets:

1. Extrapolation from previous years with adjustments for growth of the system and inflation.

2. Demand based on condition-responsive budgeting. Work responds to expected performance indictors and conditions. The method involves only necessary expenditures, but the needs are determined in the course of time.

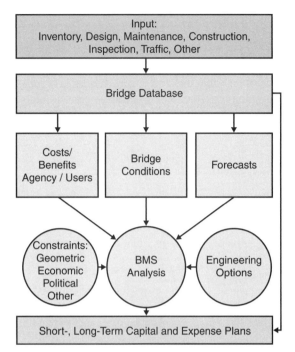

Figure 12.1   Flow of information in bridge management.

3. Budgeting based on a desired level of performance or level of service, for instance, based on a composite serviceability index.

4. Budgeting seeking an optimum level of service for the network or system of assets as a whole. As in all cases, the optimal may not be affordable.

5. Affordable budgeting, determined by administrators and elected bodies.

The five alternatives range from spending as little as possible to more or less "enlightened" optimization. The demand for decision support from a BMS varies with the adopted method. Hudson et al. (in NCHRP Report 300, 1987) and (1997) refer to practices avoiding any level of optimization as "business as usual." Executive management is not likely to use a BMS for decision support it intends to ignore. The following are extreme possibilities:

- BMSs are used for inventory only, either because other modules are inconclusive or because restructuring the database and the operation is considered too costly.

This mixture of expediency and attitude typically governs so long as the BMS capabilities remain limited, unreliable, and dependent on the quality of expertise. Eventually, BMSs (as computers before them) have become robust, refined, and available in customized form. Their value can no longer be disputed. When the FHWA relaxed the mandate to develop BMSs (Appendix 11), bridge managers did not abandon their use but, rather, sought to upgrade them.

- BMSs are used to the maximum in decision support.

This is likelier to occur when the network comprises few assets, so that project and network level priorities concur. Then the BMS can estimate (relatively accurately) the needs and consequences of individual structures within the parameters of the selected policy or the available budget. On the project level BMS can acquire the data for the subsequent decision cycle on the network level. On a network level BMSs compare the outcomes of options, strategies, or scenarios, considering different levels of investment, allocation, and performance. Some of the tools for such analysis are described in Appendix 15.

- BMSs are used to substantiate budget requests with data and analysis. The actual budgets eventually prioritize projects according to available funding.

The latter course is a realistic compromise between extreme reliance on and complete disregard for the BMS. Neither the design nor the use of the BMS decision algorithms can entirely avoid subjectivity. Chapter 2 pointed out that decision support is often limited to prioritizing limited options and minimizing hazards. The realistic response options (also enumerated in Table 9.1) fall into the general groups of do nothing, monitor, maintain, repair, and replace.

The consequences of each option (or scenario) must be quantified over comparable life cycles. Far from trivial, the do-nothing option implies quantifiable immediate savings and hypothetical increases in future risks and expenditures. The remaining options amount to various immediate costs and future benefits. BMSs should be able to compare the alternatives over a variety of planning horizons. Expert systems can model outcomes, taking into account not only deterioration models but also the history of past subjective choices under established constraints.

The effort to formalize and optimize decision making renders strategic planning (Sections 12.1 and 12.2) relatively coherent and transparent on the transportation network level. Implementation of the decisions on the level of individual projects (Section 12.3), however, introduces more variables, constraints, and uncertainties than formalized algorithms model realistically. Moreover, on the local level networks overlap. On the territory of New York City (Example 1) 600 bridges of the United States Interstate Highway network serve traffic along with 800 local bridges and a number of major toll facilities managed by the public authorities. The demand for a robust multiredundant system with a capacity roughly exceeding the traffic needs usually supersedes fragmented optimization exercises.

12.1 STRATEGIC PLANNING/ASSET MANAGEMENT

In the general hierarchy of planning for results, strategies identify and pursue long-term objectives. Tactics provide the means or tools of carrying out strategies within the more immediate future. Operations continually implement the means and use the tools. The strategic planning for the infrastructure, viewed as an integrated system, is the subject of *asset management*. A few current definitions and descriptions follow:

FHWA (1999, p. 7):

Asset management is a systematic process of maintaining, upgrading, and operating physical assets cost-effectively. It combines engineering principles with sound business practices and economic theory, and it provides tools to facilitate a more organized, logical approach to decision making. Thus, asset management provides a framework for handling both short- and long-range planning.

FHWA (2001a, p. 7):

> Asset management is a framework for making cost-effective resource allocation, programming, and management decisions.

U.K. Bridges Board (2005, p. 80):

> Asset management is a strategic approach that identifies the optimal allocation of resources for the management, operation, preservation and enhancement of the highway infrastructure to meet the needs of current and future customers.

> Asset management is the systematic and coordinated (set of) activities and practices through which an organization optimally manages its assets, and their associated performance, risks and expenditures over the lifecycle for the purpose of achieving its organizational strategic plan.

In the broader context of assets, BMSs are tactical and operational tools. Network-level tactics, however, may be project-level strategies. Most texts on preventive maintenance refer to *strategies*. NCHRP Report 483 (2003, p. 66) defines *management strategy* as follows: "A set of actions and their timing for developing, deploying, operating, and possibly disposing of a bridge or other major asset; typically stated within the context of certain experience-based rules or standards of professional practice."

Bridge managers tend to regard strategic network level-plans as external constraints imposed by economic and political considerations. Von Neumann and Morgenstern (1964, p. 2) considered a generalized theory of economics even less likely than the unified field theory sought by generations of physicists (and still at it in the twenty-first century). Hence, the authors reduced complex social phenomena to specific problems (games) with limited validity but coherent modeling (Section 1.9). Operational research (Beale, 1988) and structural reliability (Cornell, in Freudenthal, 1972) adopted that approach (Appendices 2 and 8).

Hudson et al. (1997) described the following classes of methods that could be used in priority analysis of actions in infrastructure management (Table 15.1, p. 316):

Simple subjective ranking of projects based on judgment

Ranking based on parameters, such as level of service and conditions

Ranking based on parameters with economic conditions

Optimization by mathematical programming model for year-by-year basis

Near optimization using a marginal cost-effectiveness approach

Comprehensive optimization by mathematical programming model, taking into account the effects of which, what, and when

By optimizing the outcomes of different (more or less) global scenarios, network, or top-down management runs the risk of ignoring project, or ground-up, priorities (Section 4.1.3). Banks (2002, p. 43) recounted recent developments intended to overcome the fragmentation of transportation service and political authority in the United States as follows:

In the 1960s and 1970s, the need to overcome modal fragmentation in governmental policy, planning, and finance led to the creation of multimodal departments of transportation at the federal level and in many states.

During the 1960s and 1970s, the perceived need to overcome modal and jurisdictional fragmentation in transportation planning at the local level led to the creation of Metropolitan Planning Organizations (MPOs) and the establishment of Federal Transportation Planning Regulations (FTPR).

Beginning in the 1970s with the widespread takeover of private mass transit firms by public agencies and continuing to the present time, the need to overcome the effects of jurisdictional fragmentation on transit routes, schedules, and fares has led to the creation of special-purpose regional agencies for transit planning and coordination in some metropolitan areas and assignment of these functions to MPOs in others.

In the 1980s, as a part of the federal deregulation of freight transportation, the need to overcome modal fragmentation led to repeal of provisions forbidding intermodal integration of freight firms.

Finally, a current institutional challenge is to overcome jurisdictional fragmentation in traffic control, data management, information services, and other areas affected by the ITS (Intelligent Transportation Systems) initiative. Specific areas in which the need to coordinate action is perceived include arterial signal control systems that cross jurisdictional boundaries, incident management, traffic and congestion monitoring, and provisions of real-time traffic information to the public.

All the described initiatives depend on the integration of information. Adams et al. (2005) reported on a method of integrating all bridge related data into a single Highway Structure Information System (HSIS), forming a consolidated data warehouse (see Appendix 18), for the Wisconsin DOT. D'Ignazio and Hunkins (TRR 1904, 2005, pp. 75–83) presented a framework for the integration of long-range and project planning in North Carolina. Key steps in their Comprehensive Transportation Plan (CTP) include the following:

Air quality conformity

Fiscal constraints

Land use integration

Stakeholder involvement

Multimodal integration

Modeling

Environmental considerations

Documentation

FHWA (2001e, p. 16) described *data integration* (Appendix 18) as a *workflow management* procedure. Workflow management considers business processes as sets of tasks performed in prescribed order. It combines and regulates the flow of information between various sources and participants, for instance, as shown in Tables 11.1 and 12.1.

A recently defined objective is to gain the satisfaction (or at least the understanding) of the users (e.g., the stakeholders) by engaging them in the planning process (Appendix

Table 12.1 Asset Management Outputs, Indicators, Constraints and Vulnerabilities

	Structural Engineering	Engineering Management	Asset Management	Users
Expertise	Structures Operations	Information Administration		Communication
	(product/process)	Resources Assets	Transportation	Economy
Output	Analysis Construction	Assessment Decision support		Cultural
				development
	Design Maintenance	Service Database	Industrial production	
Indicators				
Quantity	Load ratings	Structural accidents Productivity	Expenditures	Traffic
				accidents
	Surface roughness Geometry	Traffic Useful life	Pollution	
Quality	Condition ratings Productivity	Multimodal coordination	Quality of life	
	(priority rankings) Innovation	Aesthetics Satisfaction		
Standards,	Specifications Guides	Budget Administration	Policies	Laws
guidelines,	Safety QA&QC	Recommendations Accountability		Liability
constraints				
Vulnerabilities	Structural	Administrative	Economical	Political
(malfunctions)	Operational	Fiscal	Legislative	Popular

39). Management focuses on context-sensitive solutions (CSSs). In TRR 1904 (2005) Crossett and Oldham (pp. 84–92) and Rauch (pp. 93–102) report on the development of CSSs in the states of New York and Arizona. The latter authors distinguish between process- and outcome-related focus areas on the project and organizationwide levels. In each case the findings point to a strong dependence on local conditions and on the level of quality control of the process at every stage of the implementation.

FHWA (2004b) reported a pronounced shift toward integrated asset management in Australia, Canada, Japan, and New Zealand. In Japan, economic vitality, quality of life, safety, environment, and road administration are rated in terms of 17 asset performance indicators.

12.2 OPTIMIZATION

Optimization encompasses the pursuit of happiness (Section 2.2) as well as mathematical solutions minimizing the potential energy of mechanical systems (Section 5.2). The semiprobabilistic performance-based structural design specifications (Section 4.2) combine the qualitative assessment of the former and the rigorous quantification of the latter. Decision making relies to unspecified degree on intuitive and rigorous optimization and is therefore subject to constant debate and revision.

The *utility* optimized by game theory can represent first cost, cost effectiveness, or reliability. Von Neumann and Morgenstern (1964) argued that economics needed statistics, because it lacked the empirical data available in mechanics (Section 1.9); however, even mechanics could not avoid it entirely. Statistics eventually enhanced the empirical data used in engineering design (Appendix 8). Formalized optimization of decisions under uncertainty became the subject of the generally applicable *operational research*.

Operational Research

Although optimized resource allocation is the stated purpose of management, it is the least implemented among BMS capabilities. In contrast, the BMS database is continually refined and new ways are tested to maximize its usefulness. Identifying and quantifying the significant parameters of the models to be optimized are always a work in progress. A range within which available options can be considered, however, can be defined. Knowledge-based or heuristic systems attempt to maximize the engineering value of incomplete information by representing it as fuzzy sets (Zadeh and Kacprzyk, 1992; McNeill and Freiberger, 1994; Miyamoto and Frangopol, 2001). Thus, databases learn perpetually from users, whose relative ignorance grows along with their dependence on databases.

The level of expertise required of a qualified BMS user, i.e., the competence necessary in order to take bridge management decisions supported by a BMS, and BMS-related decisions, motivated by infrastructure needs, cannot be narrowly defined. Most organizations in the United States require professional license and years of relevant experience in bridge design. The standards in bridge-related data management are less stringent or nonexistent. Beale (1988, p. 3) stated:

> It is important for model builders to know a fair amount about optimization techniques, even if they can use existing computer programs to implement them. This is because of the need to compromise in the model between realism and ease of use. . . . In practice we often need to see the numerical solution to the model to help us to realize that the data are incomplete or incorrect. This makes the techniques for computing the solutions very important: to solve a real problem we may need to compute the answers to a number of alternative mathematical models, in which case we cannot afford to take too long solving any of them.

Thompson (2005) observed that the actual state is similarly minimized if life-cycle costs are minimized based on aggregate analysis, averaging the conditions of the entire stock. Yanev (1997) projected needs deterministically on the basis of worst rather than average conditions. Linear or bilinear deterministic models for bridge deterioration and preventive maintenance benefits (Examples 18–20) are relatively easy to generate, implement, verify, and adjust.

Frangopol and Liu (Miyamoto et al., 2005, pp. 57–70) treat bridge maintenance life-cycle costs (LCC) and relevant lifetime bridge performance measures as separate objectives, subject to simultaneous optimization.

> Instead of a single optimal maintenance planning solution that has a minimum expected LCC, a set of alternative maintenance solutions is produced that comprises optimized tradeoff among all conflicting objectives. . . . One can find an optimized solution that meets the specific maintenance funds available to bridge managers, particularly if the financial resources are insufficient compared to that required by the optimal maintenance solution minimizing LCC only.

The proposed algorithm assumes a somewhat homogenized network of assets.

12.3   IMPLEMENTATION

The ability to implement strongly influences managerial decisions. During implementation, optimization tends to turn into prioritization (of expenditures) or minimization (of risk or "damage"). The NCHRP has developed a number of guides and tools for managing multimodal transportation assets on project and network levels, employing to various degrees optimization techniques. Appendix 34 describes the life-cycle cost assessment sequence and the groups of significant parameters adopted by Hawk in BLCCA (NCHRP Report 483, 2003). The sensitivity of the procedure to parameter changes must be checked. Comparisons with alternative procedures are recommended. Appendices 36, 37, and 43 summarize most commonly used optimization techniques and algorithms. As in the case of structural stability, the purpose is not to present the subject but to illustrate its complexity. A BMS user may not have full understanding of the modeling assumptions and numerical operations performed by the optimization module. Thompson (in TRR 1866, 2004, pp. 51–58) proposed to facilitate the application of automated LCCA as follows:

> Part of the difficulty many agencies have with the implementation of asset management systems is that end-users of those systems are required to have a relatively sophisticated understanding of life-cycle cost analysis to understand the outputs. . . . To expect managers to become equally familiar with several incompatible analytical systems is perhaps asking too much. Separating the life-cycle cost analysis into its own separate business process sets up a locus for accountability and technical support that is friendly to upper management.

Hence, the author recommended separating decision making from decision support, as have done Mittra (1988) and Drucker (1973). A top-management responsibility (and hence vulnerability) is to optimize the dependence on automated decision support and on personal competence, as discussed in Section 4.1.3.

NCHRP Report 545 (2005, p. 12) referred to the same findings or confirmed them independently: "Existing management systems are not typically geared for use by high-level managers to support resource allocation and program tradeoff analysis. The need for this type of capability is likely to increase given the new initiatives in asset management requirements of GASB-34 (2000)."

Figure 12.2 illustrates the commonly accepted argument that maintenance costs increase throughout the life of a structure, whereas the costs of reconstruction decline. The implication is that an optimal time exists when reconstruction becomes preferable. Implementing this model encounters the following difficulties:

Maintenance costs include increasing amounts of repairs.

Annual costs can be established for maintenance but not for construction.

Benefits can be estimated for construction but not for maintenance (Examples 19–24 illustrate attempts to resolve that problem).

The "optimal" reconstruction time of a single structure cannot govern within a network.

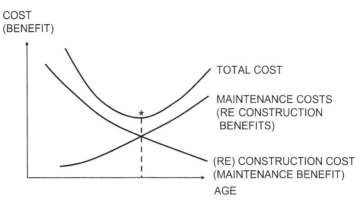

Figure 12.2 Hypothetical optimal costs (or benefits) of construction and maintenance with respect to structural age.

On the project or individual facility level, the model is feasible and can be implemented. For a network of bridges and connecting roadways, an acceptable equilibrium may be reached more or less by expediency (as in Example 18) and then adjusted to minimize annual costs.

The perceived importance of the projects and the traffic corridor considerations of the locality easily outweigh cost–benefit optimizations based on condition and vulnerability ratings alone. NCHRP Report 545 (2005) noted (among others) the following general trends in asset management attributable or relevant to the use of analytic decision support tools:

Highway "tiering" or corridor designation systems are replacing the project-centric view. Performance monitoring and investment strategy development go beyond functional classification.

Investment categories (for example, in Colorado) are organized by policy objective, as opposed to by a reactive or "worst first" approach. Investment categories, such as bridges, tunnels, pavements, rest areas, and roadside maintenance activities, are organized by policy objectives, including mobility, system quality, safety, strategic projects, and program delivery.

Analytic studies by Washington and Michigan DOTs are quoted in support of the politically difficult (although technically and economically justified) decision to prioritize assets in good condition while work on those in poor condition may be delayed.

Asset management system upgrading is common. Data warehouses are consolidating asset inventory and project information from different systems using GIS platforms.

The report developed tools for network (NT) and project (PT) asset management. The capabilities are adjustable for implementation on any local level, as described in Appendix 15. The authors supply user manuals. The method, correlating maintenance levels and bridge condition ratings, described in Example 24 is another tool that can be used on both project and network levels.

Integrated asset management has shifted the focus from individual to network performance indicators (Chapter 13). FHWA (2004b, p. 23) quoted a transportation manager from Australia as follows: "The cultural shift [towards performance orientation] has meant collecting data as part of a conscious plan to improve performance, [rather than] as audit material."

Hudson et al. (1997) have presented concisely performance indicators for both the physical state of infrastructure facilities and the services they provide.

PART III

EXECUTION: FROM SYSTEM TO STRUCTURES

"Execution" carries several connotations, of which the relevant one, consistent with *Webster's Dictionary* is "effective action." The effect may amount to the implementation of decisions, more specifically the performance of tasks and, ultimately, getting work done. Drucker (1973, p. 128) summarized: "Everything degenerates into work."

Chapter 13

Tasks and Operations

Managed work consists of tasks. Organizations are designed to carry out sets of tasks structured into a process. For the purposes of construction, Halpin and Riggs (1992) identify the following organizational hierarchy:

Organization

Project

Operation (and process)

Task

The bridge management process consists of cyclically repetitive tasks, including those of Table 4.3. They can be structured into continuous processes or discrete projects. Processes and their products must be closely monitored and reevaluated quantitatively and qualitatively (Tables 11.1 and 12.1).

The condition and performance of structures (e.g., the engineering products) are assessed according to the broad range of systems summarized in Table 9.1 and discussed in Chapter 10. Such assessments also yield clues of past managerial performance. Direct indicators of the execution of the process are harder to establish, as pointed out in Section 4.1 and 5.4.

13.1 ADMINISTRATION

At the network level, resources are allocated through administrative actions. Project-level management uses the resources to maintain services and operations. Work on publicly owned facilities is executed by permanent staff and through competitively bid contracts. Ultimately, every type of technical and administrative expertise plays critical roles.

Personnel performance is evaluated according to indicators appropriate for design, construction, maintenance, management, and fiscal operations. Standards of productivity, quality, safety, and accountability are enforced. Quality control addresses materials and products. OSHA regulates safety in the United States. Traffic safety and more general safety aspects may come under the jurisdiction of law enforcement and national security agencies. All indicators must be periodically updated.

Network-level administration is vulnerable to isolation from field operations (Section 4.1.3). Project-level expertise equally proficient in the execution of field and clerical tasks is rare. Project management systems can facilitate the processing of information on the various task levels. The PennDOT and NYS DOT BMSs shown in Appendix 16 both address this need.

Capital improvements of civil structures are typically carried out under contracts with design consultants and construction contractors. The management of contracts and projects to considerable degree determines the quality of the infrastructure. Contract bidding and letting in free-market competitive systems are regulated on national and local levels. Appendixes 20 and 21 highlight the U.S. federal and state view on that subject. NCHRP Synthesis 331 (2004) pointed out that bridge owners should develop or adopt procedures suited to their particular environment.

Bridge owners are solely responsible for the performance of accepted projects. In order to obtain assets rather than liabilities (Fig. 4.4b), they must influence the quality of the process and the product of the contracted tasks by all available means, including the following:

Preliminary stage

   Scope of work

   Contract specifications

   Request for proposal (RFP)

   Selection procedures, prequalifications, competitions, bidding, letting

Project stage

   Peer review, value engineering

   Contract supervision, progress review

   Incentive/disincentive

   QA&QC

   Withholding payment, default/termination

Completion/operation

   Final acceptance, performance evaluation

   Combinations of design, construction, maintenance, and operation contracting

In order to administrate and supervise contracts, public agencies must be technically competent. Given a technically expert staff, regularly recurring tasks, such as maintenance, are well suited for "in-house" execution. Unless the tools effective in contract management are also provided, however, responsible owners are forced to confirm the self-fulfilling prophecy that in-house work is inefficient.

13.2   QUALITY ASSURANCE AND CONTROL, PEER REVIEW

Whereas performance-based design specifications are relatively recent, the performance of tasks and services has always been under various forms of scrutiny, including direct supervision and QC&QA. The satisfactory quality of the product and the execution of the process must be ensured by regulated procedures subject to contractual agreements. Hassab (1997, p. 419) distinguished between quality of design and conformance therewith. TRC E-C037 (2002) provides a helpful glossary of quality assurance terms (Appendix 50). NCHRP Synthesis 346 (2005) applies that terminology to state highway construction practices.

As in the case of design specifications, QC&QA standards are produced by a collaboration between government agencies (OSHA, AASHTO, FHWA), professional expertise [American Society for Testing and Materials (ASTM), American Society of Civil Engineers (ASCE)], and responsible owners. QA&QC add to the first costs of a project and therefore must produce demonstrable life-cycle benefits in terms of expenditures, safety, and quality. Benefits expected to accrue over many decades are hard to demonstrate during construction, and consequently QA&QC are the subject of vigorous negotiations and occasionally, litigations. NCHRP Synthesis 346 (2005, p. 4) stated: "Starting in the 1980's . . . there was a growing perception that a duplication of testing was taking place: QC testing by the contractor and acceptance testing by the agency."

Sections 4.1 and 5.3 pointed out that duplication of effort, although routinely targeted for elimination, has the potential to enhance reliability. "Self-certification" by the contractor is efficient but "conflict of interest" must be carefully avoided. Optimizing QC&QA methods and costs is therefore a typically conflicted management problem. QC&QA must eliminate the vulnerabilities in the product and the process depicted in Table 4.5. The matrix format benefits from the relational structure favored by system management (Appendix 18). It gives to each type of expertise the opportunity to assess the products of all others (if necessary). Existing checks, control, and assurance practices can be entered as shown in Table 4.5. New ones can be introduced.

According to the definitions in Appendix 50, *quality assurance* applies to materials and products, whereas *quality control* pertains to the performance of tasks (e.g., the process). The same distinction is reflected in recent NCHRP publications which distinguish between *best practice* and *best value*. QA involves testing samples of materials and products on delivery, as during construction, or directly duplicating results, as in *peer review* of analysis and design. Field operations, such as construction and maintenance, as well as administrative ones, related to contract and fiscal management, are supervised according to prescribed QC protocols. NCHRP Synthesis 346 (2005, p. 7) quotes AASHTO (1996) and other federal publications on performance-based specifications relying on statistical methods for QA/QC sampling (according to the definitions listed in Appendix 50). QA and peer review evaluate the product on delivery, but prior to essential completion of the project. Construction operations, such as welding, concrete pouring, and painting, are subject to specifically designed and regulated QC. NCHRP Synthesis 346 (2005) points out that the options have increased for distributing various QA/QC functions between the contractor, the owner, and specialized consultants. The synthesis (p. 12) emphasizes the importance of establishing the limits to be used for acceptance during the QA process as follows:

> Making the limits too restrictive deprives the contractor of a reasonable opportunity to meet the specifications. Making them not sufficiently restrictive makes them ineffective in controlling quality. . . . Selection of the limits relates to the determination of risks. The two types of risk encountered are the seller's (or contractor's) risk, α, and the buyer's (or agency's) risk, β.

α = probability of rejecting material that is exactly at the acceptance quality level.

β = probability of accepting material that is exactly at the rejectable quality level.

A well written QA acceptance plan takes these risks into consideration in a manner that is fair to both the contractor and the agency.

To assist in the determination of the acceptable risk levels, NCHRP Synthesis 346 provided operating characteristic (OC) curves, referring to report FHWA-RD-02-095 by Burati et al. (1995).

NCHRP Report 451 (2001) developed guidelines for warranty, multiparameter, and best-value contracting as a combination of QC&QA measures (Appendix 21).

13.3 RESPONSIBILITY, ACCOUNTABILITY, AND LIABILITY

Responsible owners are accountable for the performance of the assets they manage and, by inference, for their own performance in serving these assets (particularly public ones). Fiscal procedures are relatively well defined and controlled by audits. Less specific and therefore vulnerable (Section 4.1.4) are the guidelines associated with the level of services provided by the physical assets.

NRC (1995, p. 24) stated:

Seeking to describe and measure infrastructure performance is an attempt to judge how well infrastructure is accomplishing the tasks set for the system or its parts by the society that builds, operates, uses, or is neighbor to that infrastructure. Because infrastructure is largely a public asset or resource, this judgment is typically made in a public setting. Many people are likely to be involved and reaching a consensus can be difficult. Even when one person has clearly defined responsibility for making investment or operating decisions about some element of the infrastructure, that person must be prepared for public scrutiny of his or her premises and conclusions. This public scrutiny is sometimes intense.

Tort liability was identified as a vulnerability in Section 4.1.1. Lewis (in NCHRP Synthesis 106, 1983) provided practical guidelines for minimizing tort liability. NCHRP Report 285 (1986, pp. 5–6) cautioned that liability cannot be evaded by vaguely defined service levels:

Highway agencies have been reluctant to quantify maintenance service levels because they recognize and contend that such guides specify a standard of conduct that may be used against them in tort liability cases. With the erosion of sovereign immunity, tort liability litigation has been steadily increasing and highway agencies do not want to assist litigants by establishing standards of care that define negligence per se on the part of the agency. . . . When a highway agency does not define its own maintenance service levels, the courts will define them. . . . This simply means that the agency cannot ignore the problem, but should take steps to define levels of service that are consistent with the constraints under which they are forced to operate. . . . Once levels of service are established, they must be rigorously followed.

Banks (2002, p. 30) described risk management programs intended to limit the exposure to tort liability as follows:

> Possible elements of such programs include instituting and carefully documenting accident reduction programs, training personnel in accident response and in dealing with legal procedures involved in tort cases, institution policy and procedures related to the release of information and documents, and addition of indemnity clauses in contracts to shift some risks to other parties.

> Minimizing the causes for litigation by strict adherence to the implementation of clearly defined tasks is more effective, in cost and other value, than any legal defense after an accident.

Structural hazards (Section 10.3.1) are mitigated according to that philosophy.

As all formalized procedures with considerable impact on the national budget, the assessments of structural needs (and hence conditions) are audited. The audits do not address directly the reliability and quality of inspections (Section 14.5). Rather, they verify the adherence to established procedures. The FHWA conducted such an audit of bridge inspections in 2003–2004.

13.4   DESIGN/CONSTRUCTION

Responsible owners award design and construction contracts through competitive bidding or through emergency procedures. Most projects are initiated with a RFP by the owner and are awarded to the lowest qualified bidder. A minimum of three bidders is usually required. Prequalifying conditions may be stipulated. Among a larger pool of bidders, certain systems eliminate the highest and the lowest bids from consideration. The consistency and transparency of the selection process are defined by the contract specifications. Project specifications usually derive from a generic "boiler plate" legal document. In contrast, the technical description of the contract must clearly and unambiguously state the owner's requirements for the process and the product. In seeking the best expertise and performance, the owner must comply with the following two opposed constraints (as in Section 12.2):

> The details of large projects cannot be entirely envisioned during preliminary design. The flexibility allowed by escape phrases, such as "or the equivalent," "including, but not limited to," and "as needed" easily turns to mediocrity. If the scope of work is poorly defined by the contract documents, change orders eventually transform the cost and the duration of the project beyond recognition.

> Overdefining all tasks can limit the field of qualified competitors and preclude innovation. The extreme case of a "sole source" requires special justification. Defining the performance of the product, rather than the method of achieving that performance, reduces the risk of an administrative stalemate.

In order to reconcile the conflicted demands for flexibility and control, the review and approval of special features, modifications, and additions must be clearly defined and followed. Example 8 illustrates the standard procedure adopted by a local bridge owner responsible for coordinating numerous large projects funded by various sources. Adherence is legally binding. The complexity of the process explains why the flowchart of Fig. E18.2 anticipates a lapse of 15 years between the kick-off of rehabilitation planning and beginning of work at the site. NCHRP Report 451 (2001) addressed the inherent contradiction between the near-mandatory selection of the lowest bidder and the demand for highest quality (Appendix 21).

Selection of a Design Scheme

The feasible structural solutions are becoming increasingly diverse. Appendix 14 illustrates the variety of structural types currently represented in the NBI. Competing options must be evaluated and selected, often without supporting data. They may include maintenance-intensive and low-maintenance structures, standard designs, subject to routine quality control, and innovative alternatives requiring special QA/peer review provisions. High-profile or "signature" projects (such those referred to in Section 1.5) may call for open and solicited competitions.

The *condition deterioration* models discussed in Section 10.4.5 draw on a growing database. In contrast, the *condition improvements* resulting from the various proposed measures are hypothetical (the more so for innovative designs). Expertise may be biased toward the supply of new solutions or the demand for cost-effective life-cycle service. Olsson (1993, pp. 57–59) speculated as follows (see also Example 3): "The 1990s and beyond should be the age of supply-side bridge design . . . if the demand segment of the market lets the best ideas on how to rebuild our distressed bridges see the light of application."

The author pointed out that the post–World War II European community encouraged input from the broad-based supply side, leading to major advancements in bridge technology.

It is also relevant that in certain countries bridge owners inspect bridges 10 years after completion and can assign liabilities to the contractor. Such practice encourages confidence in proposed innovations. Appendix 21 describes the NCHRP findings on the use of warranties and best-value contracting among SHAs.

Design exceptions (according to the FHWA *Federal Aid Policy Guide*) are defined by NCHRP Synthesis 316 (2003) and summarized in Appendix 51. Alternative methods of contracting are discussed in the following section and in Appendices 20 and 21. The referenced reports are intended primarily for highway construction, but some of their recommendations can apply to bridges. Compliance with governing specifications and adherence to recommended practices are linked to the eligibility for funding.

Peer Review

The preceding paragraph defined peer review as a form of QC which independently confirms results, such as analysis and design calculations. That independence is essential to the quality of design. In a chapter dedicated to lessons from failures, Waddell (1916, p. 1546) emphasized: "In every important bridge project the completed plans should be checked in detail throughout by some capable bridge engineer who is entirely disconnected from either the consulting engineer or the contractor."

Feynman (1999, p. 166) praised that approach as it was applied to computer software programming at NASA in these terms:

> The software is checked very carefully in a bottom-up fashion . . . The scope is increased step by step until the new changes are incorporated into a complete system and checked. This complete output is considered the final product, newly released. But completely independently there is an independent verification group that takes an adversary attitude to the software as if it were a customer of a delivered product. There is additional verification in using the new programs in simulations, etc. . . . To summarize, then, the computer software checking system and attitude are of the highest quality.

Not all bridge operations receive or require such scrutiny.

A bridge owner can balance in-house and outsourced design in many ways. The minimum in-house expertise must be capable of consultant selection and supervision. Retaining consultants under flexible contracts, such as for consultant support services (CSSs) or under engineering service agreements (ESAs), not to exceed preset costs and time limits, can be highly effective.

Construction

Construction predates formalized design by millennia and, consequently, project management enjoys a considerable head start over other asset management developments. Cooke and Williams (2004) present best practices in construction planning, programming, and control, departing from the management principles formulated by Henri Fayol at the turn of the twentieth century (see also Section 1.11). Its extensive background notwithstanding, construction management continues to redefine itself, as does all management. Clough (1986, p. 16) described that process as follows:

> There is no universally accepted definition for construction management and, unfortunately, it is a term that means different things to different people. In what might be considered the typical arrangement, the owner contracts with an architect-engineer for design services and with a construction manager (CM) for specified professional services. By terms of the two contracts, a non-adversary construction team is created, consisting of the owner, the architect-engineer, and the CM. The objective of this approach is to treat project planning, design, and construction as integrated tasks within a construction system. The team, by working together from project inception to project completion, attempts to serve the owner's interests in optimum fashion. By striking a balance between construction cost, project quality, and completion schedule, the team strives to produce a project of maximum value to the owner within the most economic time frame. Adherence to the established time schedule and cost budget is a prime responsibility of the CM.

The CM of the preceding outline is independent of the owner. This is not the only option when the owner is a public agency and the project involves public transportation. Whereas all publicly funded projects are awarded through competitive bidding, their supervision is conducted through a number of mechanisms. A separate bid may be let for *consultant support services*. The tasks in that contract would consist of construction supervision by a qualified consultant other than the designer.

Scope-of-work changes must be approved by all the agencies contributing to the project funding. Example 3 lists the multiple contracts and tasks associated with the reconstruction of a major bridge funded by U.S. federal, state, and local sources, all taking part in the decisions.

The owner supervises all field operations through an *engineer in charge* (EIC). A *cost manager* and a *quality surveyor* can be appointed to supervise expenditures and the compliance with quality standards, respectively, since these are two vulnerable links in the process. Software packages are available for the transmittal, storage, and retrieval of all communications relevant to a project.

The following innovative construction management practices can save time and resources if their potential vulnerabilities (Section 4.3.2) are recognized and avoided.

Design–Build (Design–Construct)
The designing consultant or the construction company is designated as the primary contractor who enters into a subcontract agreement with the other. The advantage to the owner is the elimination of the entire bidding process for construction following the design phase. The early collaboration of designer and constructor is a further advantage. Conversely, the owner loses the equivalent of a peer review exercised by an independent consultant over a contractor and vice versa.

Design–build–maintain contracts are also awarded. The winning contractor typically charges tolls for the operation of the structure over the contract period. From the owner's viewpoint, most essential in this case is to determine the optimal duration of the contract and the method of interim supervision and quality control. Contracts for 75 years have been awarded recently on important tolled facilities, for instance, the Viaduct de Millau in France.

Value Engineering
The U.S. Office of Management and Budget Circular No. A-131 (May 21, 1993) defined value engineering as "an organized effort directed at analyzing the functions of systems, equipment, facilities, services, and supplies for the purpose of achieving the essential functions at the lowest life-cycle cost consistent with required performance, reliability, quality, and safety."

The *Federal Register,* Vol. 62, No. 31 (February 14, 1997, part 627, p. 6868) updated the definition of *value engineering* as follows: "The systematic application of recognized techniques by a multidisciplined team to identify the function of a product or service, establish a worth for that function, generate alternatives through the use of creative thinking, and provide the needed functions to accomplish the original purpose of the project, reliably, and at the lowest life-cycle cost without sacrificing safety, necessary quality, and environmental attributes of the project."

The National Highway Designation Act of the U.S. Congress (1995) directed the secretary of transportation to introduce value engineering (VE) analyses on all National Highway System (NHS) project costing over U.S. $25 million.

NCHRP Synthesis 352 (2005) summarized the background and current practices and recommended value methodology (VM). The synthesis emphasized the importance of adequate training in the management of VM.

Clough (1986, p. 168) attributed the advantage of the process to the special knowledge of the contractor; however, the preceding references note that the work can also be performed by the owner. Changes to the contract documents require the owner's approval. It is up to the owner therefore to ascertain that not only the first but also the life-cycle costs are reduced by such changes.

Incentive–Disincentive

After the contractor and the owner have agreed on a reasonable date of completion, penalties for delays and bonuses for early delivery are set (usually at the same daily rate). The amount is determined by an empirical formula calculating the user costs incurred by traffic delays and detours during construction. In a densely populated urban area the incentive/disincentive can amount to U.S. $65,000/day (with a limit of three months on the incentive). The measure proved highly effective after the Northridge earthquake (1994) in California and at the East River bridges in New York City. Key to the success of the method is the selection of the term of construction. The choice becomes increasingly uncertain for long-term projects. The difficulty is exacerbated on design-built projects, where the designer and the contractor are mutually dependent. Management maintains better control of large long-term projects by breaking them down into successive contracts (as in Example 3). In such cases, competitive bidding procedures may encourage awarding consecutive contracts to different bidders.

NCHRP Synthesis 331 (2004) summarized the letting practices of U.S. DOTs (Appendix 20). FHWA (2005c) compared practices in the United States, Canada, and Europe. The researchers recommended the following:

Align team and customer goals.

Develop (project) risk assessment and allocation techniques.

Strategically apply alternative delivery methods (e.g., design–bid–build, design–build, design–build–operate, design–bid–finance–operate, private–public partnerships, other).

Enhance qualification rating process. Use qualifications in procurement. The team identified alternatives to the low-bid procurement system, including best value, qualifications without bid price, contractor rating systems, and other options.

Pilot early contractor involvement. Qualified design and construction experts can develop an open-book target pricing system jointly with the owner.

Along unit-price payments, create incentive by lump-sump payments at preset project milestones.

Along with project management introduce long-term network management. Work toward warranties and life-cycle responsibilities.

13.5 MAINTENANCE AND REPAIR

Section 4.1.4 identified maintenance as a highly vulnerable managerial responsibility. Section 11.4.1 discussed the difficulties of optimizing and/or prioritizing maintenance tasks. Traffic-safety-related maintenance tasks (such as roadway cleaning or lighting) are readily prioritized. Tasks related to structural conditions cannot be uniquely defined. NCHRP Syntheses 110 (1984) and 148 (1989) reviewed existing and recommended indicators of measuring maintenance quality. The reports remain relevant because of the fundamental questions they address and the definitions they introduce. Maintenance is evaluated according to the following three types of standards:

1. *Quality standards,* describing the results to be achieved

2. *Quantity standards,* identifying the amount of work and resources necessary to meet the quality standard or a predetermined *level of service*

3. *Performance standards,* describing a general method of performing a task, the resources required, and the rate at which the work should be performed

NCHRP Synthesis 148 (1989) described the Caltrans *level of service* guidelines, identifying the following three types of maintenance:

1. Responsive, to be handled as needed. The rapidity of the response typically defines the level of service.

2. Scheduled. The level of service is defined by the annual number of cycles (e.g., in Examples 23 and 24).

3. Planned—for instance, as part of a major maintenance or a bridge-painting program.

So long as the level of maintenance is defined in terms of the services provided by the structures, managerial performance can be assessed with some latitude. NCHRP Report 285 (1986, p. 7) pointed out that "deferred maintenance is a relative term and one must have a reference service level before differences can be evaluated." This vagueness is avoided if maintenance tasks are defined as part of the employee's contract. Then their deferral is no longer the relative term of NCHRP Report 285 (1986). The failure to perform is sought at one or another personnel level rather than in the structures.

AASHTO (1999a, p. 1-4) quoted the *hub-and-spoke* radial management scheme recommended in NCHRP Report 363 (1994). NCHRP Report 511 (2004) introduced *customer-driven benchmarking of maintenance activities.* Both are briefly described in Appendix 52. The two approaches maximize maintenance benefits by mutually complementary methods.

AASHTO (1999a, p. 1-5) recommended preventive as opposed to reactive maintenance.

NCHRP Report 511 (2004) sought to establish a process of continuous adjustment between results, as they are perceived on all levels, and the expenditure of future resources.

Section 11.4.1 and Examples 23 and 24 discuss bridge-specific and task-specific maintenance plans. BMSs remain abstract in the absence of maintenance management systems (MMSs) addressing implementation and operation. MMSs track work tickets and performance indicators. DANBRO (Appendix 16) has championed this capability for several decades.

Operation

All maintenance activities require operational capabilities. Particularly demanding are tasks associated with traffic control, such as road surface repairs and de-icing. Traffic control is essential in the operation of toll bridges. Toll operations are managed independently of other maintenance functions. In recent years most challenging has been the transition to electronic toll collection. A future challenge is likely to be the adoption and integration of more advanced methods allowing for tolling by license plate numbers and other means.

The operation of movable bridges is defined by the navigation laws and regulations (by the U.S. Coast Guard, for instance). The compliance with such regulations is rigorously enforced and must meet the standards of all supervisory organizations. Personnel must be trained and qualified by the appropriate agencies, safety and emergency regulations must be followed, and back-up staff must be available at all times.

Chapter 14

Structural Inspection and Evaluation

Field inspections (of one type or another) assess the physical vulnerabilities, conditions, and needs of the assets (Chapters 4, 10, and 11, respectively). "Inspectability" is increasingly included in the scope of design requirements. For existing bridges, inspections must adjust to the structural demand, drawing upon variations of the options described in Sections 10.4.1 and 10.4.2. Owners design the scope of inspections according to the specifics of their networks. FHWA (2005b, p. xx) reported on the variety of field inspection frequencies and scopes in Europe and South Africa. Less detailed inspections, including *daily, semiannual, monitoring, superficial, regular, routine, general,* and *annual,* are conducted at intervals of up to a year. *Detailed, principal, major, general* (again), *biennial,* and *condition evaluations* are scheduled at 2–6- and even 9-year intervals. Expertise is commensurate with the scope of the assessments.

The size and the variety of the bridge network managed by the FHWA in the United States have given raise to a flexible yet reliable set of NBIS, described in some detail herein.

14.1 NATIONAL BRIDGE INSPECTION STANDARDS

Minimum bridge inspection requirements are broadly stipulated in NBIS (FHWA, 1995b) as follows: "§650.305 (a) Each bridge is to be inspected at regular intervals not to exceed 2 years in accordance with Section 2.3 of the AASHTO Manual."

Inspection reports must meet the following NBIS requirements: "§650.309 The findings and results of bridge inspections shall be recorded on standard forms. The data required to complete the forms and the functions which must be performed to compile the data are contained in Section 3 of the AASHTO Manual."

In the above definitions, the *Coding Guide* (FHWA, 1988) refers to the AASHTO (1983) *Manual for Maintenance Inspection of Bridges* and its many updates. Also relevant are FHWA (1978, 1991). Fracture-critical and other elements, requiring special scrutiny during inspections, are addressed in a succession of manuals and related publications, such as FHWA (1986), AASHTO (2000b, 2003), and NCHRP Report 299 (1987). NBIS 23 CFR 650 refers to the interim revisions to the AASHTO *Manual for Condition Evaluation of Bridges* (2003).

The abundance of superseded and revised manuals should alert the user that these texts are never entirely conclusive for the determinations they require. The inspector's and the owner's interpretations are ultimately decisive. For example, the scope of underwater inspection is described in AASHTO (2000b) as follows: "Underwater members must be inspected to the extent necessary to determine structural safety with certainty."

Terms such as *to the extent necessary* place the responsibility for safe operation on the owner, who in turn expects the inspector to assess the bridge condition *with certainty*. Inspections are frequently governed by more than one set of directives and numerous updates. State highway departments, railroad companies, and other bridge owners issue their own inspection standards and manuals. Noteworthy are the manual by Park (1980) of the New Jersey DOT, the NYS DOT (1997) *Bridge Inspection Manual,* the *Bridge Inspection and Rehabilitation* practical guide (Silano, 1993), and *Bridge Maintenance, Inspection and Evaluation* by White et al. (1992). A handbook for railroad bridge inspection is in preparation by AREMA Committee 10 in addition to the nonregulatory Track Safety Standards 49 CFR 213, Appendix C.

14.2   SPECIAL-EMPHASIS DETAILS

Special emphasis implies *highly vulnerable*. Since a primary inspection task is to preclude failures, the details most likely to fail are specially emphasized. Inspection manuals and standards identified such details through studies of structural vulnerabilities (Chapter 4 and Section 9.2) and local conditions. To those sources of potential failures must be added the increased likelihood of overlooking important symptoms of distress as a result of poor "inspectability" of one kind or another.

The following NBIS paragraph broadly addresses unique and special features: "§650.303 (3) . . . Those bridges which contain unique or special features requiring additional attention during inspection to ensure the safety of such bridges and the inspection frequency and procedure for inspection of each feature."

According to 650.303 (e) (FHWA, 1971, 1988, 1995b), the inspector must identify the presence and document the condition of "unique or special features." Various designations are used by NBIS and other bridge management guides for structural features requiring particular attention. In keeping with the origins of highway and railroad bridge inspection programs, foremost among them is "fracture critical."

Fracture-Critical Members (FCMs)

NBIS 23 CFR 650 subpart C (FHWA, 2005a) prescribes:

In the inspection record, identify the location of FCMs and describe the FCM inspection frequency and procedures. Inspect FCMs according to these procedures.

FCM inspection: A hands-on inspection of a fracture critical member or member components that may include other non-destructive evaluation.

Fracture-critical bridge members are defined in NBIS as follows:

§650.303 (e) The individual in charge of the organizational unit that has been delegated the responsibilities for bridge inspection, reporting and inventory shall determine and designate on the individual inspection and inventory records and maintain a master list of the following: (1) Those bridges, which contain fracture critical members, the location and description of such members on the bridge and the inspection frequency and procedures for inspection of such members. Fracture critical members are tension members of a bridge whose failure will probably cause a portion of or the entire bridge to collapse.

The preceding definition clearly resonates with the Silver and Mianus Bridge failures. Harland et al. (in FHWA, 1986) assigned top priority to the inventory and inspection of fracture-critical bridge members in the now classic *Inspection of Fracture Critical Bridge Members*. This supplement to the inspection manual addressed both the general aspects of highway bridge inspection and the details prone to fatigue and fracture. Among its many recommendations are the rotation of bridge inspectors, the quality control review of inspection reports, and the call for quality in preference to quantity of inspections. Stress concentration due to welds, corrosion, poor load distribution, and so on, is described and classified according to the level of susceptibility. Examples of fracture-critical inspections are presented in TRR 1184 (1988).

NCHRP Synthesis 354 (2005) has updated the findings of past reports with recent developments, including definitions of fracture criticality, nonredundancy, effective repairs, and retrofits. Nonfatigue fractures (as at the Hoan Bridge in Milwaukee and the Bryte Bend Bridge in Sacramento) are described. Recently developed analytic models of structural reliability as a function of fracture criticality and nonredundancy are briefly discussed. Cited are bridge owners' reports on the frequencies of fracture-critical inspections, ranging from frequent visual monitoring to five-year in-depth cycles and nondestructive testing. Survey forms can be used to expand the databank.

Fatigue

Section 4.3.1. identified fatigue as a vulnerability principally (but not uniquely) responsible for metal fracture. Fatigue-prone steel details are categorized (types A through E) in most bridge design and inspection publications, notably FHWA (1986), NCHRP Report 299 (1987), and Fisher (1997). Drdacky (1992) and Forde (1999) studied specific details and offered recommendations.

A point of departure for bridge inspectors is the identification of "stress raisers" (e.g., points of stress concentration) in the as-built inventory. Hence such an inventory must be compiled and regularly updated, for instance, according to the NBI or local provisions. Where stress raiser details are numerous, a statistically meaningful sample must be regularly inspected. "Statistically meaningful" is a vague term which depends on the size and the condition of the sample space. The perception of an appropriately sized sample may change with the information gained from each new test.

Corrosion causes stress concentration and therefore exacerbates fatigue. "Corrosion fatigue" is particularly damaging to the high-strength galvanized wires of prestressing tendons (Fig. 4.69), suspenders (Fig. 14.1), and suspension cables (Fig. 4.55), but it affects a much broader range of steel structures, as shown in Fig. 4.54). NDT&E techniques (Chapter 15) are used to monitor the rate of fatigue crack propagation or, more effectively, fatigue-induced ruptures. Current efforts are directed at the potentially most helpful forecasting of fatigue crack initiation.

Figure 14.1 Broken wires in suspender.

100% Hands On

Fracture is not the only *critical* feature inspected. Bridge owners, for instance, NYS DOT (1997), define critical features as *special-emphasis details*. Inspecting engineers must certify *100% hands-on* inspections of all special-emphasis details by their professional engineering license number and signature.

NBIS 23 CFR 650 subpart C (FHWA, 2005a) defines *hands on* as "inspection within arms length of the component. Inspection uses visual techniques that may be supplemented by nondestructive testing."

Specifying that hands-on inspections must be conducted *at arms length* should not obscure the figurative significance of the term. It is intended to alert the inspectors of their responsibility without clearly defining their task. The demand for direct physical presence implies *minds, eyes,* and *ears on.* The hands-on inspections described in Example 7 acquire value only after the direct impressions have been adequately analyzed and their implications evaluated and reported. The hands-on inspections on the cable-stayed sockets shown in Figs. 14.2*a,b* require widely different competence. The significant determination most likely to result from either inspection, however, is whether or not a more meaningful nondestructive evaluation has become necessary.

Traffic impedes hands-on inspections while allowing inspectors to observe the structure's response to live loads. As a minimum, hands-on inspections must assure the safe

(*a*)

Figure 14.2 (*a*) Cable-stayed socket.

service of the inspected details until the next scheduled inspection. Sampling may become unavoidable when specially emphasized details are prohibitively numerous. The corners of bottom flange cover plates require hands-on inspections. Inspectors have no option but to inspect a statistically significant sample, selecting locations with most advanced deterioration and/or highest stress and documenting their method. Concrete deck undersides are similarly sounded for delamination (Example 16). If distress is noted, the sample is increased.

The numerous technical advisories and engineering instructions identifying new special-emphasis details and recommending the appropriate means of inspection can be reviewed on the Internet. Consequently, it becomes essential to determine which of the various specifications, directives, mandates, advisories, instructions, and recommendations apply to the inspection at hand. Two criteria help identify critical (fracture or not) candidates for hands-on inspection: redundancy and stability. The former (see discussion above) gets more attention in recent manuals. The AASHTO evaluation manuals (AASHTO 2000b; NCHRP, Project 12-46, 2000) refer to "stability" broadly, including all displacement irregularities. Structural stability, both local and global, rigorously defined as a function of compression, geometric, and material properties, must be a primary consideration in every bridge inspection (Appendix 26). Among other tasks, this implies the ability to distinguish tension from compression members, not always eminently clear, for instance, in statically indeterminate trusses.

(b)

Figure 14.2   *(b)* Socket with rubber damper and protection, Tatara Bridge, Japan.

Stability

Inspections can easily overlook potential instabilities on the false assumption that unstable members always exhibit large displacements. It is helpful to recall that the original (usually straight) configuration remains a possible mode of equilibrium as the axial compression approaches a bifurcation point. The columns in Fig. 14.3, for instance, are extended by approximately 25% due to the exposure of their footings. As a result, their theoretical buckling load is reduced by a factor of $1.25^2 = 1.5625$.

If the original columns were hinged at the top and fixed at the top of the footings and if rotation becomes possible, once the bases have been exposed, the theoretical buckling load is further reduced by a factor of $(1/0.7)^2 = 2.04$.

Figure 14.3 Exposed column footings.

The product of the two reduction factors (3.1875) is extremely conservative, for instance, ignoring the rigidity of the base blocks. Nonetheless, the stability of the columns is a realistic concern, particularly over longer periods of changing soil and load conditions.

Areas near footings are prone to accelerated corrosion (Fig. 3.8). The resulting hinge reduces the column buckling load (as in the preceding example) by a factor of up to 2.04. In redundant structures that effect may be offset by the reduction of the axial load owing to the lost section. Columns have been observed to hang by the deck rather than support it. The vulnerability is thereby transmitted to the superstructure.

The bracings ensuring the stability of steel pier columns (Figs. E7.1 and E7.6) are primary, rather than secondary, members and require 100% hands-on inspections.

Members bent by vehicular impact are often mislabeled as "buckled" (Fig. 14.4) while buckled ones are diagnosed as "impacted" (Fig. 14.5). The failing column shown in Fig. 4.73 appears to have buckled but is actually bent due to an eccentric bearing on top.

The stability of steel solid rib or box arches (Figs. 14.6a,b) and their posts (Fig. 14.7) is critical, whereas access is difficult. Design should provide added redundancy in cases where inspections are both demanding and critical.

Buckling is not a concern at masonry arches (Fig. 14.8), but crushing and instant collapse can be, as examples repeatedly demonstrate. In France, where 75% of the bridges were built before 1900, the smallest geometry changes in masonry arches are recognized as important indications of distress and monitored according to specific guidelines (LCPC, LCPC/SETRA, 1979).

Figure 14.4 Truss diagonal, bent by impact.

Connections

Connections were identified as vulnerable in Section 4.2.3. Design assumes that connections adequately transfer forces between elements or provide a release for them. Most bridge malfunctions prove these assumptions erroneous. In recently built structures, connections are the most likely to suffer from construction or design defects. Prolonged bridge use adds to the possible causes of connection failures. Inspections continually add to the list of nonperforming connections.

Connections are discrete in steel and timber structures but can be monolithic in concrete ones. Splicing and overlapping of reinforcement can be viewed as a connection's equivalent in reinforced concrete. During the Hyogo-Ken Nanbu earthquake in 1995, for instance, pier rebars at the Hanshin Expressway ruptured at the splice welds.

Figure 14.5   Buckled truss diagonal.

For inspection purposes connections are divided into rigid and articulated. Unless otherwise specified, rigid connections are rated with the structural elements. Articulated ones, such as bearings, are rated as independent elements.

Connections, rigid under regular loads, may be designed to yield during extreme events. Bearings, free to adjust to thermal fluctuations, can act as either free or restrained during earthquakes. After a fixed bearing failure at the San Francisco–Oakland Bay Bridge during the Loma-Prieta earthquake of 1989, the expansion bearing displacement capacity proved inadequate (see Figs. 4.32 and 4.33). As a temporary measure, the emergency rehabilitation replaced the 5-in. (127-mm) bearing pads with 3-ft (915-mm) brackets.

Periodic visual inspections are likely to witness neither slow (thermal) nor sudden (seismic) motion. Large displacements must therefore be deduced and anticipated. If

(*a*)

Figure 14.6 (*a*) Redundant steel rib arches and posts, New York City, 1888.

potential displacements are inadvertently restrained, for example, by poorly designed details or corrosion, these conditions must be identified as structural hazards. Connections (as all structures) are best inspected in the following sequence:

Identify all loads and displacements.

Reduce loads and displacements to their six triaxial components.

Verify that loads and displacements are adequately transmitted or released.

Understanding the design assumptions is essential for adequate evaluation. In order to recognize early symptoms of failure, inspectors must be familiar with the intended function of the connection, as in Example 7.

(*b*)

Figure 14.6   (*b*) Longest steel box tied arch, Shanghai, 2002.

Pin-and-hanger assemblies (Fig. 14.9) are the foremost fracture critical among bearings (FHWA, 1986, 2002c). Rocker and sliding metal plate bearings perform poorly during earthquakes, as well as under normal conditions, and are to be eliminated during rehabilitations (see Figs. 14.10, 14.11, A33.2, and A33.3). Inspections must quantify bearing shifts and tilts along with the temperature and other ambient conditions. In order to assess the potential hazard, the inspecting engineer must determine the range of acceptable displacements.

Inspecting bolted and welded connections requires knowledge of their construction specifications (e.g., were the welds "shop" or "field," are the bolts slip critical or galvanized). Inspections typically assume that structures are built in compliance with design. The burden of confirming this assumption falls on the inspection at the essential completion or the final acceptance of a project, discussed in a following section.

Figure 14.7   Nonredundant steel arches, posts, and floor beams.

Expansion joints

Expansion joints were discussed in Section 4.3.3 as a vulnerability of both design and construction. In Section 11.4 they are identified as a high-maintenance item. Inspection must inevitably specialize in the evaluation of expansion joints because their failures trigger the rapid deterioration of bearings, decks, pier caps, and consequently bridges as a whole. The joint inspection form in Example 19 supplies information expressly for a joint replacement and repair program.

14.3   INSPECTION TYPES

Comparisons show that the same inspection scope qualifies as routine for one bridge owner and as in-depth for another. The scope and frequency of inspections vary with the type and condition of the bridges. No inspection program should be undertaken without

Figure 14.8 Masonry arch.

a defined set of objectives, standard document format, data storage, and response capabilities. The NBI *biennial* inspections are one example. Another popular alternative combines less detailed *annual* inspections with *in-depth* inspections every five to six years. Adjustable scheduling is continually explored. Relating inspection frequency to bridge condition poses an as attractive logistic challenge but provides little benefit to management. The economy realized by avoiding inspections of relatively new bridges is offset by the loss of regularity in task performance and data input.

The categories discussed in the following paragraphs address most inspection needs. Manuals, guides, advisories, and instructions are available for each of them.

Regular (Biennial)

This is the standard bridge inspection defined by NBIS §650.305. It performs three essential functions: identify potential hazards, rate the condition, and update the inventory. Regular inspections are conducted according to local (Park, 1980; NYS DOT, 1997) and federal (FHWA, 1971, 1988, 1995); (AASHTO, 2003) bridge inspection manuals and provide reports in the format prescribed therein. The numerical outputs of these reports are annually submitted to the FHWA in a form compatible with the NBIS. New York State requires railroads to submit biennial inspection reports as well. AREMA requires annual "routine" inspections. The various inspections conducted regularly in Europe and South Africa are described in BRIME (2002) and FHWA (2005b).

Regularly scheduled inspections are often termed routine, although their execution should be anything but. The term *routine* can imply either high level of organization or

Figure 14.9 Pin-and-hanger assemblies.

mediocrity. Engineering and management must constantly resist slipping from the former into the latter. Routine performance can be checked and improved by breaking repetitive tasks down into discrete projects. Discontinuity for its own sake causes ignorance, but ignorance is an inevitable byproduct of innovation.

Bridge inspection reports are definitive technical and legal documents signed by the engineer in charge and by the QC reviewer with their professional engineering license numbers. Beginning in recent years, inspection reports are electronically generated and transmitted, introducing new procedures related to the legal status of the documentation.

Beside their findings, routine inspections must clearly document their own limitations. When structural changes between scheduled inspections cannot be adequately anticipated, further action must be recommended. Normally inaccessible but critical structural elements must be identified and scheduled for special inspections. These in-

Figure 14.10 Failed steel plate, sliding bearing.

clude foundations, channel, embedded reinforcement, encasement anchors, suspension cable wires, masonry arches, and so on.

Interim

The NBIS (FHWA, 1971, 1988, 1995) addresses the need for inspections performed between biennial inspections as follows: "650.305 (b) Certain types or groups of bridges will require inspection at less than 2-year intervals. The depth and frequency to which bridges are to be inspected will depend on such factors as age, traffic characteristics, state of maintenance, and known deficiencies. The evaluation of such factors will be the responsibility of the individual in charge of the inspection program."

Interim inspections are planned as more limited than biennials but may call for "in-depth" investigations if the need arises.

Monitoring

Monitoring inspections address specific conditions and locations. Documentation is limited to describing changes of the localized condition and recommend remedial action. Such inspections were found indispensable in New York City during the early 1990s. Yanev (in ENPC, 1994, pp. 501–516) reported annual flag incidence reaching 3000 on roughly 800 bridges with 5000 spans (Example EA46). The organization chart shown in Fig. E18.2 shows how monitoring of less critical conditions can alleviate the demand for emergency repairs without compromising the service.

Figure 14.11 Broken steel plate, fixed bearing.

Temporary Repairs

Temporary repairs mitigate imminent hazards without improving the condition of the structure. Their monitoring is a high priority for owners of previously neglected bridges under heavy-traffic demands. Typical temporary repairs are timber and steel shorings, deck steel plates (Figs. 11.7–11.11), and holes drilled at the tips of fatigue cracks (see Fig. 14.12). Sandwiching a steel girder web with bolted plates to arrest crack propagation (Figs. 4.22a,b) is, ultimately, a temporary repair, because it alters the stiffness of the member and may cause unwelcome stress concentration elsewhere in the structure. The occasionally improvised nature of such repairs places a stronger demand for judgment on the inspecting engineer.

Temporary repairs must be installed with a specified life span and a prescribed frequency of monitoring. Both are subject to field verification. If exceedingly frequent monitoring is required, the repair is ineffective. Timber shorings (Fig. 11.8) shrink, creep, rot, and split. The buckling load of a typical 14 × 14-in. (35 × 35-cm) timber post longitudinally split in half is reduced by a factor of 8. Steel shorings (Fig. 11.9) are still temporary. Their pedestals and bearings do not conform to design specifications. Temporary columns should not be considered effective without inspection of the wedging or other details at points of load transfer. An uninterrupted load path must adequately replace the failed one from the supported element to an adequate footing.

Roadway steel plates bounce and shift under traffic, crushing the concrete decks. Under typical urban traffic, the fatigue life of anchoring bolts or straps is exhausted in less than a month. Holes drilled at fatigue crack tips occasionally engender multiple crack

Figure 14.12 Fatigue crack propagating beyond 1-in. (25-mm) drilled hole.

propagation paths (see Fig. 14.12). Temporary repairs imply the mounting urgency of permanent replacement.

Special

Inspections may be *special* because of circumstances or the structure. The scope varies with the purpose. An inspection can be termed "interim" or special depending on local priorities. The expansion joint inspection, formalized according to Fig. E21.1, is considered special because of its narrowly focused purpose and irregular scheduling. The resulting reports may not fully "meet or exceed" NBI requirements. They must, however, be produced and stored in a form compatible with the standard database records.

In Depth

NBIS 23 CFR 650 Subpart C (FHWA, 2005a) contains the following definition: "In-depth inspection: Inspection within arms length of the component. Inspection uses visual techniques that may be supplemented by nondestructive testing."

The definition would benefit from a description of the expected product. It is indicative of the nature of regular biennial inspections that they can serve to prioritize rehabilitation projects but cannot be used to determine the scope of rehabilitation work. That is the task of in-depth inspections. They include field verification of as-built drawings. Destructive testing, such as coring of concrete decks, avoided during most inspections, is standard during in-depth ones. Videotaping of inspection highlights has become the practice. The NYS DOT has issued a guide for in-depth inspections. An example of an in-depth inspection is the Chicago transit rail bridge assessment (Walther and Coob, 2002).

Essential Completion

Essential completion inspections serve two purposes. First, they ascertain whether the work on the bridge is essentially complete under the current contract(s). If appropriate, a "punch list" of items to be completed is created and verified by a subsequent inspection. The scope of rehabilitation work is particularly hard to assess, because bridges are not always restored to their original condition while significant items may have been added by contract change orders.

Second, as-built field-verified drawings are entered into the bridge file. The bridge inventory is updated (or initiated if the bridge is new) and becomes a reference source for future inspections.

Periodic inspections assume that as-built conditions of new structures fully comply with design, although that is not always the case. Essential completion inspections should recognize that construction occasionally deviates from design and that design is not always flawless. Occasionally welds and bolts are substituted during construction without adequate record. Widely publicized are the pedestrian walkway failure at the Kansas City Hyatt Regency Hotel (Levy and Salvadori, 1992) in 1980, caused by an unauthorized hanger discontinuity, and the successful strengthening of the Citicorp building in New York City in the 1990s, following a construction change from welded to bolted connections.

In contrast, the strength of a bridge may not be fully restored during rehabilitation, because the live loads have been reduced. In such a case inspections should not compare the structure to the original design.

Condition ratings assigned to bridges after rehabilitation, for a variety of reasons, are rarely the highest. What constitutes a completed project can become a matter of litigious dispute. A manual of professional practice on the quality of the constructed project (ASCE, 1990) was prepared as a guideline for owners, designers, and constructors but remained only a recommendation. The lack of uniform standards in this field underscores the importance of construction supervision. Large bridges in Western Europe are inspected 10 years after construction completion by commissions representing the contractor, the owner, and government auditors in order to determine the responsibility for any nonperformance.

Complex Bridges

According to FHWA (2005a), movable, suspension, cable-stayed, and other bridges with unusual characteristics are *complex*.

From a management point of view, a bridge may be complex or require exceptional attention either because regular inspections do not capture its condition or because it provides essential services and the rating "not functioning as designed" would be unacceptable. Long spans, suspended or otherwise, are exceptional; however *long* and *exceptional* are vague terms. Some guidance is implied in the limit of 500 ft (152.5 m) adopted by AASHTO design specifications.

The number of spans, rather than the bridges, best quantifies the scope of regular inspection. Structures comprised of multiple, relatively short spans are typical for urban areas, sometimes not even perceived as bridges. Multispan bridges often have unique details, but their magnitude remains the main challenge, adequately handled by NBIS and state inspection practices.

Long spans combine critical structural details with the demand for rapid assessment of voluminous data. In the inventory they are at a disadvantage if treated as average spans. Discretizing long trusses into fictitious spans between panel points improves data management. Approaches are typically multispan structures with important features of their own.

The scope of biennial inspections had to be exceeded at New York's Williamsburg Bridge in 1988 (Example 3). Critical details, including portions of the suspension cables and the cantilever roadways, were so deteriorated that inspectors closed the bridge to all but pedestrians until a detailed evaluation could allow a reopening. The ensuing in-depth investigation resulted in a 17-year rehabilitation. A bridge-specific maintenance manual is part of the as-built (in this case, more precisely, as-rehabilitated) documentation.

Eye-Bar Chains

The Silver Bridge collapse (Example 6) discredited the internally nonredundant two eye-bar chain suspension systems; however, multiple eye-bar chains are present on many railroad and highway trusses and on a few suspension bridges (Fig. E6.1).

When the parallel eye-bars between two panel points are numerous, a single one of them is not as critical; however, the analysis must consider the probability that, if one has failed, the others might be close to failure for similar reasons, in addition to experiencing overstress. A new support system became necessary for the trusses shown in Fig. 3.6 partly because of potential overstress in the eye-bars but, to a greater degree, as a result of concern for the condition of the connecting pins. Shifts and loss of section in the eye-bars can alter the pins' mode of operation from shear to bending, increasing their vulnerability to fatigue fracture. Corrosion typically fuses the entire connection assembly into a rigid block (Fig. E6.3), completely eliminating the intended rotational release and therefore subjecting the eye-bars to bending.

Eye-bar chains and their pinned connections are typical candidates for NDT&E, particularly by ultrasonic methods (Fig. E6.3). An NDE program for a large pin-connected truss is a considerable undertaking in terms of time, equipment, expertise, and traffic constraints. If it proves impractical to test all connections, a sample must be tested for an initial estimate of the conditions. Based on that, the size of a statistically meaningful sample is determined for full testing. Expanding the testing program and providing emergency strengthening must be anticipated.

Suspension and Stay Bridges

NBI (Table A14.2) refers to 98 suspension bridges, of which 33 are considered structurally deficient and 41 are obsolete. According to NCHRP Report 534 (2004, p. 1–11) 29 suspension bridges with aerially spun parallel wire cables were constructed in the United States until 2000; 2 have prefabricated parallel wire strands, and 21 have helical strands. Most spans are greater than 700 ft (213 m). The number of cable-stayed bridges is rapidly growing. NBI contains 35 "stayed girder" bridges, including only 1 deficient and 2 obsolete. The Cincinnati–Covington Bridge (Fig. 1.37) and the Brooklyn Bridge (Figs. E1.1 and 3.5) are examples of John Roebling's trademark hybrids of suspension and stay bridges. Both are considered as suspension bridges for inventory purposes.

The details peculiar to cable-supporting systems are recognized as critical in FHWA (1986) and AASHTO (2000b, 2003). Included are the main cables and the suspenders of suspension bridges and the stays of cable-stayed ones, saddles, and anchorages (see Fig. 14.13).

The high-strength steel wires used in cable-supported bridges are grouped in parallel or in helical strands. They are usually galvanized, although there are exceptions, such as the Williamsburg Bridge in New York City (parallel wires) and the original cables on Pont de Tancarville in France (helical strands). Helical strands or wire ropes are typical for suspenders (Fig. 14.1) and stays. At the latest record holders, Akashi-Kaikyo for suspension and Tatara for cable-stayed spans (Figs. 1.40*a,b*), in Japan, parallel wire strands are used for suspenders and stays, respectively (Fig. 14.14.*a*).

The critical details of suspension and stay systems are mostly inaccessible for visual inspections (see Figs. 14.2 and E3.5). If apparent, the indications of cable deterioration are likely to be extreme, as in Figs. 14.1 and E3.4. Adequate assessment of suspension bridges must take into account the mechanisms of material degradation, the structural failure mechanisms, and the techniques of detecting malfunctions. An understanding of cable-supported structures is required. Steinman (1949) described the state of the art pertaining to a number of bridges designed up to the mid-twentieth century. Irvine (1981) and Gimsing (1983) are classic texts on suspension structures in general and bridges in particular. A number of sources, including Troitsky (1988) and Chen and Duan (1999), describe cable-supported bridge condition and performance assessment.

The developments in the design, construction, and maintenance of cable-stayed and cable-supported structures are presented in particular detail by the Association Française

Figure 14.13 Corroded eye-bars of suspension cable anchorage.

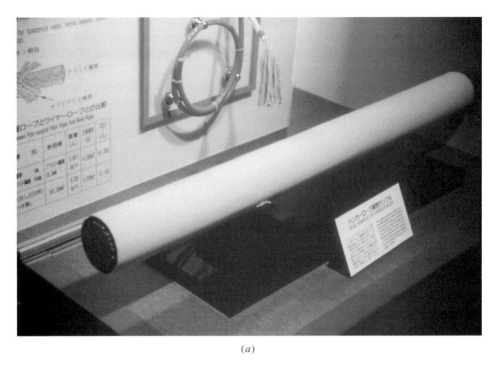

(a)

Figure 14.14 (a) Model of PWS suspender ($85 \times \phi 7$), Akashi–Kaikyo Bridge, Japan.

Pour la Construction (AFPC, 1994) and the International Association for Bridge and Structural Engineering (IABSE, 1995, 1999, 2001). Gabriel and Schlaich (IABSE, 1995, pp. 897–902) concluded their article on the robustness of stranded cables in suspended bridges as follows:

> Laymen are unable to evaluate a cable after prolonged use. Either the client's employee (bridge inspector) familiar with the whole structure and constantly present at the site or the specialized rope inspector (expert) is of particular importance. Together they will serve the safety and durability of a cable structure better than any sophisticated automatism for damage detection.

The corrosion mechanism of high-strength, highly stressed, galvanized (or not) steel is the subject to many detailed investigations reported, for example, in TRR 1654 (1999) and Stahl and Gagnon (1996). For parallel wire suspension cables, NCHRP Report 534 (2004) recommended:

(a) Periodic unwrapping of main cables (as in Fig. 11.2b)

(b) Improved stochastic and phenomenological modeling of high-strength wire and cable failure modes

(c) Development of noninvasive inspection methods for suspension cables

(*b*)

(*c*)

Figure 14.14 (*b*) Glebe Island Bridge, Sydney. (*c*) Extrados bridge under construction, Nagoya.

In 2005 the FHWA initiated such a project. To date, most common is the acoustic monitoring of wire breaks (see Chapter 15). The method provides rough estimates of strength losses in a cable but cannot identify the causes or forecast the time dependence of the process.

Pitting and stress corrosion culminate in square wire breaks. If this is detected or suspected, the cable must be unwrapped and wedged (see Fig. E3.5) in order to determine the extent of the deterioration throughout the length and the cross section. Acoustic monitoring of wire breaks is a noninvasive alternative, becoming standard practice for monitoring suspension and stay cables as well as prestressing tendons. An independent method for confirming acoustically obtained results would be very valuable (Chapter 15).

If confirmed by in-depth inspection, the deterioration of a suspension cable triggers a detailed evaluation of the bridge load-bearing capacity and life expectancy, including laboratory tests of wire samples, analysis of the as-built and as-is structure and cost–benefit assessment (Example 3). All suspenders and stays of the Brooklyn Bridge (Fig. 3.5) were replaced after one stay ruptured due to corrosion in 1981, killing a pedestrian. Virlogeux (in TRR, 1654, 1999, pp. 113–120) reported on the replacement of the main cables of Pont de Tancarville. The original cables had comprised unwrapped, nongalvanized, lock-coil, helical strands, one of which corroded to failure. The new strands are galvanized. The similar cables of Pont d'Aquitaine in Bordeaux (Fig. 5.3) were replaced by different methods (Kretz et al. in IABSE, 2006).

At intervals of 20–30 years, suspension bridge owners test sample suspenders to rupture.

Inspections should report the condition of the corrosion protection systems at cables and stays. Over the years suspension cables have used linseed oil, synthetic corrosion inhibitors, red lead paste, polyester, parallel wire wrapping, and dry air injection (Figs. 11.12 and 11.13).

NCHRP Synthesis 353 (2005) reported on the performance of corrosion protection systems of stays, including individual sheathing with corrosion-inhibiting coating and epoxy coating. The author emphasized that corrosion can be externally and internally driven and that in the latter case the corrosion protection can contribute to the problem. The following filler and blocking compounds are reviewed:

Portland cement grout

Synthetic polybutadiene resin

Metal soap hydrocarbon grease

Solidifying wax

Lightweight liquid oil

Synthesis 353 divides inspection and monitoring techniques into short and long term. Briefly reviewed are the following:

Vibration-based cable force measurements

Sound, ultrasound, laser ultrasound

Impulse radar

Infrared thermography

Magnetic methods (flux leakage, perturbation)

Magnetostriction

Radiography

Photogrametry

Other (see also Chapter 15)

Surveyed cable-stayed bridge owners report concerns about the lack of accessibility, reliable inspection, and forecasting methods. Bridge-specific inspection manuals are developed.

Unique to the suspender and stay systems is the need to monitor the response to wind. On bridges in Japan, such as the Akashi-Kaikyo and the Hakkucho Bridge in Hokkaido, real-time monitoring systems are designed along with the new structure.

The concerns and techniques associated with corrosion and loss of tension apply similarly to prestressed structures.

Prestressed Bridges

Although cable-stayed bridges (Figs. 4.7 and 14.14*b*) appear globally closest to suspension bridges, their function has a lot in common with prestressed concrete structures. Extrados bridges (Fig. 14.14*c*) represent a transition between the two.

Manuals specifically address prestressed bridge condition assessment (AASHTO, 1999a; Anglo–French Liaison Report, 1999). Most recent is the NCHRP Report 496 (2003), "Prestress Losses in Pretensioned High-Strength Concrete Bridge Girders." Loss of prestressing tension is critical to the integrity of prestressed concrete structures. The causes include shrinkage, creep, relaxation, construction imperfections, corrosion, and combinations of factors. The inaccessibility of the critical high-strength tendons and rods is a well-recognized vulnerability (Section 4.3.1). It is discussed, for instance, in the Anglo–French Liaison Report (1999) and Buderkin et al. (1991). Prestressing principles and applications to bridges are comprehensively developed by Menn (1986).

Inspections should be aware of the bridge construction sequence (pre- or posttensioned), as well as the prestressing tendons' protection systems (grouted or not). Access is often entirely obstructed. For example, the inspections of three-span bridges with inclined pier columns prestressed by tie-downs at the abutments are meaningless without the means to investigate the condition and the performance of the tie-downs.

High-precision monitoring of overall structural geometry becomes particularly important in prestressed concrete structures as well as close observation of any local indication of distress, such as cracking and pigmentation of concrete (Fig. 14.15). The crushing of the concrete at the joints between the segments in Fig. 4.68 prompted indepth inspections and eventually a number of retensioning measures (Fig. 4.69) aimed at restoring the estimated loss of prestressing forces. Wire breaks are acoustically monitored, as in cable-supported bridges.

Movable Bridges

Koglin (2003, pp. 467–8) wrote:

> The analytical methods available for evaluating bridges are limited and can, in some ways, be considered fanciful. . . . It is even more difficult to apply bridge management principles to movable bridges than to fixed ones, because almost every movable bridge is a unique entity . . .

Figure 14.15   Cracks in prestressed box girder.

The author provided a list of roughly 3780 moveable bridges operating at various times in the United States. NBI (Table A14-2) contains 874 lift, bascule, and swing bridges.

Movable bridges are classified as structurally complex by the NBIS. In the United States they must operate at the request of the U.S. Coast Guard. Owners must demonstrate their functionality periodically. Mechanical, hydraulic, electrical, interlocking, control, and other special components must be inspected by appropriately licensed mechanical and electrical engineers. Inspections are guided by the AASHTO (1998d) manual, currently in its second edition. Also essential are the design specifications of the AASHTO (1988). The cables supporting counterweights on lift bridges must be certified by the manufacturer and inspected according to the ANSI. To the vertical-lift (Fig. 4.78), bascule (Fig. 4.79), and swing bridges (Figs. 1.10, 11.1, and 1.16), discussed in AASHTO (1998d), the rarer retractile ones (Fig. 14.16) could be added for completeness. The sign warning of high voltage in Fig. 14.16 reminds of the importance of electrical and mechanical inspections. Koglin (2003, p. 31) reported the bridge as "no longer operable," but it was still active in 2006.

A number of movable bridge details are unique and prone to specific modes of failure that must be recognized and anticipated (see Section 4.6.1). Decks are typically open steel gratings (Fig. 4.21), susceptible to corrosion fatigue and fractures. Structural repairs, particularly temporary ones, must not alter significantly the weight of the movable parts.

Figure 14.16 Retractile bridge sign on fascia girder cautions of electric hazard.

461

Underwater (Diving) Inspections

The NBIS (FHWA, 1971, 1988, 1995) specified underwater inspection frequency as follows: "§650.303 (2) Those bridges with underwater members which cannot be visually evaluated during periods of low flow or examined by feel for condition, integrity and safe load capacity due to excessive water depth or turbidity . . . shall be described, the inspection frequency stated, not to exceed five years, and the inspection procedure specified."

The *Manual for Condition Evaluation* (AASHTO, 2000b) classifies underwater inspections in three levels according to their scope. The NBIS mandate above refers to a level III inspection, requiring hands-on verification of the entire underwater substructure by qualified divers. State DOTs issue specific directives for diving inspections (such as NYS DOT, September 1, 1993).

NBIS 23 CFR 650 subpart C (FHWA, 2005a) allowed shorter or longer (not to exceed 72 months) intervals between diving inspections. Scour-critical bridges (discussed in the following paragraphs) may require continuing or frequent monitoring. *Wading* is not to be confused with *diving*. Details located at the "wet line" are highly vulnerable but cyclically visible. Figure 4.63 shows timber piles rotted between 50 and 100%, entirely submerged at high tide but in full view 6 hours later.

Access to underwater bridge elements, a professional feat in itself, must be followed by collection and correct interpretation of meaningful data. Scour is particularly hard to detect even during direct access to a susceptible pier footing. Infrared camera photography allows for visual inspection, but scour can still go unnoticed. The early stages of marine borer attacks on timber piles are only detectable by sample testing.

Few consultants are qualified to perform underwater inspections in most areas and the owners have limited options for independent review of past findings. Consequently, both underwater conditions and inspections require a higher margin of safety.

Extreme Events/Emergencies

Two types of assessment are required: pre- and postevent. Preassessment considers extreme events as loading combinations (Section 4.2.4) and prepares to sustain them. Consequently this is a design function. Postassessment determines the losses and recommends remedial measures. Regardless of the level of preparation, extreme events require emergency response. To the extent that extreme events are random, adequate response requires a core team of permanent employees to be supplemented by contracted forces as needed. Personnel charged with emergency inspections must be specially trained in safety and communications by the responsible agency as well as by law enforcement, federal (FEMA), and local emergency management authorities (e.g., the Office of Emergency Management).

Regardless of prior preparation, field assignments must be preceded by clear immediate instructions. Safety of the personnel must be specifically addressed. Inspectors must be advised in detail of their responsibilities and options for remedial action, for instance, to maintain or close traffic and request emergency repairs. Emergency procedures must be documented, coordinated with related agencies, and tested. Means of contact with all responsible personnel and access to any location must be provided.

Triage

Post-extreme-event assessments must perform triage on demand. Emergency response depends on limited workforce, time, and material. For expediency, bridge owners occasionally combine the inspection and repair function by assigning engineers competent to assess the "damage," select the appropriate repair, and supervise its completion. Independent inspections should determine the useful life of emergency repairs. The bridge inventory and the condition ratings may have to be updated accordingly.

All factors relevant in an emergency inspection, including parties present, possible causes of accidents, atmospheric conditions, types and licenses of vehicles involved, must be documented in addition to the structural assessment. The most recent inspection report should be available at the site as early as possible. Repair recommendations should include measures to mitigate the cause. The most common causes for emergency inspections are briefly discussed.

Traffic Accidents

Under the bridge, vehicular impact can destroy columns and primary structures (Figs. 4.44–4.48). The nature and the location of the damage are often repetitive. A steel signpost or overhead floor beam may sustain numerous impacts with minor distortions but eventually could crack. Complacency, always a threat to bridge inspections, is particularly likely to affect investigations of traffic-induced damage as accidents recur at the same locations and create the impression of routine.

Train impact (Fig. 4.45) is much greater than vehicular one and requires more effective prevention. Even more destructive are vessel collisions (see Fig. 4.44), discussed in numerous reports, such as Gluver and Olsen (1998). Protection systems, such as crash walls, fenders, and dolphins, are inspected as part of the bridge. Traffic must be considered unsafe unless they function reliably. After the collisions shown in Figs. 4.44 and 4.45, traffic was suspended until completion of emergency repairs.

Accidents on top of the bridges typically damage railings (Fig. 14.17), traffic signs, and light standards (Fig. 14.18). Traffic must be considered unsafe until destabilized elements are removed and the railings are restored.

Local traffic departments maintain information on accident-prone locations and frequency of occurrence. Inspections of such locations should recommend appropriate mitigation, possibly including changes to the bridge geometry. After numerous collisions, bridge inspections recommended the removal of the structure of Fig. 4.47. The NYS DOT *Collision Vulnerability Manual* (NYS DOT, 1995a) is an attempt to address this hazard systematically.

Earthquakes

The ATC offers the *Field Manual for Post-Earthquake Safety Evaluation of Buildings* (ATC 20-1, 1989) and conducts courses on the subject. CalTrans has issued manuals for postearthquake structural inspections and has emergency procedures tested during the Loma-Prieta (Housner, 1990) and Northridge (January 1994) earthquakes. The Japanese experience, particularly after the Hyogo-Ken Nanbu earthquake, Kobe, January 1995, is another important source of information on postearthquake response. Numerous bearings and restrainers failed during that event. The Bay Bridge span collapsed due to a failure

Figure 14.17 Railing damaged by collision.

of both the fixed and expansion bearings (Figs. 4.32 and 4.33). Rocker and sliding steel plate bearings must be inventoried for replacement during bridge retrofits in seismic zones. The bearings of Figs. 14.10 and 14.11 did not fail in an earthquake but would have performed poorly. Such considerations inform the preevent vulnerability assessments (Section 4.2.4).

Preearthquake inspections identify vulnerable details and determine the urgency of their replacement or strengthening (Appendix 33). Figures 4.37, 4.38, 4.43, and 10.2 show retrofits of bearings included in the scope of bridge rehabilitations in a moderate seismic zone. Figures 4.35 and 4.36 depict tensioned and loose tie-downs typical in active seismic zones.

The three U.S. National Seismic Engineering Centers are continually developing hazard assessment and design manuals for new and existing bridges. One product of these projects is the network-level pre- and postevent assessment and management program package (Werner et al., 2000).

A detailed inventory allows rapid preliminary assessment of the seismic vulnerability of the stock. Information could be sorted out according to a variety of criteria, such as the ones developed by Basöz and Kiremidjian (1996) (Appendix 33), taking local priorities and estimated hazard levels into account.

Figure 14.18 Light standard damaged by collision.

Responsible owners prioritize projects according to combinations of importance and vulnerability ratings (e.g., weighted products). The seismic hazard is determined by geological surveys. Traffic counts, the availability of alternate routes, and other considerations (such as historic landmark status) are considered, not always within formalized algorithms. The structural dynamic characteristics are obtained from analytic studies and "dynamic signature" tests. Example 25 describes a dynamic signature test at the Brooklyn Bridge.

Example 25. Monitoring Dynamic Response of Brooklyn Bridge under Ambient and Forced Excitations

Ye, Fanjiang, and Yanev (in Ansari, 2005, pp. 65–72) presented a summary of the monitoring of the Brooklyn Bridge under forced excitations. Figure E25.1 shows the location of sensors on the main span of the bridge. A 2-ton dynamic actuator (Fig. E25.2) was placed consecutively at several locations along the span. Figure E25.3 shows fundamental response mode shapes.

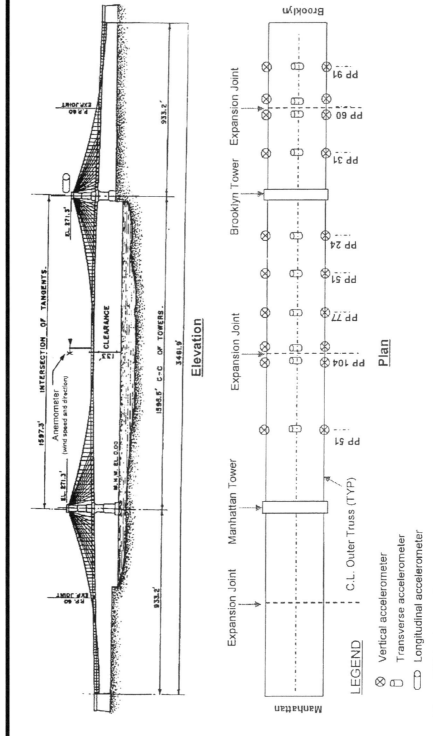

Figure E25.1 Sensors for monitoring response of Brooklyn Bridge to ambient and forced excitations.

Figure E25.2   Two-ton mass actuator on Brooklyn Bridge.

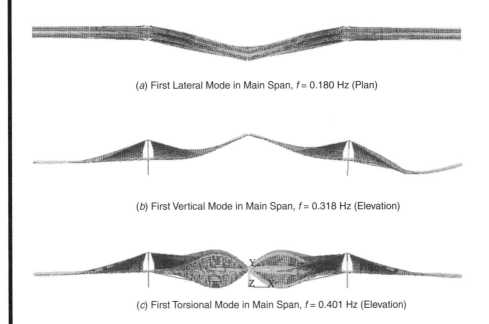

Figure E25.3   First natural mode shapes.

Floods/Hurricanes

Recent events in southeastern United States demonstrated that hurricanes and floods must be treated jointly. Floods are historically recognized as the primary cause for bridge failures. The NYS DOT *Hydraulic Vulnerability Manual* (1991) is an example of a systematic management approach to mitigating scour-related hazards. During the 1994 floods in Delaware County, New York, New York City bridge inspection teams were deployed by helicopter to inspect the affected areas. Even though emergency visual inspections can be helpful in reopening closed bridges, only underwater inspections can verify the integrity of foundations in scour-prone areas. Strong currents can undermine footings at very high speed and without warning. Scour can be monitored sonically.

NCHRP Report 489 (2003, p. 19) refers to three scour components. Long-term *aggradation and degradation* are elevation changes in the streambed of the river or waterway caused by erosion and deposition of material. *Contraction scour* is due to removal of material from the bed and the banks of a channel. *Local scour* is caused by the accelerated flow of water around the piers and abutments. Local and contraction scour can be *clear water,* (i.e., permanent) or *live bed* (i.e., cyclic). Because of the cyclic and dynamic nature of the undermining, diving inspections are not very reliable in predicting potential hazards over extended periods.

FHWA (1998) summarizes plans of action for scour-critical bridges and techniques for monitoring of scour. Fixed and portable instruments are described, including sounding rods, buried devices, sonar instruments (electronic echo sounders) and positioning systems. Briaud et al. (NCHRP Report 516, 2004) have proposed a method for predicting scour depth over time for complex piers in a contracted channel and for a given hydrograph, based on a combination of existing knowledge flume tests, numerical simulations and field verification. Some scour symptoms can be monitored sonically.

Inspections conducted during floods and wind storms may be primarily able to limit losses by closing the threatened structures to traffic.

Wind

Most bridges are designed for 100-mph wind loads (160 km/h); however, no live loads are combined with the wind loads beyond 56 mph (90 km/h). The corresponding forces are applied statically. The Tacoma Bridge collapse more than any other demonstrated the destructive dynamic nature of wind loads. The wind response of the Whitestone Bridge (Fig. E1.8) has been adjusted by numerous retrofits, the latest one still in progress at this writing.

The wind response of cable-supported structures is assessed both analytically and by testing models in wind tunnels. Ambient vibrations are continually monitored on a number of bridges in Japan, such as the Akashi-Kaikyo and the Hakucho, as reported in several publications by Y. Fujino of the University of Tokyo and by the Honshu-Shikoku Bridge Authority. The behavior of long suspenders and diagonal stays remains a matter of concern (Dallard et al., 2001) and can be continually monitored due to the increasingly available accelerometers.

Crowds

Section 4.2.4 identifies pedestrians as a source of potentially vulnerable live load. Fujino et al. (1993) have shown that pedestrian-induced excitations are not random. Field tests are therefore recommended. Pedestrian and seismic excitations vary significantly, but the studies of the induced structural response require similar expertise, instrumentation, and analysis.

In a closely monitored test, pedestrians excited a lateral response at the London Millennium Bridge (Dallard et al., 2001). Large numbers of pedestrians using the Manhattan-bound traffic lanes of the Brooklyn Bridge during the blackout on August 14, 2003, reported lateral motion at 7 p.m. Thirty minutes later the author and his colleagues could not confirm the reports at the site. Nonetheless, in the light of the findings at the Millennium Bridge (Fig. 3.1), Passerelle de Solferino (Figs. 3.2a,b), and analysis, the NYC DOT investigated the dynamic response of the bridge under simulated lateral and vertical excitations (Example 25).

Structural response can be modified by stiffening and damping. Alternatively, cautionary signs could advise the public of potentially disturbing but structurally admissible motion. Because of the sensitivity of pedestrians to motion, inspectors should report their experience of the structural response to traffic and wind along with a description of the ambient conditions. Close observation of the behavior of light posts and sign structures on bridges can be revealing. Although valuable, however, such observations do not reflect the structural performance under extreme crowd excitations.

Fire

Fire is a recognized bridge-related vulnerability (Section 4.2.4). Flammable materials, including all debris (Fig. 14.19), must be treated as hazards and immediately removed from the vicinity of bridges.

The transport of flammable materials is subject to local safety regulations. Inspections after a fire (Figs. 14.20 and 14.21) must observe the safety regulations determined by the fire department.

Figure 14.19 Illegally dumped garbage, burning and eventually compromising a bridge.

Figure 14.20   Inspection of bridge construction site after fire caused by gasoline truck traffic accident.

Sabotage

Bridge sabotage is an act of war; however, the events of September 11, 2001, in New York City (Fig. 4.53) justify peacetime concern. In that and other instances, qualified bridge inspectors provided all the key assets of training and equipment required for emergency structural assessment.

Federal and local guidelines are continually upgraded as new emergencies expose their vulnerabilities. Communication between the responsible agencies, for instance, federal and local offices of emergency management and police, fire, and transportation departments, is of primary importance. Appointing one agency as coordinator in charge of operations is a first step (Fig. 14.22).

Nonstructural Elements

Movable bridges can suffer costly malfunctions due to electric and mechanical failures and the occasional operator error. Condition evaluation in such cases is invariably an emergency due to the demand for bridge service.

Nonstructural bridge appurtenances, such as masonry cladding (Example 16), can be extremely hazardous to the public, the more so because they do not contribute to the structure's function and can be easily overlooked. Complete removal of cladding is a typical emergency measure. The maintenance needs of purely decorative elements must be included in cost–benefit analyses by design.

Figure 14.21   Spalled concrete, peeled paint, and buckled bracings caused by fire.

Closed Bridges

Closing a bridge to traffic because it has been rated "failed" resolves the urgent structural problems (while creating traffic ones for the network). A closed bridge is not a closed case, however, because, as pointed out in the section on that subject (Section 13.3), physical deterioration extends beyond the rating scale.

As a rule, dead loads (although static) amount to 90% of the design loads on vehicular bridges (beyond a certain length) and 80% on railroad ones. Combining that load with the poor condition that would have precipitated the closure can have catastrophic consequences. Figure 4.73 shows a deflected 45-ft (14-m) steel pier column with a 100% corroded web. The structure had been closed for 12 years before the condition was noticed and declared an emergency. Temporary support towers (Fig. 11.9*b*) and the strong-backs shown in Figs. 11.10*a,b* were installed.

Closed bridges frequently become dumping grounds and are more than commonly exposed to fire hazards.

Maintenance Inspections

Inspections can be incorporated into the continuous process of maintenance on unique bridges. Recurring (routine) maintenance tasks, such as spot painting, caulking, and drain cleaning, have been combined with the condition rating and record keeping required by

Figure 14.22   Response by police, fire, and emergency coordination agencies.

inspections and conducted concurrently. Certain components, such as the movable parts of movable bridges, travelers of suspension bridges, and electric equipment, may require frequent inspections (including by electrical and mechanical engineers), while others, as, for instance, the wrapped wires of suspension cables, are beyond the biennial scope. Recent designs of unique and complex bridges include bridge-specific maintenance manuals. If future inspections are able to compare field findings with maintenance records, the actual effect of maintenance on bridge condition will finally emerge.

Maintenance inspections are favored by owners of toll facilities, where both the structure's needs and the staff qualifications are highly specialized. A comparison of bridge-related expenditures and conditions suggests that maximizing toll revenues requires maintenance intensity well beyond the needs of safe traffic. As a side effect, improved maintenance minimizes the need for inspections and can absorb their performance.

Since the purpose of inspections is to obtain information, they should enjoy a certain independence, as has been recommended for all information management (see Chapter 12 on the bivalent nature of information). Separating the staff conducting assessment from that performing maintenance tasks avoids a potential conflict of interest.

14.4   PERSONNEL

The basic qualifications of inspectors required by the NBIS are summarized in Appendix 53.

A civil engineering license is not uniformly mandatory, reflecting the two approaches to assessment discussed in Section 10.4. Rating/descriptive evaluations require experts (e.g., licensed engineers) who can qualify and quantify both routine and unprecedented findings. Defect/action reports can be prepared by experienced technicians given clear and comprehensive instructions (as by an expert system). Both methods suffer from the shortages of their main assets, the experts in the former and the expert systems in the latter.

As in the case of hazard identification (Section 10.3) and mitigation (Section 11.2), the inspector's ability to identify critical conditions is a last resort, when prevention has failed.

Expert systems (Appendix 42) integrate the bridge database with existing experience, service requirements, and repair options. Information can be treated as fuzzy sets and modeled by neural networks. Portable computers facilitate on-site data access and processing. "Heuristic" guides can prompt the inspector to seek and identify catalogued defects. These new developments will enhance inspection quality if implemented without relaxing personnel qualification requirements.

The partial failures of Example 7, the lost bearing in Fig. 4.31, the cracked truss chord in Fig. 3.7, or the damaged pier in Fig. 4.44 must be promptly assessed by qualified engineers capable of assuming responsibility for decisions affecting traffic and workforce deployment. Periodical inspections lose significance if they do not reflect engineering expertise. A database can be homogeneous, yet impenetrable if it has been mechanically obtained.

Safety and Equipment

All bridge inspection manuals instruct inspectors to comply with safety rules and regulations "as required." This is unavoidably vague, as safety rules are often revised by a number of responsible agencies with occasionally overlapping jurisdiction.

The Occupational Safety and Health Administration (OSHA), U.S. Department of Labor, is the federal authority on construction work standards. The standards are periodically updated. Bridge inspection differs from bridge construction in the level of exertion and the time spent at locations requiring safety precautions. The extent of that difference is ill-defined but significant, because it is argued that inspection is not the type of labor OSHA always envisions. An example is the requirement that anyone working higher than 6 ft (1.835 m) from the ground should be tethered or otherwise protected (OSHA, 1995, p. 3). The rigid enforcement of this requirement would effectively abolish the widespread and mostly uneventful inspections of bridge abutments by ladders. The existing ambivalence makes it imperative to define clear safety rules for any specific inspection operation as well as a line of responsibility for their implementation and adherence.

Safety rules must be reviewed at the start of an assignment regardless of past experience. This is particularly important in emergencies. All inspections must be supervised by a duly assigned, responsible, qualified, and licensed (for the purposes intended) professional.

Example 26 presents an inventory of the standard safety gear and inspection equipment supplied to the in-house bridge inspection unit of a transportation agency.

Example 26. Standard Inspection Equipment

Inspection equipment cannot be standardized across the entire profession. Particular operations must adopt, maintain, and upgrade their own selection. A possible list of standard inspection equipment is presented in Table E26.1.

The proper use of equipment during work must be rigorously supervised. Failure to maintain the equipment in good order is a hazardous breach of safety procedures. Expiration dates for safety gear must be monitored in order to obtain replacements. Inclusion of certain items can be the subject of vigorous discussion and investigation. For instance, ear plugs are standard for iron workers; however, they were found to be potentially hazardous for inspectors, who must be aware of all warning signals during work. The insulating rubber mat shown in Fig. E26.1 is intended to facilitate work near an energized third rail. Inspectors, however, are not trained to install it, and consequently it should not be part of their equipment. Rather, inspectors should treat all third rails as energized and should be trained to step safely over them. In addition, railroad flagmen should be requested to provide and install the mat if extended work in the vicinity of a third rail is anticipated.

Figure E26.1   Insulating rubber mat used over energized third rail during inspection. Courtesy of A. Leyco.

Table E26.1   Standard Bridge Inspection Equipment

Inspector Equipment	Inspection Team Equipment	Inspection Van Equipment
Boots, knee high	Hand-held computer	Tool chest
Dust masks (disposable)	Portable phone/radio	Clip boards
Overalls	Telephone directory	Flashlight (3 D cell)
Hard hat with liner	Lap-top computer	Fire extinguisher
Rain hat and jacket	Photo cameras, digital, 35 mm, infrared	First-aid kit
Work gloves, long cuff	Borescope	Flags (3)
Work gloves, unlined	Hand compass	Step ladder, 6 or 8 ft
Work gloves, lined	Optical distance meter	Traffic cones (10)
Chipping hammer	Telescopic survey rod 25 ft	Broom
Belt (2 drop-forged D-rings)	Dye-penetrant kit	
Deceleration lanyards	Lantern	**In trucks:**
Flashlight (2 D cell)	D-meter with test block	Generator
Safety reflector vest	Marking paint spray	Oil for generator
Level 9 (magnetic)	Screwdriver sets	Extension ladder, 32 ft
Tool bags (24 in.)	Sledge hammer (8 pounds)	Extension ladder, 24 ft
Class III body harness	Thermometer	Extension ladder, 16 ft
Lanyards	Spray penetrating oil	Shovel
Bridge inspection manuals	Binoculars	Push broom
Technical advisories for Inspection manuals	Vernier calipers	Dust pan and sweep broom
Emergency procedure instructions/contacts	Wrenches, 12 in.	Water cooler
Respirators and filters (OSHA)	Tool pouch	Bolt cutter
Clip boards	Lumber crayons	Flood lights
Safety goggles	Spray paint	Approved safety gasoline can
	Awl	Traffic cones
	Calipers	Variable message signs (optional)
	Drafting equipment	Arrow boards
	Hacksaw	
	Hacksaw blades (extra)	
	Paint scraper	
	Inspection mirror	
	Level, 24 in.	
	Pliers, 8 in.	
	Plumb bob	
	Pocket knife	
	Ruler, 25 or 30 ft (metal)	
	Ruler, 100 ft (fiberglass)	
	Scraper blades (extra)	
	Snips	
	Wire brush	
	Folding ruler, 8 ft	
	Rope ½ in. with 100-ft coil	
	Maps	

Traffic Management

Most inspection accidents are traffic related. Traffic conditions, rules, and needs vary for different communities and municipalities. Temporary embargos are often imposed on the traffic closures required for inspections. Transportation agencies normally have construction (and other operation) coordination centers, authorizing lane closures and issuing permits. Lane closures in heavy traffic (Fig. 14.23) may require recourse to specialized professionals. Inspection personnel must regularly attend the traffic management courses offered by local law enforcement or transportation agencies.

Night Work

Night inspection may be required due to lack of daytime access, for instance, over railroad tracks, or in emergencies. All personnel must have adequate vision for the tasks. Lights powered by generators must be available. Infrared vision aids are available. Light-reflecting safety gear is mandatory. The responsible inspector must determine if the findings are satisfactory or, alternatively, if daytime inspection remains necessary.

Railroad Track Work

"Track work" in this case consists of inspecting bridges over railroad tracks. Such work must comply with OSHA *and* the safety standards of the particular railroad company. All personnel working on the tracks must complete a training course offered annually by the respective company and obtain a certificate. No bridge inspection is to be conducted from train tracks without the supervision of railroad personnel ("flagmen"). Before proceeding with the inspection, the engineer in charge must obtain a clear statement

Figure 14.23 Lane closure for inspection with bucket truck.

of responsibility for the safety of his or her staff from a responsible railroad representative.

Roadway bridge owners secure the services of railroad personnel through force account agreements. Alternatively, the responsibilities and expenditures relevant to the safety of both operations can be shared according to a long-term interagency agreement.

On mixed-use bridges, coordination involves multiple tasks and all responsible owners. The bearings shown in Figs. 4.30, 4.31, 4.61, and 14.11 support train tracks on structures also carrying vehicular traffic and managed by a vehicular transportation agency. Each owner must inspect the primary structure; however, their respective design, maintenance and operation standards, and procedures differ. On the rare occasions when this has been possible, as in Figs. 4.5 and 4.6, ownership has been divided during rehabilitation, simplifying the chain of responsibility.

Access Equipment

All access equipment, such as bucket trucks, snoopers, scissor lifts, and ladders, must be appropriately certified for the work intended. A bucket truck may be appropriate for inspection but not for repair. Aluminum ladders are inadequate in the proximity of power lines. Thirty-foot (10-m) boom trucks can be used by qualified bridge inspectors after completing a course of instruction. Other equipment, such as 80-ft (25-m) boom trucks (Fig. 14.23) and snoopers (see Fig. 14.24), require a licensed operator. Bucket trucks operated from barges (see Fig. 14.25) are easily swayed by waves. Vessels should be kept at a safe distance. Inspections conducted from waterborne vessels in a navigable channel must comply with U.S. Coast Guard regulations.

Contemporary design is required to provide access for bridge inspection to all hands-on or critical locations. Travelers or fixed inspection platforms are built along with new bridges or added to older ones. Such nonstructural appurtenances must be inspected along with the rest of bridge elements and rated, most appropriately, as "utilities." Their safe operation must be certified by qualified mechanical and electrical engineers.

Because of their relatively light cross sections, maintenance and inspection platforms are the first to lose structural integrity under corrosion attack. Fatal inspection accidents have been caused by unsafe platforms. "Pans" built under bridge decks for collecting debris may not be designed for any live loads and are often unsafe. In the absence of access platforms and adequate equipment, scaffolding must be erected by a qualified contractor.

Personal Safety Gear

All personal safety gear (Example 26), including helmets, light-reflecting jackets, harnesses, lanyards, belts, goggles, flotation vests, and respirators, must be certified, periodically inspected, and replaced according to specifications. Incomplete or inadequate safety equipment is cause for work cancellation. It is up to the team leaders and their supervisors to ensure the proper use of all safety equipment. Whenever possible, inspectors must be secured at elevations higher than 6 ft (1.83 m) above ground, as shown in Fig. 14.26. The inspection shown in Fig. E7.3 does not comply with several safety requirements.

Mountain-climbing techniques are effective on a number of bridges where scaffolding would be the only alternative. Successful use of mountain-climbing techniques for bridge inspections was reported, for instance, at the New River Gorge arch in West Virginia.

(*a*)

Figure 14.24 Inspection with snooper.

Frostbite is a common inspection hazard in winter. Gloves, clothing, and shoes must provide adequate protection in low temperatures and high winds.

Portable computers are standard for many bridge inspections. In addition to the inspector's computer skills, managers must ascertain that operating this equipment in field conditions does not pose new hazards. With the availability of portable telephones, procedures must be established for prompt communication in the event of accidents. Most inspection reports are electronically generated and transmitted. Data filing and management must be adapted to the new and less cumbersome technology.

Environmental Hazards
Inspectors may come in contact with toxic materials, particularly waste, illegally disposed under bridges. The inspection personnel periodically receive instructions on operating in

(b)

Figure 14.24   (*Continued*)

a potentially toxic environment. When such a hazard is suspected, the area must be immediately vacated and declared unsafe. The sanitation department must be notified. Respirators must be used in the presence of pigeon droppings, frequently abundant under bridges, and in cellular abutment structures. Heavy shoes provide protection against sharp objects (sneakers are never appropriate).

Inspections can also be the source of toxic waste, such as lead paint chippings under a bridge. The debris generated during hands-on inspections are not likely to contaminate the environment significantly; however, a lead disposal protocol usually exists for the specific area and must be followed.

14.5   INSPECTION RELIABILITY AND QUALITY (QC&QA)

Section 4.5 identified inspections as highly vulnerable, because their tasks require advanced analytic and physical capabilities. The inspecting engineers must simultaneously move, observe, and think correctly and reliably. After evaluating the predominantly visual inspections conducted under the mandate for biennial bridge inspections, FHWA (2001b) gave a relatively low assessment to that capability. The report concluded that "visual inspections alone are not likely to detect or identify the specific types of defects for which the inspection is prescribed." In contrast, in-depth inspections produced superior findings. The observed inspection deficiencies fall into the groups of quality and reliability.

Figure 14.25   Inspection with bucket truck from barge.

Inspections must maintain the reliability of their product (e.g., the assessments) and maintain a quality of operation as a process. Inspection reliability measures the adequacy of the product as it has been designed, primarily in the following aspects:

Extent to which the prescribed methods can evaluate the targeted conditions

Significance of the targeted conditions

Quality measures the level of performance or the extent to which inspections attain their potential reliability.

In a study performed for PennDOT, Purvis and Koretzky (in TRR 1184, 1988, pp. 10–21) made the following distinction between quality control and assurance:

QC is the enforcement, by supervisor, of procedures that are intended to maintain the quality of a product or service at or about a specified acceptable level. QC of the in-

Figure 14.26 Work on suspension cable with fall protection.

spection at PennDOT's bridges is a daily operational function performed within each district for designated staff members under the supervision of the district engineer.

QA is the verification or measurement of the level of quality of a sample product or service generally by a third party organization. The sampling must be sufficiently representative to permit a statistical correlation with the whole group. The findings are compared against accepted standards to determine whether specified procedures are followed. . . . Statewide bridge inspection QA activities are the responsibility of the BMS Division, Bureau of Bridge and Roadway Technology.

Management is responsible for maintaining appropriate QC&QA. The safety of the performing personnel is a critical vulnerability of the process (Sections 4.6) subject to independent QC (Section 14.4). It influences significantly the quality of the assessments (e.g., the product).

Quality: Management and Qualifications

A detailed definition of the tasks and training of the staff can considerably improve the reliability of the findings. Under any circumstances, the term *routine* is too vague to serve as a professional reference.

The inspection process adopted by the NYC DOT (Example 27), consistently with New York State policies, ensures QC&QA of all scheduled inspections (e.g., biennial, interim, monitoring) by requiring a valid professional engineering (PE) license for the performance of the following two key tasks:

Team leader (TL) in charge of and physically present during inspections

QC engineer reviewing the inspection reports

The TL and QC engineers must cosign the final report and affix their PE license numbers.

Example 27. Bridge Inspection Process

Bridge inspections primarily evaluate, rate, and describe structural conditions and potential hazards. Examples 18 and EA46 show these to be the main sources of information on the bridge management process. The relational flowcharts of Figs. E27.1 and E27.2 illustrate the management process of bridge inspections and the

BDS CONCEPT MODEL

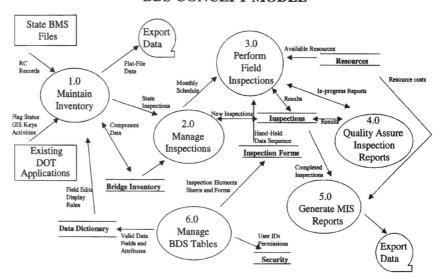

Figure E27.1   Relational flowchart of bridge data management system (BDS).

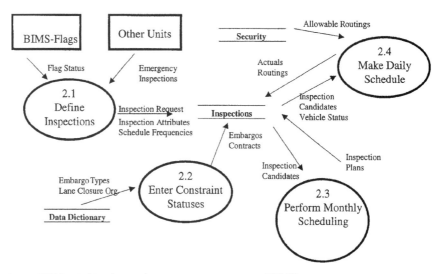

Figure E27.2   Bridge inspection management system (BIMS).

information they generate. Tables E27.1 and E27.2 show checklists that can be used to verify the completeness of the flowcharts. Figure E27.3 illustrates the flow of information between an in-house inspection unit and the units responsible for maintenance and rehabilitation.

Table E27.1   Checklist Corresponding to Bridge Data Management System (BDS) Flowhart of Fig. E27.1

Process (Inspect) → / Product (Data) ↓		Import	Maintain	Manage	Perform/ Generate	QC/QA	Export	Secure
State BMS	Data records	Y						Y
	Inspections	Y						Y
	Flags	Y						Y
Inventory	Inventory	Y	Y	Y	Y	Y	Y	Y
	Element data	Y	Y	Y	Y	Y	Y	Y
	Data dictionary	Y	Y	Y	Y		Y	Y
	Flags	Y	Y	Y	Y	Y	Y	Y
	GIS	Y	Y	Y	Y		Y	Y
	Applications	Y	Y	Y	Y	Y	Y	Y
Inspections	Schedules		Y	Y	Y			
	Forms	Y	Y				Y	Y
	Reports			Y	Y	Y	Y	Y
	Resources	Y	Y	Y		Y		Y
	Equipment	Y	Y	Y		Y		Y
Data management	BDS tables	Y	Y	Y	Y	Y	Y	Y
	MIS reports		Y	Y	Y	Y	Y	Y
	Software	Y	Y	Y		Y		Y
	Hardware	Y	Y	Y		Y		Y

Table E27.2   Bridge Inspections According to Flowchart in Fig. E27.2

Process → / Product ↓		Inpsect/ Monitor	Evaluate	Report	Assure QC	File	Import/ Export	Manage/ Train
Resources/ Equipment	Field				Y		Y	Y
	Office				Y		Y	Y
Structure		Y	Y	Y	Y			
BMS	Inspecting/ monitoring report	Y	Y	Y	Y	Y	Y	
	Flag	Y	Y	Y	Y	Y	Y	Y
	Repairs	Y	Y	Y	Y	Y		
	Postings		Y			Y		
	Foreast $/schedules		Y	Y	Y	Y	Y	Y
	Inventory					Y	Y	Y
Personnel	Field	Y	Y	Y		Y	Y	Y
	Office		Y		Y	Y	Y	Y

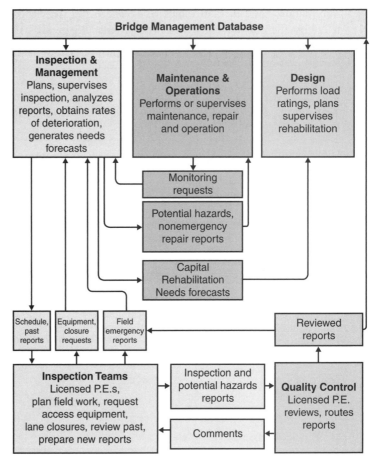

Figure E27.3   Flow of information between inspection and management, maintenance and operations, and design and construction.

NBIS 23 CFR 650 subpart C stipulates the following QC&QA measures:

Assure systematic quality control (QC) and quality assurance (QA) procedures to maintain a high degree of accuracy and consistency in the inspection program.

Include periodic field review of inspection teams, periodic bridge inspection refresher training for program managers and team leaders, and independent review of inspection reports and computations.

The standards require a TL to be always present during inspections; however, TLs are not required to be licensed engineers (Appendix 53).

Field supervision of safety operations (e.g., Q.A) is indispensable. Bridge owners regularly field verify the compliance with the adopted standard operating procedures (SOMs) by checklists.

Efficiency/Productivity

Inspection productivity evolves under the contradictory demands for quantity and quality. More than other tasks, inspections create the misleading impression that quantity compensates for quality. The manager must determine and maintain the highest average productivity that does not compromise the reliability of the findings.

Productivity is usually measured in inspector/span/day. It depends on the type and condition of the structures, the climate, the access, qualifications, experience, and other factors. Inspection teams typically consist of the TL, one or more assistants (civil engineers), and the required equipment operators.

The efficiency of evaluating structural conditions should not be confused with the rate of generating inspection reports. A more elaborate (and possibly slower) field investigation, taking full advantage of the methods in Fig. 4.4a and Table 9.1, may produce a more accurate and ultimately faster assessment of the structural condition.

Inspectors increasingly use portable computers (Chapter 15). The intent is to complete the reports on site and to transmit urgent findings if necessary. The quality of the output is likely to benefit from this inevitable upgrade of the operation before the quantity can show any effect. The transition from a paper-and-photography to a computer-based digital record of the findings requires considerable administrative and physical adjustments. The more experienced the inspectors, the longer may be the adjustment to the new technology. Unexpected drawbacks of the portable equipment, such as the glare of the screen, weight, battery life, shoulder straps, and writing recognition, can sidetrack an otherwise well-planned system overhaul. Field testing of proposed hardware prior to a final selection identifies potential production drawbacks.

The staff must regularly attend inspection and safety refresher courses. Rotation of personnel on bridges may slightly reduce productivity but amounts to a valuable peer review.

Bridges can be inspected by the owner or, as in most transportation departments, by consultants. Authorities in charge of toll bridges (which are often complex, unique, and essential structures) find the NBIS type of biennial inspections too general and design bridge-specific *maintenance inspections*. The subject in the preceding section indicated that duplication of effort is thus eliminated; however, the independence of the tasks and the implied peer review is lost. Despite their close cooperation, inspection and maintenance cultivate different and complementary types of expertise.

Reliability and Technological Enhancements

Advances in NDT&E (Chapter 15), data acquisition, and transmittal (by telephone, radio, satellite) enhance structural condition assessment in the following two ways:

By providing new information about the structural behavior

By processing the information faster and more reliably

Bridge management must upgrade the standard hardware, software, and expertise to take full advantage of the emerging condition assessment technologies. The task requires optimization. Not all innovations are directly applicable; others must be designed to the specific needs of the users. Tools previously considered exceptional, because of their cost and complexity, have become regular "off-the-shelf" products. Some of them are discussed in the following chapter.

Chapter 15

New Technologies and BMS

The new and more readily available technologies for NDT&E and for health monitoring of structural behavior (SHM) are transforming the field of structural assessment. With the gained capability for data acquisition come the vulnerabilities of misinterpretation and mismanagement. Structural engineers are challenged to obtain data cost-effectively and to interpret them correctly. Infrastructure management, in turn, must obtain life-cycle benefits from relatively untested smart materials and instrumentation systems (Yun and Spencer, 2005).

15.1 NONDESTRUCTIVE TESTING AND EVALUATION

So far regular inspections tend to recommend NDT&E applications beyond their own scope; however, technology advances and increased demands for quantified condition assessments are changing that practice. Nationally, the FHWA Nondestructive Evaluation Validation Center (NDEVC) is the source of information on available techniques and their performance. Worldwide, a number of international events are held to discuss NDE developments annually and various websites carry related information.

Section 8 of AASHTO (2003) *Guide Manual for Condition Evaluation* is dedicated to nondestructive load testing of the following types:

Diagnostic Tests. Included are measurements of load effects on bridge members, primarily for calibration of analytic models.

Proof Tests. The bridge response is measured under specific loads. A distinction is made between bridges that lend themselves to analytic load rating and those that cannot be so rated.

Dynamic Tests. Included are weigh-in-motion, dynamic response (primarily for stress level and fatigue evaluations) and vibration (e.g., dynamic signature) tests.

The tests are expected to complement the following assessments:

Load distribution, including unintended composite action, continuity/fixity, participation of secondary, nonstructural members, and the deck

Unknown or low-rated components

Deteriorated or damaged members

Fatigue

Dynamic allowance

The FHWA (2002c) bridge inspection manual dedicated its final Chapter (13) to "advanced inspection techniques" for supplementing the evaluations obtained by visual inspections and for inspection of otherwise inaccessible members. Hearn (in TRC 498, 2000, C-1) proposed NDE as a means for determination of condition states used in BMS, such as PONTIS (Appendix 44).

The theoretical basis for NDT applications to material quality control was addressed by Halmshaw (1987) and Collacott (1985). Hull and John (1988) describe NDT methods for the detection of defects in metals.

The following general NDT categories were identified by Agbabian and Masri (1988):

Visual

Radiological

Ultrasonic

Magnetic

Electrical

Penetrant flaw detection

Acoustic emission

Others

Collacott (1985) divided NDE into primary, secondary, and tertiary, depending on whether the sought parameters are related to the primary function of the structure, to a fault that might lead to a failure thereof or directly to a failure symptom.

The American Society for Non-Destructive Testing (ASNDT) has produced a large number of important publications. ASTM (Bush and Baladi, 1989) and CRC Press (Malhorta and Carino, 1991) specify a number of nondestructive methods. Specifically for bridges, the FHWA has conducted a number of strategic highway research projects (SHRPs) to develop and assess NDT&E techniques. One of many examples is the FHWA Demonstration Project No. 84, "Corrosion Detection Equipment." Recent manuals summarizing the theory and applications of NDE include Hellier et al. (2001), Shull (2002), and Gonkang Fu (2005).

Based on the type of wave propagation involved, NDE methods can be grouped into mechanical, sonic, electromagnetic, optical, X-ray, nuclear, and thermal. Specific applications include strain gauging, corrosion surveying by half-cell potential, magnetic particle evaluation of welds, ultrasonic evaluation of steel, acoustic emission and other methods for evaluation of fatigue crack propagation, infrared thermography for concrete decks, nuclear measuring of water–cement ratio in fresh concrete, impact-echo thickness and flaw evaluation, fiber-optic sensors for concrete, radar survey for roadway surfaces, dynamic characteristic method, long-range remote monitoring using electronic clinometers, and ultrasonic testing. Pile and foundation testing are gaining interest.

The Canadian Network of Centers of Excellence has placed a course on SHM by Bisby and Briglio (2004) on the website www.isiscanada.com.

The abundance of information adds to the need for a sound understanding of each method's range of application. A brief description of most common nondestructive techniques follows with examples of their applications.

Enhanced Methods of Observation

Enhanced methods of observation, such as dye penetrant testing and thickness meters have long been common practice; however, bridge inspectors find the equipment excessively sensitive or cumbersome. Steel thickness meters, for instance, are largely underused despite the importance of the information they provide, primarily because of their need for repeated calibration and the unrealistic demand for surface preparation. Dye penetrant testing can determine the size of fatigue cracks, once their location is suspected. The application of the dye penetrant requires some expertise, particularly in locating the tip of the examined crack (Fig. 15.1).

The capabilities of electronic borescopes are constantly improving and thus previously inaccessible locations can be viewed and photographed. Training is required before the relatively expensive equipment can be effectively used in the field. High-resolution digital photography is combined with advanced software into quick-time virtual reality

Figure 15.1   Dye penetrant test revealing crack in steel plate girder.

(QTVR) and panoramic image creation utilities for recording of field observations. This is a noteworthy database enhancement, so long as it does not become a substitute for hands-on inspections by qualified engineers.

Surveying

Laser technology lends to surveying high accuracy and excellent field performance. Laser beam distance gauges have replaced the telescopic poles and facilitated the essential (but previously deficient) bridge inspection task of measuring clearances. Using such equipment to monitor changes in the bridge geometry can be extremely useful. Essential is the appropriate choice of monitored locations and frequency of sightings as well as adequate record keeping (including, e.g., ambient temperature). Typical applications include long-term monitoring of movable bridges for indications of approach settlement and temperature effects.

Geographic Information and Global Positioning Systems

FHWA (2001e, p. 13) defined GIS as follows: "A computer-based tool used to gather, transform, manipulate, analyze, and produce information related to the surface of the earth."

In its modern form, information related to the surface of the earth is provided by satellites and incorporated into GPSs. Currently GPSs can monitor structural displacements with accuracy within 5 mm in the horizontal plane and are attempting to match that in the vertical plane. Results can be transmitted by telephone to any computer station.

The "spatial data" obtained from GPSs is useful both at the structural level, for monitoring displacements, and on the network (or GIS) level, for managing traffic flow, identifying alternative routes, priority corridors, and emergency management. In TRR 1889 (2004) Huang et al. (pp. 54–62) and Tsai et al. (pp. 21–34) report on pavement distress data collection and pavement rehabilitation planning based on GISs.

Load Deflection Tests

Load testing by calibrated vehicles is gaining recognition as a method of bridge load rating (NCHRP Report 12-46, 2000). On a number of occasions bridges have been subjected to controlled loadings, for instance, with hydraulic jacks (Fig. 15.2) or by calibrated trucks, and the response has been measured with extensometers and strain gauges, as in Figs. 15.3 and 15.4. Load cells are used during the replacements of suspenders on suspension bridges. Strain gauges were used to evaluate the live-load response of the eye-bars at the Franklin Square truss of the Brooklyn Bridge approach in New York City (Fig. 3.6), while X-ray diffraction gave estimates of the total load. Total load on a steel member can also be evaluated by electromagnetic methods (Schwesinger and Wittmann, 2000).

With the improved accuracy and accessibility of nondestructive measurements, load testing is increasingly used as a means of load rating. Fu (in Frangopol, 1999b) distinguished three types of load testing:

Proof Load Testing. The structure is subjected to a predetermined load enveloping (with a margin of confidence) expected service loads. Elastic response must be demonstrated for a satisfactory outcome.

Figure 15.2   Load test with hydraulic jacks during construction.

Diagnostic Load Testing. The structure is examined under incrementally increasing loads below the operating rating values in order to model its properties. A proof load may eventually be reached.

Weigh-in-Motion Testing. Moses and Ghosn (1985) formulate procedures for instrumenting, calibrating, and analyzing structures in order to measure their elastic response under multiple and repeated traffic loads. Dynamic and static traffic loads can be monitored continually.

All load test methods imply detailed instrumentation of the structure in the field and independent calibration of material properties in a controlled environment.

Strain and Displacement Gauges

In a functioning structure, strain gauges determine mainly the response to transient loads, such as those due to traffic and temperature variation. Since the amplitude and frequency of live load cycles are essential in fatigue life estimates, gauges have been employed in a number of in-depth and special inspections of steel structures supporting automobile and train traffic (Walther and Koob, 2002; ATLSS, 1995). In exceptional cases strain gauging may yield data on the structural response to the total load. Such opportunities arose on both the Manhattan and Williamsburg Bridges in New York City during the reanchoring of suspension cable strands. The new anchorage rods (Fig. 15.3) were strain gauged before the load transfer and the progress of the operation was monitored.

Figure 15.3 Prestressing anchorage rods.

Most common are the electric resistance gauges. The exposed reinforcing bars in Fig. E16.6 were strain gauged in order to demonstrate that they still respond to live loads. The results allowed the structure to remain open to traffic despite the low condition rating, based on visual inspections.

The relative difficulty in attaching resistance strain gauges to the tested surface has prompted a search for more robust alternatives. Acoustic gauges can be attached to steel by magnets (Fig. 15.4*a*) or by clamps and bolts, eliminating the sensitive gluing process.

Inevitably, such gauges have a larger base, averaging measurements over a finite length. As a result, they can be regarded as "displacement" rather than "strain" gauges and used when that limitation is acceptable. Figure 15.4*b* shows a cracked anchorage monolith, monitored by gauges with a 6-in. (150-mm) base.

Tilt Meters

Tilt meters are inexpensive and reliable and yield important information. The tilt meter shown in Fig. 15.5 provided online data on the torsion of a bridge during the global stiffening operation illustrated in Fig. 4.23.

Fiber-Optic Sensors

Fiber-optic sensors reliably measure displacement, temperature, and pressure. Due to their exceptionally adaptable length and durability, they are frequently embedded in prestressing tendons, although other applications are also reported. Results are readily transmitted

(a)

Figure 15.4 (a) Acoustic strain gauge.

(*b*)

Figure 15.4 (*b*) Displacement gauge monitoring crack in anchorage monolith.

via a local data acquisition unit by mobile or fixed line telephone connection for remote computer monitoring.

X-ray Diffraction

X-ray diffraction is used for the determination of residual stress in steel. When the stress distribution across the section of a structural member is relatively uniform or well known, this method can estimate the stresses in a loaded structure. Realistic results were obtained in New York City at the Williamsburg Bridge anchorage eye-bars (Fig. 15.6) and on the Brooklyn Bridge Franklin Square truss (Fig. 3-5). Stresses in posttensioning tendons were measured at the La Guardia Airport runways, owned by the Port Authority of New York & New Jersey (Carfagno et al., in IABSE, 1995, pp. 201–206).

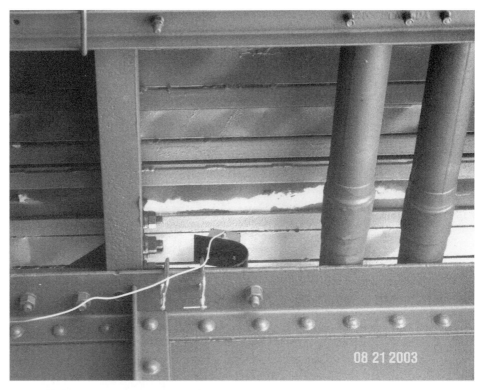

Figure 15.5 Tilt-meter monitoring bridge torsional movement.

The method can be used to compare the stress levels in structural members intended to carry equal loads, for instance, two legs of a bridge pier tower. A reliable calibration of a stress-free sample, comparable to the in situ tested material, is highly desirable but not always possible. A good knowledge of the elastic and plastic stress–strain relationships in steel (Noyan and Cohen, 1987) is indispensable.

Acoustic Emission (AE)

Acoustic emission (AE) is used to detect fatigue or corrosion–fatigue cracks in steel bridge members. The State of Virginia Transportation Council reported results of acoustic monitoring of steel bridge members (Lozev et al., 1997). Once crack locations are known, AE can possibly monitor their propagation. Main difficulties arise from two sources:

Vehicular traffic cannot generate reliably meaningful response from the structure. Train traffic has proven more effective.

At the level of the significant acoustic response noise is hard to filter.

Monitoring of wire breaks is widely used for prestressed and cable-supported structures. The process can be alternatively conducted online or triggered by the monitored events.

Figure 15.6   X-Ray diffraction test on suspension bridge anchor eye-bar.

Acoustic detection of corrosion in inaccessible high-strength wires has been attempted.

Ultrasonic Testing

Ultrasonic waves can detect nonuniformities in metals, for instance, in nickel alloy steel pins (diameter 0.4 m, length 2.0 m approximately) with exposed face sections. The challenge is to distinguish cracks from surface roughness. The California DOT reported self-compensating techniques for improved results. The method was used at the Williamsburg Bridge in the mid-1990s for evaluating the condition of nickel alloy steel pins as part of the structural rehabilitation. A calibration allowing for visual verification of the sonograms was demonstrated on test pins. The results were found satisfactory. A statistically significant sample of the pins and eye-bars on the Queensboro Bridge (Figs. E1.2 and E6.2) in New York City was inspected by that method with an accelerated scanning.

Corrosion Sensor

An early corrosion sensor, developed at the ATLSS Center, Lehigh University, consists of a sample probe implanted in normally inaccessible areas of a structure, for instance, under the wrapping of the suspension bridge cable, where the level of corrosion is of interest. An estimate can be made of the ambient corrosion by measuring the resistance of the probe to direct current. Corrosion sensors of this or other types would be highly desirable implants under the protective wrappings or the grouting of cable-supported high-strength steel tension members. The FHWA is conducting research on the subject.

Half-Cell Potential

Potential readings were used to assess the corrosion of deck-reinforcing bars on several bridges prior to their demolition. When adequately applied, the method yielded accurate results. Currently similar techniques are used to evaluate the effect of deck-waterproofing membranes. One reservation in applying this method is that corrosion in rebars embedded in concrete decks is usually localized in certain areas, while other areas remain free of it. The erratic nature of the results can (sometimes unduly) cast doubt on the procedure.

Dynamic Signature Tests

Dynamic signature investigations measure structural natural frequencies and compare them to theoretical values or earlier measurements. The discrepancies between measured and analytically obtained values can indicate modeling errors or structural damage. The method has had some success in components with strongly pronounced lowest vibration modes, such as the vertical suspenders of the George Washington Bridge where tests were conducted by the Port Authority of New York & New Jersey and Columbia University in the 1980s.

The dominant natural frequencies of bridges are of great interest during analyses of seismic performance. Response to ambient excitations is recorded by a number of transducers located on the structures and synchronized by satellite. Natural modes and frequencies are thus provided for the dynamic analysis for seismic, wind, and pedestrian load response (see Example 25). It is noted that computerized structural modeling tends to underestimate the stiffness of large truss structures. One possible explanation is that a number of rotation and translation release devices in the investigated structures act rigidly under the relatively small traffic excitations.

Wenzel and Pichler (2005) presented a comprehensive up-to-the-moment treatment of ambient vibration monitoring.

Ultraviolet Rays for Detection of Alkali–Silica Gel in Concrete

Guthroe and Carey (in TRR 1668, 1999, pp. 68–71) described a geochemical method for identifying alkali–silica reaction gel.

Alkali–silica reaction in concrete can be quickly detected by applying 5% uranil acetate solution to a sample. Alkali–silica gel exhibits a yellow pigmentation under ultraviolet light. The method is not quantitative but helps to determine if alkali–silica reaction is the reason for spalling, for instance, in a concrete deck or pier.

Infrared Thermography

Thermal methods can measure material characteristics and imperfections. Temperature is measured by thermometers. Emission is measured by infrared radiation detectors. The heat flux can be determined by measuring changes of temperature with respect to time. Readings of radiation or temperature from many points can constitute a thermal image. Thermal imaging systems are mounted on vehicles and used to inspect roadway surfaces for delaminations.

Duke and Warfield (in TRR 1347, 1992) evaluated the use of infrared thermography for detecting delaminations in reinforced-concrete bridge substructure elements. They pointed out that interpretation of the images is critical.

Magnetic Flux Leakage

The magnetic flux leakage (MFL) method measures changes in an induced magnetic field, for instance, in the proximity of otherwise inaccessible prestressing tendons. The method was reviewed at the FHWA Fairbank Validation Center. Field tests suggest considerable potential, for instance, in investigating painted steel members deformed by impact. The Honshu-Shikoku Bridge Authority (HSBA, 2002) successfully inspected suspender ropes at the Inoshima Bridge by this method.

The combination of different methods is particularly promising. Magnetostriction is used with acoustic emission; GPSs are used jointly with strain gauges and fiber-optic sensors.

15.2 STRUCTURAL HEALTH MONITORING

As NDE technologies advance, the lists of available capabilities become obsolete within months. The on line and intermittent surveillance of environmental and structural conditions have evolved into SHM (Aktan et al., 2001), changing the relationship between structural design and assessment. The *International Journal on Structural Health Monitoring,* published quarterly since July 2002, and *Structure and Infrastructure Engineering* (since March 2005) are among many informative professional periodicals on the subject. The annual conferences held by the SPIE and the International Conferences on Structural Health Monitoring and Intelligent Infrastructure (the first in Tokyo, November 2003, the second in Shanghai, 2005) are representative of the intensified exchange of information.

The tools of health monitoring are consistent with the *predictive maintenance* of mechanical and electronic equipment, as formulated by Mobley (1990). In that formulation maintenance can be cost ineffective, unless it is based on quantified models of structural response. The tactic of predictive maintenance has limited application to bridges and their networks, where components are not as readily replaceable.

So far, most successful NDT&E applications have either maintained a bridge in operation, thus benefiting the traveling public and the local economy to an inestimable degree, or avoided potentially fatal accidents by identifying unsafe structures (another incalculable cost). This mode of use pertains during the "liability" stage of the structural life cycle. At that point NDT&E is often expected to determine levels of potential hazard. While that purpose may indeed be served, it must be approached with caution. NDT&E methods have occasionally indicated no significant structural distress, but existing codes and safety considerations have overruled in favor of emergency remedial work (as in the support arch shown in Fig. 3.5). Surveying alone, for instance, does not fully guarantee the safety of a deteriorating brick masonry arch. The absence of AE signals is not a conclusive proof that wires are not breaking in a suspension cable.

At the "asset" stage (Fig. 4.4*b*), NDT&E can extend the bridge life by recommending optimal maintenance levels. Monitoring structural parameters for serviceability rather than "health" also obtains life-cycle cost benefits. At many bridges in Japan (Aktan et al., 2001), the Tsing-Ma in Hong Kong (Fig. E2.4) and the future Woodrow Wilson in Washington, D.C. (Aktan et al., 2001), health-monitoring systems are designed along with the structure. Interest is growing in monitoring of maintenance effectiveness. Certain maintenance tasks depend on advanced sensor systems. Anti-icing systems can be equipped with weather-dependent control.

NDT&E is also incorporated at the "project" stage, while the bridge is in construction and design. A design alternative can become feasible only because of advanced health-monitoring capabilities but should be evaluated on the merit of its lifecycle cost/ benefits. NDT&E must always be preceded by analysis. Anticipated results and the appropriate response actions must be defined. Data will invariably be wasted, unless it is part of a systematic structural evaluation plan. The independent monitoring of different structural response parameters, such as strain and acoustic emission, strain and displacement, and acceleration and velocity, provides a welcome redundancy. Approval of any proposed technique by ASTM, ASCE, and AASHTO is essential.

The inspections described herein, and the NDT&E technologies they can employ, are primarily oriented toward steel and concrete bridges. As carbon, glass, and other materials gain application, management will have to adapt to their needs.

The validation and calibration of NDT&E techniques for use, compatible with design specifications, are the current topics of greatest interest to bridge managers and NDE experts. At meetings in 2004 (Ansari, 2005) and 2005 (Juhn et al. 2005), the International Society for Health Monitoring of Intelligent Infrastructure (ISHMII), the Validation Center at FHWA, the National Science Foundation (NSF) , SPIE, and the Asian-Pacific Network of Centers for Research in Smart Structures Technology (ANCRiSST), among others, undertook to improve the correlation of NDT&E capabilities and the needs of project-level users.

Smart Materials, Structures, and Systems
S. C. Liu (Yun and Spencer, 2005, pp. 3–11) defined the terms as follows:

> Smart materials have multiple functionalities mimicking sensory nerve systems or actuating systems of human or other bio-living bodies.

> Smart structures and systems (SSS) should be capable of adapting to changing requirements for functionality and external excitations, be able to perform self-calibration. self-diagnosis, self-control and self-healing, and ultimately achieve minimum life-cycle cost.

The intended smart materials, structures and systems present new demands on management expertise and intelligence. Kong et al. (Liu and Spencer, 2005, pp. 837–848) identify the following familiar key elements of a smart bridge management system (SBMS):

> multiple health indices system, integrating assessments based on discrete health indexing of various resolutions;

quantification of material and structural degradation as a function of various effects, including maintenance;

cost–maintenance interaction;

calibration of health condition based on inspection and monitoring results, with continuous upgrading.

Example 28 illustrates a current effort to correlate the needs for and the availability of NDT&E options.

Example 28. Needs for and Capabilities of Nondestructive Testing and Evaluation (NDT&E)

Yanev (in Juhn et al. 2005, pp. 29–43) reported on an effort by the International Society for Health Monitoring of Intelligent Infrastructures (ISHMII), FHWA, and NSF to correlate NDT&E capabilities with the demands of bridge managers. The objective is to establish a correspondence between significant events, measurable effects, and appropriate techniques, as shown in Fig. E28.1.

The task is envisioned as a continuing open-ended compilation of a NDT&E applications database. Achieving redundant monitoring by different methods or techniques is highly desirable. Check lists such as the ones shown in Tables E28.1–E28.3 can help summarize the demand for and supply of monitoring capabilities.

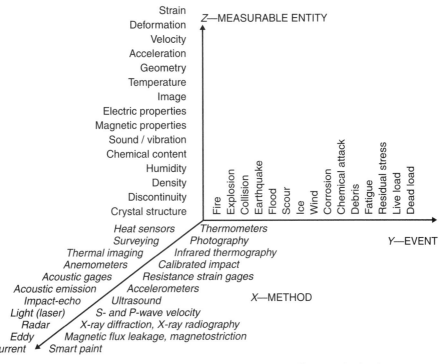

Figure E28.1 Monitoring methods (techniques), measurable effects, and related events.

Table E28.1 NDT&E Methods and Measurable Physical or Chemical Entities

Method/Technique	Strain	Deformation	Velocity	Acceleration	Temperature	Geometry	Image	Electric Properties	Sound/Vibration	Chemical Content, Humidity	Magnetic Properties	Density	Discontinuities	Crystal Structure
Thermometers					Y									
Heat sensors					Y							Y	Y	
Photography		Y				Y	Y							
Surveying		Y				Y								
Infrared thermography	Y	Y		Y			Y							
Thermal imaging	Y				Y	Y								
Calibrated impact	Y	Y		Y	Y	Y						Y		
Anemometers			Y		Y				Y					
Resistance strain gauges	Y	Y	Y											
Acoustic strain gauges	Y	Y												
Accelerometer	Y			Y										
Acoustic emission	Y			Y	Y				Y					
Ultrasound						Y	Y						Y	
Wave velocity (S&P)													Y	
Impact echo										Y		Y		
Light (laser)						Y	Y							
Radar														
X-ray diffraction	Y													Y
X-ray radiography													Y	
Magnetic flux leakage, magnetostriction											Y		Y	
Eddy current											Y		Y	
Smart paint		Y											Y	

Table E28.2 Bridge-Related Events and Measurable Entities

Event	Strain	Deformation	Velocity	Acceleration	Temperature	Geometry	Image	Electric Potential	Sound/Vibration	Chemical Content Humidity	Magnetic Properties	Density	Discontinuities	Crystal Structure
Fire	Y	Y		Y	Y		Y		Y					
Explosion	Y	Y	Y	Y			Y							
Collision	Y	Y				Y								
Earthquake	Y	Y		Y			Y							
Scour	Y	Y	Y	Y	Y	Y	Y		Y			Y	Y	Y
Live loads	Y	Y		Y	Y	Y	Y							
Wind	Y	Y		Y										
Corrosion								Y	Y	Y	Y			Y
Fatigue	Y			Y					Y		Y			Y
Dead load	Y			Y										
Residual stress														Y
Void and crack												Y	Y	Y
Debris		Y			Y	Y	Y					Y		
Ice					Y					Y				
Chemical attack								Y		Y	Y	Y		Y

Measurable Entities

501

Table E28.3   NDT&E Methods and Bridge-Related Events

Method/Technique	Fire	Explosion	Collision	Earthquake	Flood	Scour	Live Loads	Wind	Corrosion	Fatigue	Dead Load	Residual Stress	Void & Crack	Debris	Ice	Chemical Attack
Thermometers	Y	Y													Y	Y
Heat sensors	Y	Y													Y	Y
Photography	Y	Y					Y							Y		
Surveying			Y			Y										
Infrared thermography													Y			
Thermal imaging										Y			Y			
Calibrated impact													Y			
Anemometers								Y								
Resistance strain gauges			Y	Y		Y	Y	Y		Y						
Acoustic strain gauges			Y	Y		Y	Y	Y								
Accelerometer		Y	Y	Y				Y								
Acoustic emission									Y	Y			Y			
Ultrasound						Y										
Impact echo													Y			
Wave velocity (S&P)													Y			
Light (laser)																
Radar																
X-ray diffraction												Y				
X-ray radiography													Y			
Magnetic flux leakage, magnetostriction										Y			Y			
Eddy current										Y			Y			
Smart paint							Y			Y			Y			

The tables represent works in progress. "Y" indicates a potential match.

The independent input of technology developers, researchers, and potential users of NDT&E can be matched. Related entries along the three axes of Fig. E28.1 can be grouped and addressed by the same methods. For example, accelerations and velocities are relevant to some degree to earthquakes, collisions, explosions, winds, floods, and scour.

15.3 EXPERT SYSTEMS

Expert systems (ESs) "narrow and prioritize the inspection terms" in order to "improve the reliability and efficiency of bridge inspection" (Mizuno et al., p. 112, in Miyamoto and Frangopol, 2001); see Chapter 9 and Appendix 47. Also known as knowledge-based systems, they update a priori semideterministic deterioration models and, consequently, are *dynamic*. New advances in this field are following in the steps of earlier developments in computerized structural analysis. While the traditional "stiffness analysis" of elastic structures was facilitated to the point that technicians can input data of the geometry, designing the problems and interpreting the results remain highly specialized and sensitive engineering activities.

Expert systems must solve the inverse problem of quantifying the parameters of a known model. Redesigning the model, rather than optimizing its parameters, somewhat strains the capabilities of system identification methods. Bridge management ESs must still rely on experts to determine the adequacy of adopted models. Expert systems reach beyond structural condition assessment and extend to bridge management in general. Appendices 42 and 43 outline general aspects of ESs stemming from the evolution of artificial intelligence (AI).

15.4 INTELLIGENT TRANSPORTATION SYSTEMS

Similarly to HMSs (Section 15.2), ITSs are the result of technological advances that allow improved information acquisition and processing. At an international workshop in New York City in 1996 presentations on the subject were made by transportation managers of Berlin, Boston, Brussels, Chicago, Houston, London, Los Angeles, Madrid, New York, Paris, Philadelphia, Rome, and Toronto. The U.S. national ITS program has the following objectives:

To improve the safety of the surface transportation system

To increase the operational efficiency and capacity of the surface transportation system

To reduce energy and environmental costs associated with traffic congestion

To enhance present and future productivity

To enhance the personal mobility, convenience, and comfort of the surface transportation system

The serviceable condition of the transportation infrastructure network is clearly instrumental to the feasibility of any ITS application. Consequently ITSs and BMSs must be linked.

15.5 BMS MANAGEMENT

A defining decision related to a BMS is whether to develop the system locally or to "customize" an available (off-the-shelf) one. A custom-made BMS can be developed in-house or under contract. Several decades of experience have taught bridge managers to explore both options and their combinations.

Developing a custom-made BMS is a long-term commitment involving initial investment, life-cycle maintenance costs, and state-of-the-art expertise. Information management systems (MISs) require periodic upgrades. Mittra (1988) suggested 3-year intervals (as is the case for design specification updates). Larger system overhauls entail the hazards described in Section 4.1.7.

Modular design with generally compatible software allows owners to enhance their operating systems with new packages. Shepard (2005) reported the successful implementation of PONTIS at the California DOT by this approach. The software and hardware systems benefit from a relative independence. (The present review similarly contains its topics in independent sections, so that each of them can grow out of date at its own pace.)

The introduction of portable computers to bridge and roadway inspections exemplifies the difficulties of development. In the early 1990s a number of transportation agencies developed software and purchased portable computers. The primary incentive was to

(a)

Figure 15.7 Lap-top, touch-screen, and office computers with inspection report software.

facilitate inspection report writing and transmittal. In that spirit, the software was designed to replicate existing inspection forms. The hardware was selected primarily for durability and simplicity. During the ensuing implementation, inspectors found the screen-type hand-held computers awkward and illegible, particularly under the glare of sunlight. The process did not accelerate overnight. Keyboards were not entirely avoided in the field, as handwriting recognition proved unreliable. The experience has demonstrated that development and implementation expenditures are inevitable, regardless of the selected hardware and software.

Over the following decade, software and hardware were repeatedly updated. Paper forms and film processing became obsolete. At this writing, inspection reports are generated on laptop computers (usually stationed in the inspector's vehicles) and transmitted by e-mail or on CD-ROMs. Emergency reports can be transmitted by telephone. Portable

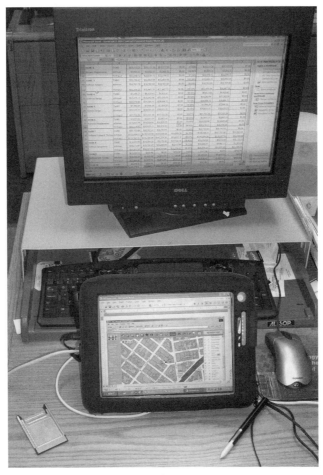

(b)

Figure 15.7 (*Continued*)

(hand-held) computers became "ruggedized," incorporated digital photography and voice recognition (a still-underutilized development). Off-the-shelf and custom-made software is becoming increasingly available and accessible.

The competition between laptop (Fig. 15.7) and portable computers (Fig. 15.8) for field data acquisition and report generation has resulted in their complementary use. This may be viewed initially as duplication but is liable to become a standard redundancy as equipment becomes more readily available.

Developing a customized management system can be justified by the size and the uniqueness of the bridge stock. With some adjustments, an off-the-shelf product may be the optimal choice for smaller networks. Many intermediate options exist. Example E27 shows a bridge inspection data management system (BIMS) of a local owner supplementing state-managed federally funded biennial inspections.

The owner may retain a consultant to develop the system and train in-house staff for future maintenance. Alternatively, the contract may stipulate a maintenance responsibility until the anticipated upgrade. Finally, a major bridge owner can maintain an in-house fully staffed computer center. In that case, the weakest and therefore most maintenance-intensive link is not the software or the hardware but the support staff.

As in all other operations, duplication can improve reliability but at a higher cost. The owners of tolled facilities, who enjoy dedicated funds, occasionally duplicate tasks. Consultants perform work, which is compared to results obtained in-house. A BMS can be effectively developed and maintained by a consultant in collaboration with the owner's bridge management unit when both are independently qualified to complete the project (as observed by Feynman at NASA; see Section 13.4). Eliminating duplication of effort, on the other hand, is the favorite priority of fiscally responsible management (Section 4.1). Consequently, budget requests for BMS maintenance and development must be justified by detailed and soundly substantiated life-cycle cost–benefit analyses.

Figure 15.8 Hand-held devices for digital data input, voice and photo options.

Chapter **16**

Conclusion

A conclusion might be appropriate for a well-organized text but not for its subject. Consequently, in a risky disregard for Pascal's advice to begin a book with the last thought to occur to the writer, this one ends where the logic and the purpose formulated in the introduction were able to take it. An open-ended treatment leading from fundamental principles and historic background to most recent applications and unresolved questions seems appropriate for topics that have been under development for more than 2500 years, with no final pronouncement either envisioned or desired. As presented, the material might prove instructive to users with diverse interests while avoiding a falsely definitive tone.

Engineering texts demonstrate predetermined conclusions concisely. The best literature surprises readers and authors alike. As a result, the two types of writing are mutually exclusive, the few exceptional borderline examples notwithstanding. The management of civil infrastructure contains a similar dichotomy. The established practices rely on formal methods, systematic procedures, and rigorous solutions, all of which the best professionals aim to exceed. Engineers and managers can neither be confined by a rigidly defined process nor indulge in unconstrained improvisation. Just as engineering draws on theory and empiricism, management combines the opposed approaches of centralized optimization and decentralized iterative adjustments. Theory remains above the conflict by offering only decision support. Practice constantly questions the value of that support.

The reviewed ideas and practices may suggest with some justification that everything has been said and revised repeatedly. By the last count, the reference list contains 516 titles. Many are discussed in the 53 appendixes and 30 examples. None should be adopted without verification or evaluation. Taking, as much as possible, into account the relevant precedents, engineers and managers, particularly those who aspire to be both, must formulate and solve their problems, as have done their predecessors. The quality of the decision support they can expect will depend on the quality of the questions they are prepared to address.

The commitment to rationalized reasoning has established Aristotle and Descartes as the precursors of optimized decision making. Descartes described his motivation as follows (1976, p. 174): "I am convinced that if all the truths that I have searched by demonstrations had been taught to me during my youth, and if I had had no difficulty learning them, I would have perhaps known no others, and even less would I have acquired the habit and the facility, which I think I have, of finding new ones, when I apply myself to the search."

For those who have similar ambitions, Descartes drew examples from infrastructure management (1976, pp. 102–103):

So it is that the buildings designed and achieved by a single architect are usually more beautiful and better organized than those readjusted by many, using old structures, built for other purposes. Thus the ancient towns, which begin as mere settlements, and become, in the course of time, great cities, are typically so poorly organized, compared to the regular fortresses that an engineer plans according to his vision. One often finds as much or more art in the individual edifices, than in their arrangement, here a big one, there a small one, and with streets so winding and uneven, that one might say they have been designed by chance, rather than by the will of men using reason. And if one considers that at all times certain officials have taken care of particular buildings, so that they could

Figure 16.1 Holding the bridge by reins and riding it into the city . . .

serve the public, one realizes that it is hard to produce fine accomplishments by working on the structures of others.

Since the writing of this Discourse, existing designs and methods have been both refined and abandoned in favor of new and older ones. The life of cities, built according to grand visions, has survived only in accidental spontaneous developments, i.e., those "designed by chance." Master plans have swept the details of old tenements while granting historic landmark status to others. Overreaching designs have been rejected by voting majorities.

Parkinson (1957, p. 60) was to write, more than half-seriously: "It is now known that a perfection of planned layout is achieved only by institutions on the point of collapse."

In the *New York Times Magazine* of November 27, 2005, page 28, C. Caldwell asked: "Were the French riots (of November, 2005) produced by Modern architecture?"

Pascal (1962, Section 619 [394]) cautioned: "All their principles are true, skeptics, stoics, atheists, etc., but their conclusions are false, because the opposite principles are also true."

Descartes conceded (1976, p. 105): "It is true that we do not envision tearing down all the town houses for the sole purpose of redoing them in a different way and rendering the streets more beautiful."

In retrospect, most revealing is the opinion of Leibnitz that analytic geometry owes more to the genius than to the method of Descartes. Recent developments in parallel

Figure 16.2   . . . And into the future.

computational techniques offer intriguing alternatives to sequential reasoning. J. Bailey (1996, p. 220) opined, ". . . The Cartesian side of things was left with nothing to do but falsely take credit for having shown the way. . . . Everything, including thought itself is up for reconsideration."

Operating in the uncertain margin between genius and formalized decision support, managers must set their objectives according to their vision, seek out and create the tools appropriate for their reasoning. The present text is an example of one among many possible such exercises. In an interview with the *Wall Street Journal* (November 6, 1991, p. 1) this writer compared descending on the cables of an East River suspension bridge to holding the structure by the reins and riding it into the city (Fig. 16.1). Obtaining and applying the resources that will take the bridges and their users into a designed but nonetheless uncertain future are even more exhilarating (Fig. 16.2).

References

AASHTO (1983). *Manual for Maintenance Inspection of Bridges,* including revisions from *Interim Specifications for Bridges,* 1984, 1985, 1986, 1987–1988, 1989, 1990, American Association of State Highway and Transportation Officials, Washington, DC.

AASHTO (1987). *Manual for Bridge Maintenance,* American Association of State Highway and Transportation Officials, Washington, DC.

AASHTO (1988). *Standard Specifications for Movable Bridges,* American Association of State Highway and Transportation Officials, Washington, DC.

AASHTO (1989). *Guide Specifications for Strength Evaluation of Existing Steel and Concrete Bridges,* American Association of State Highway and Transportation Officials, Washington, DC.

AASHTO (1996). *Quality Assurance Guide Specification,* AASHTO Quality Assurance Task Force, American Association of State Highway and Transportation Officials, Washington, DC.

AASHTO (1998a). *Load and Resistance Factor Bridge Construction Specifications* (LRFD), 2nd ed., American Association of State Highway and Transportation Officials, Washington, DC.

AASHTO (1998b, Interim 2002). *Guide for Commonly Recognized (CoRe) Structural Elements,* American Association of State Highway and Transportation Officials, Washington, DC.

AASHTO (1998c). *Maintenance Manual—1998,* American Association of State Highway and Transportation Officials, Washington, DC.

AASHTO (1998d). *Movable Bridge Inspection, Evaluation and Maintenance Manual,* American Association of State Highway and Transportation Officials, Washington, DC.

AASHTO (1999a). *The Maintenance and Management of Roadways and Bridges,* American Association of State Highway and Transportation Officials, Washington, DC.

AASHTO (1999b, 2000a). *Interim Guide Specifications for Seismic Isolation Design,* 2nd ed., American Association of State Highway and Transportation Officials, Washington, DC.

AASHTO (2000b). *Manual for Condition Evaluation of Highway Bridges,* 2nd ed., American Association of State Highway and Transportation Officials, Washington, DC.

AASHTO (2001). *Pavement Preventive Maintenance Guidelines,* American Association of State Highway and Transportation Officials, Washington, DC.

AASHTO (2002). *Standard Specifications for Highway Bridges,* 17th ed., American Association of State Highway and Transportation Officials, Washington, DC.

AASHTO (2003). *Guide Manual for Condition Evaluation and Load and Resistance Factor Rating of Highway Bridges,* 3rd ed., American Association of State Highway and Transportation Officials, Washington, DC.

AASHTO (2004). *Load and Resistance Factor Bridge Design Specifications (LRFD),* 3rd ed., American Association of State Highway and Transportation Officials, Washington, DC.

ACI, American Concrete Institute (1992). *Manual of Concrete Inspection,* Publication SP-2(92), ACI.

511

Adams, T. M., and P. R. M. Sianipar (1995). *Project and Network Level Bridge Management, Proceedings of the Transportation Congress,* San Diego, CA, October 22–26, 1995, pp. 1667–1681.

Adams, L. H., F. D. Harrison, and A. Vandervalk (2005). Issues and Challenges in Using Existing Data and Tools for Performance Measurement, *TRB Conference Proceedings 36: Performance Measures to Improve Transportation Systems,* Transportation Research Board, Washington, D.C.

Adeli, H., Ed. (1988). *Expert Systems in Construction and Structural Engineering,* Chapman & Hall, London and New York.

Adey, B. T., and R. Hajdan (2005). Potential Use of Inventory Theory to Bundle Interventions in Bridge Management Systems, Paper 05-0185, 84th Annual Meeting, Transportation Research Board, Washington, D.C.

AFPC (1994). *Cable-Stayed and Suspension Bridges,* Proceedings Vol. I and II, Association Française pour la Construction, Bagneux, France.

Agbabian, M. S. and Masri, S. F., Eds. (1988). *Nondestructive Evaluations for Performance of Civil Structures,* University of Southern California, Los Angeles, CA.

Aktan, E., et al. (2001). Health Monitoring of Long Span Bridges, National Science Foundation Workshop, University of California, Irvine.

Alexander, M. J. (1974). *Information System Analysis,* Science Research Associates, Chicago.

Ang, A. H.-S., and D. De Leon (2005). Modeling and Analysis of Uncertainties for Risk-Informed Decisions in Infrastructure Engineering, *Structure and Infrastructure Engineering,* Vol. 1, No. 1, pp. 19–31.

Ang, A. H.-S., and W. H. Tang (1975, 1984). *Probability Concepts in Engineering Planning and Design,* Vols. I and II, Wiley, New York.

Anglo–French Liaison Report (1999). *Post-Tensioned Concrete Bridges,* Thomas Telford, London.

Ansari, F., Ed. (2005). *Sensing Issues in Civil Structural Health Monitoring,* Springer, Dordrecht, The Netherlands.

AREMA, American Railroad Engineering and Maintenance Association (2001). *Manual for Railway Engineering,* AREMA, Washington, DC.

Aristotle (1941). *The Basic Works,* Random House, New York.

Aristotle (1943). *On Man in the Universe,* L. R. Loomis, Ed., Walter J. Black, Roslyn, NY.

Armytage, W. H. G. (1976). *A Social History of Engineering,* Westview, Boulder, CO.

ASCE, American Society of Civil Engineers, New York (1990). *Quality in the Constructed Project,* Manuals and Reports on Engineering Practice 73.

ASCE (1993). J. L. Gifford, D. R. Uzarski, and S. McNeil, Eds., *Infrastructure Planning and Management, Proceedings,* June 21–23, 1993, Denver, CO.

ASCE (1997). Saito M., Ed., *Infrastructure Condition Assessment: Art, Science and Practice, Proceedings,* Boston, MA., August 25–27.

Asimov, I. (1990). *The Complete Short Stories,* Vol. 1, Doubleday, New York.

ATC (1983). Seismic Retrofitting Guidelines for Highway Bridges, Report No. FHWA/RD-83/007, FHWA Applied Technology Council, U.S. Department of Transportation, Washington, DC.

ATC (1989). *Field Manual: Postearthquake Safety Evaluation of Buildings,* Applied Technology Council, Redwood City, California.

ATLSS, Advanced Technology and Large Structural Systems (1995). *Williamsburg Bridge Orthotropic Deck Prototype,* Lehigh University Center for ATLSS, Bethlehem, PA.

Babaei, K., and N. Hawkins (1993). Bridge Retrofit Planning Program, Report No. WA-RD 217.1, Washington State DOT, Olympia, WA.

Bailey, J. (1996). *After Thought, the Computer Challenge to Human Intelligence,* Basic Books, Harper Collins Publishers, New York.

Bailey, S. F. (1996). *Basic Principles and Load Models for the Structural Safety Evaluation of Existing Bridges,* École Polytechnique Fédérale de Lausanne, Lausanne, Switzerland.

Lord Baker (1978). *Enterprise versus Bureaucracy,* Pergamon Press, Oxford.

Baker, M., et al. (2003). Bridge Software—Validation Guidelines and Examples, NCHRP Report 485, National Research Council, Transportation Research Board, Washington, DC.

Banks, J. H. (2002). *Introduction to Transportation Engineering,* McGraw-Hill, New York.

Barker, R. M., and J. A. Puckett (1997). *Highway Bridges,* Wiley, New York.

Barlow, R. E., F. Proschan, and L. C. Hunter (1965). *Mathematical Theory of Reliability,* Wiley, New York.

Barrow, J. D. (1991). *Theories of Everything,* Clarendon, Oxford.

Barzun, J. (1959). *The House of Intellect,* Harper & Brothers, New York.

Basöz, N., and A. S. Kiremidjian (1996). Risk Assessment for Highway Transportation Systems, Report No. 118, The John A. Bloom Earthquake Engineering Center, Department of Engineering, Stanford University, Stanford, CA.

Bažant, Z. P., and L. Cedolin (1991). *Stability of Structures,* Oxford University Press, New York.

Beale, E. M. L. (1988). *Introduction to Optimization,* Wiley, New York.

Bell, D. (1973). *The Coming of Post-Industrial Society,* A Venture in Social Forecasting, Basic Books, New York.

Berlin, I. (1996). *The Sense of Reality, Studies in Ideas and their History,* Farrar, Straus and Giroux, New York.

Bieniek, M., et al. (1989). Preventive Maintenance Management System for the New York City Bridges, Report of a Consortium of Civil Engineering Departments of New York City Colleges and Universities, the Center of Infrastructure Studies, Columbia University, New York.

Billington, D. P. (1983). *The Tower and the Bridge,* Basic Books, New York.

Bisby, L. A., and M. B. Briglio (contributor) (2004). An Introduction to Structural Health Monitoring, Canadian Network of Centres of Excellence, *ISIS Education Module Journal,* available at www.isiscanda.com.

Bjerrum, J., et al. (2006). Internet-Based Management of Major Bridges and Tunnels Using DANBRO+, *Proceedings, Conference on Operation, Maintenance and Rehabilitation of Large Infrastructure Projects, Bridges and Tunnels,* Copenhagen, May 15–17, IABSE, Zurich.

Boller, A. P. (1885). *Construction of Iron Highway Bridges,* for the use of town committees, Wiley, New York.

Boresi, A. P. and O. M. Sidebottom (1985). *Advanced Mechanics of Materials,* Wiley, New York.

Born, M. (1968). *My Life and My Views,* Charles Scribner's Sons, New York.

Boudon, R. (1968). *A Quoi Sert la Notion de "Structure"?* Gallimard, Paris.

Bremauld, P. (1998). *Markov Chains,* Springer-Verlag, New York, Berlin and Heidelberg.

BRIME (2002). *Bridge Management in Europe,* B. Godart, Ed., Laboratoire Central des Ponts et Chaussées, Paris.

Brock, D. S., S. M. Levy, and L. L. Sutcliffe, Jr., Eds. (1986). *Field Inspection Handbook,* McGraw-Hill, New York.

Brown, S., Ed. (1995). *Forensic Engineering,* ISI Publications, Humble, TX.

Brühwiler, E., and B. Adey (2005). Improving the Consideration of Life-Cycle Costs in Bridge Decision-Making in Switzerland, *Structure and Infrastructure Engineering,* Vol. 1, No. 2, pp. 145–157.

Bucher, C. and H. A. Pham (2005). On Model Updating of Existing Structures Utilizing Measured Dynamic Responses, *Structure and Infrastructure Engineering,* Vol. 1, No. 2, pp. 135–143.

Burati, J. L., W. C. Bridges, and S. A. Ackerman (1995). Evaluation of Quality Assurance Programs for Bituminous Paving Mixtures, Final Project Report, FHWA-SC-95-02, Clemson University, Clemson, SC.

Burdekin, F. M., et al. (1991). Non-Destructive Methods for Field Inspection of Embedded or Encased High Strength Steel Rods and Cables, CAPCIS, NCHRP 10-30 (3), University of Manchester, United Kingdom.

Burke, M. P. (1993). Integral Bridges: Attributes and Limitations, Paper No. 930104, 72nd Annual Meeting, Transportation Research Board, Washington, DC.

Burke, M. P. (1994). Semi-Integral Bridges: Movements and Forces, Paper No. 940051, 73rd Annual Meeting, Transportation Research Board, Washington, DC.

Burke, M. P. (2000). Warning! LRFD May Be Hazardous to Your Bridge's Health, 17th Annual International Bridge Conference, Engineers' Society of Western Pennsylvania, Pittsburgh, PA.

Busa, G. D., et al. (1985). A National Bridge Deterioration Model, Report No. SS-42-U5-26, U.S. Department of Transportation Research and Special Programs Administration, Transportation Systems Center, Cambridge, MA.

Bush, J. and G. Y. Baladi (1989). *Nondestructive Testing of Pavements and Backcalculation of Moduli,* ASTM 89-38726, American Society for Testing and Materials, New York.

Button, M. R., J. C. Colman, and R. L. Mayes (1999). Effect of Vertical Ground Motions on the Structural Response of Highway Bridges, Technical Report MCEER-99-0007, Multidisciplinary Center for Earthquake Engineering Research, State University of New York, Buffalo, NY.

Cady, P. D. and R. E. Weyers (1984). Deterioration Rates of Concrete Bridge Decks, *Journal of Transportation Engineering,* Vol. 110, No. 1, pp. 34–44.

Calgaro, J. A. (2004). L'Ingenieur et le Gestionnaire, editorial, *Le Pont,* No. 18, December 2004, Toulouse, France.

Calgaro, J. A., and R. Lacroix, Eds. (1997). *Maintenance et Reparation des Ponts,* Presses de l'Ecole National des Ponts et Chaussées, Paris.

Calgaro, J. A., and M. Virlogeux (1988). *Analyse Sructurale des Tabliers de Ponts,* Presses Ponts et Chaussées, Paris.

Camo, S. (2004, January). The Evolution of a Design, *Structural Engineer,* Vol. 4, No. 1, pp. 32–37.

Caro, R. A. (1974). *The Power Broker,* Alfred A. Knopf, New York.

Casti, J. L. (1990). *Searching for Certainty,* William Morrow and Co., New York.

CBO, Congressional Budget Office (1991). *How Federal Spending for Infrastructure and Other Public Investments Affects the Economy,* U.S. Government Printing Office, Washington, DC.

Chang, C. M. (2005). *Engineering Management,* Challenges in the New Millennium, Prentice-Hall, Upper Saddle River, NJ.

Chang, S. E., et al. (1996). Estimation of the Economic Impact of Multiple Lifeline Disruption, Technical Report NCEER-96-0011, NCEER, State University of New York, Buffalo, NY.

Charnes, A., and W. W. Cooper (1959). Chance-Constrained Programming, *Management Science,* Vol. 5, p. 73.

Chen, W. F., and L. Duan, Eds. (1999). *Bridge Engineering Handbook,* CRC Press, Boca Raton, FL.

Chen, W. F., and E. M. Lui (1987). *Structural Stability,* Prentice-Hall, Englewood Cliffs, NJ.

Chopra, A. (1995). *Dynamics of Structures,* Prentice-Hall, Englewood Cliffs, NJ.

Cleland, D. I., and D. F. Kocaoglu (1981). *Engineering Management,* McGraw-Hill, New York.

Clifton, J. R. (1991). Predicting the Remaining Life of Concrete, NISTRIP 4712, National Institute of Standards and Technology, U.S. Department of Commerce, Washington, DC.

Clough, R. W., and J. Penzien (1993). *Dynamics of Structures,* 2nd ed., McGraw-Hill, New York.

Clough, R. H. (1986). *Construction Contracting,* Wiley, New York.

Clough, R. H., et al. (2000). *Construction Project Management,* Wiley, New York.

Collacott, R. A. (1985). *Structural Integrity Monitoring,* Chapman & Hall, London.

Commission for the Investigation of the Schoharie Bridge Collapse (1987). Final Report, New York State Department of Transportation, Albany.

Considere, A. (1903). *Experimental Researches on Reinforced Concrete,* McGraw Publishing, New York.

Cooke, B., and P. Williams (2004). *Construction Planning, Programming & Control,* Blackwell Publishing, Oxford.

Cremona, C., Ed. (2003). *Application des Notions de Fiabilité à la Gestion des Ouvrages Existants,* Presses de l'École Nationale de Ponts et Chaussées, Paris.

Dallard, P., et al. (2001). London Millennium Bridge: Pedestrian-Induced Lateral Vibration, *ASCE Journal of Bridge Engineering,* Vol. 6, No. 6, pp. 412–417.

Daniels, H., W. Kim, and J. L. Wilson (1989). Recommended Guidelines for Redundancy Design and Rating of Two-Girder Steel Bridges, Report No. 319, National Research Council, Transportation Research Board, Washington, DC.

Das, P., D. M. Frangopol, and A. S. Nowak (1999). *Current and Future Trends in Bridge Design, Construction and Maintenance,* Thomas Telford, London.

Date, C. J. (1973, 1983). *An Introduction to Database Systems,* Vols. I and II, Addison-Wesley, Reading, MA.

De Finetti, B. (1974). *Theory of Probability,* Vol. 1., Wiley, New York.

De Gramo, E. P., J. R. Canada, and W. G. Sullivan (1979). *Engineering Economy,* Macmillan, New York.

Den Hartog, J. P. (1949). *Strength of Materials,* McGraw-Hill, New York.

Descartes, R. (1976). *Discours de la Methode,* Livres de Poche, Paris.

Dewey, J. (1929). *The Quest for Certainty, a Study of the Relation of Knowledge and Action,* Minton, Balch & Co., New York.

Diwekar, U. (2003). *Introduction to Applied Optimization,* Kluwer Academic, Norwell, MA.

Dowrick, D. J. (1977). *Earthquake Resistant Design,* Wiley, New York.

Drdácký, M. (1992). editor, Lessons from Structural Failures, Aristocrat, Telč.

Drucker, P. F. (1954). *The Practice of Management,* Harper & Row, New York.

Drucker, P. F. (1968). *The Age of Discontinuity,* Harper & Row, New York.

Drucker, D. (1973). *Management,* Harper & Row, New York.

Drucker, P. F. (1995). *Managing in a Time of Great Change,* Truman Talley/Dutton, New York.

Duke, J. C., and S. C. Warfield (1992). Evaluation of Infrared Thermography as a Means of Detecting Delaminations in Reinforced Concrete Bridge Substructure Elements, Transportation Research Record No. 1347. National Academy Press, Washington, D.C.

Eary, D. F., and G. E. Johnson (1962). *Process Engineering for Manufacturing,* Prentice-Hall, Englewood Cliffs, NJ.

Einstein, A. (1921). *Geometry and Experience,* lecture before the Prussian Academy of Science, January 27, 1921, Springer, Berlin.

Einstein, A. (1941). The Common Language of Science, *Advancement of Science,* Vol. 2, No. 5.

Einstein, A. (1950). "On the General Theory of Gravitation," *Scientific American,* Vol. 182, No. 4.

Einstein, A., and L. Infeld (1942). *The Evolution of Physics,* Simon & Schuster, New York.

Emmons, P. H. (1993). *Concrete Repairs and Maintenance Illustrated,* R. S. Means, Kingston, MA.

ENPC Presses de l'École National des Ponts et Chaussées (1994). *Maintenance of Bridges and Civil Structures,* ENPC, Paris.

ENR (May 24, 2004). Team Looking at Temporary Bracing for Cause of Collapse, *Engineering News Record,* Vol. 252, No. 21, McGraw-Hill Construction, New York.

ENR (May 31, 2004). Focus on Construction of Columns at Airport, *Engineering News Record,* Vol. 252, No. 22, McGraw-Hill Construction, New York.

ENR (July 12, 2004). Questions Raised over Steel Roof Structures, *Engineering News Record,* Vol. 253, No. 2, p. 10, McGraw-Hill Construction, New York.

ENR (Sept. 27, 2004). Planned, Collapsed Terminals Featured at Peer Review, *Engineering News Record,* Vol. 253, No. 12, p. 16, McGraw-Hill Construction, New York.

Federal Infrastructure Safety Program (1994). Infrastructure in the 21st Century Economy, Interim Report Vol. 2, Institute for Water Resources, U.S. Army Corps of Engineers, Alexandria, VA.

Feld, J. (1968). *Construction Failure,* John Wiley & Sons, Inc., New York.

FEMA (Federal Emergency Management Agency) 403 (2002). *World Trade Center Building Performance Study,* ASCE, Reston, VA.

Feynman, R. (1998). *The Meaning of It All,* Perseus Books, Reading, MA.

Feynman, R. (1999). *The Pleasure of Finding Things Out,* Perseus Books, Cambridge.

FHWA (1979). *Bridge Inspector's Training Manual 70,* Federal Highway Administration, U.S. Department of Transportation, Washington, DC.

FHWA (1986). *Inspection of Fracture Critical Bridge Members,* FHWA-IP-86-26, Federal Highway Administration, U.S. Department of Transportation, Washington, DC.

FHWA (1987). *The Pennsylvania Bridge Management System,* FHWA-PA-86-036-84-28A, February, 1987, Harrisburg, PA, Federal Highway Administration, U.S. Department of Transportation, Washington, DC.

FHWA (1991). *Bridge Inspector's Training Manual/90,* R. A. Hartle et al., FHWA-PD-91-015, Federal Highway Administration, U.S. Department of Transportation, Washington, DC.

FHWA (1994). *Life Cycle Cost Analysis,* A Policy Discussion Series, No. 12, Nov. 1994, Federal Highway Administration, U.S. Department of Transportation, Washington, DC.

FHWA (1995a). *Seismic Retrofitting Manual for Highway Bridges,* I. G. Buckle and I. M. Friedland, Eds., FHWA-RD-94-052, NCEER, CUNY, Buffalo, Federal Highway Administration, U.S. Department of Transportation, Washington, DC.

FHWA (1971, 1988, 1995b) *Recording and Coding Guide for the Structure Inventory and Appraisal of the Nation's Bridges,* FHWA-PD-96-001, Federal Highway Administration, U.S. Department of Transportation, Washington, DC.

FHWA (1997). *Federal-Aid Policy Guide, Part 625—Design Standards for Highways,* Federal Highway Administration, U.S. Department of Transportation, Washington, DC.

FHWA (1998). *Scour Monitoring and Instrumentation,* FHWA-SA-96-036, Federal Highway Administration, U.S. Department of Transportation, Washington, DC.

FHWA (1999). *Asset Management Primer,* Federal Highway Administration, U.S. Department of Transportation, Washington, DC.

FHWA (2000). *Primer: GASB 34,* Government Accounting Standards Board's Statement 34, Office of Asset Management, Federal Highway Administration, U.S. Department of Transportation, Washington, DC.

FHWA (2001a, August). *Data Integration Primer,* Office of Asset Management, Federal Highway Administration, U.S. Department of Transportation, Washington, DC.

FHWA (2001b). *Reliability of Visual Inspection for Highway Bridges,* Vols. I and II, FHWA-RD-01-020 and FHWA-RD-01-021, Federal Highway Administration, U.S. Department of Transportation, Washington, DC.

FHWA (2001c). *Performance of Concrete Segmental and Cable-Stayed Bridges in Europe,* FHWA-PL-01-019, American Trade Initiatives, Alexandria, VA, Federal Highway Administration, U.S. Department of Transportation, Washington, DC.

FHWA (2001d). *Data Integration Glossary,* U.S. Department of Transportation, FHWA-IF-01-017, Federal Highway Administration, U.S. Department of Transportation, Washington, DC.

FHWA (2002a). *Seismic Retrofitting Manual for Highway Structures,* MCEER, Buffalo, NY, Federal Highway Administration, U.S. Department of Transportation, Washington, DC.

FHWA (2002b). *Guidelines for Detection, Analysis, and Treatment of Materials-Related Distress in Concrete Pavements,* Vol. 2: *Guidelines Description and Use,* FHWA-RD-01-164, Federal Highway Administration, U.S. Department of Transportation, Washington, DC.

FHWA (2002c). *Bridge Inspector's Reference Manual* (*BIRM*), Vols. I and II, FHWA-NHI-03-001, Federal Highway Administration, U.S. Department of Transportation, Washington, DC.

FHWA (2004a). *National Bridge Inspection Standards,* 23 CFR Part 650, Docket No. FHWA-2001-8954, pp. 74419–74439, *Federal Register,* Vol. 69, No. 239, Dec. 14, 2004, Federal Highway Administration, U.S. Department of Transportation, Washington, DC.

FHWA (2004b). *Transportation Performance Measures in Australia, Canada, Japan, and New Zealand,* International Technology Exchange Program, FHWA-PL-05-001, Federal Highway Administration, U.S. Department of Transportation, Washington, DC.

FHWA (2005a). *Specifications for the National Bridge Inventory,* Draft, May 16, 2005, Federal Highway Administration, U.S. Department of Transportation, Washington, DC.

FHWA (2005b). *Bridge Preservation and Maintenance in Europe and South Africa,* International Technology Exchange Program, FHWA-PL-04-007, Federal Highway Administration, U.S. Department of Transportation, Washington, DC.

FHWA (2005c). *Construction Management Practices in Canada and Europe,* International Technology Exchange Program, FHWA-PL-05-010, Federal Highway Administration, U.S. Department of Transportation, Washington, DC.

FHWA (2005d). *Transportation Asset Management in Australia, Canada, England, and New Zealand,* International Technology Exchange Program, FHWA-PL-05-019.

Fisher, J. W. (1984). *Fatigue and Fracture in Steel Bridges,* Case Studies, Wiley, New York.

Fisher, J. W., G. L. Kulak, and I. F. C. Smith (1977). A Fatigue Primer for Structural Engineers, ATLSS Report No.97-11, Lehigh University, Bethlehem.

Fisk, E. R., and R. R. Rapp (2004). *Introduction to Engineering Construction and Inspection,* Wiley, New York.

Florman, S. C. (1987). *The Civilized Engineer,* St. Martin's Press, New York.

Forde, M., Ed. (1999). *Structural Faults + Repair, Proceedings, International Conferences,* Engineering Technics Press, Edinburgh.

Frangopol, D., Ed. (1998). *Optimal Performance of Civil Infrastructure Systems,* American Society of Civil Engineers, Structural Engineering Institute, Reston, VA.

Frangopol, D., Ed. (1999a). *Case Studies in Optimal Design and Maintenance Planning of Civil Infrastructure Systems,* American Society of Civil Engineers, Structural Engineering Institute, Reston, VA.

Frangopol, D., Ed. (1999b). *Bridge Safety and Reliability,* American Society of Civil Engineers, Structural Engineering Institute, Reston, VA.

Frangopol, D. M., and H. Furuta, Eds. (2001). *First International Workshop on Life-Cycle Cost Analysis and Design of Civil Infrastructure Systems,* American Society of Civil Engineers, Structural Engineering Institute, Reston, VA.

Frank, T. (2005). The President's Man, *N.Y. Times Book Review,* February 27, p. 18.

Freudenthal, A. M. (1972). *International Conference on Structural Safety and Reliability, Proceedings,* Pergamon, Oxford, United Kingdom.

Freyssinet, E. (1993). *Un Amour Sans Limite,* Editions du Linteau, Paris.

Fujino, Y., et al. (1993). Synchronisation of Human Walking Observed During Lateral Vibration of a Congested Pedestrian Bridge, *Earthquake Engineering and Structural Dynamics,* Vol. 22, pp. 741–758.

Fukuyama, F. (1992). *The End of History and the Last Man,* Avon Books, New York.

Fukuyama, F. (1999). *The Great Disruption,* Simon & Schuster, New York.

Furuta, H., et al. (2003). Performance Measures for Bridge Management, *Proceedings of the International Workshop on Structural Health Monitoring of Bridges,* September 1–2, 2003, Kitami Institute of Technology, Japan Society of Civil Engineering, Kitami, Japan, pp. 181–186.

Galbraith, J. K. (1967). *The New Industrial State,* Houghton Mifflin, Boston.

Galbraith, J. K. (1973). *Economics & the Public Purpose,* Houghton Mifflin, Boston.

Galbraith, J. K. (1977). *The Age of Uncertainty, A History of Economic Ideas and Their Consequences,* Houghton Mifflin, Boston.

Gans, D., Ed. (1991). *Bridging the Gap,* Van Nostrand Reinhold, New York.

GASB 34 (2000). Office of Asset Management, U.S. Department of Transportation, Washington, DC.

George, C. S. (1968). *The History of Management Thought,* Prentice-Hall, Englewood Cliffs, NJ.

Gertsbakh, I. (2000). *Reliability Theory,* with Application to Preventive Maintenance, Springer-Verlag, Berlin.

Gibble, K., Ed. (1986). *Management Lessons From Engineering Failures,* American Society of Civil Engineers, Reston, VA.

Gies, J. (1963). *Bridges and Men,* Grosset & Dunlap, New York.

Gimsing, N. J. (1983). *Cable Supported Bridges,* Wiley, New York.

Gluver, H., and D. Olsen, Eds. (1998). *Ship Collision Analysis,* A. A. Balkema, Rotterdam.

Godfrain, J. (2003). *Les Ponts, le Diable et le Viaduc,* Le Jardin des Livres, Paris.

Golabi, K., P. Thompson, and W. A. Hyman (1992). *Pontis Technical Manual,* Optima, Cambridge Systematics, Cambridge.

Goltz, J. D., Ed. (1994). The Northridge, California Earthquake of January 17, 1994: General Reconnaissance Report, Technical Report No. 94-0005, Buffalo, NY.

Gómez-Ibáñez, J. A., and J. R. Meyer, Eds. (1993). *Going Private,* Brookings Institution, Washington, DC.

Gongkang, F. Ed. (2005). *Inspection and Monitoring Techniques for Bridges and Civil Structures,* Woodhead Publishing, Cambridge, England.

Gordon, J. E. (1978). *Structures or Why Things Don't Fall Down,* Da Capo Press, New York.

Gourmelon, J.-P. (1988). Matiere a Réflexion pour Une Politique de Gestion des Ponts Suspendus par Temps froid, *Bulletin Liaison Laboratoire Central des Ponts et Chaussées,* Jul.–Aug., No. 158, pp. 105–107, Paris.

Gramet, C. (1966). *Highways over Waterways,* Abelard-Schuman, London.

Graves, R. (1992). *The Greek Myths,* Penguin Books, London.

Günther, H. P., Ed. (2005). *Use and Application of High-Performance Steels for Steel Structures,* International Association for Bridge and Structural Engineering, Zürich.

Halmshaw, R. (1987). *Non-Destructive Testing,* Edward Arnold, London.

Halpin, D. W. and L. S. Riggs (1992). *Planning and Analysis of Construction Operations,* Wiley, New York.

Hambly, E. C. (1976). *Bridge Deck Behavior,* Wiley, New York.

Harding, J. E., G. A. R. Parke and M. J. Ryall (1990, 1993, 1997, 2000, 2005). *Bridge Management 1, 2, 3, 4, 5,* Thomas Telford, London.

Harriss, J. (1975). *The Tallest Tower,* Houghton Mifflin, Boston.

Hartle, R. A., et al. (1991). *Bridge Inspector's Training Manual/90,* FHWA-PD-91-015, Federal Highway Administration, Washington, DC.

Hassab, J. C. (1997). *Systems Management,* CRC Press, Boca Raton, FL.

Hays, W. L. (1994). *Statistics,* 5th ed., Harcourt Brace College Publishers, Fort Worth, TX.

Hegel, G. F. (1989). *Science of Logic,* Humanities Press International, Inc., Atlantic Highlands, N.J.

Heisenberg, W. (1958). *Physics and Philosophy, the Revolution in Modern Science,* Harper & Row, New York.

Hellier, C. J., et al. (2001). *Handbook of NDE,* McGraw-Hill, New York.

Hersey, P. (1985). *Situational Selling,* Center for Leadership Studies, Escondido, CA.

Heyman, J. (1996). *Elements of the Theory of Structures,* Cambridge University Press, Cambridge.

Heyman, J. (1998). *Structural Analysis: A Historical Approach,* Cambridge University Press, Cambridge.

Hodge, Ph. G. (1981). *Plastic Analysis of Structures,* Robert E. Krieger Publishing, Malabar, FL.

Holmes, O. W. (1895). *The Complete Poetical Works,* Houghton Mifflin, Riverside Press, Boston.

Hopkins, H. J. (1970). *A Span of Bridges,* Praeger, New York.

Horne, M. Z., and Merchant, W. (1965). *The Stability of Frames,* Pergamon, London.

Housner, G. W., Chairman (1990). Competing Against Time, The Governor's Board of Inquiry on the 1989 Loma-Prieta Earthquake, Office of Planning and Research, State of California.

HSBA, Honshu-Shikoku Bridge Authority (2002). Nondestructive Testing of Hanger Rope, *Newsletter on Long-Span Bridges,* No. 11, p. 1, March.

Hudson, R., R. Haas, and W. Uddin (1997). *Infrastructure Management,* McGraw-Hill, New York.

Hull, B., and V. John (1988). *Non-Destructive Testing,* Macmillan Education, Macmillan, London.

IABSE (1995). *Extending the Lifespan of Bridges,* Symposium, San Francisco, International Association for Bridge and Structural Engineering, Zurich, Switzerland.

IABSE (1999). *Cable Stayed Bridges—Past, Present and Future,* Malmö, International Association for Bridge and Structural Engineering, Zurich, Switzerland.

IABSE, (2000). International Association for Bridge and Structural Engineering (2000). *Structural Engineering for Meeting Urban Transportation Challenges, 16th Congress,* Lucerne, IABSE, Zurich, Switzerland.

IABSE (2001). *Cable-Supported Bridges—Challenging Technical Limits,* Conference, Seoul, International Association for Bridge and Structural Engineering, Zurich, Switzerland.

IABSE (2006). *Operation, Maintenance and Rehabilitation of Large Infrastructure Projects, Bridges and Tunnels,* May 15–17, 2006, Copenhagen, Denmark.

Imbsen, R. A., et al. (1997). Structural Details to Accommodate Seismic Movements of Highway Bridges and Retaining Walls, Technical Report NCEER-97-0007, National Center for Earthquake Engineering Research, State University at Buffalo, NY.

Inman, D. J., et al., Eds. (2005). *Damage Prognosis for Aerospace, Civil and Mechanical Systems,* Wiley, Chichester, West Sussex, England.

Ireson, W. G., and C. F. Coombs, Jr., Eds. (1988). *Handbook of Reliability Engineering and Management,* McGraw-Hill, New York.

Irvine, H. M. (1986). *Structural Dynamics,* Unwin Hyman, London.

Irvine, M. (1981). *Cable Structures,* Dover Publications, New York.

ISO (1995). *Guide to the Expression of Uncertainty in Measurement,* 2nd Ed., International Organization of Standardization, Geneva.

Jay, A. (1994). *Management and Machiavelli,* Pfeiffer & Company, San Diego.

Jones, D. A. (1992). *Principles and Prevention of Corrosion,* Macmillan Publishing Company, New York.

Juhn G.-H., C.-B. Yun, and B. F. Spencer, Jr., Eds. (2005). *Smart Infrastructure Technology for Maintenance of Infrastructure,* Proceedings, Korea Infrastructure Safety & Technology Corporation (KISTEC), Smart Infrastructure Technology Center (SISTeC), Korea Advanced Institute of Science and Technology (KAIST), and the Asian-Pacific Network of Centers for Research in Smart Structures Technology (ANCRiSST), Gyengju, Korea.

Kant, I. (1965). *Critique of Pure Reason,* St. Martin's Press, New York.

Karnakis, E. (1997). *Constructing a Bridge,* MIT Press, Cambridge, MA.

Kelly, J. M. (1993). *Earthquake-Resistant Design with Rubber,* Springer-Verlag, London.

Kierkegaard, S. (1849, 1980). *The Sickness unto Death,* Princeton University Press, Princeton, NJ.

Kline, M. (1953). *Mathematics in Western Culture,* Oxford University Press, New York.

Kline, M. (1980). *Mathematics, the Loss of Certainty,* Oxford University Press, New York.

Kodur, V. K. R., and T. Z. Harmathy (2002). Properties of Building Materials, *FPE Handbook of Fire Protection Engineering,* 3rd Ed., National Fire Protection Association.

Koglin, T. L. (2003). *Moveable Bridge Engineering,* John Wiley & Sons, Inc., Hoboken, NJ.

Kretz, T., et al. (2006). "Haute surveillance et evaluation de l'aptitude au service du pont suspendu d'Aquitaine," pp. 13–32, *Bulletin des Laboratoires des Ponts et Chaussées,* Vol. 260, Paris.

Kulkami, R. B. (1984). *Dynamic Decision Model for Pavement Management System,* TRR No. 997, Transportation Research Record, National Research Council, Washington DC, pp. 11–18.

Larsen, A. and S. Esdahl, Eds. (1998). *Bridge Aerodynamics,* A. A. Balkema, Rotterdam.

LCPC, Laboratoire Central des Ponts et Chaussées (1979). Ausculation, Surveillance Renforcee, Haute Surveillance, Mesures de Securite immediate ou de Sauvegarde, Instruction Technique du 19 Octobre 1979, Paris.

LCPC/SETRA (1979). *Instruction Technique pour la Surveillance et l'Entretien des Ouvrages d'Art,* Laboratoire Central des Ponts et Chaussées/Service d'Etudes Techniques des Routes et Autoroutes, Direction des Routes, Paris.

Leonardo da Vinci (1935). *Notebooks,* E. McCurdy, Ed., Empire State Book Co., New York.

Leonhardt, F. (1980). *Bridges,* MIT Press, Cambridge, MA.

Levy, M., and M. Salvadori (1992). *Why Buildings Fall Down,* W. W. Norton & Company, New York.

Livesley, R. K., and D. B. Chandler (1956). *Stability Functions for Structural Frameworks,* Manchester University Press, Manchester.

Llanos, J. (1992). *La Maintenance des Ponts Routiers,* Approche économique, Presses de l'École Nationale de Ponts et Chaussées, Paris.

Llanos, J., and B. Yanev (1991). Models of Deck Deterioration and Optimal Repair Strategies for the New York City Bridges, Proceedings, 2nd Civil Engineering Automation Conference, pp. 1–28, New York.

Lucas, Jr., H. C. (1985). *The Analysis, Design and Implementation of Information Systems,* McGraw-Hill, New York.

Mach, E. (1956). The Economy of Science, in *The World of Mathematics,* Vol. 3, Simon and Schuster, New York, pp. 1787–1795.

Mahorta, V. M., and N. J. Carino (1991). *CRC Handbook on Nondestructive Testing of Concrete,* CRC Press, Boca Raton, Fla.

Mallet, G. P. (1994). *Repair of Concrete Bridges,* State of the Art Review, Thomas Telford, London.

Mander, J. B., et al. (1996). Response of Steel Bearings to Reversed Cyclic Loadings, Technical Report NCEER-96-0014, NCEER, University of Buffalo.

McCullough, D. (1972). *The Great Bridge,* Avon Books, New York.

McCullough, D. (1977). *The Path Between the Seas,* Simon & Schuster, New York.

McNeill, D. and P. Freiberger (1994). *Fuzzy Logic,* Simon & Schuster, New York.

Melchers, R. E. (1987). *Structural Reliability Analysis and Prediction,* Wiley, New York.

Menn, C. (1986). *Prestressed Concrete Bridges,* Birkhauser Verlag, Wien.

MIL-Hdbk-472 (1984). *Maintainability Prediction,* Naval Forms and Publications Center, Philadelphia, PA.

Minor, J., K. R. White, and R. S. Busch (1988). *Condition Surveys of Concrete Bridge Components, User Manual,* Report 312, National Research Council, Transportation Research Board, Washington, DC.

Mintzberg, H. (1979). *The Structuring of Organizations,* Prentice-Hall, Englewood Cliffs, NJ.

Mittra, S. S. (1988). *Structured Techniques of System Analysis, Design, and Implementation,* Wiley, New York

Miyamoto, A. and D. Frangopol, Eds. (2001), Second International Workshop on Life-Cycle Cost Analysis and Design of Civil Infrastructure Systems, Proceedings, University of Yamaguchi, Ube, Japan.

Miyamoto, A., A. Sarja, and T. Rissanen, Eds. (2005). *Lifetime Engineering of Civil Infrastructure,* Yamaguchi University, Ube, Japan.

Mladjov, R. (2004). The Most Expensive Bridge in the World, *Modern Steel Construction,* Vol. 44, No. 9, pp. 53–56.

Mobley, R. K. (1990). *An Introduction to Predictive Maintenance,* Van Nostrand Reinhold, New York.

Morcous, G. H., et al. (2002). Modeling Bridge Deterioration Using Case-Based Reasoning, *Journal of Infrastructure Systems,* Vol. 8, No. 3, pp. 86–95.

Moses, F. and M. Ghosn (1985). A Comprehensive Study of Bridge Loads and Reliability, Report FHWA/OH-85/005, Department of Civil Engineering, Case Western Reserve University, Cleveland, OH.

Mullen, C. L. and A. S. Cakmak (1997). Seismic Fragility of Existing Conventional Reinforced Concrete Highway Bridges, Technical Report NCEER-97-0017, SUNY, Buffalo.

National Commission Final Report on the Terrorist Attacks upon the U.S. (2004), *The 9/11 Report,* W. W. Norton & Company, New York and London.

NCHRP Digest 232, National Cooperative Highway Research Program Report, National Research Council, Transportation Research Board, Washington, DC.

NCHRP Digest 234 (1993). Bridge Rating Though Nondestructive Load Testing, Project 12-28(13)A, National Research Council, Transportation Research Board, Washington, DC.

NCHRP Project 2-17 (3) (1994). Macroeconomic Analysis of the Linkages Between Transportation Investments and Economic Performance (M. E. Bell and T. J. McGuire), National Research Council, Transportation Research Board, Washington, DC.

NCHRP Project 12-46 (2000). Manual for Condition Evaluation and Load Resistance Factor Rating of Highway Bridges, Final Draft, National Research Council, Transportation Research Board, Washington, DC.

NCHRP Report 141 (1989). Bridge Deck Joints (M. Burke), National Research Council, Transportation Research Board, Washington, DC.

NCHRP Report 222 (1980). Bridges on Secondary Highways and Local Roads: Evaluation and Replacement, National Research Council, Transportation Research Board, Washington, DC.

NCHRP Report 243 (1981). Rehabilitation and Replacement of Bridges on Secondary Highways and Local, Roads, National Research Council, Transportation Research Board, Washington, DC.

NCHRP Report 273 (1984). Manual for the Selection of Optimal Maintenance Levels of Service, National Research Council, Transportation Research Board, Washington, DC.

NCHRP Report 285 (1986). Evaluating Alternative Maintenance Strategies, National Research Council, Transportation Research Board, Washington, DC.

NCHRP Report 292 (1987). Strength Evaluation of Existing Reinforced Concrete Bridges, National Research Council, Transportation Research Board, Washington, DC.

NCHRP Report 293 (1987). Methods of Strengthening existing Highway Bridges, National Research Council, Transportation Research Board, Washington, DC.

NCHRP Report 299 (1987). Fatigue Evaluation Procedures for Steel Bridges, National Research Council, Transportation Research Board, Washington, DC.

NCHRP Report 300 (1987). Bridge Management Systems, National Research Council, Transportation Research Board, Washington, DC.

NCHRP Report 304 (1988). Determining Deteriorated Areas in Portland Cement Concrete Pavements Using Radar and Video Imaging, National Research Council, Transportation Research Board, Washington, DC.

NCHRP Report 312 (1988). Condition Surveys of Concrete Bridge Components—User's Manual, National Research Council, Transportation Research Board, Washington, DC.

NCHRP Report 319 (1989). Recommended Guidelines for Redundancy Design and Rating of Two-Girder Steel Bridges, National Research Council, Transportation Research Board, Washington, DC.

NCHRP Report 333 (1990). Guidelines for Evaluating Corrosion Effects in Existing Steel Bridges, National Research Council, Transportation Research Board, Washington, DC.

NCHRP Report 377 (1994). Life-Cycle Cost Analysis for Protection and Rehabilitation of Concrete Bridges Relative to Reinforcement Corrosion, National Research Council, Transportation Research Board, Washington, DC.

NCHRP Report 363 (1994). Role of Highway Maintenance in Integrated Management Systems, National Research Council, Transportation Research Board, Washington, DC.

NCHRP Report 403 (1998). Guidance for Estimating the Indirect Effects of Proposed Transportation Projects, National Research Council, Transportation Research Board, Washington, DC.

NCHRP Report 406 (1998). Redundancy in Highway Bridge Superstructures, National Research Council, Transportation Research Board, Washington, DC.

NCHRP Report 437 (2000). Collection and Presentation of Roadway Inventory Data, National Research Council, Transportation Research Board, Washington, DC.

NCHRP Report 447 (2001). Testing and Inspection Levels for Hot-Mix Asphaltic Concrete Overlays, National Research Council, Transportation Research Board, Washington, DC.

NCHRP Report 451 (2001). Guidelines for Warranty, Multi-Parameter, and Best Value Contracting, National Research Council, Transportation Research Board, Washington, DC.

NCHRP Report 454 (2001). Calibration of Load Factors for LRFD Bridge Evaluation, National Research Council, Transportation Research Board, Washington, DC.

NCHRP Report 458 (2001). Redundancy in Highway Bridge Substructures, National Research Council, Transportation Research Board, Washington, DC.

NCHRP Report 466 (2002). Desk Reference for Estimating the Indirect Effects of Proposed Transportation Projects, National Research Council, Transportation Research Board, Washington, DC.

NCHRP Report 467 (2002). Performance Testing for Modular Bridge Joint Systems, National Research Council, Transportation Research Board, Washington, DC.

NCHRP Report 483 (2003). Bridge Life-Cycle Cost Analysis (BLCCA), National Research Council, Transportation Research Board, Washington, DC.

NCHRP Report 485 (2003). Bridge Software—Validation Guidelines and Examples, National Research Council, Transportation Research Board, Washington, DC.

NCHRP Report 489 (2003). Design of Highway Bridges for Extreme Events, National Research Council, Transportation Research Board, Washington, DC.

NCHRP Report 495 (2003). Effect of Truck Weight on Bridge Network Costs, National Research Council, Transportation Research Board, Washington, DC.

NCHRP Report 496 (2003). Prestress Losses in Pretensioned High-Strength Concrete Bridge Girders, National Research Council, Transportation Research Board, Washington, DC.

NCHRP Report 505 (2003). Review of Truck Characteristics as Factors in Roadway Design, National Research Council, Transportation Research Board, Washington, DC.

NCHRP Report 511 (2004). Guide for Customer-Driven Benchmarking of Maintenance Activities, National Research Council, Transportation Research Board, Washington, DC.

NCHRP Report 516 (2004). Pier and Contraction Scour in Cohesive Soils, National Research Council, Transportation Research Board, Washington, DC.

NCHRP Report 517 (2004). Extending Span Ranges of Precast Prestressed Concrete Girders, National Research Council, Transportation Research Board, Washington, DC.

NCHRP Report 519 (2004). Connection of Simple-Span Precast Concrete Girders for Continuity, National Research Council, Transportation Research Board, Washington, DC.

NCHRP Report 525 (2004–2005), Guide for Emergency Transportation Operations, National Research Council, Transportation Research Board, Washington, DC.

NCHRP Report 534 (2004). Guidelines for Inspection and Strength Evaluation of Suspension Bridge Parallel-Wire Cables, National Research Council, Transportation Research Board, Washington, DC.

NCHRP Report 538 (2005). Traffic Data Collection, Analysis, and Forecasting for Mechanistic Pavement Design, National Research Council, Transportation Research Board, Washington, DC.

NCHRP Report 543 (2005). Effective Slab Width for Composite Steel Bridge Members, National Research Council, Transportation Research Board, Washington, DC.

NCHRP Report 545 (2005). Analytical Tools for Asset Management, National Research Council, Transportation Research Board, Washington, DC.

NCHRP Synthesis 106 (1983). Practical Guidelines for Minimizing Tort Liability, National Research Council, Transportation Research Board, Washington, DC.

NCHRP Synthesis 110 (1984). Maintenance Management Systems, National Research Council, Transportation Research Board, Washington, DC.

NCHRP Synthesis 148 (1989). Indicators of Quality in Maintenance, National Research Council, Transportation Research Board, Washington, DC.

NCHRP Synthesis 153 (1989). Evolution and Benefits of Preventive Maintenance Strategies, National Research Council, Transportation Research Board, Washington, DC.

NCHRP Synthesis 238 (1997). Performance Measurement in State Departments of Transportation, National Research Council, Transportation Research Board, Washington, DC.

NCHRP Synthesis 284 (2000). Performance Survey on Open-Graded Friction Course Mixes, National Research Council, Transportation Research Board, Washington, DC.

NCHRP Synthesis 319 (2003). Bridge Deck Joint Performance, National Research Council, Transportation Research Board, Washington, DC.

NCHRP Synthesis 327 (2004). Cost-Effective Practices for Off-System and Local Interest Bridges, National Research Council, Transportation Research Board, Washington, DC.

NCHRP Synthesis 316 (2003). Design Exception Practices, National Research Council, Transportation Research Board, Washington, DC.

NCHRP Synthesis 330 (2004). Public Benefits of Highway System Preservation and Maintenance, National Research Council, Transportation Research Board, Washington, DC.

NCHRP Synthesis 331 (2004). State Highway Letting Program Management, National Research Council, Transportation Research Board, Washington, DC.

NCHRP Synthesis 333 (2004). Concrete Bridge Deck Performance, National Research Council, Transportation Research Board, Washington, DC.

NCHRP Synthesis 345 (2005). Steel Bridge Erection Practices, National Research Council, Transportation Research Board, Washington, DC.

NCHRP Synthesis 346 (2005). State Construction Quality Assurance Programs, National Research Council, Transportation Research Board, Washington, DC.

NCHRP Synthesis 352 (2005). Value Engineering Applications in Transportation, National Research Council, Transportation Research Board, Washington, DC.

NCHRP Synthesis 353 (2005). Inspection and Maintenance of Bridge Stay Cable Systems, National Research Council, Transportation Research Board, Washington, DC.

NCHRP Synthesis 354 (2005). Inspection and Management of Bridges with Fracture Critical Details, National Research Council, Transportation Research Board, Washington, DC.

NCHRP Synthesis of Highway Practice, National Research Council, Transportation Research Board, Washington, DC.

Neal, B. G. (1956, 1981). *The Plastic Methods of Structural Analysis,* Chapman & Hall, London and New York.

Neale, B. S., Ed. (2001). *Forensic Engineering,* Thomas Telford, London.

Newman, W. H. and C. E. Summer, Jr. (1961). *The Process of Management,* Prentice-Hall, Englewood Cliffs, NJ.

Newmark, N. M. and E. Rosenblueth (1971). *Fundamentals of Earthquake Engineering,* Prentice-Hall, New York.

Nims, D. K., et al. (1989). Collapse of the Cypress Street Viaduct as a Result of the Loma Prieta Earthquake, Earthquake Engineering Research Center, University of California, Berkeley.

Northouse, P. G. (1997). *Leadership, Theory and Practice,* SAGE Publications, Thousand Oaks, London and New Delhi.

Noyan, C., and J. B. Cohen (1987). *Residual Stress,* Springer-Verlag, Berlin.

NRC, National Research Council (1990). Truck Weight Limit, Issues and Options, Special Report 225, Transportation Research Board (TRB), Washington, D.C.

NRC (1993a). Concrete Bridge Protection, Repair, and Rehabilitation Relative to Reinforcement Corrosion: A Method Application Manual, Strategic Highway Research Program, SHRP-S-360, TRB, Washington, DC.

NRC (1993b). Eliminating or Minimizing Alkali-Silika Reactivity, SHRP-C-343, TRB, Washington, DC.

NRC (1995). Measuring and Improving Infrastructure Performance, Board on Infrastructure and the Constructed Environment, National Academy Press, Washington, DC.

NYC DOT (1992). Bridges and Tunnel, Annual Condition Report, Department of Transportation, New York City.

NYS DOT (1991). Hydraulic Vulnerability Manual, Department of Transportation, New York State, Albany, NY.

NYS DOT (1993a). Mini-Decision Support System, Bridge Management System, New York State Department of Transportation, Albany, NY.

NYS DOT (1993b). Overload Vulnerability Manual, Department of Transportation, New York State, Albany, NY.

NYS DOT (1993c). Steel Details Vulnerability Manual, Structures Design and Construction Division, Department of Transportation, New York State, Albany, NY.

NYS DOT (1995a). Collision Vulnerability Manual, Structures Design and Construction Division, Department of Transportation, New York State, Albany, NY.

NYS DOT (1995b). Seismic Vulnerability Manual, Structures Design and Construction Division, Department of Transportation, New York State, Albany, NY.

NYSBA/HSBA New York State Bridge Authority/Honshu-Shikoku Bridge Authority (2002), Third International Bridge Operator's Conference, Awaji Island, Japan, May 16–17, 2002.

NYS DOT, New York State Department of Transportation (1997), *Bridge Inspection Manual,* NYS DOT, Albany, NY.

O'Connor, C. (1971). *Design of Bridge Superstructures,* Wiley, New York.

O'Connor, D. S. and W. A. Hyman (1989). Bridge Management Systems, Report FHWA-DP-71-0R, U.S. Department of Transportation, Washington, DC.

OECD (1981). *Bridge Maintenance,* Organization for Economic Cooperation and Development, Paris.

OECD (1992). *Bridge Management,* Organization for Economic Cooperation and Development, Paris.

OHBDC (1983, 1993). *Ontario Highway Bridge Design Code,* 2nd and 3rd eds., Ontario Ministry of Transportation and Communications, Toronto, Ontario.

Olsson, N. (1993). Supply-Side Solutions, *Civil Engineering,* April 1993, pp. 57–59.

OSHA (1995). Fall Protection in Construction, Occupational Safety and Health Administration, U.S. Department of Labor, Washington, D.C.

Paine, T. (1945). *Selected Works,* Duell, Sloan and Pearce, New York.

Park, S. H. (1980). *Bridge Inspection and Structural Analysis,* NJDOT, Trenton, NJ.

Park, W. R., and D. E. Jackson (1984). *Cost Engineering Analysis,* Wiley, New York.

Parkinson, C. N. (1957). *Parkinson's Law and Other Studies in Administration,* Houghton Mifflin, Boston.

Pascal, B. (1962). *Pensées,* Editions du Seuil, Paris.

Patterson, W. D. O., and T. Scullion (1990). Information Systems for Road Management: Draft Guidelines on System Design and Data Issues, Technical Paper INU77, Infrastructure and Urban Development, The World Bank, Washington, DC.

PBQ&D, Parsons Brinkerhoff Quade & Douglas Inc. (1993). *Seismic Awareness: Transportation Facilities, A Primer for Transportation Managers,* U.S. Department of Transportation, Washington, DC.

Pearl, J. (1990). Bayesian and Belief-Function Formalisms for Evidential Reasoning: A Conceptual Analysis, in G. Shafer, and J. Pearl, Eds., *Readings in Uncertain Reasoning,* Morgan Kaufmann, San Mateo, CA, pp. 540–574.

Persy, J.-P., and A. Raharinaivo (1987). Etude de la Rupture par Temps Froid d'Éléments en Acier Provenant d'Un Pont Suspendu, *Bulletin Liaison Laboratoire Central des Ponts et Chaussées,* Nov.–Dec., No. 152, pp. 49–53, Paris.

Petroski, H. (1992). *To Engineer Is Human,* Vintage Books, Random House, New York.

Petroski, H. (1993). Predicting Disaster, *American Scientist,* Vol. 81, Mar.–Apr., pp. 110–113.

Petroski, H. (1994). Success Syndrome: The Collapse of the Dee Bridge, *Civil Engineering,* Vol. 64, No. 4, pp. 52–55.

Petroski, H. (1995). *Engineers of Dreams,* Alfred A. Knopf, New York.

Phillips, K. (2004). *American Dynasty,* Viking, Penguin Group, New York.

PIARC (1996). International Seminar on Bridge Engineering and Management in Asian Countries, World Road Association, Jakarta, Indonesia.

Picon, A. (1992). *L'Invention de l'Ingenieur Moderne,* Presses de l'École des Ponts et Chaussées, Paris.

Plato (1942). *The Works of Plato,* translator, B. Jowett, Dial Press, New York.

Plato (1956). *The Great Dialogues,* translator W. H. D. Rouse, New American Library, New York.

Plutarch. *The Lives of the Noble Grecians and Romans,* translator J. Dreyden, Random House, New York.

Post, N. M. (2005). Mysteries of Building Codes, *Engineering News Record,* July 18, 2005, pp. 26–29, McGraw Hill Construction, New York.

Priestley, M. J. N., F. Seible, and G. M. Calvi (1996). *Seismic Design and Retrofit of Bridges,* Wiley, New York.

Pritchard, B. (1992). *Bridge Design for Economy and Durability,* Thomas Telford, London.

Raheja, D. G. (1991). *Assurance Technologies,* McGraw-Hill, New York.

Rahman, S., and M. Grigoriu (1994). A Markov Model for Local and Global Damage Indices in Seismic Analysis, Technical Report NCEER-94-0003 (NSF), NCEER, CUNY, Buffalo.

Raiffa, H., and R. Schlaifer (1961). *Applied Statistical Decision Theory,* Harvard University Press, Boston, MA.

Ramberg, G. (2002). *Structural Bearings and Expansion Joints,* International Association for Bridge and Structural Engineering, Zurich, Switzerland.

Ratay, R., Ed. (2005). *Structural Condition Assessment,* Wiley, Hoboken, NJ.

Ratay, R. T., Ed. (2000). *Forensic Structural Engineering Handbook,* McGraw-Hill, New York.

Reier, S. (1977). *The Bridges of New York,* Quadrant Press, New York.

Roads & Bridges (Aug. 2003). Vol. 41, No. 8.

Roads & Bridges (Dec. 2003). Vol. 41, No. 12.

Roberts, J. (1991). Recent Advances in Seismic Design and Retrofit of California Bridges, *Proceedings, Third U.S. National Conference on Lifeline Earhquake Engineering,* Los Angeles, pp. 52–64.

Robison, R. (1988). The Williamsburg: Rehab After All, *Civil Engineering,* Sept. 1988, ASCE, New York.

Rojahn, C., et al. (1997). Seismic Design Criteria for Bridges and Other Highway Structures, Technical Report NCEER-97-0002 (FHWA), NCEER, CUNY, Buffalo.

Rolfe, S. T., and J. M. Barsom (1987). *Fracture and Fatigue Control of Structures,* 2nd ed., Prentice-Hall, Englewood Cliffs, NJ.

Romains, J. (1940). *Sept Mysteres du Destin de l'Europe,* Editions de la Maison Francaise, New York.

Ross, S. S. (1984). *Construction Disasters,* McGraw-Hill Book Company, New York.

Rossi, P. H., H. E. Freeman, and S. R. Wright (1979). *Evaluation, a Systematic Approach,* Sage Publications, Beverly Hills and London.

Russell, B. (1985). *The Philosophy of Logical Atomism,* Open Court, La Salle, IL.

Ryall, M. J. (2001). *Bridge Management,* Butterworth Heinemann, Oxford.

Salvadori, M. (1980). *Why Buildings Stand Up,* W. W. Norton, New York and London.

Salvadori, M. With R. Heller (1963). *Structure in Architecture, the Building of Buildings,* Prentice-Hall, Englewood Cliffs, NJ.

Santayana, G. (1928). *The Life of Reason,* Charles Scribner's, New York.

Saul, J. R. (2004). *On Equilibrium,* Four Walls Eight Windows, New York.

Schlaich, J., and H. Scheef (1982). *Concrete Box-Girder Bridges,* International Association for Bridge and Structural Engineering, Zurich.

Schneider, J. (1997). *Introduction to Safety and Reliability of Structures,* International Association for Bridge and Structural Engineering, Zurich.

Schopenhauer, A. (1942). *Complete Essays,* Wiley, New York.

Schwesinger, P., and F. H. Wittmann, Eds. (2000). *Present and Future of Health Monitoring,* Aedificatio Publishers, Freiburg, Germany.

Scott, R. (2001). *In The Wake of Tacoma,* ASCE Press, Reston, VA.

Servan-Schreiber, J.-J. (1967). *Le Defi Americain,* Denoël, Paris.

Servan-Schreiber, J.-J. (1991). *Passions,* Diffusion Hachette, Rungis, France.

SETRA/LCPC (1975). Defauts Apparents des Ouvrages d'Art en Beton, Service d'Etudes Techniques des Routes et Autoroutes/Laboratoire Central des Ponts et Chaussées, Ministere de l'Equipment, Paris.

SETRA/LCPC (1981). Defauts Apparents des Ouvrages d'Art Metalliques, Service d'Etudes Techniques des Routes et Autoroutes, Ministere de l'Equipmen/Laboratoire Central des Ponts et Chaussées, Paris.

SETRA/LCPC (1982). Defauts Apparents des Ouvrages d'Art en Maçonnerie, Service d'Etudes Techniques des Routes et Autoroutes/Laboratoire Central des Ponts et Chaussées, Ministere de l'Equipment, Paris.

SFPE (2000). *Engineering Guide to Performance Based Fire Protection Analysis and Design of Buildings,* Society of Fire Protection Engineers, Bethesda, MD.

Shanley, F. R. (1957). *Strength of Materials,* McGraw-Hill, New York.

Shanley, F. R. (1960). *Weight-Strength Analysis of Aircraft Structures,* 2nd ed., Dover Publications, New York.

Shenk, D. (1997). *Data Smog,* Harper Collins, New York.

Shepard, R. W. (2005). Bridge Management Issues in a Large Agency, *Structure and Infrastructure Engineering,* Vol. 1, No. 2, June, pp. 159–164.

Shepard, R. W., and M. B. Johnson (2001). Health Index: A Diagnostic Tool to Maximize Bridge Longevity, Investment, *TR News,* Vol. 215, July–Aug., pp. 6–11, Washington, DC.

Shepherd, R., and D. Frost, Eds. (1995). *Failures in Civil Engineering: Structural, Foundation and Geoenvironmental Studies,* American Society of Civil Engineers, Reston, VA.

Shinozuka, M., Ed. (1995). The Hanshin-Awaji Earthquake of January 17, 1995: Performance of Lifelines, Technical Report NCEER-95-0015, Buffalo, NY.

Shull, P. J., Ed. (2002). *NDE, Theory, Techniques and Applications,* Marcel Dekker, Weston, CT.

Sibly, P., and A. S. Walker (1977). Structural Accidents and Their Causes, *Proceedings of the Institution of Civil Engineers,* Vol. 62, Part 1, pp. 191–208, London.

Stahl, F. L., and C. P. Gagnon (1996). Cable Corrosion in Bridges and Other Structures, American Society of Civil Engineers, New York.

Silano, L. G., Ed. (1993). *Bridge Inspection and Rehabilitation,* Wiley, New York.

Sinha, K., and T. F. Fwa (1987). *On the Concept of Total Highway Management,* TRR 1229, Transportation Research Board, National Research Council, Washington, DC.

Stark, R. (2005). *The Victory of Reason,* Random House, New York.

Steiner, G. A., et al., Eds. (1982). *Management Policy and Strategy,* Macmillan Publishing, New York.

Steinman, D. B. (1909). The Hudson Memorial Bridge, Thesis for Engineering Degree, Columbia University, New York.

Steinman, D. B. (1945). *The Builders of the Bridge,* Harcourt, Brace, New York.

Steinman, D. B. (1949). *A Practical Treatise Suspension Bridges,* Wiley, New York.

Stidger, R. W. (2004). How Agencies Manage Their Tight Budgets, *Better Roads,* Vol. 74, No. 3, pp. 26–28, Des Plaines, IL.

Sun-Tzu (1994). *The Art of War,* Barnes & Noble, New York.

Talese, G. (1970). *Fame and Obscurity,* World Publishing, New York.

Taly, N. (1998). *Design of Modern Highway Bridges,* McGraw-Hill, New York.

Taylor, C., and E. Van Marcke, Eds. (2002). *Acceptable Risk Processes,* Lifelines and Natural Hazards, ASCE Press, Reston, VA.

Thoft-Christensen, P., and M. J. Baker (1982). *Structural Reliability Theory and Its Applications,* Springer-Verlag, Berlin.

Thoft-Christensen, P. D., and Y. Murotsu (1986). *Application of Structural Systems to Reliability Theory,* Springer-Verlag, Berlin.

Thompson, P. D. (2005). Markovian Bridge Deterioration: Developing Momdels from Historical Data, *Structure and Infrastructure Engineering,* Vol. 1, No. 1, pp. 95–91.

Timoshenko, S. (1936). *Theory of Elastic Stability,* McGraw-Hill, New York.

Tocqueville, Alexis de (2000). *Democracy in America,* HarperCollins, New York.

Tobin, J. (2001). *Great Projects,* Free Press, New York.

Toffler, A. (1980). *The Third Wave,* William Morrow and Company, New York.

Tonias, D. E. (1995). *Bridge Engineering,* McGraw-Hill, New York.

TRB, Transportation Research Board (1990). *Truck Weight Limits,* Special Report 225, National Research Council, Washington, DC.

TRB (1992). *Data for Decisions: Requirements for National Transportation Policy Making,* Special Report 234, National Research Council, Washington, DC.

TRB (2001). *Maintenance Management,* Conference Proceedings 23, National Academy Press, Washington, DC.

TRC No. 1350 (1992). *Hydrology and Bridge Scour,* National Research Council, Transportation Research Board, Washington, DC.

TRC No. 423 (1994). *Characteristics of Bridge Management Systems,* National Research Council, Transportation Research Board, Washington, D.C.

TRC No. 498 (2000). *Eighth International Bridge Management Conference,* Vols. I and II, National Research Council, Transportation Research Board, Washington, DC.

TRC No. E-C037 (2002). *Glossary of Quality Assurance Terms,* National Research Council, Transportation Research Board, Washington, DC.

TRC No. E-C049 (2003). *Ninth International Bridge Management Conference,* National Research Council, Transportation Research Board, Washington, DC.

Troitsky, M. S. (1988). *Cable-Stayed Bridges,* Van Nostrand, New York.

Troitsky, M. S. (1994). *Planning and Design of Bridges,* Wiley, New York.

TRR, Transportation Research Record, National Research Council, Transportation Research Board, National Academy Press, Washington, DC. TRR No. 1083 (1986). *Pavement and Bridge Maintenance.*

TRR No. 1113 (1987). *Bridge Maintenance, Corrosion, Joint Seals, and Polymer Mortar Materials.*

TRR No. 1124 (1987). *Transportation Needs, Priorities, and Financing.*

TRR No. 1183 (1988). *Systematic Approach to Maintenance.*

TRR No. 1184 (1988). *Structures Maintenance.*

TRR No. 1351 (1992). *Construction Quality and Construction Management.*

TRR No. 1389 (1993). *Innovations in Construction.*

TRR No. 1490 (1995). *Management and Maintenance of Bridge Structures.*

TRR No. 1654 (1999). *Construction: Pavement, Bridge, Quality Control/Quality Assurance, and Management*

TRR No. 1668 (1999). *Concrete in Pavements and Structures.*

TRR No. 1697 (2000). *Maintenance and Management of Bridges and Pavements.*

TRR No. 1866 (2004). *Maintenance and Management of Pavement and Structures.*

TRR No. 1877 (2004). *Maintenance Management and Services.*

TRR No. 1889 (2004). *Pavement Management, Monitoring, Evaluation, and Data Storage.*

TRR No. 1901 (2005). *Bituminous Binders.*

TRR No. 1904 (2005). *Highway Facility Design.*

U.K. Bridges Board (2005). *Management of Highway Structures, Code of Practice,* Department of Transport TSO, London.

U.K. Highway Agency/LCPC (1999). *Post-tensioned Concrete Bridges,* Thomas Telford, London.

Vaicaitis, R., et. al (1999). Preventive Maintenance Management System for the New York City Bridges, Report of a Consortium of Civil Engineering Departments of New York City Colleges and Universities, the Center of Infrastructure Studies, Columbia University, New York.

Valéry, P. (1941). *Tel Quel,* Gallimard, Paris.

Valéry, P. (1945). *Regards sur le Monde Actuel,* Gallimard, Paris.

Van Der Zee, J. (1986). *The Gate,* Simon and Schuster, New York.

Veshosky, D. (1992). Life-Cycle Cost Analysis Doesn't Work for Bridges, *Civil Engineering,* Vol. 62, No. 6, p. 6.

Veshosky, D., et al. (1994). Comparative Analysis of Bridge Superstructure Deterioration, *Journal of Structural Engineering,* Vol. 120, No. 7, pp. 1223–2136.

Vesikari, E. (1988). *Service Life of Concrete Structures with Regard to Corrosion of Reinforcement,* Technical Research Centre of Finland, Espoo.

Vick, S. G. (2002). *Degrees of Belief,* ASCE Press, Reston, VA.

Vincentsen, L. J. and J. S. Jensen, Eds. (1998). *Operation and Maintenance of Large Infrastructure Projects,* A. A. Balkema, Rotterdam.

Virlogeux, M. (1999). *Replacement of the Suspension System on the Tancarville Bridge,* Paper No. 99-0604, TRR No. 1654, Transportation Research Board, National Research Council, National Academy Press, Washington, DC, pp. 113–120.

Von Karman, T., and M. A. Biot (1940). *Mathematical Methods in Engineering,* McGraw-Hill, New York.

Von Neumann, J., and O. Morgenstern (1964). *Theory of Games and Economic Behavior,* Princeton University Press, Wiley, New York.

Waddell, J. A. L. (1916). *Bridge Engineering,* Wiley, New York.

Waddell, J. A. L. (1921). *Economics of Bridgework,* Wiley, New York.

Wadia-Fascetti, S., et al. (2002). *Subsurface Sensing for Highway Infrastructure Condition Diagnostics,* Transportation Research Board, Washington, DC.

Walther, R. A., and M. J. Koob (2002). Condition Assessment of Chicago's 100 Year-Old Elevated Mass Transit System, Stahlbau, #71, No. 2, Feb., 2002, Berlin, pp. 117–124.

Wearne, P. (1999). *Collapse, When Buildings Fall Down,* TV Books, New York.

Wenzel, H., and D. Pichler (2005). *Ambient Vibration Monitoring,* Wiley, New York.

Werner, S. D., et al. (2000). A Risk-Based Methodology for Assessing the Seismic Performance of Highway Systems, Technical Report MCEER-00-0014, MCEER, State University of New York, Buffalo.

White, K. R., J. Minor, and K. N. Derucher (1992). *Bridge Maintenance, Inspection and Evaluation,* Marcel Dekker, New York.

Williamsburg Bridge Technical Advisory Committee (1988). Technical Report to the Commissioners of Transportation of the City and State of New York, Howard Needles Tammen & Bergendorf, June 30.

Xanthakos, P. P. (1994). *Theory and Design of Bridges,* Wiley, New York.

Xanthakos, P. P. (1996). *Bridge Strengthening and Rehabilitation,* Prentice-Hall, Englewood Cliffs, NJ.

Yanev, B. (1989). The Elusive Engineering Style, *Journal of Professional Issues,* Vol. 115, No. 4, pp. 418–421.

Yanev, B. (1994). Emergency Repair Needs Assessment for the New York City Bridges, in *Maintenance of Bridges and Civil Engineering Structures,* Presses de l'École Nationale des Ponts et Chaussées, Paris, pp. 501–513.

Yanev, B. (1997). Life-Cycle Performance of Bridge Components in New York City, *Recent Advances in Bridge Engineering,* Proceedings of the US–Canada–Europe Workshop on Bridge Engineering, EMPA Switzerland and Columbia University, New York, pp. 385–392.

Yanev, B. (1998). The Management of Bridges in New York City, *Engineering Structures,* Vol. 20, No 11, pp. 1020–1026.

Yanev, B. (2003). Management for the Bridges of New York City, *International Journal of Steel Structures,* Vol. 3, No. 2, pp. 127–135. The Korean Society of Steel Construction, Seoul.

Yanev, B., and R. B. Testa, (2001). Maintenance Level Assessment for New York City Bridges, *Second International Workshop on Life-cycle Cost Analysis and Design of Civil Infrastructure Systems,* Yamaguchi, Japan, pp. 83–92.

Yanev, B., and D. Tran (1997). Prioritizing New York City Bridges According to Earthquake Hazard Criteria, *Economic Consequences of Earthquakes: Preparing for the Unexpected, Proceedings,* NCEER-SP-0001, University of Buffalo, Buffalo, NY, pp. 155–166.

Yao, J. T. P., and H. Furuta (1986). Probabilistic Treatment of Fuzzy Events in Civil Engineering, *Journal of Probabilistic Mechanics,* Vol. 1, No. 1, pp. 58–64.

Yao, J. T. P., and J. M. Roesset (2001). Suggested Topics for a Curriculum in Infrastructure Management, *Public Works Management & Policy,* Vol 5, No. 4, pp. 308–317.

Yun, C.-B., and B. F. Spencer, Jr., Eds. (2005). Advanced Smart Materials and Smart Structures Technology, Proceedings of the Second International Workshop, Gyengju, July 21–24, 2005, Technopress, Daejeon, Korea.

Zadeh, L., and J. Kacprzyk, Eds. (1992). *Fuzzy Logic for the Management of Uncertainty,* Wiley, New York.

Zokaie, T., T. A. Osterkamp, and R. A. Imbsen (1991). Distribution of Wheel Loads on Highway Bridges, NCHRP Report 12-26, Transportation Research Board, Washington, DC.

Appendixes

APPENDIX 1. "THE DEACON'S MASTERPIECE OR THE WONDERFUL ONE-HOSS SHAY" BY OLIVER WENDELL HOLMES (1895, p. 158)

The poem, published in 1858, perfectly described the concepts applied in structural optimization and engineering in general, as noted by Petroski (1992, p. 29). In a delightfully tongue-in-cheek tone appropriate for a poet who happens to be the Harvard professor of anatomy and the father of a future Supreme Court justice, O. W. Holmes (Fig. A1.1) explained his "perfectly intelligible conception, whatever material difficulties it presents" as follows: "Observation shows us in what point any particular mechanism is likely to give way. In a wagon, for instance, the weak point is where the axle enters the hub or nave. When the wagon breaks down, three times out of four, I think, it is at this point that the accident occurs. The workman should see to it that this part should never give way; then find the next vulnerable place, and so on, until he arrives logically at the perfect result attained by the deacon."

For his "one-hoss shay" the Deacon develops equal-reliability, performance-based specifications as follows:

> Now in building of chaise, I tell you what,
> There is always *somewhere* a weakest spot,—
> In hub, tire felloe, in spring or thill,
> In panel, or crossbar, or floor, or sill,
> In screw, bolt, thoroughbrace,—lurking still,
> Find it somewhere you must and will,—
> Above or below, or within or without,—
> And that's the reason, beyond a doubt,
> That a chaise *breaks down,* but doesn't *wear out.*
> . . . .
> Fur . . . 't's mighty plain
> That the weakes' place mus' stan' the strain;
> 'N' the way t' fix it, uz I maintain,
> Is only jest
> T'make that place uz strong uz the rest.

After a century and a day:

Figure A1.1 Oliver Wendell Holmes (1809–1894), Hall of Fame, the Bronx.

> There are traces of age in the one-hoss shay,
> A general flavor of mild decay,
> but nothing local, as one may say.

Then the shay abruptly turns to dust.

> You see, of course, if you're not a dunce,
> How it went to pieces all at once,—
> All at once, and nothing first,—
> Just as bubbles do when they burst.

The analogy is rigorous since bubbles are optimal uniform stress shapes. As a warning to optimization and failure analyses, the poem remains valid. The term *vulnerability* is employed as in Chapter 4 herein. The mention of axle bearings brings comparable bridge details to mind. The somewhat flippant ending casts a doubt over the entire method of reasoning: "Logic is logic. That's all I say."

Figure A1.2 Oliver Wendell Holmes, Jr. (1841–1935).

Possibly in response to that parental caution, Oliver Wendell Holmes, Jr. (Fig. A1.2) concluded: "The life of the law has not been logic, it has been experience" (Fig. A1.3).

In contrast to the law, engineering builds on logic, confirmed by experience. Unique experience suggests a *possible* useful life of 100 years. What should the *design* life be? When did the shay become unsafe and by what standard?

If the process were considered as a chain of independent tasks, such as construction, use, and maintenance, each carrying a 10% likelihood of failure, the resulting cumulative

Figure A1.3 Inscription under monument of Fig. A1.2.

reliability would be unacceptable (see following Appendices). To that process must be added inspections, clueless on the mode of failure and therefore not reliable beyond 50%.

This view renders the uniform-strength shay unmanageable, which may explain why there has been little incentive to build it. Imperfect but predictable structures would be easily competitive and, hence, are likelier to be considered by life-cycle management.

APPENDIX 2.   BAYESIAN STATISTICAL DECISION THEORY AND RELIABILITY-BASED DESIGN

Cornell (in Freudenthal, 1972, pp. 47–66) presented the application of Bayesian statistical theory to structural reliability as follows:

Analysis of Uncertainty

Statistical uncertainty in parameters is handled, not through confidence statements, but in a manner parallel to the probabilistic uncertainty inherent in the stochastic model of the physical mechanism. Techniques are suggested for extracting or assessing both judgmental probabilities and the relative benefit or utility of various levels of structural performance. . . . Statistical decision theory introduces two new concepts which can be discussed somewhat separately. The first is statistical and is related to the analysis of uncertainty in parameters. The second is decision analysis and is concerned with quantifying the benefits and expected benefits of the various possible outcomes of a decision.
. . .

In decision theory any uncertain factor may be treated as a random variable. In addition to the familiar random variable, that is, the output of a stochastic model of a physical process, we may now treat as random variables the parameters of a deterministic or stochastic model, and even the model itself. Over and above the inherent, physical, or modeled randomness, the engineer is always uncertain about the values of his model's parameters. This can be called statistical (as opposed to probabilistic) uncertainty in that it becomes smaller as more observations of the process become available to provide more reliable parameter estimates. In addition, there always remains a fundamental uncertainty in the model itself. The material does not actually behave in a linear manner; fatigue life may follow a Weibull or gamma distribution (or, more likely, some as yet unnamed distribution). Our concern here is with the consistent and parallel treatment of all these sources of uncertainty in order that they are properly included in the choice among engineering designs. . . .

A major contribution of Bayesian statistical decision theory is the procedure for coupling professional information and the information contained in . . . statistical data. The vehicle is Bayes' theorem. In words, this theorem says that the posterior probability, $P[H_i|A]$, of hypothesis i, given observation A, is proportional to the prior probability $P[H_i]$, times the likelihood of A, given H_i, $P[H_i|A]$:

$$P[H_i|A] \propto P[A|H_i]P[H_i] \qquad \text{[A2.1]}$$

. . .

Decision analysis requires the assignment of a measure of preference, relative value, or *utility* to each action-outcome pair $u(a_i, \theta_j)$. The numerical value of the utility assignment may be simply a monetary measure, but in general utility must be assigned in a particular way to ensure that the prescribed decision criterion is valid. That criterion is that the action or design be chosen to maximize the expected utility.

Formally, Bayesian decision theory states that the decision maker choose the action a_0 from among the set of alternative actions $a_1, a_2, \ldots, a_m$, such that the expected utility, given an action $E[u|a_i]$, be maximized. The expectation is over $\theta_1, \theta_2, \ldots, \theta_n$, the set of possible events given a_i:

$$E[u|a_i] = \sum_{j=1}^{n} u(a_i, \theta_j) \, P[\theta_j|a_i] \qquad [A2.2]$$

Expected value maximization is an obvious criterion when the decision represents one in a succession of many, similar small decisions. Its use in major, one-of-a-kind decisions, such as those normally encountered in structural design, requires further discussion. Expected monetary value is surely not a reasonable decision criterion in all such cases. This is true even if one manages to express difficult-to-quantify events, such as loss of professional prestige or human life, in dollars (say, in the dollars the profession or society is willing to pay to avoid such a loss, as suggested by Dr. E. Rosenblueth in private communication). . . . It is unreasonable to think that most firms would consider a coin toss a good risk in the light of such potential payoffs. . . .

Applications of decision theory in structural design have been reported by many authors. . . . It is not always true that these authors' positions have been explicitly Bayesian in the definition of the probabilities or that the expected value criterion has been justified on any but an intuitive basis, or applied to values other than monetary. The concept of explicitly balancing higher initial cost versus higher risk of poor performance has long been an attractive advantage of the probabilistic approach to structural design. As a realistic, practical approach, however, it has been criticized for failure to include all sources of uncertainty and for the limitations of the expected monetary value criterion. Formal Bayesian decision theory removes these objections.

APPENDIX 3. "THE MACHINE THAT WON THE WAR," I. ASIMOV (1990)

The supercomputer is universally celebrated for an ultimate victory over the enemy. Privately, the young chief programmer, the middle-aged chief database manager (e.g., the Oracle), and the senior chief executive voice some reservations. The chief programmer invokes the principle now familiar as "garbage in–garbage out." He boasts of the complex preprocessing that crunched the vast and meaningless information arriving from all corners of the universe into coherent data input. The Oracle complains that the output that he had to interpret was worthless. He admits to revising it, with obsolete hardware but clearly superior managerial wisdom, in order to produce the invaluable recommendations submitted to the chief executive. The latter, significantly more exhausted than the rest, proclaims both efforts inconclusive. He reveals the "ancient computer" that has backed up his critical decisions. It is a coin that, when flipped, produces random heads-and-tails outcomes.

APPENDIX 4. CONDITIONAL PROBABILITY

Ang and Tang (1975) develop the subject as follows:

> A. *Conditional Probability (p. 43).* The conditional probability $P(E_1|E_2)$ of event E_1 given E_2 is denoted as

$$P(E_1|E_2) = \frac{P(E_1E_2)}{P(E_2)} \tag{A4.1}$$

B. *Multiplication Rule of Probabilities for Joint Events (p. 47)*. The probability $P(E_1E_2)$ of a joint event E_1E_2 is

$$P(E_1E_2) = P(E_1|E_2)P(E_2)$$

or

$$P(E_1E_2) = P(E_2|E_1)P(E_1)$$

Statistically independent events are mutually exclusive; the occurrence of one precludes the occurrence of the other as follows:

$$P(E_1|E_2) = P(E_1) \tag{A4.2}$$

and

$$P(E_2|E_1) = P(E_2)$$

The probability $P(E_1E_2 \cdots E_n)$ of a joint event $E_1E_2 \cdots E_n$ is

$$P(E_1E_2) = P(E_1)P(E_2) \cdots P(E_n) \tag{A4.3}$$

where $E_1, E_2, \ldots, E_n$ are statistically independent events.

C. *Theorem of Total Probability (p. 52)*. The total probability $P(A)$ of an outcome, depending on the probability of n mutually exclusive and collectively exhaustive events $E_1, E_2, \ldots, E_n$ is

$$P(A) = P(AE_1) + P(AE_2) + \cdots + P(AE_n) \tag{A4.4}$$

where

$$E_1 \cup E_2 \cup \cdots \cup E_n = S \tag{A4.5}$$

From Eq. A4.2, it follows that

$$P(A) = P(A|E_1)P(E_1) + P(A|E_2)P(E_2) + \cdots + P(A|E_n)P(E_n) \tag{A4.6}$$

Ang and Tang (1975) describe the probability of A as an average probability weighted by the probabilities of E_i.

D. *Bayes's Theorem (1764)*. The theorem, due to Thomas Bayes (1701–1761), is formulated by Ang and Tang (1975, p. 56) and Thoft-Christensen and Baker (1982, p. 17), as follows:

$$P(E_i|A) = \frac{P(A|E_i)\ P(E_i)}{\sum\limits_{j=1}^{n} P(A|E_j)\ P(E_j)} \tag{A4.7}$$

where $E_1, E_2, \ldots, E_n$ are n mutually exclusive events in a sample space and A is an event in the same sample space.

According to Ang and Tang (1975) the Bayes theorem determines a "reverse probability"; i.e., if A occurred, what is the probability that E_i also occurred?

(Vick, 2002, p. 37) paraphrases the theorem as shown below:

$$p[\text{failure}|\text{indicator}] = \frac{p[\text{indicator}|\text{failure}] \times p[\text{failure}]}{\{p[\text{indicator}|\text{failure}] \times p[\text{failure}]\} + {} \atop {} + \{p[\text{indicator}|\text{no failure} \times p[\text{no failure}]\}}$$

where $p[\text{cause}]$ = probability that cause occurs

$\quad\quad p[\text{effect}]$ = probability that effect occurs

$p[\text{effect}|\text{cause}]$ = conditional probability of effect given cause

$p[\text{cause}|\text{effect}]$ = conditional probability of cause given effect

Pearl (1990, p. 541) identified the following "three flags of Bayesian probability":

1. It reasons backward from evidence to hypothesis.

2. It accepts subjective evaluations.

3. It builds a total model of the situation, creating an edifice in which all probabilities add up to 100%, rather than assessing individual frequencies in isolation.

Ang and Tang (1975) define the latter condition by Eq. A4.5.

Hays (1994, p. 47) explains:

In and of itself, Bayes's theorem is in no sense controversial. However the question of its appropriate use has, in years past, been a focal point in the controversy between those who favor a strict relative-frequency interpretation of probability and those who would admit a subjective interpretation as well. The issue emerges clearly when some of the probabilities figuring in Bayes's theorem are associated with *states of nature* or with nonrepetitive events. . . . It is usually difficult to give meaningful relative frequency interpretations to probabilities for such states or *one-time* events.

APPENDIX 5.   UNCERTAINTY

Wadia-Fascetti et al. (2002), consistent with ISO (1995), refer to the following three types of uncertainty associated with engineering products and processes:

Ignorance. The significant parameters defining the as-built and as-is state cannot be fully known. Included are quantitative measurements of structural, material behavior, remaining useful life, vulnerabilities, potential hazards, material properties such as fatigue life, amount and effect of repair, and maintenance work. Ignorance, for instance, the absence of quantitative measurements, is compensated by deterministic analysis resulting in qualitative (but vague) expert judgments or opinions.

Randomness. Events may be unstable and have an indeterminate outcome. Material properties such as yield, ultimate strength, fracture toughness, and chemical resistance vary. Random phenomena lend themselves to stochastic analysis. The causes for bridge deterioration and accidents contain randomness that can be statistically modeled given ample data.

Vagueness. According to B. Russell (Section 1.4), vagueness affects every attempt at a precise definition. Particularly susceptible are qualitative assessments, for example, condition ratings, load ratings, remaining useful life, redundancy, safety, reliability, vulnerability, potential hazards, and socioeconomic constraints. Vaguely defined bridge or element conditions can be represented by fuzzy sets (Frangopol and Furuta, 2000). Genetic algorithms and neural network models (Miyamoto and Frangopol, 2001) consider vagueness, producing ranges of possibilities with perceived likelihood, dependent on the data and the models.

McNeill and Freiberger (1994, pp. 187–188) list various combinations of vagueness with other uncertainties resulting in fuzziness, including the following:

Nonspecificity. Ambiguity or lack of informativeness. A one-to-many relation between statement and possible meaning. Can be addressed by crisp set theory. This definition appears similar to ignorance.

Dissonance. Pure conflict. Treated as Bayesian probability of one statement being correct as opposed to another.

Confusion. Pure and potential conflict. There is conflict and the meaning of the data is unclear. Treated by "possibility" theory.

Fuzziness. Vagueness, for example, to what degree does a term apply. Treated by fuzzy-set theory, which considers Bayesian probability (e.g., randomness) as a subset.

For the purposes of Bayesian probability, Thoft-Christensen and Baker (1982, p. 6) distinguish physical, statistical, and model uncertainties. Melchers (1987), recognizing that predecessor, organizes uncertainties into phenomenological, decision, modeling, prediction, physical, statistical, and human (error and intervention). Any one of the latter seven groups can contain the former three in numerous combinations of somewhat fuzzy distinction.

Ang and De Leon (2005) identify the following two types of uncertainty:

Aleatory. A nondeterministic property of natural randomness modeled by random variables.

Epistemic. An inability to correctly represent a possibly deterministic reality, particularly significant in risk-informed decisions.

Aleatory uncertainty corresponds to the randomness in the preceding classification, whereas epistemic uncertainty is related to ignorance.

APPENDIX 6.   QUANTITATIVE MANAGEMENT TECHNIQUES

George (1968, p. 157) described quantitative management techniques according to their main characteristic and areas of application (in alphabetical order):

Decision theory, including performance organization, learning theory, cybernetics, suboptimization	Objective determination, planning, conflict resolution, job estimates
Experimental design	All predictive models
Game theory	Timing and pricing in competitive market, military strategy
Information theory	Data processing system design, organization analysis, market research
Inventory control	Economic lot size and inventory control
Linear programming	Assignment of equipment and personnel, scheduling input–output analysis, transportation routing, product mix, allocation processes
Probability theory	All areas (author may be considering Bayesian probability)
Queuing theory	Inventory, traffic control, telephone trunking systems, scheduling of services, radio communications, etc.
Replacement theory	Replacement of equipment in response to failure and deterioration
Sampling theory	Quality control, simplified accounting and auditing, market research
Simulation theory (Monte Carlo methods)	System reliability evaluation, profit planning, logistic system studies, inventory and personnel needs
Statistical decision theory	Estimation of model parameters in probabilistic methods
Symbolic logic	Circuit design, legal inference, contract consistencies

Also mentioned are the significant contributors. The emphasis is on application. Thus the fundamental theoretical developments are considered along with derivative techniques. The main objective is to quantify the previously intuitive management.

APPENDIX 7.   STRUCTURAL RELIABILITY

The reliability of civil engineering structures is a special case of system reliability analysis. Hudson et al. (1997, p. 240) wrote: "Data regarding failure rate versus time are usually the basis of reliability predictions for infrastructure components. . . . Six generic trends of failure probability against age, for a variety of electrical and mechanical items, have been reported by Moubray."

The six patterns show different time dependence during the initial, middle, and final stages of their respective life cycles. The rapid deceleration of initial failures (e.g., a "shakedown" period) followed by a long uneventful plateau and ending with a swiftly increasing failure rate is known as the "bathtub" curve.

For general purposes (mostly applicable to electronic systems), Barlow et al. (1965, p. 5) formulate probabilistic reliability along with availability, interval availability, efficiency, and effectiveness as follows: "Unfortunately, the definitions given in the literature are sometimes unclear and inexact, and vary among different writers. . . . We shall define mathematically a single generalized quantity which, when appropriately specialized, will yield most of the fundamental quantities of reliability theory."

That generalized quantity of the preceding quote is the *expected gain* $G(t)$. It is associated with the *state* of a system at time t, described by vector-valued random variables $X(t)$.

Over a time interval $a \leq t \leq b$, the expected gain $G(t)$ has the following average $H(a,b)$ with respect to a weight function $W(t)$:

$$H(a,b) = \int_a^b G(t)W(t) \tag{A7.1}$$

In terms of the expected gain $G(t)$ and average expected gain $H(a,b)$, the authors define or refer to the following five "basic quantities arising in reliability theory":

1. *Reliability* $= G(t)$. The probability of a device performing its purpose adequately for the period of time intended under the operating conditions encountered (without need of repair).

2. *Pointwise Availability.* The probability that the system will be able to operate within the tolerances at a given instant in time (allowing repair).

3. *Interval Availability.* The expected fraction of a given interval of time that the system will be able to operate within the tolerances (allowing repair). The authors equate this quantity with *efficiency.*

4. *Limiting Interval Availability.* The expected fraction of time in the long run that the system operates satisfactorily.

5. *Interval Reliability.* The probability that at a specified time the system is operating and will continue to operate for an interval of duration, say x (without the benefit of replacement).

Thoft-Christensen and Baker (1982, Chapter 4) develop reliability for load factor (ultimate-strength) design of engineering structures subjected to random events.

A *safety factor* implies that a structure would be "safe" until conditions are aggravated by the amount of that factor. The authors (p. 1) reject that approach on the following grounds: "It is now widely recognized that some risk of unacceptable structural performance must be tolerated."

In order to estimate that risk, reliability is defined (p. 8) as "the *probability* that a structure will not attain each specified limit state (ultimate or serviceability) during a specified *reference period.*"

The reliability function $R_T(t)$ is the probability that a system will still be operational at time t:

$$R_T(t) = 1 - F_T(t) \tag{A7.2}$$

where the failure distribution $F_T(t)$ is a function of the random time T to failure.

If the density function f_T of the time to failure were known, then

$$R_T(t) = 1 - \int_0^t f_T(\tau)\, d\tau = \int_t^\infty f_T(\tau)\, d\tau \tag{A7.3}$$

If $\lim_{t \to \infty} [t R_T(t)] \to 0$, then the *expected life* of the system is

$$E[T] = \int_0^\infty R_T(t)\, dt \tag{A7.4}$$

Assuming that both the design load (demand) and structural resistance (supply) have normal distributions F_S and F_R, respectively, the structural reliability $\check{R}$ is the probability that the structure will survive under the load as follows:

$$\check{R} = 1 - P_f = 1 - \int_{-\infty}^\infty F_R(x)\, f_S(x)\, dx \tag{A7.5}$$

or

$$\check{R} = 1 - P_f = 1 - \int_{-\infty}^\infty [1 - F_S(x)]\, f_R(x)\, dx \tag{A7.5a}$$

where P_f = probability of failure
f_S = density function corresponding to F_S
f_R = density function corresponding to F_R

Thoft-Christensen and Baker (1982, p. 73) emphasize that, although Eqs. A7.5 and A7.5a yield numerically identical results, they represent two different assumptions, the former pertaining to the strength distribution and the latter to the distribution of the loads in the failing structure. The probability of failure can be modeled in terms of the distribution of the resistance R (Eq. A7.6) or as the distribution of the load S (or Q) (Eq. A7.6a) in a failing structure. Here, P_f is not equal to the area where the two distributions overlap:

$$P_f = \int_{-\infty}^\infty F_R(x)\, f_Q(x)\, dx \tag{A7.6}$$

$$P_f = \int_{-\infty}^\infty [1 - F_Q(x)]\, f_R(x)\, dx \tag{A7.6a}$$

Figure A7.1 illustrates the reliability model. Recent LRFD texts denote the demand of the load effects by Q rather than S.

For a linear safety margin and normally distributed independent variables, the *reliability index β* (Thoft-Christensen and Baker, 1982, p. 89) is the number of standard deviations σ_M by which the mean μ_M exceeds zero, or

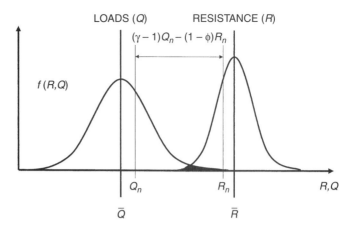

Figure A7.1 Normally distributed supply R and demand Q according to load factor design of structures.

$$\beta = \frac{\mu_M}{\sigma_M} = \frac{\mu_R - \mu_S}{(\sigma_R^2 + \sigma_S^2)^{1/2}} \qquad (A7.7)$$

where μ_R, μ_S, μ_M = normally distributed functions R, S, and M, respectively

σ_R, σ_S, σ_M = standard deviations of distributed functions R, S, and M, respectively

R = resistance, random variable with normal distribution function F_R

S = load effect, random variable with normal distribution function F_S

M = failure or limit state function, $= R - S$

Defining μ_R/μ_S as a *safety factor* does not imply that R and S (or Q) are constants. Barker and Puckett (1997, Section 3.46) develop a similar *safety index* β for normal and lognormal distributions.

When the failure function is not linear, the reliability index β can be estimated by Monte Carlo simulation (MCS) or by the iterative linearization procedures known as FORM (first-order reliability method) or SORM (second-order reliability method), described by Ang and Tang (1984).

A target reliability index $\beta = 3.5$ has been used to calibrate LRFD AASHTO (1998a), implying a probability of failure of 0.0233% for new structures. For older structures β is estimated closer to 2.5, suggesting a failure probability of 0.621%. Both values are based on assumptions about the probability of failure of individual structural components in nonredundant systems. The calibration of β takes into account existing satisfactory design loads as described by Nowak (in Frangopol, 1999b, Chapter 7) and NCHRP Report 454 (2001). The result is a recommendation for strength and load factors (Section 4.2.2). Attempts are made to optimize β with respect to cost; however, life-cycle cost estimates are highly speculative. Ghosn and Moses develop a procedure accounting for redundancy in NCHRP Report 406 (1998) (Appendix 17). Ghosn, Moses, and Wang (in NCHRP Report 489, 2003) calibrate β for extreme events.

APPENDIX 8. OPTIMIZATION

In his lectures at the Imperial College, London, from 1967 to 1985, Beale (1988) defined the subject as follows:

1.1 Introduction to optimization

Optimization involves finding the best solutions to a problem. Mathematically, this means finding the minimum or maximum of a function of n variables, $f(x_1, \ldots, x_n)$, say, where n may be any integer greater than zero. The function may be unconstrained or it may be subject to certain constraints on the variables of the function, say $g(x_1, \ldots, x_n) = b_i$ for $i = 1, \ldots, m$. . . . The functions $f(\mathbf{x})$ and $g(\mathbf{x})$ usually have some real physical meaning (for example total cost and profit or capacity and demand restrictions).

Although optimization is used in various branches of applied mathematics and statistics, it is particularly associated with operational research. Before discussing what operational research involves, we will spend a few minutes distinguishing it from statistics.

We can argue that statistics is concerned with trying to understand what is happening (or what might happen) in an uncertain world full of apparently random phenomena, and that operational research is concerned with deciding what to do about it. . . .

Operational research is therefore concerned with *decision making*. This can be an instinctive, or at least intuitive, process, but this is not always a satisfactory way to make decisions, particularly those made on behalf of other people, for example by a government department or a commercial organization. We may therefore approach the situation more methodically, and list the alternative possible decisions and their respective advantages and disadvantages. We may then go further and quantify them, and this leads us to make what is generally known as a *mathematical model* of the situation requiring a decision.

Mathematical modeling is at the heart of operational research. This . . . means analyzing some logical structure that is as simple as possible while still representing the essence of the problem faced by the decision maker. . . .

There are three things . . . typical of operational research models. The first is that we do *not* try to make the model as *realistic as possible*. . . . Just as the problem for which the model is developed is one of finding the best compromise between partially conflicting objectives, so the art of model building itself is one of finding the best compromise between realistically representing the situation and being able to collect data easily and draw conclusions from the results.

The next thing to note is that the use of the model involves *optimization*. . . . (Such) problems involve quantitative variables. . . . These problems may require numerical techniques for finding maxima or minima, as well as skill in model building. . . .

The third point to be made about . . . many models is that their value is not so much that they give the best answer to the problem—which, of course, is only true to the extent that the model is valid—it is much more that the model provides a convenient framework for constructive thought about the problem. . . . This may lead us on to more elaborate versions of the model if we are dissatisfied with the alternatives offered to us by the simple model.

APPENDIX 9. PROBABILITY

Probability is usually defined as the likelihood of an uncertain outcome. The type of uncertainty should suggest the appropriate method of probability assessment.

Stochastic analysis estimates likelihood of random outcomes according to an assumed probability distribution best fitting available data. Ang and Tang (1975, Chapter 3) list a number of random-variable distributions suitable for engineering purposes, including the normal (or Gaussian), standard normal, lognormal, binomial, geometric, Poisson, Gumbel, and exponential. The authors derive probability distributions, discuss their validity, and use them for parameter estimation. Some of the basic terminology is quoted from Chapter 3 of the text.

If X is a random variable, its probability distribution can always be described by its *cumulative distribution function (CDF)*, which is

$$F_X(x) \equiv P(X \leq x) \tag{A9.1}$$

where x are all values of X.

For a discrete random variable X, the *probability mass function (PMF)* is $P(X = x)$ for all x. If the PMF is $p_X(x_i) \equiv P(X = x_i)$, the distribution function of X is

$$F_X(x) = P(X \leq x) = \sum_{\text{all } x_i \leq x} P(X = x_i) = \sum_{\text{all } x_i \leq x} p_X(x_i) \tag{A9.2}$$

For a *continuous random variable* X the probability P in the interval $(a, b]$ is defined in terms of the *probability density function (PDF)* $f_X(x)$ as follows:

$$P(a < X \leq b) = \int_a^b f_X(x)\, dx \tag{A9.3}$$

The corresponding *distribution function* is

$$F_X(x) = P(X \leq x) = \int_{-\infty}^x f_X(\xi)\, d\xi \tag{A9.4}$$

Hence

$$f_X(x) = \frac{dF_X(x)}{dx} \tag{A9.5}$$

The *central limit* theorem is stated as follows: If a population has finite variance σ^2 and a finite mean μ, the distribution of sample means from samples of N independent observations approaches the form of a normal distribution with variance σ^2/N and mean μ as the sample size N increases. When N is very large, the sampling distribution of $\bar{x}$ is approximately normal.

The mean μ or expected value $E(x_i)$ of a sample x_i is defined as

$$\mu = E(x_i) = \frac{\sum_{i=1}^N x_i}{N} \tag{A9.6}$$

where x_i is the sample.

The variance σ^2 is defined as

$$\sigma^2 = \sum_{i=1}^{N} \frac{(x_i - \mu_x)^2}{N} \tag{A9.7}$$

where σ is the standard deviation.

The normal, or Gaussian, distribution, referred to in the central limit theorem, is symmetric and representative of the widest range of phenomena (see Examples 12 and 18). As pointed out by Thoft-Christensen and Baker (1982, Chapter 9.3), it is particularly helpful that linear operations on a Gaussian process result in another Gaussian process. The distribution proposed by Weibull in 1939 as "the simplest mathematical expression of the appropriate form" has proven particularly suitable for modeling the behavior of deteriorating structures. The density functions f_X of these frequently used distributions are as follows:

- *Normal (Gaussian)*

$$f_X(x) = \frac{1}{\sigma(2\pi)^{1/2}} \exp\left[-\frac{1}{2}\left(\frac{x - \mu}{\sigma}\right)^2\right] \quad -\infty < x < \infty \tag{A9.8}$$

  where μ = mean of variate
  σ = standard deviation of variate

- *Standard Normal*

$$f_S(s) = \frac{1}{(2\pi)^{1/2}} \exp\left(-\frac{s^2}{2}\right) \quad -\infty < s < \infty \tag{A9.9}$$

  where

$$\mu = 0 \quad \sigma = 1.0$$

- *Weibull*

$$f(t; \lambda, \alpha) = \lambda \, \alpha \, t^{a-1} \exp\left[-(\lambda t)^\alpha\right] \tag{A9.10}$$

  where $t \geq 0$, $\lambda > 0$ is a scale parameter, and $\alpha > 0$ is a shape parameter.

- *Poisson*

$$P(X_t = x) = \frac{\nu t^x}{x!} e^{-\nu t} \quad x = 0, 1, 2, \ldots \tag{A9.11}$$

  where X_t = number of occurrences in time or space interval t
  ν = mean occurrence rate

  The mean or expected value and the variance are

$$\mu(X_t) = \sigma^2(X_t) = \nu t \tag{A9.11a}$$

- *Exponential.* If events occur according to a Poisson process, then the time T_1 until the first occurrence has an exponential distribution. According to Eq. A.9.11

$$P(T_1 > t) = P(X_t - 0) = e^{-\nu t} \tag{A9.12}$$

where $(T_1 > t)$ = no event occurs in time t

T_1 = first occurrence *and* recurrence time (since occurrences of an event in nonoverlapping time intervals in a Poisson process are statistically independent)

$$F_{T1}(t) = P(T_1 \leq t) = 1 - e^{-\mu} \qquad (A9.12a)$$

where $F_{T1}(t)$ = distribution function of T_1

$$f_{T1}(t) = \frac{dF}{dt} = \nu e^{-\mu} \qquad t \geq 0 \qquad (A9.12b)$$

where $f_{T1}(t)$ = density function

If ν = const, then the mean recurrence time for a simple Poisson process is

$$\mu(T_1) = \frac{1}{\nu} \qquad (A9.12c)$$

• *Bernoulli Sequence and Binomial PMF.* The following conditions must be satisfied:

1. Each trial has only two possible outcomes, e.g., occurrence and nonoccurrence. (This is known as bivalence, in contrast with the polyvalence of fuzzy sets.)

2. The probability of occurrence of the event in each trial is constant.

3. The trials are statistically independent.

Ang and Tang (1975, p. 107) point out that if the operational conditions of a fleet of equipment are statistically independent and the probability of malfunction is the same, the condition of that fleet constitutes a Bernoulli sequence. Events are statistically independent and the annual probability of exceeding a prescribed threshold is constant. If the probability of occurrence of an event in each trial is p and that of nonoccurrence is $1 - p$, then the probability of exactly x occurrences among n trials in a Bernoulli sequence is obtained by the binomial PMF as follows:

$$P(X = x) = \frac{p^x (1 - p)^{n-x} n!}{x!(n - x)!} \qquad x = 0, 1, 2, \ldots, n \qquad (A9.13)$$

where p is a parameter.

APPENDIX 10.   UPPER AND LOWER BOUND THEOREMS OF PLASTIC FRAME ANALYSIS

According to Neal (1956, 1981) the plastic collapse theorems were formulated by Gvozdev in 1936, Horne in 1950, and Greenberg and Prager in 1952. His formulation is as follows (pp. 48–49):

Static Theorem. If there exists any distribution of bending moment throughout a frame which is both safe and statically admissible with a set of loads λ, the value of λ must be less than or equal to the collapse load factor λ_c.

Kinematic Theorem. For a given frame subjected to a set of loads λ, the value of λ which corresponds to any assumed mechanism must be either greater than or equal to the collapse load factor λ_c.

Uniqueness Theorem. For a given frame and set of loads λ, if there is at least one safe and statically admissible bending moment distribution in which the plastic moment occurs at enough cross sections to produce a mechanism, the corresponding load factor will be the collapse load factor λ_c.

In sum:

Static conditions	$\lambda \leq \lambda_c$
Kinematic conditions	$\lambda \geq \lambda_c$
Collapse	$\lambda = \lambda_c$

Because static and kinematic solutions approach λ_c from below and from above, respectively, they are also referred to as the *lower* and *upper bound* methods (e.g., by Hodge, 1970, p. 20). The upper bound approach envisions the failure mechanism (e.g., the form), whereas the lower bound is concerned with the supply of strength (e.g., the content). To general safety considerations, the upper/lower bound analogy contributes the following reminders:

- Failures occur at the *lowest critical combination of demands.*
- The strength reserve (or "overdesign") of performing structures depends on the failure mode.
- Strengthening one weak point does not increase the overall safety of a system.
- Critical combinations of vulnerabilities match or exceed the demands a structure can sustain safely. (Hence, Murphy's Law is a lower-bound formulation.)

APPENDIX 11.   HIGHLIGHTS OF HISTORY OF U.S. NATIONAL BRIDGE INVENTORY (NBI)

1968   The Federal-Aid Highway Act introduces nationwide bridge inventory.

1971   The National Bridge Inspection Standard (NBIS) mandates biennial inspections. The first *Coding Guide,* the *Bridge Inspector's Training Manual 70,* and the AASHTO *Manual for Maintenance and Inspection of Bridges* are issued.

1973   274,000 bridges on the Federal-Aid Highway System are inventoried.

1978   The Surface Transportation Assistance Act extends NBIS to all publicly used vehicular bridges longer than 20 ft (6.1 m). The inventory includes 577,000 bridges, an estimated 97% of the total.

1988   The Uniform Relocation Assistance Act introduces fracture-critical and underwater inspection procedures (86% of the 592,000 bridges on the inventory are over water). Second edition of the *Coding Guide* is published.

1991   The Intermodal Surface Transportation Efficiency Act (ISTEA) mandates the use of BMSs, along with pavement, safety, congestion, public transportation, and intermodal management systems.

FHWA sponsors the development of PONTIS—a BMS able to satisfy the needs of any agency in charge of a bridge network.

NCHRP (National Cooperative Highway Research Program) of the TRB (Transportation Research Board) develops BRIDGIT—a BMS for smaller networks or local systems.

1994   AASHTO revises the *Manual for Condition Evaluation of Bridges.* FHWA revises the *Coding Guide* (primarily in compliance with the metric conversion). FHWA (Nov. 1994) recommends life-cycle cost–benefit considerations.

1995   The National Highway System Act (NHS) downgrades the requirement for BMSs to a recommendation. The act mandates that states conduct life-cycle cost analysis (LCCA) for federally funded projects with costs greater than U.S. $25 million.

NCHRP sponsors the development of software, e.g., BLCCA (NCHRP Report 483, 2003), compatible with PONTIS and BRIDGIT for that purpose.

1998   The Transportation Equity Act of the Twenty-First Century (TEA-21) rescinds the LCCA mandate of the NHS (1995) but proposes criteria for integrated asset management (FHWA, 2001c). TEA-21 charges the Office of Asset Management at the U.S. DOT with the task of integrating the diverse infrastructure databases nationwide. The Information Systems (IS) infrastructure becomes a target of advanced management and design.

2003   FHWA issues the *Bridge Inspector's Reference Manual.* AASHTO issues the *Bridge Condition Evaluation Manual.*

2004   FHWA conducts a survey in preparation of a new *Inventory Guide* (2005), subject to AASHTO approval.

APPENDIX 12.   FIRST COSTS

In Chapter 53, "True Economy in Design," Waddell (1916, p. 1182) made the following statement:

> The great majority of bridge designers believe that the most economic structure is the one for which the first cost is a minimum; and from the contractor's prejudiced point of view this is correct, because his interest generally lies in securing the contract for the work regardless of all other considerations than his own profits; but from the purchaser's point of view that structure is the most economic which will do the work required of it for as long a time as necessary with the least possible expenditure for operation, maintenance, and repairs, all these desiderata being obtained with the smallest practicable initial cost of construction.
>
> In making an economic comparison of two or more designs for any proposed structure there are two methods of procedure, either of which is correct and satisfactory.
>
> The first is to find for each case what sum of money at the governing rate of interest will produce an income just sufficient to defray the average annual cost of operation, maintenance, repairs, and all other regular necessary expenditures, and add this amount to the total initial cost of the structure. The sum will be the "equivalent total first cost"; and if the designs be all satisfactory, and the proposed structures of practically equal life, that structure for which the equivalent total first cost is the least is the most economic.

The other method is to assume several future dates, preferably those at which certain large expenditures would probably have to be made for renewals or repairs of perishable portions, and compute the grand total cost to each date for each proposed structure under the assumption that it is then put into perfect condition, and allowing standard compound interest not only on the first cost but also on all annual expenditures. A comparison of these grand total costs at the several dates adopted will indicate clearly which is the most economic structure.

APPENDIX 13.   NETWORK AND PROJECT BRIDGE MANAGEMENT

NCHRP Report 300 (1987, p. 6) distinguished between the network- and project-level spheres of competence as follows: The project level treats each bridge on an individual basis for inspection, maintenance, repair, and/or rehabilitation (MR&R) needs. Once network-level decisions are made on priorities and funding, then a detailed evaluation of each selected bridge must follow at the project level, including:

- Detailed structural engineering analyses
- Distress type extent and severity of critical component
- Estimated remaining life
- Rate of deterioration
- Condition of secondary components
- Cost of design life of alternative MR&R treatments
- Availability of funds
- Essentiality of bridge to public
- Impact of repairs on traffic flow
- Related bridge or highway work nearby
- Type and size of bridge
- Load-carrying capacity of bridge
- Projected future use of bridge
- Historical significance of bridge

The activities associated with network-level planning and programming include the following:

- Automate data entry, editing, storage, and management.
- Summarize global network structural and functional conditions.
- Establish candidate project lists.
- Prioritize and select among the various MR&R actions for all candidate bridges in the system and identify resource requirements.
- Develop life-cycle cost estimates.
- Optimize the various alternatives.
- Evaluate funds and resource allocation alternatives.
- Develop outputs specifically related to bridge posting and load permit routing.

- Develop MR&R action schedules and cost data.
- Ensure that standards of optimal safe maintenance levels are followed.
- Ensure uniform reporting of inventory and inspection information.
- Report historical expenditures for different types of work (dollars, manpower, materials).
- Report historical changes of condition of plant and inventory as well as predicting effectiveness of global maintenance strategies.

APPENDIX 14.   U.S. NATIONAL BRIDGE INVENTORY (NBI) AND PROPOSED NBI SPECIFICATIONS

The *Coding Guide* (FHWA, 1995b, p. viii) defines the U.S. NBI as follows: "The aggregation of structure inventory and appraisal data collected to fulfill the requirements of the National Bridge Inspection Standards (NBIS) that each State shall prepare and maintain an inventory of all bridges to the NBIS."

The *Specifications* (FHWA, 2005a, pp. 6–9) propose 33 new or amended definitions among which is the following: "An aggregation of State and Federal agency bridge and roadway records maintained by the Federal Highway Administration (FHWA). The NBIS requires each state [and federal agency] to prepare a bridge inventory, which must be submitted to the FHWA in accordance with this Specification when requested (usually annually)."

The original definition distinguishes between "inventory" and "appraisal," or between the as-built structure that could be defined from documents (if available) and its as-is condition which must be subject to periodic reassessment.

NBIS 23 *Code of Federal Regulations* (CFR) 650, Subpart C, specifies the following structural inventory and appraisal (SI&A) requirements:

§650.315 (a) Each State or Federal agency must prepare and maintain an inventory of all bridges subject to NBIS. Certain SI&A data must be collected and retained by the State or Federal agency for collecting by FHWA as requested. A tabulation of this data is contained in the SI&A sheet distributed by the FHWA as part of the "Recording and Coding Guide for the Structure Inventory and Appraisal of the Nation's Bridges:" (December 1995), together with subsequent interim changes or the most recent version. Report the data using FHWA established procedures as outlined in the Recording and Coding Guide.

(b) For routine, in depth, fracture critical, underwater, damage and special inspections, enter the SI&A data into the State or Federal agency inventory within 90 days of the date of inspection for State or Federal agency bridges and within 180 days of the date of inspection for all other bridges.

(c) For existing bridge modifications that alter previously recorded data and for new bridges, enter the SI&A data into the State or Federal agency inventory within 90 days after the completion of the work for State or Federal agency bridges and within 180 days after the completion of the work for all other bridges.

(d) For changes in load restriction or closure status, enter the SI&A data into the State or Federal agency inventory within 90 days after the change in status of the structure for State or Federal agency bridges and within180 days after the change in status of the structure for all other bridges.

The NBI comprises the following "records": "Data which has been coded according to the Guide for each structure carrying highway traffic or each inventory route which goes under a structure. These data are furnished and stored in a compact alphanumeric format on magnetic tapes or disks suitable for electronic data processing."

The definition appropriately allows for technological updates, to be expected in a national infrastructure. FHWA (2005a) emphasizes the distinction between bridge and roadway records.

Bridges are defined according to the AASHTO transportation glossary and 23 CFR, Section 650.301, as follows:

A structure including supports erected over a depression or an obstruction, such as water, highway or railway, and having a track or passageway for carrying traffic or other moving loads, and having an opening measured along the center of the roadway of more than 20 feet (610 cm) between undercopings of abutments or spring lines of arches, or extreme ends of openings for multiple boxes; it may also include multiple pipes, where the clear distance between openings is less than half of the smaller contiguous opening.

FHWA (2005a) defines *highway bridges* as follows: "A bridge that carries a public road, which is any road or street under the jurisdiction of and maintained by a public authority and open to public travel. Bridges that carry only pedestrians, railroad tracks, pipelines, or other types of non-highway passageways are not highway bridges."

Approximately 150,000 railroad bridges and as many pedestrian bridges are managed by the respective responsible owners. For the years 1992 and 2004, the breakdown of highway bridges (including culverts) by material and type is shown in Tables A14.1 and A14.2.

FHWA (2005a) proposes the following definition of *culvert:* "A structure designed hydraulically to take advantage of submergence to increase hydraulic capacity. Culverts, as distinguished from bridge type structures, are usually covered with embankment and

Table A14.1 Bridge Type by Material (NBI)

Bridge by Material	1992			2004		
	Total	Deficient	Obsolete	Total	Deficient	Obsolete
Concrete	170,711	18,265	22,790	168,346	14,389	20,505
Concrete continuous	58,331	4,348	8,445	77,040	5,009	8,128
Steel	165,430	62,927	27,374	143,682	37,887	26,708
Steel continuous	43,151	5,411	7,617	48,612	4,377	8,719
Prestressed concrete (PC)	76,238	3,139	8,043	104,313	3,973	10,574
PC continuous	9,386	226	649	18,636	300	1,211
Wood	44,673	23,107	4,603	29,660	11,122	4,013
Masonry	1,959	520	557	1,857	432	521
Aluminum/iron	944	287	85	1,295	219	96
Other	1,701	506	273	461	51	73
Total	572,524	118,736	80,436	593,902	77,759	80,548

Table A14.2 Bridges by Main Structural Type (NBI)

Year	1992			2005		
Structure Type	Total	Deficient	Obsolete	Total	Deficient	Obsolete
Slab	73,974	8,628	10,120	78,677	6,792	9,545
Stringer/multibeam or girder	261,648	71,774	40,934	248.970	42,478	40,370
Girder and floor beam system	11,342	5,153	2,242	7,922	2,754	1,878
Tee beam	37,816	5,141	8,909	36,657	4,795	7,902
Box beam or girders (multiple)	34,273	1,347	4,567	46,337	2,448	5,445
Box beam/girders (single or spread)	4,662	261	430	7,525	289	910
Frame (except culverts)	4,557	1,055	1,200	5,010	329	1,310
Orthotropic	263	15	18	446	90	155
Truss deck	1,082	575	184	772	288	171
Truss thru	23,383	16,458	3,353	13,105	7,708	2,449
Arch deck	8,197	2,356	2,548	7,106	1,650	2,243
Arch thru	435	146	119	387	82	106
Suspension	110	47	41	98	33	41
Stayed girder	15	1	1	36	2	2
Movable lift	143	56	42	170	44	73
Movable bascule	525	196	139	475	132	154
Movable swing	251	121	66	229	102	65
Tunnel	89	8	59	68	5	40
Culvert	101,066	3,426	4,241	123,376	2,863	5,523
Mixed types	331	87	57	1,322	295	163
Segmental box girder	43	0	3	169	7	9
Channel beam	3,817	468	393	13,724	2,167	1,113
Other	4,506	1,416	772	2,649	556	695
Total	572,528	118,735	80,438	595,230	75,909	80,362

are composed of structural material around the entire perimeter, although some are supported on spread footings with the streambed serving as the bottom of the culvert. Culvert that meet the NBIS definition of a bridge are considered bridges.''

Inventory route is defined as (FHWA, 2005a) ''the route for which the applicable inventory data is to be recorded. The inventory route may be on the structure or under the structure. Inventories usually proceed from west to east and from south to north.''

FHWA (2005a) proposes: ''A highway for which applicable roadway data is recorded in the NBI. A highway bridge will always have one inventory route carried on the bridge and may have zero, one or multiple inventory routes passing under it. An inventoried non-highway bridge (such as a railroad bridge over a highway) will not carry an inventory route on the bridge, but will have one or more inventory routes passing under it.''

Thus, although the NBI maintains bridge and roadway records, information of non-highway bridges is anticipated.

Defense items and public roads are also defined. FHWA (2005a) defines the *Strategic Highway Network* (*STRAHNET*) as follows: ''A system of public highways that is a key deterrent in United States strategic policy. It provides defense access, continuity, and emergency capabilities for movements of personnel and equipment in both peace and

Table A14.3   Bridge Type: Feature Carried (FHWA, 1988)

Type	Feature Carried	Type	Feature Carried
1	Highway	6	Overpass structure at an interchange or second level of a multilevel interchange
2	Railroad	7	Third level (interchange)
3	Pedestrian exclusively	8	Fourth level (interchange)
4	Highway, railroad	9	Building or plaza
5	Highway, pedestrian	0	Other

Table A14.4   Bridge Type: Feature Crossed (FWHA, 1988)

Type	Feature Crossed	Type	Feature Crossed
1	Highway, with or without pedestrian	6	Highway, waterway
2	Railroad	7	Railroad, waterway
3	Pedestrian exclusively	8	Highway, waterway, railroad
4	Highway, railroad	9	Relief for waterway
5	Waterway table	0	Other

war. It includes the entire 73,025 km Interstate System and 25,215 km of other important public highways."

The 20-ft (6.10-m) lower limit excludes culverts. They require different expertise and may fall under the purview of other responsible owners, for instance, the environmental protection agencies. Tables A14.3 and A14.4 show the NBI designations for features carried and crossed by bridges.

APPENDIX 15.   ANALYTICAL TOOLS FOR ASSET MANAGEMENT

NCHRP Report 545 (2005), developed by Cambridge Systematics, PB Consult, and System Metrics Group, reviewed asset management practices in California, Florida, Massachusetts, Maryland, Michigan, Montana, New York, Ohio, South Carolina, and Wisconsin. The findings suggested that user-friendly adaptable what-if tools can benefit the managers of the reviewed and other localities if they provide the following four views:

- *Budget view*—allowing the user to explore the relationship between the average level of investment over time and the value of a single performance measure for a selected asset and portion of the network.

- *Targeting view*—allowing the user to determine the average annual expenditure required to reach a target performance measure value over a selected time frame.

- *Dashboard view*—allowing the user to survey several different performance indicators at once and to explore their sensitivity to overall budget levels and the allocation of budget across assets, geographic areas, and portions of the network.

- *Allocation view*—allowing the user to define different resource allocation scenarios and to see graphs that compare their performance impact over time.

Two tools are developed to provide these views on different levels:

A. Network tool (NT), analyzing investment versus performance across categories for the highway mode, satisfying the following functional requirements:

  1. Accept investment and performance data from asset management systems.

  2. Provide capability for interactive what-if trade-off analysis.

  3. Display results in graphical views.

  4. Produce standard reports.

B. Project trade-off tool (PT), limited to a functional spreadsheet proof-of-concept system, satisfying the following functional requirements:

  1. Allow user definition of performance measures and analysis categories.

  2. Accept project information.

  3. Accept system-level information.

  4. Calculate impacts of a set of projects on system performance and expenditures.

  5. Provide interactive interface for adjusting projects in the program.

  6. Provide summary reports and graphs of program performance impacts.

The PT output includes the following:

- Expenditures by budget category for a budget scenario allowing filtering by geographic and network category
- Before and after performance measure values (compared to targets, if established) for a single budget scenario or comparison across budget scenarios (in tabular form)
- Expenditures by project type for a budget scenario, allowing filtering by geographic and network category
- Amount of work by type compared to established work targets for a budget scenario or comparison across budget scenarios (in tabular form)
- List of projects selected for a given budget scenario organized by program category

Detailed subrequirements and definitions are provided. Future functionality enhancements include the following:

- Analysis of subsets of the project list
- Autoaggregation for baseline measures and indicators
- Accommodation of annual budget constraints
- Accommodation of "plug" program items
- Accommodation of multiple sets of budget categories

Improvements of the asset management analytic tools are recommended in the following areas:

- Definition of preservation strategies
- Full assessments of costs and benefits of alternative investments
- Resource allocation decisions

- Management of other assets
- Monitoring and feedback support

The report is accompanied by user guides for the software.

APPENDIX 16.   BRIDGE MANAGEMENT SYSTEM (BMS)

Figures A16.1–A16.7 show examples of BMS flowcharts.

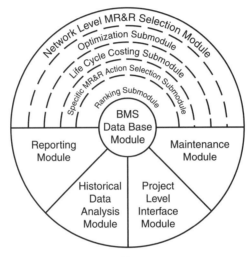

Figure A16.1   Essential BMS modules (FHWA).

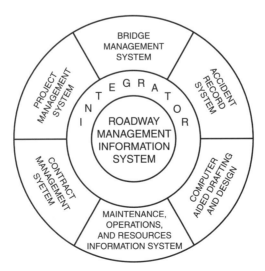

Figure A16.2   PennDOT BMS.

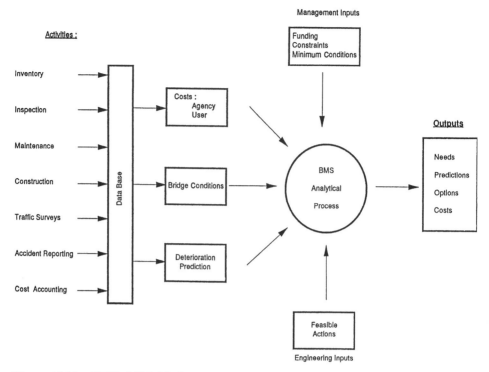

Figure A16.3   OECD (1992) BMS.

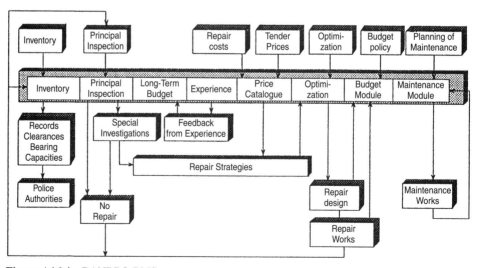

Figure A16.4   DANBRO BMS.

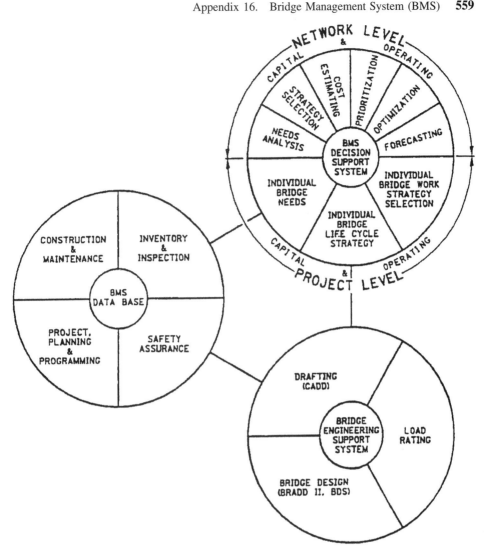

Figure A16.5   NYS DOT BMS.

NCHRP Report 300 (1987) identified the following "minimum basic modules" of a BMS (p. 9):

- Database
- Network-level maintenance rehabilitation and replacement (MR&R) selection

  Ranking
  Specific MR&R action selection
  Life-cycle costing
  Optimization

- Maintenance
- Historical data analysis

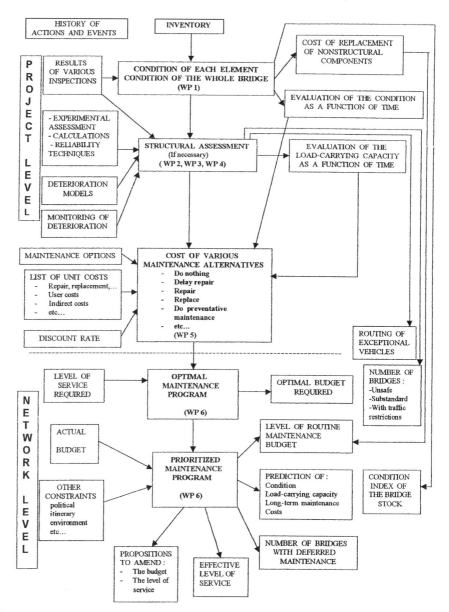

Figure A16.6   BRIME (2002) BMS.

- Project-level interface
- Reporting

Bjerrum et al. (2006) report that the bridge management system of the Danish Road Directorate (DANBRO; Fig. A16.4) is upgrading to a new Internet-based client–server system (DANBRO+) with the following modules:

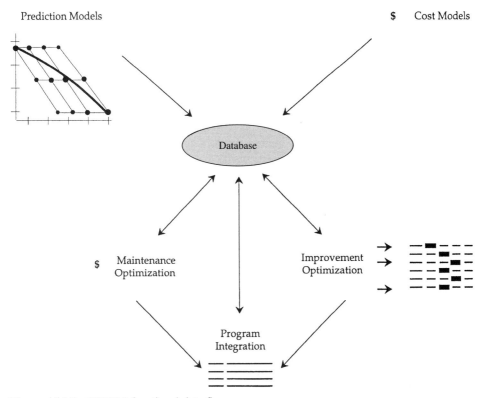

Figure A16.7   PONTIS functional data flow.

- Introduction
- Administration
- Inventory
- Management tools
- Traffic management
- Maintenance and repair/rehabilitation
- Equipment and materials

APPENDIX 17.   BRIDGE RELIABILITY ACCOUNTING FOR REDUNDANCY

Barlow et al. (1965, Chapter 6) consider network redundancies of two types:

- Parallel (redundant units operate simultaneously in parallel)
- Standby (redundant units stand by as spares and are used successively for replacement)

The optimal allocation of redundancy is obtained under one or more constraints (such as cost, volume, weight). The approach is better suited for networks than bridges.

Cremona (2003, Chapter 12.5) defined the reliability of parallel and series systems as follows: A system is a set of failure elements. If a structural element is not a failure element, the structure is not a system by this definition.

A failure mechanism is a subset of elements which, by their failure, cause the failure of the system. The system failure therefore results from the failure of all the elements in one mechanism.

In a *series* system all failure mechanisms consist of a single element. A *parallel* system has a single failure mechanism.

Ghosn and Moses (in NCHRP Report 406, 1998) link structural redundancy to reliability in terms of the following limit states, distinct from the ones listed in LRFD (AASHTO, 2004) (Appendix 25 herein):

Member failure	Determined by elastic analysis
Ultimate	Ultimate load capacity and formation of collapse mechanism
Functionality	Maximum deflection equal to 1% of span length
Damaged condition	Ultimate capacity with one main load-carrying component removed

The capacities at each of the above limit states are LF_1, LF_u, LF_f, and LF_d, respectively. At each limit state, target system reliability indices $\Delta\beta$ are set such that

$$\Delta\beta_u = \beta_{ult} - \beta_{member} \qquad \Delta\beta_f = \beta_{funct} - \beta_{member} \qquad \Delta\beta_d = \beta_{damaged} - \beta_{member}$$

where β is defined by Eq. A7.6.

Lognormal distributions are assumed, based on empirical evidence of structural performance. Gumbel (or other) distribution functions which may be more representative of the random variables involved must be converted to equivalent normal distributions *at the point where the failure is most likely to occur* (NCHRP Report 406, 1998, p. 12) by a "level 2 reliability" program. Level 2 is used in the sense proposed by Thoft-Christensen and Baker (1982) cited above.

With respect to the loss of one main member, the system reserve ratios R_u, R_f, and R_d for the *ultimate, functionality,* and *damage* limit states are

$$R_u = \frac{LF_u}{LF_1} \qquad R_f = \frac{LF_f}{LF_1} \qquad R_d = \frac{LF_d}{LF_1}$$

Deterministic values are computed for the capacity of typical structures at each limit state in appendices outside the scope of the main report. A direct redundancy check is developed based on required load factor ratios in Table 2 in NCHRP Report 406 (1998).

The reserve of nonredundant members $(R - D)$ should be increased by a factor φ_{red}^{-1}. The redundancy factor φ_{red} is defined as

$$\varphi_{red} = \min (r_1 r_u, r_1 r_f, r_1 r_d) \qquad (A17.1)$$

where $r_1 = LF_1/LF_{1,\,req} = (R_{provided} - D)/(R_{req} - D)$
$r_u = R_u/R_{u,\,req} = R_u/(LF_u/LF_1)_{req} = R_u/1.30$

$$r_f = R_f/R_{f,\text{req}} = R_f/(\text{LF}_f/\text{LF}_1)_{\text{req}} = R_f/1.10$$
$$r_d = R_d/R_{d,\text{req}} = R_d/(\text{LF}_d/\text{LF}_1)_{\text{req}} = R_d/0.50$$
$$\text{LF}_{1,\text{req}} = (R_{\text{req}} - D)/L_{\text{HS-20}} \text{ for AASHTO HS-20 live load}$$
$$R_{\text{provided}} = \text{provided member capacity}$$
$$R_{\text{req}} = \text{required member capacity}$$
$$D = \text{dead load}$$

The procedure can be applied to all structural members or to critical ones. The redundancy of a bridge superstructure can be evaluated by incremental nonlinear structural analysis allowing critical members to fail. The following system redundancy factor φ_s is introduced:

$$\gamma_d D_n + \gamma_l L_n(1 + I) = \varphi_s \varphi R' \tag{A17.2}$$

Here, φ_s^{-1} replaces η_i of the basic LRFD equation (Eq. 4.1 herein). Tables of system factors φ_s are developed for various bridges based on target average reliability indices.

Liu et al. (in NCHRP Report 458, 2001) apply the same approach to bridge piers.

The proposed procedure clearly defines the contributions of stochastic and deterministic analyses to the final results. It is up to the designer or manager to select the most critical members. Critical loading conditions (with respect to limit states) can be added to the left side of Eq. A17.2 if the member reserve ratio r_1 of Eq. A17.1 and the reliability indices β are adjusted accordingly.

AASHTO (2004) accounts for redundancy by the η factor in Eq. 3.1 above as follows:

For loads for which γ_i max is appropriate

$$\eta_i = \eta_D \eta_R \eta_I \geq 0.95 \tag{A17.3a}$$

For loads for which γ_i min is appropriate

$$\eta_i = \frac{1}{\eta_D \eta_R \eta_I} \leq 1.0 \tag{A17.3b}$$

where γ_i = load factor, statistically based multiplier applied to force effects
$\quad \eta_i$ = load modifier, relating to ductility, redundancy, and operational importance
$\quad \eta_D$ = factor relating to ductility:
$\qquad$ = 1.05 for nonductile components and connections in strength limit state
$\qquad$ = 1.00 for conventional designs and details complying with specifications
$\qquad \geq 0.95$ for components and connections for which additional ductility-enhancing measures have been specified beyond those required by specifications
$\qquad$ = 1.00 for all other limit states
$\quad \eta_R$ = factor relating to redundancy:
$\qquad \geq 1.05$ for nonredundant members in strength limit state
$\qquad$ = 1.00 for conventional levels of redundancy in strength limit state
$\qquad \geq 0.95$ for exceptional levels of redundancy in strength limit state
$\qquad$ = 1.00 for all other limit states
$\quad \eta_I$ = factor relating to operational importance:

≥ 1.05 for important bridges in strength limit state
$= 1.00$ for typical bridges in strength limit state
≥ 0.95 for relatively less important bridges in strength limit state
$= 1.00$ for all other limit states

APPENDIX 18.   DATA INTEGRATION

Data integration is the process of combining or linking numerous data sets from different sources, such as bridge, highway, railroad, and lifeline management systems. The *Data Integration Primer* (FHWA, 2001c) defined two database structures as follows:

Fused Database (e.g., Data Warehouse, Fig. A18.1). Selected data are gathered from various sources, filtered, and exported to a centralized database, replicating the data at other locations. A common user interface is used for the relevant subsets of all component databases following specified rules of data fusion.

Interoperable Database (e.g., Federated or Distributed System). This is a collection of separate and possibly diverse interpreting database systems over multiple sites linked through a computer communication network. Integrated (e.g., diverse) data standards and distributed processing capabilities are typical.

FHWA (2001c, pp. 17, 18) compared fused and interoperable databases as shown in Tables A18.1 and A18.2.

The comparison identifies the typical differences between centralized (top-down) nonredundant and decentralized (ground-up) redundant structures.

Within the global (fused or interoperable) database environment, data are managed by system, for example, a database management system (DBMS). FHWA (2001e) defined DBMS as "a collection of programs that enables information to be sorted in, modified, and extracted from a database." DBMS can be structured according to one of the following models:

Flat File. Data records have no structured interrelationship. Computer space is minimized, but applications must be familiar with the structure.

Hierarchical. Records form a family tree, each retaining a single owner. Typical of the first mainframe database management systems, this model cannot relate to real structures.

Table A18.1   Data Integration Alternatives

Interoperable (federated, distributed) database				*Fused data (data warehouse)*			
Pavement database	Roadway inventory	Structures database	Other databases	Pavement condition	Inventory location	Culvert structures	Maintenance history
				Pavement database	Roadway inventory	Structures database	Other databases
Multiple independent data servers				Single data server			

Table A18.2 Comparison of Fused and Interoperable Databases, Primer (FHWA, 2001c, p. 18)

Characteristic	Fused Database (Data Warehouse)	Interoperable Database
Data servers	One (central)	Multiple (distributed)
Data replication	Yes	No
Advantages	Easy to manage and control databases; maximum data processing power (quick access)	Can keep data in independent locations and file servers (autonomy of sites)
		No reliance on single site that can become point of failure (e.g., redundancy)
	Processing capacity	Transparency
	Security	Unified data description (e.g., no need to know data models)
		Access to resources on computer network
Disadvantages	Slow and costly implementation	Demanding and costly maintenance of global data model
	Read-only data, no online updates	Changes in data export protocol require rebuilding
	Storage demands	Demands for database management procedures (access and updates)

Relational. Data form sets of tables (relations) in which they can be accessed, extracted, or reassembled without reorganizing the original arrangement. Each table (relation) contains one or more data categories in columns. Each row contains a unique instance of data for the categories defined by the column.

Object Oriented. Data are units of information to be operated on by object codes (sequences of computer instructions). This model merges the data and the code into a single object.

The flowchart and table descriptions of bridge inspection in Example 26 are roughly analogous to the hierarchical and relation models.

Beyond the data models, the FHWA (2001c) discusses standards, reference systems, metadata, data dictionary, computer communication requirements, software, hardware, staffing, and data management requirements.

FHWA (2001e) is a concise glossary defining 100 data integration terms, including the following:

Application. Short for application program, which is a software program designed to perform a specific function directly for the user or for another software program.

Asset Management (AM). AM is a strategic approach to managing transportation infrastructure. It requires integrated data and information to make comprehensive decisions.

Atomic Data. Atomic data are items containing the lowest level of detail, as opposed to *aggregate data,* which comprise rollups of atomic data items.

Computer Network. A group of two or more computer systems linked together for the purpose of communications or application distribution. In *local-area networks* (*LANs*) computers are geographically close. In *wide-area networks* (*WANs*) computers are further apart and connected by telephone or radio.

Data.

Extraction: Process of reading one or more sources of data and creating new representation of data.

Mapping: Process of assigning source data element to target data element.

Migration/transformation: Process of converting data from one format to another. The need for migration arises when converting to a new computing or database management system that is incompatible with the current one.

Mining: Process of extracting previously unknown, valid, and actionable information or relationships from large databases.

Modeling: Method used to define and analyze data requirements, supporting business process of an organization.

Partitioning: Process of physically or logically dividing data into segments, facilitating maintenance and access (as in *relational* database management systems).

Scrubbing: Filtering, merging, decoding, and translating source data to create validated data for *data warehouse.*

Warehouse: Collection of databases designed to support management decision making.

Information System Architecture. An information system framework that describes:

1. Business rules: functions a business performs and information it uses

2. System structure: definitions and interrelationships between applications and products

3. Technical specifications: the interfaces, parameters, and protocols used by product and applications

4. Product specifications: standards pertaining to elements of technical specifications and application of vendor-specific tools and services in developing and running applications

Hassab (1997, p. 26) lists the following *basic structures* of organizations and databases:

Hierarchical. Depicted by a tree where each node, as a work center with its subsidiary parts, is linked to one superior, and all pre-set links among the nodes form no cycles. A cycle is a series of links that begin and end at the same node, without need for backtracking on any link. Connectivity from node to node follows the links to integrate the functions of the organization. Connectivity may be physical or logical.

Network. Depicted by a tree with added cycles to it and/or multiple superiors at any node.

Relational. Depicted by a table where the attributes of the parts are listed and no pre-set pointers are established. Each part has at least one multiple attribute which forms the key (pointer) to it. Connectivity of the parts is done on-line through dynamically linking unique attribute(s) that fulfill the sought after organizational function.

Although the author stresses that "there is no single best way to structure organizations," the relational structure appears to provide the highest level of adaptability to specific and possible dynamic needs. FHWA (2001c) provides a similar summary of data management system structures. The matrix form used throughout this text for representing two-dimensional interactions, such as management and engineering, process and product, and so on, is a variation of the relational structure.

APPENDIX 19.   PRIVATIZATION

Gómez-Ibáñez and Meyer (1993) summarized the significant experience in privatizing transportation facilities and services in a number of countries, including Britain, France, Spain, Mexico, Chile, and the United States. The authors identified three basic types of privatization (p. 1):

- Sale of existing state-owned enterprise.

Such sales have been largely motivated by the "widespread belief that the private sector is inherently more efficient than the public sector."

- Use of private financing and management rather than public for new infrastructure development.

Infrastructure privatization is often motivated by a desire to tap new sources of funds to supplement the constrained resources of the public sector. Efficiency may still be claimed as an important advantage as the private sector is often thought to build infrastructure cheaper or faster than public counterparts. . . . Unlike many other government services, infrastructure can often be supported by charges levied on users. Privatization offers the potential for financing infrastructure without overt increases in taxes; in many ways privatization can take an activity off the political agenda.

- Outsourcing (contracting out to private vendors) public services previously provided by government employees

Outsourcing of existing state-owned enterprises or infrastructure facilities offers "the prospect of immediate financial gain to government," on condition that "the to-be-privatized enterprise can generate operating income in excess of that needed to cover operating expenses and finance expected new investment needs."

The authors cautioned (pp. 4–5): "Privatization creates winners and losers. . . . Critics point out that a newly privatized state enterprise may have few incentives to be efficient or market oriented if it operates in a monopoly or uncompetitive market. . . . Critics also argue that private contractors may hold back on the quantity or quality of services they render unless their performance is monitored, and the costs of such monitoring could offset any savings in efficiency."

A. M. Howitt (in Gómez-Ibáñez and Meyer, 1993, p. 254) defined two types of efficiency as follows:

Technical efficiency concerns how effectively a firm or an industry uses its resources. A firm is technically efficient if it produces its output by using the least-cost production techniques and input combinations.

Allocative efficiency concerns whether the firm or industry is producing the appropriate output given society's competing needs for other goods and services. Allocative efficiency implies rules for both pricing and investment in new capacity.

According to Howitt (in Gómez-Ibáñez and Meyer, 1993, pp. 264–265): "Privatization is likely to lead to a revaluation of assets when capacity is expanded. . . . Users are likely to perceive revaluing old investments as unfair. . . . It is no surprise, therefore, that most often wholly new transportation infrastructure, rather than existing facilities, is privatized. Virtually all of the highway privatization programs in the United States, Europe, and the developing world have been of new facilities."

The author concluded: "When privatization involves losses, someone will have to make up the shortfall. The fundamental issue is who. There are two basic options: other private enterprises or the public sector. Often the two sources are used in combination, especially when the prospects for profitability are low, but each has advantages and disadvantages."

Gómez-Ibáñez and Meyer (1993, p. 8) summarized the conditions necessary for successful privatization as follows:

- Competition in the markets in which privatized firms buy and sell is highly desirable.

- Privatization is easier to effect, all else equal, when the efficiency gains from privatization are fairly large, that is, when the private sector is for some reason or another inherently more efficient than the public sector.

- Privatization is easier to implement when there are not too many redistributions or transfers linked with it.

- Privatization works best when associated with fewer controversial consequences such as environmental concerns or general opposition to economic development of growth (the toll road experience in the United States is cited as an example).

- Privatization is easier when the activity or service approximately covers its costs, neither requiring significant government subsidy nor generating significant surplus. If the revenues are far below expectations, the organization in charge would have to rely on government subsidies, whereas if they are exceedingly high, the government will attempt to tap into them for other purposes. Again examples are derived from the U.S. toll experience.

Subsequent to this publication, railway services were privatized in the United Kingdom with results that can be described as mixed and still hotly debated.

To the question "Should the autoroutes (in France) be privatized?" professor C. Saint-Etienne, president of France-Strategie, responded on the front page of *Le Monde,* September 29, 2005, as follows:

This operation cannot be analyzed independently of the desirable transportation policy of France, the creation of powerful French concession-construction groups, capable of intervening in a 25 member Europe, and, furthermore, the macroeconomic impact of a highway construction program that the country urgently needs.

The sale of the State's share makes no sense unless the French government intends, by this transaction, to establish major construction-concession organizations in order to design an ambitious program, building modern infrastructure capable of improving the quality of the global transportation system of the country, creating employment and adding significantly to the potential for economic growth.

This motivation appears similar to the one inspiring the public but autonomous transportation authorities in the United States. The George Washington Bridge, built and operated by the Port Authority of New York and New Jersey, is highlighted in Example 2.

Banks (2002) wrote:

The popularity of toll financing [in the U. S.] has fluctuated with the perceived adequacy of fuel tax revenues. In almost all cases, toll financing has been used for especially expensive facilities that were expected to be particularly attractive to users, thus making up for their higher costs. [p. 39]

Recent legislation has allowed public participation in toll projects operated by private firms under franchise. This is in contrast to the more traditional arrangement in which toll facilities are constructed and operated by public agencies or special district authorities.

In the case of privatization of functions such as engineering and maintenance, there has never been any uniform policy as to which functions are done in house by public agencies and which are contracted out. . . . In the current climate favoring privatization, some of the agencies that formerly did most functions in-house have increased their use of outside contractors. [p. 43]

APPENDIX 20.   STATE HIGHWAY LETTING PROGRAM MANAGEMENT

In NCHRP Synthesis 331 (2004) S. D. Anderson and B. C. Blaschke, along with a panel of state highway officials, reviewed the letting practices of state highway agencies (SHAs) and requirements they must meet as an essential component of the Statewide Transportation Improvement Program (STIP). The report identified the following key terminology:

- *Project Development.* A series of processes (e.g., planning, programming, design, and construction) that convert a highway transportation need into a completed facility that satisfies the need.
- *Viable Project.* A scope and concept of work with identified limits, meeting a transportation need(s) and consistent with long-range plans.
- *Letting Program Process.* A series of steps that uses the products of the planning and programming functions as a basis for authorizing and controlling the stages of project development and for establishing the time schedules for letting of proj-

ects. The process also incorporates the management of the time table for the flow of projects from initial development authorization through letting.

- *Letting Schedule.* A document that lists projects and specific dates on which the projects will be bid for construction (month, day, year). Typically, it includes projects that will be let in a period of one year or less.

- *Letting.* A function that includes advertisement of the proposed construction projects, receipt of bids, and opening and reading of the bids in a public setting.

Project development comprises the key stages of planning, programming, advanced planning and preliminary design, final design, letting, award, and construction (subject to modifications by the respective SHAs). The components of each stage are described in detail.

The letting program process commences at the completion of the project development and consists of a pool of projects. NCHRP 331 recommended a pipeline analogy. The viable projects enter a system of filters representing authorizations, budget constraints, and "on/off" valves controlling the project flow rate. The main factors influencing the project scheduling for letting are the design completion, the funding, and the constraints. The factors causing changes in the letting program are divided into two groups as follows:

A. Funding and/or cost, environmental/clearance, right-of-way, project scope

B. Utilities, design completion, schedule constraints, project priority, integracy/coordination, plan accuracy/project status

SHA bidding practices, particularly related to bid under- and overruns, are described. Practices for data sharing of letting program information between SHAs are recommended.

APPENDIX 21. WARRANTY, MULTIPARAMETER, AND BEST-VALUE CONTRACTING

NCHRP Report 451 (2001) by S. D. Anderson and J. S. Russell provides guidelines for warranty contracting as a response to the prevalent trends among state highway agencies (SHAs) to downsize, outsource, and improve efficiency and quality under the requirements of competitive bidding. The six issues identified as critical (to highway projects) are as follows:

Selection criteria

Bidding system

Agency resources

Risk allocation

Bonding requirements

Quality aspects

An extensive screening procedure identifies the following three alternative contracting methods:

1. *Warranty Contracting.* Warranty specifications are performance based and emphasize the quality of the product. The warranty is a guarantee of the integrity of a product and of the maker's responsibility for the repair or replacement of deficiencies. A warranty is an absolute liability on the part of the warrantor, and the contract is void unless it is strictly and literally performed.

    Warranties shift some of the postconstruction performance risk to the contractor. They appear most applicable to small- or medium-size, not overly complex projects. Quality is noted to improve. Project completion time may increase.

2. *Multiparameter Bidding and Contracting.* Multiparameter bidding is defined by the formula

$$A + B + \frac{I}{D} (+Q) \tag{A21.1}$$

where A = sum bid by contractor
   B = number of days bid by contractor $\times$ road user cost
I/D = incentive/disincentive
   Q = biddable quality parameter

    The method is compatible with the low-bid system. It can be applied to reconstructions, rehabilitations, and remediations in urban areas with high-traffic volumes and significant user costs. Completion time is typically reduced.

3. *Best-Value Contracting.* The method awards a contract based on price, technical excellence, management capability, past performance, personnel qualifications, and other factors. The selection of a relatively costlier proposal is justified by technical and managerial advantages. Parts of the process may be adapted to the low-bid systems, others depend on local procurement laws. The contractor may be required to present a plan and a schedule for the entire project during the evaluation stage.

APPENDIX 22. EMERGENCY MANAGEMENT

NCHRP Report 525 (2005) is prepared by S. Lockwood (PB Consult), J. O'Laughlin (PB Farradyne), and D. Keever and K. Weiss (Science Applications International Corporation) as Volume 6 of the Surface Transportation Security Project, which in turn is designed to assist transportation agencies in adopting the National Incident Management System (NIMS). The report identifies the following seven key forces defining the needs of state DOTs and their public safety partners in the face of potential emergencies and incidents:

1. Highway incidents and traffic-related emergencies are a major cause of delay and safety problems.

2. The broad and growing array of hazards that involve highways directly or indirectly has varying implications for response.

3. State DOTs and local government transportation departments are not clearly focused on emergency transportation operations (ETOs).

4. There is no clear "best practice" that is widely accepted.

5. New technology is available that could support improved ETO.

6. There is limited institutional commitment to traffic incident and related emergency operations as part of ETO.

7. Significant highway performance improvement opportunities are being missed.

The report identifies the following five general categories of ETO events, along with the corresponding transportation-related emergency characteristics and ETO scope implications:

1. Planned activities
   - Special events
   - Work zone
   - Amber alert
   - Crime control
   - Civil disturbance
2. Traffic incidents
   - Breakdown
   - Crash (major/minor)
   - HAZMAT release
3. Weather related
   - Fog
   - Snow and ice
   - Wildland fire
   - Utility fire
   - Rock/mud/avalanche
4. Natural disaster
   - Earthquake
   - Hurricane
   - Tornado
   - Flood
5. Terrorism/weapons of mass destruction (WMDs)

Typical vulnerabilities are identified. The following four strategies are recommended, along with detailed tactical procedures for their implementation:

1. Make hazard-specific/proactive preparations.

2. Develop and implement coordinated protocols, procedures, and training.

3. Deploy advanced technology/equipment.

4. Measure/benchmark performance against best practice.

The resources for the implementation of the recommended practices must be addressed beyond this report and represent another critical vulnerability.

APPENDIX 23. LINKAGES BETWEEN TRANSPORTATION INVESTMENTS AND ECONOMIC PERFORMANCE

The subject has been addressed in numerous publications sponsored by the U.S. Government through the Congressional Budget Office, the Bureaus of Labor and Economic Analysis, and the Departments of Commerce and Transportation. Bell and McGuire [in NCHRP Project 2-17 (3), 1994, p. 3] summarized the current positions and the needs for future research in this field as follows:

> The major conclusion of this review is that infrastructure investments have a modest positive effect on the nation's private economic activity. (Consumption benefits of infrastructure are not considered. Therefore, conclusions about the value of transportation systems to the nation based on these narrowly focused studies may represent a lower bound estimate of the economic impact of transportation investments.) . . . The conclusion that infrastructure is a productive input simply acknowledges what many believe—roads, airports, water, and other core infrastructure services are important ingredients in a modern, productive economy. The conclusion that infrastructure is productive, however, does not mean that further investments in infrastructure, including transportation, are necessarily the best means of increasing the nation's output. The opportunity costs of public infrastructure investments must be considered in such decisions. Other alternatives, such as greater investments in human capital or in private capital must be considered to determine the best means for expanding the economy.

Dalenberg and Eberts (also in NCHRP Project 2-17) identified two basic approaches for measuring public infrastructure at any level of aggregation:

- Perpetual inventory method (PIM): The measure of capital under the PIM is the sum of the value of past capital purchases adjusted for depreciation and discard.

- Physical measures reflecting the quantity and quality of all pertinent structures and facilities.

The former method relies on economic indicators, the latter on engineering assessments. The authors, expert economists, state the advantages of PIM as follows (p. 85):

> The advantage of PIM is twofold. The Bureau of Economic Analysis (BEA) uses this technique to estimate public capital and private capital at the national level. The established methodology offers benchmarks and other depreciation and discard schedules that can be used to construct state-level estimates. In addition, since most analyses of the effect of public infrastructure on economic activity are based on a neoclassical production function, current input capital should be measured as the maximum potential flow of services available from the measured stock. The perpetual inventory method yields such a measure by using a depreciation function that reflects the decline in the assets' ability to produce as much output as when it was originally purchased.

The PIM relies on two assumptions:

- The purchase price of a unit of capital, which is used to weigh each unit of capital put in place, reflects the discounted value of its present and future marginal products. This assumption is met in a perfectly competitive market (thus possibly excluding state and local governments executing certain projects).

- A constant proportion of investment in each period is used to replace old capital (depreciation). This assumption implies the use of accurately estimated asset's average useful life, discard rate, and depreciation function, i.e., data obtained by engineering assessments.

Aschauer (in Federal Infrastructure Strategy Program, 1994) sought an estimate of the output elasticity of public capital Θ_{KG} as follows:

$$\Theta_{KG} = \frac{K^G}{Y} \frac{\partial Y(\cdot)}{\partial K^G} \tag{A23.1}$$

The aggregate output of the economy Y is modeled by the following aggregate production function:

$$Y = A\ f(LKK^G) \tag{A23.2}$$

where A = total factor or multifactor productivity, reflecting existing state of technology
L = labor force
K = private-sector capital force (typically restricted to business equipment and structures)
K^G = stock of public infrastructure capital

The author discussed the following approaches seeking a causal relationship between public capital and productivity:

- Disaggregation of public capital into functional categories
- Simultaneous equation modeling
- Granger causation techniques
- Cost function estimation

Chapter III-3 of NCHRP Project 2-17 (3) (1994) modeled the individual utility function u_i of the median voter as follows:

$$u_i = u_i(x_i, q_i) \tag{A23.3}$$

where u_i is maximized subject to the constraint

$$Y_i = px_i + t_i qG \tag{A23.4}$$

where G = total quantity of public goods and services
Y_i = median voter's income
t_i = tax liability per unit of G
q = unit cost of public good
p = price of private good $\times$ (assumed to be the numeraire)

The quantity of public good g_i assumed by the individual citizen is modeled as

$$g_i = N^\alpha G \qquad (A23.5)$$

where N = number of people sharing public good

α = "publicness" or congestion parameter such that $0 \leq \alpha \leq 1$

$\alpha = 0$ signifies purely public good (e.g., no congestion)

$\alpha = 1$ implies private good

Based on the maximization of utility, the authors obtain a standard demand function and hence public transportation demand functions. They conclude: "The demand for transportation investments is much more sensitive to personal income than traditionally thought, once adjustments are made for simultaneous equations bias and missing variables. . . . Results suggest that the demographic trends (of urbanization and the aging of the infrastructure), which are generally beyond the control of state and local officials, will impact the demand for investment in individual modes of transportation in different, but important, ways."

APPENDIX 24.   SYSTEM DEVELOPMENT

Mittra (1988, Chapter 2.2) identified the following phases of structured system development, along with their end products:

1. Problem definition and feasibility study, including scope and objectives of the study, description of the proposed system, high-level data flow diagram, and feasibility issues

2. System analysis, including deficiencies of existing system; functional capabilities of the proposed system; detailed data flow diagrams, data dictionary for processes, data flows, and data stores; and cost–benefit analysis

3. Preliminary system design, including flowcharts of proposed system, input and output of proposed system, screen formats, and alternative solutions and recommendations

4. Detailed system design, including record and file structure, auxiliary storage estimate, schema design (if applicable), data communication network and traffic volume (if applicable), structure charts, program flowcharts or input processing output chart, equipment specification, personnel selection, detailed cost estimates, and implementation plan

5. System implementation, maintenance, and evaluation, including structured coding; testing plan of programs; user training; documentation manuals, backup, recovery, and audit trail procedures; maintenance procedures for computer operations staff; and evaluation plan

APPENDIX 25.   LOADING COMBINATIONS AND LIMIT STATES

From AASHTO (2002), Table 3.22.1A, Eq. 3-10:

$$\text{Group}(N) = \gamma[\beta_D D + \beta_L(L + I) + \beta_C CF + \beta_E E + \beta_B B + \beta_S SF + \beta_W W +$$
$$+ \beta_{WL} WL + \beta_L LF + \beta_R(R + S + T) + \beta_{EQ} EQ + \beta_{ICE} ICE] \qquad (A25.1)$$

where N = group number
     γ = load factor, =1 or working stress (ASD)
     β = coefficient
     D = dead load
     L = live load;
     I = live-load impact
     E = earth pressure
     B = buoyancy;
     W = wind load on structure
   WL = wind load on live load, 100 psf (4.8×10^{-3} MPa)
    LF = longitudinal force from live load
   CF = centrifugal force
     R = rib shortening
     S = shrinkage
     T = temperature
   EQ = earthquake
    SF = stream flow pressure
  ICE = ice pressure

From LRFD, 3rd ed., Table 3.4.1-1, *Limit state* is defined as a condition beyond which the bridge or component ceases to satisfy the provisions for which it was designed (p. 1-2). The following limit states are described:

> The *service limit state* provides certain experience-related provisions that cannot always be derived solely from strength or statistical considerations.
>
> The *fatigue limit state* is intended to limit crack growth under repetitive loads to prevent fracture during the design life of the bridge.
>
> Extensive distress and structural damage may occur under the *strength limit state,* but overall structural integrity is maintained.
>
> *Extreme-event limit states* are considered to be unique occurrences whose return period may be significantly greater than the design life of the bridge.

Limit states are achieved through the following loading combinations:

Strength I	Basic load combination relating to the normal vehicular use of the bridge without wind
Strength II	Load combination relating to the use of the bridge by owner-specified special design vehicles, evaluation permit vehicles, or both without wind
Strength III	Load combination relating to the bridge exposed to wind velocity exceeding 90 km/h (55 mph)
Strength IV	Load combination relating to very high dead load–live load force effect ratios
Strength V	Load combination relating to normal vehicular use of the bridge with wind of 90 km/h (55 mph) velocity
Extreme event I	Load combination including earthquake

Extreme event II Load combination relating to ice load, collision by vessels and vehicles, and certain hydraulic events with a reduced live load other than that which is part of the vehicular collision load

Service I Load combination relating to the normal operational use of the bridge with 90-km/h (55-mph) wind and all loads taken at their nominal values. Also related to deflection control in buried metal structures, tunnel liner plate, and thermoplastic pipe and to control crack width in reinforced-concrete structures. This load combination should also be used in the investigation of slope stability.

Service II Load combination intended to control yielding of steel structures and slip of slip-critical connections due to vehicular live load

Service III Load combination relating only to tension in prestressed concrete structures with the objective of crack control

Fatigue Fatigue and fracture load combination relating to repetitive gravitational vehicular live load and dynamic responses under a single design truck having the axle spacing specified in Article 3.6.1.4.1, that is, 9000 mm (30.0 ft) between the 145.0-kN (32.0-kip) axles

Force effects are determined for the above loading combinations by Eq. 3.4.1-1 (p. 3-6) as follows:

$$Q = \sum \eta_i \gamma_i \, Q_i \qquad \text{(A25.2)}$$

where η = load modifier

γ_i = load factors

Q_i = force effects due to loads specified below

Permanent	**Transient**
DD = downdrag	BR = vehicular breaking force
DC = dead load of structural components and nonstructural attachments	CE = vehicular centrifugal force
	CR = creep
	CT = vehicular collision force
DW = dead load of wearing surfaces and utilities	CV = vessel collision force
	EQ = earthquake
EH = horizontal earth pressure load	FR = friction
EL = accumulated locked-in effects resulting from construction process, including secondary forces from posttensioning	IC = ice load
	IM = vehicular dynamic load allowance
	LL = vehicular live load
	LS = live-load surcharge
ES = earth surcharge load	PL = pedestrian live load
EV = vertical pressure from dead load of earth fill	SE = settlement
	SH = shrinkage
	TG = temperature gradient
	TU = uniform temperature
	WA = water load and stream pressure
	WL = wind on live load
	WS = wind load on structure

APPENDIX 26. STRUCTURAL STABILITY

The supply of elastic response extends to the yield stress and strain of the material. Up to that point the two are related according to Hooke's law as follows:

$$\sigma = E\varepsilon \tag{A26.1}$$

where σ = stress (force/area)
E = Young's elastic modulus (force/area)
ε = strain

The demand is expressed in terms of force according to Newton's second law as follows:

$$F = ma \tag{A26.2}$$

where F = force
m = mass (density $\times$ volume)
a = acceleration (distance/time^2)

Force (or stress) can be viewed as the cause in Eq. A26.1 and as the effect in Eq. A26.2; however, both relationships are linearly proportional. By taking into account the internal flexural energy of the system, Leonard Euler (1707–1783) gave it the option to change (qualitatively) its shape, rather than to continue to deform in the direction of the applied load.

Bending Induced by Compression

Euler obtained the deflected shape of linear elastic bodies in 1744. In 1826 Navier extended this relationship to a general "small-deflexion bending theory . . . leading to the basic elastic equation" (Heyman, 1998, p. 97):

$$M = -EI\frac{d^2y}{dx^2} \tag{A26.3}$$

where the bending moment M (force $\times$ length) in an elastic member is directly proportional to its deflection y normal to the axial coordinate x.

In 1757 Euler used his method to determine the critical concentrated axial compression P_{cr} (or P_e) under which an initially straight, simply supported column could deflect as follows:

$$P_{cr} = \left(\frac{\pi}{KL}\right)^2 EI_y \tag{A26.4}$$

where L = column length
$I_y < I_z$ = section moments of inertia (length^4)
E = Young's modulus of elasticity (force/length^2)
K = effective length factor (=1 for simply supported column)

At $P \geq P_{cr}$ the original straight shape is no longer the only equilibrium configuration. Salvadori (1963, p. 90) memorably explained that beyond the critical load the column finds it "easier" to bend than to shorten any further. That critical shortening is obtained from Eq. A26.4:

$$\Delta_{cr} = \frac{(\pi \, r_y)^2}{L} \qquad \text{(A26.4a)}$$

where Δ_{cr} = critical axial column deformation (length)
$\quad r_y = (I_y/A)^{1/2}$, cross-sectional radius of gyration (length)
$\quad A$ = cross-sectional area (length^2)

According to Eq. A26.4a, all elastic straight columns of given original dimensions buckle at the same Δ_{cr}, independently of material strength and external load. Since force equilibrium is at the center of all engineering analysis, it is worth noting that, regardless of strength, the original shape of a structure will not tolerate deformation beyond a critical limit. Unstable distortions always assume shapes of a higher order than stress analysis envisions. Axially compressed columns buckle in bending or torsion. Deflected beams buckle in torsion. Torsion warps thin-wall sections and they buckle locally, as plates. Othmar Ammann attributed the failure of the Quebec bridge to a lack of knowledge about torsional buckling. Gies (1963) drew parallels to the earlier Tey and Ashtabula failures.

Equations A26.3 and A26.4 set the norm for most engineering analysis of elastic structures with the concise and sufficiently accurate assumption that deformations are *small,* implying the following:

$$\left[1 + \left(\frac{dy}{dx}\right)^2 \right]^{3/2} \approx 1$$

Shanley (1957, p. 574) concluded that, given the acceptable accuracy of the above approximation, "large deformations are largely of academic interest."

A distinction must be made between the theoretical validity of this statement and the definition of *large* in practice. Large displacements are routine in cable-supported structures. In any structure small deflections can be amplified into large under given circumstances. Relatively small deflections or distortions can be significant to axial compression. Similarly, small distortions can matter in bending. The elastic response that qualifies as small in a frame may be neither elastic nor acceptable in a masonry arch. When comparing structures with significantly different rates of deformation, the term *elastic* is often (confusingly) replaced by *rigid,* as in *rigid frame.*

Materials only approximate the linear elastic springs of the theoretical models and only within limits. The assumed force distributions must be correlated with the actual imperfect and deformed shapes that structures assume over time. Deformations consistent and inconsistent with the elastic (small) response to the applied load must be treated differently. The qualitative distinctions between the various uncertainties involved in probabilistic analysis (Appendix 5) are analogous.

Torsion Induced by Compression

Equation A26.4 assumes that the stability of a straight column is lost in bending. Another possibility is torsion. Equation A26.5 (Bažant and Cedolin, 1991, Chapter 6) describes torsion as Eq. A26.3 describes bending:

$$M_t = T_{SV} + T_W = GJ \frac{d\theta}{dx} - EI_W \frac{d^3\theta}{dx^3} \tag{A26.5}$$

where M_t = applied torque (force $\times$ length)

I_W = warping moment of inertia (or warping constant) (length^6)

G = elastic shear modulus (force/length^2)

J = section torsional constant (length^4)

θ = torsion angle (rad)

A member resists torsion in a combination of simple or Saint-Venant (T_{SV}) and warping (T_W) modes, each contributing according to the shape of the cross section, defined by J and I_W, respectively. A simply supported straight column buckles in torsion at the following critical load $P_{\theta, \, cr}$:

$$P_{\theta, cr} = \frac{A\,(GJ + EI_w\,\pi^2/L^2)}{I_y + I_z} \tag{A26.6}$$

where I_y, I_z = bending section moments of inertia

$A = \int_A dA$

A column will buckle in bending rather than torsion if $P_{cr\theta} > P_{cr}$. From Eqs. A26.4 and A26.6, that condition can be expressed as

$$\frac{GJ}{EI_y} > \left(\frac{\pi}{L}\right)^2 \left(\frac{I_y + I_z}{A} - \frac{I_w}{I_y}\right) \tag{A26.7}$$

For typical cross sections designed to resist bending, inequality A26.7 is likely to pertain. Hence, most columns buckle in bending, although, particularly with imperfections, the two buckling modes may interact.

Torsion Induced by Bending

Beams subjected to bending can buckle in torsion. The critical moment M^0_{cr1} of a simply supported beam under bending moments M^0 applied at both ends (Bažant and Cedolin, 1991, p. 387) is

$$M^0_{cr1} = \frac{\pi}{L} \left\{ EI_y \left[GJ + EI_w \left(\frac{\pi}{L}\right)^2 \right] \right\}^{1/2} \tag{A26.8}$$

Bending does not induce buckling in all cross-sections. Chen and Lui (1987, p. 325) quote the approximate solution of Kirby and Nethercott as follows:

$$M^0_{cr1} = \frac{\pi}{L} \left\{ \frac{EI_y \left[GJ + EI_w \left(\frac{\pi}{L}\right)^2 \right]}{1 - I_y/I_x} \right\}^{1/2} \tag{A26.9}$$

where M^0_{cr1} is real only if $I_y < I_z$. Approximate formulas are available for variable loads and cross sections.

Thin Plates

The critical in-plane load N_{cr} uniformly distributed along the edge b of a simply supported thin rectangular plate with sides $a \geq b$ and thickness h (Bažant and Cedolin, 1991, p. 433) is

$$N_{cr} = \left(\frac{\pi}{b}\right)^2 \left(\frac{mb}{a} + \frac{a}{mb}\right)^2 \frac{Eh^3}{12(1 - \nu^2)} \tag{A26.10}$$

where ν = Poisson ratio
   $m = a/b$ for minimum N_{cr}

Critical loads for other boundary conditions are available in all texts on plate analysis and design. For free edges a, the term $1 - \nu^2 \approx 0.91$ (for metals) would be the only difference between Eqs. A26.10 and A26.4. Significantly, the width b, rather than the height a, governs N_{cr}. Increasing the width b of a plate in order to distribute a constant load N over greater area is not a solution.

For the critical shear load $N_{xy, cr}$ along simply supported edges a and b, Bažant and Cedolin (1991, p. 436; Fig. A26.1) derive the following approximation:

$$N_{xy,cr} = \pm \frac{9}{32} \pi^4 ab \left(\frac{1}{a^2} + \frac{1}{b^2}\right)^2 \frac{Eh^3}{12(1 - \nu^2)} \tag{A26.11}$$

Figure A26.1   Plate girder buckled in shear (Laboratoire Central des Ponts et Chaussées, Paris).

If N_{cr} of Eq. A26.11 is exceeded, strips of narrowing width $c < b/2$ adjacent to the simply supported edges a remain roughly in plain and continue to resist the load until yield is reached. Von Karman obtains the width c from equating the yield stress f_y and the critical load as follows:

$$N_{cr} = \frac{(\pi/c)^2 \, Eh^3}{12(1 - \nu^2)} = hf_y \qquad (A26.12)$$

Under this approximation, the ultimate concentrated load P_{ult} resisted by a simply supported plate would depend on its thickness h. Substituting $f_y = 36$ ksi (250 MPa), $E = 29,000$ ksi (200,000 MPa), and $\nu = 0.3$, P_{ult} is obtained as

$$P_{ult} = 2chf_y = \pi h^2 \left(\frac{Ef_y}{3(1 - \nu^2)}\right)^{1/2} = 1942h^2 \quad \text{kips} \qquad (4280h^2 \quad \text{kN}) \quad (A26.13)$$

Bažant and Cedolin (1991, p. 448) point out that rectangular plates have considerable postbuckling strength in shear because they act as rectangular trusses with one tension diagonal. The diagonal buckling waves can be observed as they constantly switch directions on the side walls of moving trailer trucks.

In contrast with one-dimensional columns and beams, plates have considerable postbuckling reserve. This could be explained by noting that, after plates buckle under compression or shear, parts of them retain their original straight shape and continue to carry loads up to the material limit, albeit by a modified mechanism.

The postbuckling strength reserve of plates is widely exploited in design. The cutouts in the webs of truss members (Fig. A26.2) are examples. It is also recognized that plate buckling, even if not critical, can initiate other failure modes. A number of AISC design formulas prescribe web and flange thickness–depth ratios based on the above model. Girder sections where plate buckling may govern are designated as *noncompact*. Stiffening and bracing requirements are specified. LRFD (3rd ed.) discusses stability requirements for steel in Sections 6.9 and 6.10.

The design of a section can be balanced so that local and global buckling occur simultaneously. In a multiredundant structure, a buckled slender member is relieved of the critical load and can show little distress apart from some deflection.

Superposition and Amplification

Most failures result from combinations of independently sustainable effects. LRFD, (AASHTO, 2004, Section C4.6.1.1) states: "Simultaneous torsion, moment, shear, and reaction forces and the attendant stresses are to be superimposed as appropriate. The equivalent beam idealization does not alleviate the need to investigate warping effects in steel structures. In all equivalent beam idealizations, the eccentricity of the loads should be taken with respect to the centerline of the equivalent beam."

The term *as appropriate* is often used in specifications in order to invoke the responsibility and competence of the user. In this case it is a reminder that combinations of diverse loads and geometric imperfections easily exhaust the limitations of linear elastic structural analysis and, consequently, invalidate direct superposition.

Figure A26.2 Typical truss member cutouts.

Torque and Compression

The combined effect of torque M_t and axial load P has implications beyond linear structural analysis. Bažant and Cedolin (1991, Section 1.10) obtain the following solution for a simply supported column:

$$\frac{P}{P_{cr}} + \left(\frac{M_t}{M_{cr}^0}\right)^2 = 1 \qquad (A26.14)$$

where M_t = torque applied to column

$$M_{cr}^0 = 2\pi \frac{EI}{L}$$

Beyond the surprising absence of torsional stiffness from Eq. A26.14, the authors note that "the problem is not as simple as it seems." Under certain conditions, the torque is path dependent and therefore nonconservative. The static solution becomes "a priori illegitimate and dynamic analysis is required." As it happens, in this particular case the solution would be the same. Time-dependent pressure, exciting bending and torsional oscillations with similar frequencies, results in the nonconservative "flutter," causing failure in aircraft wings and suspension bridges.

Bending and Compression

Beam columns carry bending and axial loads simultaneously. Linear elastic stress–strain analysis should allow superposition of the effects. Stability does not.

Under an axial load P, the deflection y_1 of a column with initial imperfection a (or eccentricity Δ) is amplified as follows (Timoshenko, 1936, p. 31):

$$y_1 = \frac{a}{1 - P/P_{cr}} \tag{A26.15}$$

The same equation pertains if the initial deflection a in the axially loaded column is caused by a bending moment (Timoshenko, 1936, p. 33). In a straight column under an axial load P, the factor $1/(1 - P/P_{cr})$ amplifies the initial eccentricity Δ or reduces the bending stiffness k (Shanley, 1957, p. 565). Example EA26 illustrates the reduction of lateral stiffness under axial load.

Bažant and Cedolin (1991) review design provisions for P–Δ effects from the secant formula to the above amplification factor. Deformation amplification can be taken into account by iterative "second-order" structural analysis (Bažant and Cedolin, 1991, p. 27). Within limits, beam columns subjected to both axial and flexural loads are designed by first-order analysis with amplified moments.

Bažant and Cedolin (1991, p. 388) combine M_{cr}^0 of Eq. A26.8 with an eccentric axial load P to obtain the following amplified effect:

$$M_{cr1}^P = M_{cr1}^0 \left[\left(1 - \frac{P}{P_{y,cr}} \right) \left(1 - \frac{P}{P_{\theta,cr}} \right) \right]^{1/2} \tag{A26.16}$$

LRFD (AASHTO, 2004, Section C4.5.3.2.1) refers to large-deflection theory as a "synergistic effect of interaction . . . (between compressive axial forces and out-of-straightness, causing) the apparent softening of the component, i.e, a loss of stiffness." As a first approximation, LRFD (Section 4.5.3.2.2b) recommends the single-step moment magnification in beam columns as follows:

$$M_c = \delta_b M_{2b} + \delta_s M_{2s} \tag{A26.17a}$$

$$f_c = \delta_b f_{2b} + \delta_s f_{2s} \tag{A26.17b}$$

where $\delta_b = C_m/(1 - P_u/\varphi P_e) \geq 1.0$
$\qquad \delta_s = 1/(1 - \Sigma P_u/\Sigma \varphi P_e)$
$\qquad P_u = $ factored axial load
$\qquad P_e = P_{cr}$ of Eq. A26.4 with effective length factor K according to Section 4.6.2.5
$\qquad \varphi = $ resistance factor of Eq. 4.1 for axial compression
$M_{1b}, M_{2b} = $ smaller and larger end moments on compression member due to factored gravity loads that result in no appreciable side sway, calculated by conventional first-order elastic frame analysis, > 0

M_{2s} = moment on compression member due to factored lateral or gravity loads that result in side sway Δ, greater than $1_u/1500$, calculated by conventional first order elastic frame analysis, > 0

$C_m = 0.6 + 0.4M_{1b}/M_{2b}$ for $\Delta = 0$ and $C_m = 1.0$ for $\Delta \neq 0$

f_{2b} = stress corresponding to M_{2b}

f_{2s} = stress corresponding to M_{2s}

If the deformation of the structure results in a significant change in force effects, the effects of deformation shall be considered in the equations of equilibrium. The effect of deformation and out-of-straightness shall be included in stability analyses and large deflection analyses.

The past traditional boundary between small- and large-deflection theory becomes less distinct as bridges and bridge components become more flexible due to advances in material technology, the change from mandatory to optional deflection limits and the trend toward more accurate optimized design. (AASHTO, 1998a)

The designer defines "significant," the manager selects the "optimized design." The specifications can only remind of these responsibilities. Equation A26.17a is similar in AASHTO (2002) and LRFD (3rd ed.); however, the latter adds Eq. A26.17b and the commentary.

Frames

For the effect of *constant* axial forces on structural stiffness Clough and Penzien (1993, p. 171) introduce *geometric stiffness coefficients* and show a number of ways to incorporate them in finite-element analysis.

Bridge piers are typically frames of very simple geometry; however, their performance is critical to the structural integrity and can be inelastic nonlinear under extreme loads. Bažant and Cedolin (1991, p. 53) point out that the axial loads in frames significantly influence not only the stability of the carrier but also that of the adjacent members. Frame analysis takes that phenomenon into account by using flexibility matrices, which are the inverse of the stiffness matrices. Also called stability functions, they are discussed in detail by Horne and Merchant (1965). The combination of buckling and vibration effects can be significant, as discussed in a subsequent section herein. That problem is addressed, for instance, in Livesley and Chandler (1956).

LRFD (AASHTO, 2004, Section 4.6.3.5) requires a refined plane or space frame analysis of trusses, including:

Composite action with deck system

Continuity among the components

Force effect due to component weight, change of geometry due to deformation and axial offset at panel points

In- and out-of-plane buckling under above effects

The use of large-deflection analysis is recommended for long-span arches and suspension bridges normally considered beyond the scope of the specifications.

Statics and Dynamics

The fundamental model of structural dynamics is an undamped single-degree-of-freedom oscillator under harmonic excitations. The ratio of the dynamic to static displacement of such an oscillator is defined, recently by Chopra (1995, p. 65), as the *displacement response factor R_d*:

$$R_d = \frac{u_o}{(u_{st})_o} = \left[1 - \left(\frac{\omega}{\omega_n} \right)^2 \right]^{-1} \tag{A26.18}$$

where u_o = dynamic displacement
$(u_{st})_o$ = static displacement
ω = forcing function frequency
ω_n = oscillating system natural frequency

Karman and Biot (1940) refer to the same term as *resonance factor*. The amplification depends on the ratio of the forcing function frequency to the natural frequency of the oscillator and (in the absence of damping) tends to infinity as that ratio approaches unity. A more realistic *dynamic magnification factor D* (Clough and Penzien, 1993) includes the viscous damping ζ as follows (p. 38):

$$D = [(1 - \beta^2)^2 + (2\zeta\beta)^2]^{-1/2} \tag{A26.19}$$

where $\beta = \omega/\omega_n$
ζ = viscous damping ratio

Equation A26.18 is similar to Eq. A26.15. Both excitations and deflections are modeled by harmonics. Actual dynamic loads are random and frequently sudden. Chopra (1995) reminds us that sudden impact is amplified by a factor of 2, contributing to the systematic failure of rigid restrainers at expansion joints. LRFD (AASHTO, 2004) shows awareness of that amplification by stipulating a dynamic factor of 75% for deck joints, where repeated impact is guaranteed.

For equivalent single-degree-of-freedom analysis, LRFD (AASHTO, 2004, Section 4.7.4) defines structural periods and their limitations. Section 4.7.4.3.3 states: "Multimode spectral analysis shall be performed for bridges in which coupling occurs in more than one of the three coordinate directions within each mode of vibration."

When inelastic behavior is likely, time–history analysis is prescribed. Clough and Penzien (1993, p. 191) caution that geometric stiffness matrices with constant terms cannot consider dynamic variations of the axial loads. Under a time-dependent axial load $P(t)$, the natural frequency ω_n of a single-degree-of-freedom undampened oscillator depends on the stiffness $k(t)$, which is also time dependent and must be modified according to the equation

$$\omega_n(t) = \left(\frac{k(t)}{m} \right)^{1/2} = \omega_n \left(1 - \frac{P(t)}{P_{cr}} \right)^{1/2} \tag{A26.20}$$

A simplified illustration is provided in Example EA26.

Under simultaneous lateral and vertical excitations, the dynamic amplification factor R_d of Eq. A26.18 becomes a function of a variable stiffness. Comparable vertical and lateral natural frequencies (only likely in very long spans) can become a source of in-stability. Dynamic response can be obtained by the application of Mathieu functions. It is not taken into account by typical elastic frame analysis software. The AASHTO stan-dard quasi-static seismic loadings simulate transverse and longitudinal ground motions. The effect of vertical ground motions on highway bridges is discussed in Button et al. (1999).

Example EA26. Lateral Stiffness under Axial Load

A structure consisting of an infinitely rigid column with length L rigidly connected to two colinear springs each with axial stiffness $c/2$ is subjected to axial (P) and transverse (F) loads independently and simultaneously, as shown in Fig. EA26.1:

1. $P \neq 0; F = 0: \Sigma M_O = 0 \rightarrow P_{cr} L \sin \alpha = cD^2 \sin \alpha \rightarrow P_{cr} = cD^2/L$, where $c/2$ is the axial spring stiffness (force/displacement)

2. $P = 0; F \neq 0: \Sigma M_O = 0 \rightarrow FL \cos \alpha = cD^2 \cos \alpha \sin \alpha \rightarrow$
$F = c(D/L)^2 L \sin \alpha = k \Delta H$

where $k = c(D/L)^2 = P_{cr}/L$ is the lateral stiffness of structure
$\Delta H = L \sin \alpha$ is the lateral deflection of structure (length)

3. $P \neq 0; F \neq 0: \Sigma M_O = 0 \rightarrow FL \cos \alpha + PL \sin \alpha = cD^2 \sin \alpha \cos \alpha$
$\rightarrow F = P_{cr} \sin \alpha [1 - P/(P_{cr} \cos \alpha)] \rightarrow F \approx k (1 - P/P_{cr}) \Delta H$
where $\cos \alpha \approx 1$ for small deflections.
$k (1 - P/P_{cr}) \approx$ modified lateral stiffness under axial load, as in Eq. A26.15.
$k (1 - P/P_{cr}) \rightarrow 0$ as $P \rightarrow P_{cr}$

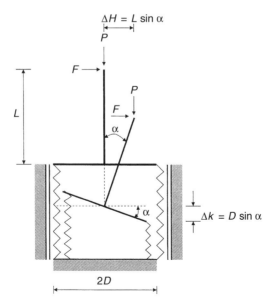

Figure EA26.1 Lateral stiffness reduction under axial load.

APPENDIX 27. EFFECTIVE SLAB WIDTH FOR COMPOSITE ACTION

The traditional AASHTO recommendation for width of the reinforced-concrete bridge deck contributing to the compression flange of interior girders was based on empirical estimates of the shear lag in the deck slab. LRFD (AASHTO, 2004) modified the formula (denoted by an asterisk). The effective slab width for interior girders of all types of composite bridge superstructures, except orthotropic decks and segmental concrete structures, is the least of the following:

1. Effective span length/4
2. Average depth of slab × 12 + max (web thickness, top flange width/2)*
3. Average spacing of adjacent beams

NCHRP Report 543 (2005) proposed a new definition of effective width intended to reconcile results of stress analysis by the finite-element method (FEM) and bending moments obtained by equivalent beam analysis as follows:

$$b_{ef} = \frac{C_{slab}}{A} = \frac{C_{slab}}{0.5\, t_{slab}\,(\sigma_{max} + \sigma_{min})} \tag{A27.1}$$

where A = area of equivalent compressive block for simple beam theory

t_{slab} = total structural slab thickness

σ_{max} = maximum compressive stress at extreme compression fiber of slab, obtained by FEM

σ_{min} = minimum compressive stress at bottom of slab, obtained such that resultant compressive force C_{slab} obtained by beam theory and FEM the same

C_{slab} = total or resultant compressive force in slab, obtained by Eq. A27.2

$$C_{slab} = \sum_{i=1}^{n} \sigma_i\, \text{Area}_i \tag{A27.2}$$

where Area_i = element cross-sectional area

i = element number

σ_i = longitudinal stress in element i

The maximum compressive stress at the extreme compression fiber according to beam theory is obtained as

$$\sigma_{max,\, \text{Beam theory}} = \frac{M_{FEM}}{S_{top,\, \text{Beam theory}}} \tag{A27.3}$$

where M_{FEM} = bending moment at specific section, obtained by FEM

$S_{top,\, \text{Beam theory}}$ = elastic section modulus for extreme compression fiber

APPENDIX 28. LIVE-LOAD DISTRIBUTION FACTORS

- AASHTO (2002, 3.23.2.3.1.5): Concrete deck, four or more steel or concrete beams (interior), two or more design lanes loaded:

$$\text{Distribution factor} = \frac{S}{4.0 + 0.25S} \qquad 6.0 < S \leq 14.0 \text{ ft}$$

$$= \frac{S}{5.5} \qquad S \leq 6.0 \qquad \text{(A28.1)}$$

In SI units

$$\text{Distribution factor} = \frac{S}{(1.2 + 0.25S)} \qquad 1.8 < S \leq 4.3 \text{ m}$$

$$= \frac{S}{1.7} \qquad S \leq 1.8$$

(subject to multiple presence factors)

- AASHTO (2004, Table 4.6.2.2.2b-1): Distribution of live loads *per lane* for moment in interior beams (multiple presence factors included):

$$\text{Distribution factor} = 0.075 + \left(\frac{S}{9.5}\right)^{0.6} \left(\frac{S}{L}\right)^{0.2} \left(\frac{K_g}{12.0Lt_S^3}\right)^{0.1}$$

$$\text{SI units} = 0.075 + \left(\frac{S}{2900}\right)^{0.6} \left(\frac{S}{L}\right)^{0.2} \left(\frac{K_g}{Lt_S^3}\right)^{0.1} \qquad \text{(A28.2)}$$

where $3.5 \leq S \leq 16.0$ ft or $1100 \leq S \leq 4900$ mm spacing of primary members
$20 \leq L \leq 240$ ft or $6000 \leq L \leq 73\,000$ mm span length
$4.5 \leq t_S \leq 12.0$ in. or $110 \leq t_s \leq 300$ mm depth of concrete slab
$K_g = n\,(I + A\,e_g^2)$ longitudinal stiffness parameter (in.^{4} or mm^4)
A = area of cross section (in.^{2} or mm^2)
e_g = distance between centers of gravity of basic beam and deck (in. or mm)
I = moment of inertia of beam (in.^{4} or mm^4)

APPENDIX 29.   SUPERSTRUCTURE DEFLECTIONS

- AASHTO (17th ed., 2002), Sections 8.9.3 (concrete) and 10.6.2 (steel): "Members having simple or continuous spans preferably should be designed so that the deflection due to the service load plus impact (HS 20) should not exceed 1/800 of the span, except on bridges in urban areas used in part by pedestrians whereon the ratio preferably shall not exceed 1/1000. For cantilevers, the ratios are 1/300 and 1/375, respectively."

- LRFD (AASHTO, 2004):
Section 2.5.2.6: "Bridges should be designed to avoid undesirable structural or psychological effects due to their deformations. While deflections and depth limitations are made optional, except for orthotropic plate decks, any large deviation from past successful practice regarding slenderness and deflections should be cause for review of the design to determine that it will perform properly."

Section 2.5.2.6.2: Provides a method for deflection calculations and, in the absence of any other, reverts to the numerical values of AASHTO (2002) above.

APPENDIX 30. AASHTO DESIGN LIVE LOADS

The specifications for NBI (FHWA, 2005a) propose the following live-load designations (item L-1):

Code	Metric	English
00	Unknown	Unknown
01	M 9	H 10
02	M 13.5 1 axle × 110 kN + 1 axle × 26 kN at 4.3 m	H 15 1 axle × 24.0 kips + 1 axle × 6.0 kips at 14 ft
03	MS 13.5 2 axles × 110 kN at 4.3–9.0 m + 1 axle × 26 kN at 4.3 m	HS 15 2 axles × 24.0 kips at 14–30 ft + 1 axle × 6.0 kips at 14 ft
04	M 18 1 axle × 145 kN + 1 axle × 35 kN at 4.3 m	H 20 1 axle × 32.0 kips + 1 axle × 8.0 kips at 14 ft
05	MS 18 2 axles × 145 kN at 4.3–9.0 m + 1 axle × 35 kN at 4.3 m	HS 20 2 axles × 32.0 kips at 14–30 ft + 1 axle × 8.0 kips at 14 ft
06	MS 18 + Mod	HS 20 + Mod
07	Pedestrian	Pedestrian
08	Railroad	Railroad
09	MS 22.5 2 axles × 182 kN at 4.3–9.0 m + 1 axle × 44 kN at 4.3 m	HS 25 2 axles × 40.0 kips at 14–30 ft + 1 axle × 10.0 kips at 14 ft
10	HL 93	HL 93
11	> MS 22.5/HS 25 (AASHTO trucks only)	

The AASHTO design truck (AASHTO, 2004, p. 3–21) is MS 18/HS 20.

In addition, AASHTO specifies a design tandem consisting of a pair of 25.0-kip (110-kN) axles spaced 4.0 ft (1200 mm) apart. The uniformly distributed lane load simulating continuous traffic is 0.64 KLF (9.3 kN/m), applied transversely over a width of 10 ft (3m).

States designate local "Permit Design Vehicles."

APPENDIX 31. IMPACT FACTORS

Impact factors (IM) are empirically determined multipliers augmenting the maximum static deflections caused by the empirically selected design vehicles such that

$$\text{IM} = \frac{D_{\text{dyn}}}{D_{\text{sta}}} \tag{A31.1}$$

Empiricism in this case strongly relies on statistics. Barker and Puckett (1997, p. 157) emphasize that the *dynamic amplification* of structural deflection (rather than the *impact*) varies widely depending on the location of the design vehicle. The dynamic load allowance (DLA) is based on extreme deflections, not necessarily on the largest IM that Eq. A31.1 obtains.

Current specifications recommend the following provisions:

- AASHTO, 2002, p. 21:

$$I = \frac{50}{L + 125} (\%) \leq 30\%$$

  where L is the length in feet of the span portion loaded to produce maximum stress in the member.

- AASHTO, 1998a, p. 3–27:

$$IM = 33\%$$

  except

$$IM = 15\% \text{ for fatigue and fracture limit state}$$

$$IM = 75\% \text{ for deck joints, all limit states}$$

Thus supplemented, live loads LL + I or LL + IM are further amplified by a factor reflecting the uncertainty of their *static* magnitude as follows:

AASHTO, 2002, p. 31 $\gamma \times \beta_{LL} = 1.3 \times 1.67$ (load factor combination I)

$\gamma \times \beta_{LL} = 1.3 \times 2.20$ (load factor combination IA)

AASHTO 1998a, p. 3-11 $\gamma_{LL,IM} = 0.50$ to 1.75, except

$\gamma_{EQ} \leq 1.0 \ (=0.50)$ for earthquake effect on live loads the latter issue is not resolved (p. 3-12)

Live loads are subject to reduction in the case of multiple-lane use; however, it is pointed out that the LRFD (2nd ed.) distribution factors include the effect of multiple presence.

NCHRP Report 12-46 (2000) lists design, legal (types 3, 3S2, 3-3, lane, state), and permit loads. NCHRP Report 495 (2003) developed a new algorithm for evaluating the "significant damage" on bridges due to trucks with increasing weight. The report (p. 20, Eq. 2.3.3.1) refers to the live-load factor formula, proposed by NCHRP Report 12-46 and based on statistical evaluation of site-specific data, as follows:

$$\gamma_L = \frac{1.8 \ (2W^* + 1.41t^{ADTT} \ \sigma^*)}{240} \tag{A31.2}$$

where γ_L = live-load factor
 W^* = average truck weight for top 20% of truck weight histogram (TWH)
 σ^* = standard deviation of top 20% of TWH
 t^{ADTT} = 2–4.5 (typically), a factor depending on annual average daily truck traffic

For projected changes in the truck weight limit over an expected life of 75 years, NCHRP Report 495 (2003, p. 22) adjusts t^{ADTT}, modifying the live-load factor γ_L as

$$\gamma_L = \frac{1.75(2W^* + 6.9\ \sigma^*)}{265} \qquad (A31.2a)$$

For concentrated loads, the *Ontario Highway Bridge Design Code* (OHBDC, 1983) recommended an impact factor ranging between 0.20 and 0.40 (for structures with first-mode natural frequencies from 1 to 6 Hz) and 0.30 for higher and 0.20 for lower frequencies. Taly (1998, Section 3.3.1) pointed out that OHBDC (1993) has reverted to a DLA of 0.4, 0.3, and 0.25 for one, two, and three axles, respectively.

For railroad bridges where tolerance to fatigue is lower and dynamic effects are higher, AREMA recommends the following impact factors I for diesel and electric locomotives:

Span Length	Impact Factor (%)	SI
< 80 ft (25 m)	$100/S + 40 - 3L^2/1600$	$30.5/S + 12.2 - 3L^2/150$
≥ 80 ft (25 m)	$100/S + 16 - 600/(L - 30)$	$30.5/S + 5 - 180/(L - 9)$

where S = spacing of girders, trusses, or floor beams, ft (m)
L = span length, ft (m)

The maximum recommended value is $I = 60\%$. For concrete structures AREMA recommends $I = 100/(1 + \text{dead load}/\text{live load})$.

APPENDIX 32. SEISMIC DESIGN CRITERIA FOR BRIDGES AND OTHER HIGHWAY STRUCTURES

A survey was sponsored by the FHWA and conducted by the National Center for Earthquake Engineering Research (NCEER) and the Applied Technology Council (ATC). As part of the findings, Rojahn et al. (1997) tabulated (Table 4-6, pp. 78–96) a comparison of specifications for seismic design of bridges, including the following sources:

- AASHTO, 2002, Division I-A
- LRFD AASHTO (1998a)
- Caltrans
- ATC-32
- Transportation Corridor Agencies (TCA)
- New Zealand
- Eurocode 8
- Japan

The reviewed provisions included the following:

1. General: performance criteria, design philosophy, design approach
2. Seismic loading: return period, geographic variation, importance considerations, side effects, damping, duration

3. Analysis: selection guidelines, equivalent static, elastic dynamic, inelastic static, inelastic dynamic, directional uncertainty, load combinations

4. Seismic effects: design forces, ductile components, nonductile components, displacements, minimum seat width

5. Concrete design: columns—flexure, shear, spirals (confinement), anchorages; superstructure and column joints, caps, piers, footings, superstructure, shear keys

6. Column splices

7. Steel design

8. Foundation design: spread footings, pile footings, liquefaction

9. Miscellaneous design: restrainers, base isolation, active/passive control

The comparison highlighted the issues deemed by all bridge designers as significant to structural safety in seismic zones. Emphasized were provisions preventing catastrophic failures, redundancy and ductility, elastic and inelastic design, dynamic analysis, design detailing, soil properties, the use of one- and two-step design approaches, and estimates of return periods.

APPENDIX 33. PRIORITIZATION OF SEISMIC VULNERABILITIES

Caltrans

Caltrans prioritizes bridge retrofitting according to a *level 1 risk analysis* consisting of the following steps (Roberts, 1991):

1. Identify major faults with high event probabilities (priority 1 faults).

2. Develop attenuation relationships at faults identified in step 1.

3. Define the minimum ground acceleration capable of causing severe damage to the bridge structure.

4. Identify all the bridges within high-risk zones defined by the attenuation model of step 2 and the critical acceleration boundary of step 3.

5. Prioritize the bridges at risk by summing weighted bridge structural and transportation characteristic scores.

A long-term risk factor is computed based on macro- and micro-components as follows:

Macrocomponents	Microcomponents
Load factor (seismicity)	Magnitude, acceleration, duration (long, intermediate, short), soil (high, low risk)
Structural factor (vulnerability)	Number of hinges, number of columns per bent, year of construction, outriggers, etc.
Social factor (importance)	On lifeline, multilevel, ADT, route type, detour length, etc.

Structural vulnerability is considered mainly as a function of the following factors:

Components/elements	Relevant factors
Bearings	Support skewness, bearing type, support length
Columns, piers, footings	Shear and flexural capacity as function of effective column length, column bent type, reinforcement ratio, transverse and longitudinal reinforcement, skewness
Abutments	Abutment type, settlement of fill at abutment, skewness
Foundations	Soil conditions, acceleration, discontinuity of superstructure, skewness, redundancy

The procedure is designed for zones of high seismic risk and densely populated urban areas, heavily dependent on transportation. Consequently geological and inventory data govern, whereas structural deterioration is not essential.

Basöz and Kiremidjian (1996)

The authors review essential prioritization procedures, such as those of ATC-6-2 (1983, California and Washington) and develop a vulnerability assessment combining results of seismic hazard and structural fragility analyses, as shown in Fig. A33.1.

The flowchart is an exemplary combination of the tools used by most seismic prioritization procedures. Fragility analysis is a probabilistic procedure which typically does not account for structural deterioration (Mullen and Cakmak, 1997).

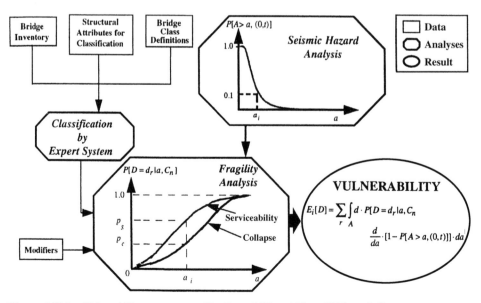

Figure A33.1 Vulnerability assessment (Basöz and Kiremidjian, 1995, p. 3-6).

NYS DOT

In 1990 the NYS DOT issued an engineering instruction supplementing the AASHTO design specifications for seismic loadings. The instruction was updated as EI 92-046 in 1992. The criteria for seismic retrofit activities were identified as follows:

Seismicity of site

Functional importance

Structure type and details

Scope of rehabilitation project

Guidelines were prescribed for seismic retrofit measures. For example, the rehabilitations described in Example 7 were planned independently but had to add seismic retrofitting to their scope.

Design rock acceleration was set at 0.19g. Detailed seismic analysis for bridges with two or more spans was prescribed. Steel sliding bearings (Fig. 14.10) were banned from use. Bearing area was increased, anchorage bolts were upgraded, reinforcement lap splices were detailed, and continuity was recommended, for instance, by splicing the primary members, as in Fig. A33.2, whenever possible.

In 1995 the NYS DOT issued a seismic vulnerability manual, revised in 2002. The manual introduced a procedure for prioritizing seismic strengthening and retrofit of bridges consisting of the following 3 steps:

Screening. The inventory is screened for seismically vulnerable bridge characteristics, including importance, and bridges are sorted into four preliminary susceptibility groups.

Figure A33.2 Providing span continuity by splicing of steel girders during rehabilitation.

Classifying. In an order determined by the screening, bridges are investigated in detail. Their classification score is determined and, accordingly, they are grouped into the high, medium, or low seismic vulnerability classes. The classification score is the product of the structure's vulnerability (*V*) and the seismic hazards (*E*).

Vulnerability Rating. The purpose is to create a uniform rating suitable for prioritization. A *likelihood-of-failure* score is assumed according to the vulnerability class and a *consequence-of-failure* score is added according to the type of expected failure and the level of exposure.

The vulnerability score (*V*) is informed by the following deterministic reasoning (NYS DOT, 1995, p. 3-2):

Although the performance of a bridge is based on the interaction of all its components, it has been observed in past earthquakes that certain bridge components are more vulnerable to damage than others. These are: (a) connection bearings and seats, (b) piers, (c) abutments, and (d) soils. Of these, bridge bearings seem to be the most economical to retrofit. For this reason, the vulnerability score to be used in the classification process is determined by examining the connections, bearings and seat details separately from the remainder of the structure. Connections refer to whether the superstructure is continuous or interrupted by joints. A separate vulnerability score V_1 is calculated for these components.

The vulnerability of the piers, abutments, and soils forms a score V_2. The overall score for the bridge can be equal to 0, V_1, or V_2 depending on the seismic performance category (SPC) and the criticality of the structure.

Vulnerability scores range from 0 to 10. The manual offers step-by-step instructions intended to assist the user in the considerable engineering judgment required for implementing the procedure.

The seismic hazard addressed by the manual is moderate and it is concurrent with significant steel detail and deterioration vulnerabilities. Consequently, the methods and objectives are combined as follows:

Deterministic prescriptive procedure and probabilistic risk assessment

Retrofitting for normal-service conditions and extreme events

Certain bearings, targeted as vulnerable, underwent emergency retrofits, because of their poor condition and, later, permanent rehabilitation, as shown in Figs. A33.3*a*, *b*, *c*.

NBI Specification (FHWA, 2005a)

The screening and evaluation are conducted according to FHWA (1995a). Seismic vulnerability implies that bridge components can collapse or lose function in the design earthquake. The following seismic vulnerability codes are proposed:

U No seismic evaluation performed.

8 Bridge is not vulnerable to seismic collapse or loss of function because the location has a seismic acceleration coefficient ≤ 0.09.

(*a*)

(*b*)

Figure A33.3   (*a*) Vulnerable bearing. (*b*) Temporary retrofit.

(c)

Figure A33.3   (c) Rehabilitation.

7 Bridge location has a seismic acceleration coefficient greater than 0.09, and bridge has been determined to meet current seismic performance standards. Any deficiencies in the original seismic design have been retrofitted to the standard for new design.

6 Bridge superstructure has been determined to perform satisfactory in the design earthquake and needed retrofits completed. Substructure or foundation components do not meet current seismic design standards but should prevent the loss of function of the bridge.

5 Bridge superstructure has been determined to perform satisfactorily in the design earthquake and needed retrofits completed. Performance of substructure or foundation components is unknown.

4 Bridge is seismically vulnerable. Only partial retrofit of bridge has been completed. Additional retrofit is necessary to provide satisfactory performance in the design earthquake.

3 Bridge is seismically vulnerable. No components have been retrofitted.

2 Bridge is seismically vulnerable. Structural distress or damage has occurred from previous earthquake.

1 Bridge is seismically vulnerable. Structural damage from an earthquake has made the bridge or one of its components unstable and failure is imminent or has occurred.

REDARS 2 Software (FHWA-MCEER Project 094, S. Werner et al., 2000)

The REDARS 2 software is part of the deliverables under the program sponsored by FHWA through the Multidisciplinary Center for Earthquake Engineering Research (MCEER) for improving the reliability and safety of transportation structures under seismic loadings. The objective is to estimate potential losses from a hypothetical earthquake and actual alternatives after the event. The risk-based methodology developed by Werner et al. (2000) is used. Originally calibrated for the New Madrid Fault area, the software is becoming generic and applicable to any densely populated area. Transportation networks are considered jointly with lifelines (Chang et al., 1996) and other seismically vulnerable assets.

Input must include bridge and roadway topology and attributes, origin-destination (O-D) zones and trip tables, economic loss data, and National Earthquake Hazard Reduction Program (NEHRP) site soil conditions.

Users have the following options:

Select deterministic or probabilistic analysis of losses under arbitrary earthquake events.

Compare losses with retrofitted and current structural conditions.

Estimate congestion-dependent trip demands.

Preliminary (first-order) applications allow the user to decide if more elaborate analysis and a more detailed database are justified.

The challenge for developers and users of such software is to use data from existing inventories and management systems while adding more details as they become available from other sources or are obtained for this specific purpose.

Example 14 describes a preliminary deterministic scanning of an existing bridge inventory intended to identify vulnerable structural elements and critically important structures in an urban bridge network.

APPENDIX 34.   BRIDGE LIFE-CYCLE COST ANALYSIS (BLCCA)

NCHRP Report 483 (2003) described the basic steps of the BLCCA process (p. 21) as follows:

- Characterize the bridge and its elements.
- Define planning horizon, analysis scenario(s), and base case.
- Define alternative bridge management strategies.
- Specify/select appropriate deterioration models and parameters.
- (Re)estimate costs:
  - Agency, routine maintenance
  - User, work related, other
  - Vulnerability, agency and user
- Calculate net present values.
- Review results.
- Select preferred strategy or modify strategy and reestimate costs.

The primary purpose of the BLCCA is to identify a management strategy with least total life-cycle cost (TLCC). Because of the uncertainty in estimating TLCC, there are several ways in which the indicators can identify a "best" alternative, for example, least total present value of all costs, least agency cost, and others. In addition, the analysis may use the expected value (mean) of such a parameter or a value associated with another probability level (e.g., the value of TLCC such that there is estimated to be only a 20% probability that a lower value will occur, the mean representing the 50% level).

Significant parameters are grouped as follows (NCHRP Report 483, p. 32)

- Total bridge life-cycle cost metrics and parameters influencing computation of life-cycle cost
- Descriptors of actions, deterioration, hazards, and consequences
- Agency costs
- User costs
- Vulnerability costs
- Value metrics for decision making

APPENDIX 35.   HAMBLY'S PARADOX

Heyman (1998, p. 154) illustrates the limitations of simplistic assessments of structural strength reserve, as well as redundancy, by Hambly's paradox. The paradoxical structures are two chairs A and B supported on three and four legs, respectively. Both chairs must resist a load Q by a structural strength $R = 1.2Q$, where Q is applied symmetrically and R is equally divided among the chair's legs. Chair A's legs will have strength equal to $0.4Q$, and those of chair B will have strength of $0.3Q$, as shown in Fig. A35.1.

If surface imperfections exceed the elastic shortening of the legs under the design load, one of chair B's legs would be unloaded. Load Q would be resisted by a capacity of $0.9Q$. If the chair, having lost its symmetry, rocks, the load may end up resisted by only two of the legs, having a strength equal to $0.6Q$.

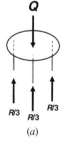
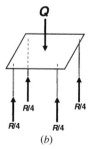

Figure A35.1   Hambly's paradox.

Because of its nonredundancy, chair A is not susceptible to the uncertainties of the support conditions. If, on the other hand, the uncertainty consisted of a 17% variation in the strength of the legs and if three legs were likely to be affected, but not four, the worst-case scenario would be as follows:

Chair A $\qquad 0.4 \times 3 \times 0.83Q = 0.996Q$

Chair B $\quad 0.3 \times (3 \times 0.83 + 1)\, Q = 1.047Q$

Under lateral loads, uneven load distribution to the supports becomes a certainty. Figure E7.6 shows a four-legged pier tower which, as a result of a malfunctioning articulation device at the top, was supporting all loads with only two of the legs. Upon considering numerous realistic scenarios related to inelastic behavior and load redistribution, Heyman (1998, p. 161) concluded: "Under these conditions it is meaningless to ask for a calculation of the "actual" state of a structure; that state is an accidental product of the reaction between the structure and its environment and can indeed change as a result of unpredictable events."

Cremona (2003, p. 233) attributed a similar statement to Streletsky dating to 1928.

APPENDIX 36.   OPTIMIZATION MODELS

Cleland and Kocaoglu (1981, Chapter 8) describe *prescriptive* optimization models and applications belonging to the following general categories:

Classical Optimization. The optimization problem is represented by a continuous differentiable function. Minima and maxima are sought. Solutions can pertain to structural weight minimization, optimal storage and ordering costs, and optimal size of containers.

The range of modeling demands frequently exceeds the scope of *classical optimization*. The following four categories represent *mathematical programming*:

Linear Programming (*LP*). The objective function and the constraints are expressed as linear mathematical statements and the decision variables have continuous values. Applications include crane configurations for maximum load capacity, water resource systems, control of traffic flow, competitive bidding, manpower assignment, equipment location, resource allocation, and production planning.

Nonlinear Programming. If the objective function and the constraint equations are nonlinear, a number of methods may be suitable. *The gradient method* requires a differentiable objective function. *Direct-search* methods do not employ derivatives but require continuous objective functions. Equality constraints can be incorporated in the objective function by means of Lagrange multipliers. Inequality constraints can be modeled by penalty functions.

Integer Programming. Integer programming applies to problems involving discrete variables or outcomes (e.g., yes/no).

Dynamic Programming. Time considerations and sequences of decisions in time become inputs or decision variables. Such are the problems considered by game theory, equipment replacement policies, economic planning, reliability theory, control theory, and stochastic processes. The multistage decision process requires a definition of the stages, an objective of the process, and return (recursive) functions that indicate how the decisions affect the stages.

In contrast to the *prescriptive* models, *descriptive* models perform *simulation*. Whereas the prescriptive models are analytic in form and function, simulations can have analytic representation but essentially seek to match the optimized process by analogy:

Discrete Simulation. The Monte Carlo technique simulates the probability of events by randomly generated numbers. Probability distributions are obtained for the behavior of the system. It has applications in operational decisions for inventory problems, waiting lines, financial risk analysis, and so on.

Continuous Simulation. System components are identified and interrelated by logical functional relationships that describe their impact on each other. As a change is introduced into one or more of the components, that change is followed through the total system. The result is a directional shift in the system's behavior. The changes are observed over long periods in order to determine the long-range implications of major changes that can be introduced by external conditions or external policies. Social, urban, political, and energy systems can be studied by this approach.

APPENDIX 37. NUMERICAL OPTIMIZATION

Diwekar (2003, p.1) introduced the subject as follows: "As a system becomes more complicated involving more and more decisions to be made simultaneously and becoming more constrained by various factors, some of which are new to the system, it is difficult to take optimal decisions based on a heuristic and previous knowledge."

The feasible alternative is numerical optimization.

Numerical Optimization

A general optimization problem can be stated as follows:

$$\text{Optimize } Z = z(x) \qquad \text{(A37.1)}$$
$$\underset{x}{}$$

subject to

$$h(x) = 0 \qquad \text{(A37.2)}$$

$$g(x) \leq 0 \qquad \text{(A37.3)}$$

The goal of an optimization problem is to determine the decision variables x that optimize the objective function Z while ensuring that the model operates within established limits enforced by the equality constraints h (Eq. A37.2) and inequality constraints g (Eq. A37.3).

The model simulates the phenomena and calculates the objective function and constraints. The information is utilized by the optimizer to calculate a new set of decision variables. This iterative sequence is continued until the optimization criteria pertaining to the optimization algorithm are satisfied.

The author refers to some of the many software codes available for numerical optimization. Optimization algorithms mainly depend upon the type of optimization problems.

Types of Optimization Problems

Optimization problems can be divided into the following broad categories depending on the type of decision variables, objective function(s), and constraints.

- *Linear Programming (LP).* The objective function and constraints are linear. The decision variables involved are scalar and continuous.
- *Nonlinear Programming (NLP).* The objective function and/or constraints are nonlinear. The decision variables are scalar and continuous.
- *Integer Programming (IP).* The decision variables are scalars and integers.
- *Mixed-Integer Linear Programming (MILP).* The objective function and constraints are linear. The decision variables are scalar; some of them are integers while others are continuous variables.
- *Mixed-Integer Nonlinear Programming (MINLP).* A nonlinear programming problem involving integer as well as continuous decision variables.
- *Discrete Optimization.* Problems involving discrete (integer) decision variables. This includes IP, MILP, and MINLP.
- *Optimal Control.* The decision variables are vectors.
- *Stochastic Programming or Stochastic Optimization.* Also termed optimization under uncertainty. In these problems, the objective function and/or the constraints have uncertain (random) variables. Often involves the above categories as subcategories.
- *Multiobjective Optimization.* Problems involving more than one objective. Often involves the above categories as subcategories.

Optimization involves three steps: (1) understanding the system, (2) finding a measure of system effectiveness, and (3) degree-of-freedom analysis and applying a proper optimization algorithm to the solution.

Optimization under Uncertainty

Cleland and Kocaoglu (1981, Chapter 10) discuss decision-making methodologies under uncertainty and under risk. In the former case the authors apply the maximin, maximax, and minimax Laplace and Hurwicz principles. In the case of "risk," the authors adopt the Bayesian approach (Appendix 2). Since the latter approach allows for updating critical information, it is clearly the one of choice.

Diwekar (2003, Chapter 5) summarized optimization under uncertainty as follows:

> The literature on optimization under uncertainties very often divides the problems into categories such as *wait and see, here and now,* and *chance constrained optimization.* . . . Both *here and now* and *wait and see* problems require the representation of uncertainties in the probabilistic space and then the propagation of these uncertainties through

the model to obtain the probabilistic representation of the output. Many problems have both here and now, and wait and see problems embedded in them. The trick is to divide the decisions into these two categories and use a coupled approach.

The probabilistic or stochastic modeling iterative procedure consists of:

1. Specifying the uncertainties in key input parameters in terms of probability distribution
2. Sampling the distribution of the specified parameter in an iterative fashion Frangopol and Liu (Miyamoto et al., pp. 57–70) summarize the problem-dependent mathematical programming for optimization purposes as follows (p. 60):

In order to solve optimization problems with multiple and usually conflicting objective functions, one has to convert, with nontrivial inconvenience, the original multi-objective problem into a series of equivalent single-objective optimization problems, which in total produce a set of solutions that displays optimal tradeoff among all competing objectives. . . . One approach is the epsilon-constraint method, which keeps only one objective at a time and converts the other objectives into constraints. The difficulty of this method lies in the determination of constraint values, especially the upper and lower limits. The other approach is the weighted sum method, which forms a single composite objective function as a weighted sum of the original multiple objectives, using a specified set of weight coefficients. Because traditional optimization methods usually search for the optimized solution on a point-by-point basis, multiple algorithm runs are needed to generate optimized tradeoff solutions.

Alternative numerical tools capable of handling multiple conflicting objectives directly and simultaneously are the genetic algorithms (GA, Appendix 43 herein).

APPENDIX 38.   MINIMIZING LIFE-CYCLE COSTS OF CATASTROPHIC STRUCTURAL FAILURES

Ang et al. (in Frangopol, 1998, pp. 1–16) estimate the cost of earthquake damage to concrete buildings; however the probabilistic model applies to catastrophic failures in general. Trade-offs are obtained between initial and expected damage costs. A target reliability, minimizing the total expected life-cycle cost $E[C_T]$, subjected to the constraint of acceptable balanced risk of fatality $r_F \leq r_{F0}$ is determined:

$$E[C_T] = C_1 + E[C_D^0] \tag{A38.1}$$

where C_1 = initial cost of structure
C_D^0 = cumulative damage cost in present worth, including direct and indirect losses

The present worth $E[C_D^0]$ of C_D^0 is estimated by introducing the assumptions that future earthquakes represent a Poisson process, occurrences are statistically independent, and structures are repaired after every significant event:

$$E[C_D^0] = \int_0^L E[C_D] \left(\frac{1}{1+q}\right)^t \nu \, dt = \lambda \nu L E[C_D] \tag{A38.2}$$

where ν = annual mean occurrence rate of earthquakes with significant intensities

L = expected structural life

$\lambda = [1 - \exp(-\alpha L)]/(\alpha L)$ is discount factor

$\alpha = \ln(1 + q)$

q = annual discount rate

The expected damage cost $E[C_D]$ is expressed as a function of the structural damage level x:

$$E[C_D] = \sum_i \int_{y,\min}^{y,\max} \int_0^\infty C_{Di}(x) f_{X|Y}(x) f_Y(y) \, dx \, dy \tag{A38.3}$$

where X = structural damage level

Y = expected maximum ground intensity conditional on occurrence of earthquake

$C_{Di}(x)$ = cost function for damage component i

$f_{X|Y}(x)$ = probability density function of X conditional on $Y = y$

$f_Y(y)$ = probability density function of Y at site

For buildings, the direct losses include the costs of repair or replacement, loss of content, and cost of lives and injury. Each of these depends of the level of structural damage. Various relationships between the building collapse rate and fatality rate are discussed. The cost function of for life savings is expressed as

$$C_F = r_F N_0 V_F \tag{A38.4}$$

where $r_F = r_0 (p_{fc})^n$ is expected fatality rate

p_{fc} = collapse probability of structure

r_0 = expected fatality rate for $p_{fc} = 1$

n = 1.6 accounts for nonlinearities between collapse probability and fatality rate

N_0 = number of occupants

V_F = value of life saved

Nonfatal injuries are estimated similarly.

APPENDIX 39. GENERAL CATEGORIES OF PERFORMANCE INDICATORS FOR DECISION MAKING

NCHRP Synthesis 238 (1997, p. 9) recommended a national transportation performance monitoring system (NTPMS). The general categories of indicators to be incorporated in the system were proposed in TRB Special Report 234 (1992), as shown in Table A39.1. In addition to indicators of the supply and demand for transportation services, the indicators evaluate safety and personal security, access and mobility, service delivery, and cost as well as transportation impacts on economic growth, national security, environmental quality and land use, and energy consumption.

Table A39.1 NTPMS Data Attributes and Descriptors (TRB Special Report 234, 1992)

Supply	System	Providers
	General characteristics	General characteristics
	Coverage	Financial condition
	Physical condition	
	Fare or fee structure	
	Elasticity of supply	
Demand	User characteristics	
	Activity levels	
	Flows	
	Elasticity of demand	
Performance	Safety and personal security	
	Access and mobility	
	Service delivery: level, efficiency, and quality	
	Cost	
Impacts	Economic growth	
	National security	
	Environmental quality and land use	
	Energy use	

APPENDIX 40.   CONDITION RATING SYSTEMS

NBI Ratings (FHWA, 1995b, p. 37, original, 1971, Fig. A16-1 herein)

According to the NBIS (FHWA, 1995) all bridge inventory components and, ultimately, the bridges receive integer numerical condition ratings from 0 to 9 based on a mostly visual comparison of the as-is structure to a presumed as-built state. The ratings are intended to reflect general, rather than local, conditions of decks, superstructures, and substructures (inventory items 58, 59, and 60, respectively) as follows:

N NOT APPLICABLE

9 EXCELLENT

8 VERY GOOD—No problems noted

7 GOOD—Some minor problems

6 SATISFACTORY—Structural elements show some minor deterioration.

5 FAIR—All primary structural elements are sound but may have minor section loss, cracking, spalling, or scour.

4 POOR—Advanced section loss, deterioration, spalling, or scour

3 SERIOUS—Loss of section, deterioration, spalling, or scour have seriously affected primary structural components. Local failures are possible. Fatigue cracks in steel or shear cracks in concrete may be present.

2 CRITICAL—Advanced deterioration of primary structural elements. Fatigue cracks in steel or shear cracks in concrete may be present or scour may have

removed substructure support. Unless closely monitored it may be necessary to close the bridge until corrective action is taken.

1 "IMMINENT" FAILURE—Major deterioration or serious loss present in critical structural components or obvious vertical or horizontal movement affecting structure stability. Bridge is closed to traffic but corrective action may put back in light service.

0 FAILED—Out of service, beyond corrective action

The *Coding Guide* (1995b) similarly rates channels and culverts (items 61 and 62). FHWA (2005a) has proposed the following amendment to the condition rating scale:

N Not applicable

8 Very good

7 Good

6 Satisfactory

5 Fair

4 Poor

3 Serious

2 Critical

1 Failure

Detailed descriptions are provided for the physical characteristics corresponding to each rating for decks, superstructures, substructures, culverts, and stream channels.

Serviceability Appraisal (FHWA, 1995b, p. 45)

The NBIS *appraises* serviceability (inventory item 62) from 0 to 9 relative to a hypothetical new structure, built to current standards, based on the ratings of structural evaluation, deck geometry, horizontal and vertical underclearances, waterway adequacy, and approach roadway alignment (inventory items 67, 68, 69, 71, and 72, respectively). The ratings are defined as follows:

N Not applicable

9 Superior to present desired criteria

8 Equal to present desirable criteria

7 Better than present minimum criteria

6 Equal to present minimum criteria

5 Somewhat better than minimum adequacy to tolerate being left in place as is

4 Meets minimum tolerable limits to be left in place as is

3 Basically intolerable requiring high priority of corrective action

2 Basically intolerable requiring high priority of replacement

1 This value of rating code not used

0 Bridge closed

Tables 1 through 3B of the *Coding Guide* (FHWA, 1995b) link the serviceability appraisals to levels of average daily traffic (ADT) and bridge geometry, minimizing their subjectivity. That information must therefore be regularly updated (ADT inventory data are notoriously unreliable).

Maintenance Rating (FHWA, 2002c, Section 4.2.4)

The *Bridge Inspector's Reference Manual* (FHWA, 2003) proposes a maintenance rating on a scale consistent with the FHWA condition rating and serviceability appraisal, as shown in Table A40.1.

Table A40.1   Maintenance Rating (FHWA, 2003)

Urgency Index	Maintenance Immediacy of Action	Inspection Type of Report
9	No repairs needed	Note in inspection report
8	No repairs needed; list specific items for special attention during next inspection cycle	Note in inspection report
7	No immediate plans for repair; examine possibility for increased level of inspection	Note in inspection report
6	Repair by end of next season; add to scheduled work	Note in inspection report
5	Place in current schedule; current season, first reasonable opportunity	Special notification to superior
4	Priority; current season, review work plan for relative priority, adjust schedule if possible	Special notification to superior
3	High priority; current season, as soon as can be scheduled	Special notification to superior
2	Highest priority; discontinue other work if required, emergency basis or emergency subsidiary actions if needed (post, one-lane traffic, no trucks, reduced speed, etc.)	Verbally notify superiors immediately and confirm in writing
1	Facility closed for repair	—

New York State Department of Transportation (NYS DOT, 1997, original 1982, Fig. A16.5 herein)

The conditions of all structural elements in every span, the components they belong to, and the entire bridges are rated from 1 to 7 as follows:

9 NOT ACCESSIBLE

8 NOT APPLICABLE

7 NEW

6 Shade between 5 and 7

5 MINOR DETERIORATION BUT FUNCTIONING AS ORIGINALLY DE-
SIGNED

4 Shade between 3 and 5

3 SERIOUS DETERIORATION OR NOT FUNCTIONING AS ORIGINALLY DE-
SIGNED

2 Shade between 1 and 3

1 TOTALLY DETERIORATED OR FAILED

Detailed instructions accompany the above ratings for all elements and for the bridge as a whole. For certain bridge elements, for instance, the primary member and the deck, the condition rating reflects the overall state for the rated span. For bearings, on the other hand, the worst one in the span is rated.

Written comments, photographs, and sketches are required for all ratings that are lower than 5, changed by more than one point, or improved. Repair recommendations are encouraged, but not strongly emphasized. The resulting inspection reports are primarily descriptive with two very significant exceptions. One is the definition of the rating 3—not functioning as designed. By assigning a rating equal or lower than 3, an inspector qualitatively changes the structural assessment. The other is the requirement that, independently of all ratings, the inspecting engineer must identify and immediately report all structural hazards, essentially relieving the numerical ratings of the responsibility for decisions related to prompt remedial actions.

The New York State ratings are numerically converted to NBI ratings. The critical rating of 3 (NYS DOT) corresponds generally to 4 on the NBI scale.

Commonly Recognized (CoRe) Structural Elements

FHWA (2005a, p. 6) proposed the following definition of CoRe structural elements: "A set of structural elements that are commonly used in highway bridge construction and encountered on bridge safety inspections and have been endorsed by AASHTO. The CoRe elements provide a uniform basis for data collection for BMS, enable the sharing of data between States and agencies, and allow a uniform translation of data to the Deck, Superstructure, Substructure and Culvert Condition Ratings."

The following CoRe structure elements were identified by AASHTO (1998b, p. 18):

Superstructure: Girder (open, box, closed web), stringers, through truss (with or without bottom chord), deck truss, arch, cable (nor embedded), floor beam, pin and hanger assembly.

Substructure: Column/pile, pier wall, abutment, submerged pile cap/footing, pier cap, culvert.

The recognized structural materials are steel (painted and unpainted), prestressed concrete, reinforced concrete, and timber. Arches of other materials are also recognized.

Four or five condition states are described and quantified in terms of departure from the as-built condition. For painted steel and reinforced concrete the condition states as built (1) to corroded and spalled over 25% of the area with a potential for reduced strength (5). For unpainted steel and prestressed concrete the condition states are four, implying a lesser level of internal reserve.

Feasible actions are recommended for each state, including cleaning, recoating, rehabilitation, and replacement. Reanalysis is implied. The ratings can be used with the PONTIS BMS package (Fig. A16.7 herein).

Conversion of CoRe to NBI ratings

Hearn and Frangopol (in TRC 423, 1994, pp. 122–129) obtain the purely qualitative numeric *NBI* ratings from the quantified CoRe condition states by two methods as follows:

- *Weighted Average.* The method uses the formula (see also A41.1)

$$\text{NBI} = \Sigma M_i F_i \tag{A40.1}$$

where NBI = NBI condition rating computed from BMS data (real number from 9 to 0)
M_i = mapping constant for BMS condition state i
F_i = fractional quantity of bridge element reported in condition state i

- *Table Driven.* The method is similar to the NBI sufficiency rating described in Appendix 41. Mapping constants $M_{i,j}$ are established as shown in Table A40.2. The resulting NBI rating is an integer (from 9 to 0).

The mapping constants for both methods (M_i, $M_{i,j}$) are calibrated to minimize error. The two methods should obtain ratings differing by less than 0.5 on the NBI scale.

Table A40.2   Conversion of CoRe to NBI Condition Ratings

CoRe	NBI
$P1 \geq M1,9$; $P1 + P2 \geq M2,9$; $P1 + P2 + P3 \geq M3,9$; and $P1 + P2 + P3 + P4 \geq M4,9$	9
$P1 \geq M1,8$; $P1 + P2 \geq M2,8$; $P1 + P2 + P3 \geq M3,8$; and $P1 + P2 + P3 + P4 \geq M4,8$ etc.	8
$P1 \geq M1,0$; $P1 + P2 \geq M2,0$; $P1 + P2 + P3 \geq M3,0$; and $P1 + P2 + P3 + P4 \geq M4,0$	0

Note: P_i = percentages of quantities.

Hearn (in Frangopol, 1999b)

The author proposes condition ratings locating bridge elements in one of the following stages of service life: *protected, exposed, vulnerable, attacked,* and *damaged.* The stages are distinct if they can be identified by physical evidence (progressing over time) and if corresponding response actions exist.

A system for estimating the transition of elements from one state to the next is recommended by Hearn (in TRC 498, 2000).

American Railroad Engineering and Maintenance of Way Association (AREMA, 2001)

The Federal Track Safety Standards 49 CFR 213 (Appendix C) provide nonregulatory guidelines for railroad bridge inspection. The *Manual for Railway Engineering* (AREMA,

2001) contains instructions for inspections of steel, concrete and masonry, and timber bridges, independently of the detailed guidelines on rail welding and pit inspections. Inspection findings are documented in a form fundamentally different from the NBIS. The bridge is discretized into components: track, substructure, piers, primary elements, and retaining walls. As in the NBIS, there is no reference to individual spans. In contrast, however, there are no numerical condition ratings. Rather, the manual prompts the inspector to comment on anticipated defects and to report their extent. The latter determines the magnitude and urgency of the recommended repair. Excerpts from the inspection guidelines (AREMA, 2001) follow:

> Concrete and masonry bridges
> ....
> 2. Piers and Abutments
>   Material (brick, stone, concrete):_____
>   Condition of backwall (plumb, clearance of structure):_____
>   Condition of bridge seat:_____
>   ....
>   b. Concrete:
>   Cracks (location, size, description):_____
>   Condition of reinforcing (exposed, corroded-location):_____
>   Condition of waterline:_____
>   ....
>   Steel bridges
>   ....
> 8. Corrosion
>   Loss of section from corrosion, noting exact location and extent of such action, with measurement of remaining section if members are badly corroded, paying close attention to loss of metal in girder and beam flanges and webs, and parts of lateral bracing system.
>   Distortion caused by rust between rivets and built-up members.
>   Damage to overhead structures from engine blast in spans.
>   Pockets at bearing locations and at bottom of bearing stiffeners.

The above reports are defect oriented. Prompt remedial action is implied. Inspections are annual. The AREMA Manual and the track safety standards are annually updated. This relatively austere practice shows the influence of nineteenth-century railroad bridge collapses, such as Ashtabula Creek in 1876 and the Forth of Tay in 1879 (Petroski, 1994). The absence of recent incidents appears to confirm its merit.

Service d'Etudes Techniques des Routes et Autoroutes (SETRA), Laboratoire Central des Ponts et Chausées (LCPC), France

Inspections on the French national highway bridge network combine the rating/descriptive and defect/action-oriented format. Regular and emergency inspections of every bridge structure, including masonry, steel, reinforced and prestressed concrete, refer to the technical instructions of LCPC/SETRA (1979). All anticipated defects, such as spalling, cracks, corrosion, etc., are described, illustrated and assigned identification numbers in structure-specific guides (SETRA/LCPC, 1975, 1981, 1982). Identification numbers are also assigned to the locations on the structure, where such defects can occur, span by span. Defects are rated on a variable scale (two or three levels, depending on their type) and quantified by a code designation. Comments, sketches, and photographs are attached, as in most reports.

The (LCPC SETRA, 1979) manuals for reinforced-concrete, prestressed, and other structures subscribe to the AREMA approach. All anticipated defects are described, explained, and illustrated. The inspection must determine their presence and scope. An independent study (in IABSE 1995, pp. 407–412) estimated the reconstruction needs for the national highway bridge network of France in 1994. Bridges were classified into the following six categories:

1 Good condition
2 Showing defects of equipment or protection components or minor structural defects without urgent need of repair
2E As above but with urgent need of repair in order to prevent more advanced structural deterioration
2S As above with urgent need of repair in order to guarantee safety of road user
3 Bridges with structural damage
3U As above but with urgent need of repair

DANBRO, Denmark

The Bridge Management System of Denmark presented by Andersen and Lauridsen in TRC 423 (1994, pp. 55–61) and by Lauridsen in Vincentsen and Jensen (1998, pp. 49–62) (Fig. A16.4 herein) has been in operation for more than two decades and exports its expertise worldwide. It seeks to minimize data collection by the defect/action method. Routine inspections (every two weeks) issue work tickets for remedial action. General inspections (from one to six years apart) assign condition ratings limited to a maximum of four levels. Special inspections are envisioned for structural or economic reasons. The system is particularly suited for the relatively small number of bridges in relatively good condition.

BRIME (2002)

The Bridge Management in Europe Report (BRIME 2002, Chapter 4, Fig. A16.6 herein) illustrates a possible relationship between six bridge ratings, the corresponding load rating requirements, and the appropriate analytic models and methods. The provided flowchart (p. 72) also lends itself to tabular form, as shown in Table A40.3.

BRIME (2002) definitions are as follows:

Assessment. A set of activities used to determine the safe load-carrying capacity of an existing structure.

Load-Carrying Capacity. The traffic load that can be carried in combination with other loads/actions where appropriate (ULS).

Table A40.3 Load Models and Calculation Principles (Fig. 4.7, BRIME, 2002, p. 72)

Assessment Level	Strength + Load Model	Calculation of Load Effects	Type of Analysis
0	Accepted into BMS without formal assessment; implies records permit assessed capacity to be assigned and condition of the structure is not a cause of concern. This is an optional (proxy) level of assessment.	None specifically for the assessment	As in records. If bridge designed to current standards, capacity factor may be >1 for assessment loading. If calculated, the assessment level could be 2.
1	Strength and load models as devised for assessment if available. Otherwise use design standards.	Simple for level 1	Preferable semiprobabilistic method of analysis using partial factors at ULS as main criterion with SLS where this may have effect on integrity.
2	Material properties based on records and standards	Refined for levels ≥ 2[a]	Semiprobabilistic analysis.
3	Material properties based on in situ testing.	Refined for levels ≥2	
4	Bridge-specific loading based on in situ obaservatons or code values of load for special circumstances	Refined levels ≥2	Semiprobabilistic analysis; modified partial factors.
5	Strength model including probability distribution for all variables and full traffic simulation	Refined for levels ≥2	Probabilisitc analysis; full reliability analysis

[a] Use partial factors at ULS as main criterion with SLS where this may have effect on integrity or as specified by technical authority

Assessment Live Loading. A traffic-loading model used for bridge assessment, which may be less onerous than design loading.

Bridge-Specific Live Loading. A traffic-loading model using measured site-specific data (SLS).

Corrosion Stages of Galvanized High-Strength Wires

NCHRP Report 534 (2004, p. 1-17) referred to the following corrosion stages (or grades):

Stage 1: spots of zinc oxidation on the wires

Stage 2: zinc oxidation on the entire wire surface

Stage 3: spots of brown rust covering up to 30% of the surface of a 3–6 in. (75–150-mm) length of wire

Stage 4: brown rust covering more than 30% of surface of 3–6 in. (75–150-mm) length of wire

The rating is entirely defined by the description. Remaining useful life cannot be inferred, although it is considered hazardously limited at stage 4.

Taiwan Bridge Management

Table A40.4 shows a four-level condition rating system with corresponding time frame of responses proposed for bridge management developed at Taiwan Central University.

Table A40.4 Condition Ratings and Corresponding Urgency of Remedial Actions (Taiwan Central University)

Degree	Extent	Relevancy	Urgency
0: No such item	0: Can't assess	0: Can't assess	0: Can't assess
1: Good	1: <10%	1: Minor	1: Routine
2: Fair	2: <30%	2: Small	2: Within 3 years
3: Bad	3: <60%	3: Medium	3: Within 1 year
4: Severe	4: Over 60%	4: Major	4: Immediately

APPENDIX 41.   BRIDGE CONDITION RATINGS

Hearn (in Frangopol, 1999b) defines as *priority indexes* bridge condition ratings of the following general form (as in Eq. A40.1):

$$PI = \sum_i K_i F_i(a,b,c, \ldots) \qquad (A41.1)$$

where PI = priority index
K_i = weight of ith deficiency
F_i = ith deficiency
$a,b,c, \ldots$ = attributes of deficiency

NBIS (FHWA, 1995, original 1972b Appendix B)

The overall sufficiency ratings of bridges on the national highway network is defined as follows:

$$0 \leq \text{sufficiency rating} = S_1 + S_2 + S_3 - S_4 \leq 100\% \qquad (A41.2)$$

where S_1, S_2, S_3, and S_4 are described in Table A41.1.

A system of weights, rules, and charts defines the computations of $S_{1, 2, 3, 4}$ using inspection-generated condition ratings, design, geometric, traffic and designation inventory data, and computed load ratings.

Table A41.1 NBI Sufficiency Rating (FHWA, 1988)

0 ≤ S_1 ≤ 55%, *Structural* *Adequacy* *and Safety*	*0 ≤ S_2 ≤ 30%,* *Serviceability and* *Functional Obsolescence*	*0 ≤ S_3 ≤ 15%,* *Essentiality for* *Public Use*	*0 ≤ S_4 ≤ 13%,* *Special Reductions* *(if $S_1 + S_2 + S_3 \geq 50\%$)*
	Lanes on structure	Detour length	Detour length
	Average daily traffic	Average daily traffic	Traffic safety features
	Structure type, main		Structure type, main
	Bridge roadway width		
	Appr. roadway width		
	VC over deck		
Superstructure	Deck condition		
Inventory rating	Structural evaluation		
	Deck geometry		
Substructure	Underclearances		
	Waterway adequacy		
	Appr. roadway		
	alignment		
Culvert	Defense highway	Defense highway	
	designation	designation	

States using their own condition rating systems annually submit a numerically converted set of the ratings to FHWA.

New York State DOT

A. Bridge Condition

An overall bridge condition rating R is obtained by the formula

$$R = \frac{\sum_{i=1}^{n} R_i W_i}{\sum_{i=1}^{n} W_i} = \sum_{i=1}^{n} R_i k_i \qquad (A41.3)$$

where $i = 1, \ldots, n$ are bridge elements considered significant to overall bridge condition as shown in Table E23.2
  R_i = worst condition ratings to be found on bridge for each i element
  W_i = weights, e.g., importance factors assigned to each R_i as shown in Table E23.2
  k_i = normalized values of W_i

Table E23.2 shows the shortest useful life observed in New York City for each of the n bridge elements of Eq. A41.3. That equation uses the lowest element condition ratings to occur in any of the bridge spans, thus rating a hypothetical "worst span." Graphically such a condition rating history is represented by the concave line in Fig. 10.1. For better estimates of rehabilitation needs, NYS DOT (1997) introduced separate condition ratings for each span.

B. Span Condition Index (NYS DOT, 1993)

A span condition index (SCI) is defined as

$$SCI = \frac{10 \; CCI_{Super} + 8 \; CCI_{Deck} + 5 \; CCI_{Sub}}{10 + 8 + 5} \tag{A41.4}$$

where CCI_{Super} = [8 (primary member) + 2 (secondary member)]/(8 + 2)
CCI_{Deck} = structural deck rating
CCI_{Sub} = min [CCI_{Abut}, CCI_{Pier}]
CCI_{Abu} = [7 (stem) + 2 (backwall) + 1 (pedestal/seat)]/(7 + 2 + 1), or
CCI_{Abu} = [9 (backwall) + 1 (pedestal/seat)]/ 9 + 1) (if no stem is rated)
CCI_{pier} = [5 (cap beam) + 4 (solid stem or (column) +
        1 (pedestal)]/(5 + 4 + 1)

California Transportation Department (Caltrans)

Caltrans, an early cosponsor of PONTIS, supplemented the CoRe element rating system (FHWA, 1998) by a "health index" for its 12,656 bridges (Shepard and Johnson, 2001). The index relates bridge inspection data to an assumed asset value of the structure or the network, directly expressing structural deterioration in monetary loss as follows:

$$HI = \frac{\Sigma \; CEV}{\Sigma \; TEV} \times 100 \; (\%) \tag{A41.5}$$

where TEV = TEQ × FC
CEV = $\Sigma(QCS_i \times WF_i)$ × FC
WF = [1 − (condition state − 1)(1/state count − 1)] (weight factor for con-
      dition state)
HI = health index
CEV = current element value
TEV = total element value
TEQ = total element quantity
FC = failure cost of element
QCS = quantity in condition state

Equation A41.5 supplements the federal sufficiency rating of Eq. A41.2. It aptly illustrates the role of condition ratings in pursuing bridge assessment to its management objective.

The federal sufficiency rating (Eq. A41.2), the New York State bridge condition rating (Eq. A41.3), the Caltrans health index (Eq. A41.5), and the SETRA condition classification (Appendix 40) assign to bridges unique condition ratings on defined scales. The formulas are a matter of choice and some owners use more than one. New York State inspectors must assign to the bridge a "general recommendation" integer rating between 1 and 7 independently of the condition rating in Eq. A41.3. Yanev and Chen (in TRR 1389, 1993) found the two ratings in consistent agreement. It may be argued that, despite their limitations, different bridge condition ratings generally agree with each other because they all converge toward an objectively existing bridge condition.

Frangopol and Das (in Das et al., 1999, pp. 45–58)

The authors propose a bridge condition rating, and indeed management policy, based on the reliability index β (Appendix 7) as follows:

Reliability Index β	State
<4.6	1
4.6–6	2
6–8	3
8–9	4
> 9	5

Finish Road Administration (Finra)

FHWA (2005b, p. 13) reported on the *repair index* KTI computed for each of the bridges managed by Finra as follows:

$$\text{KTI} = \max \text{Wt}_i \times C_i \times U_i \times D_i) + k \, \Sigma(\text{WT}_j \times C_j \times U_j \times D_j) \quad \text{(A41.6)}$$

where Wt = weight (importance) of damaged structural part

$\quad$ C = condition of structural part, rated from 0 (like new) to 4 (very poor)

$\quad$ U = urgency of repair (e.g., 2 years, $U = 10$; 4 years, $U = 5$; indefinite, $U = 1$)

$\quad$ D = class (severity) of damage, e.g., mild (1), moderate (2), serious (4), or very serious (7)

$\quad$ k = weighting factor for damage summation (default value 0.2)

$\quad$ i = worst defect

$\quad$ j = other defects

"Structural parts" can be elements or components, including the following:

Structural Part	Wt
Substructure	0.7
Edge beam	0.2
Superstucture	1.0
Overlay	0.3
Other surface structure	0.5
Railings	0.4
Expansion joints	0.2
Other equipment	0.2
Bridge site	0.1

KTI is adjusted for average daily traffic (ADT) by a factor ranging from 1.15 (for ADT > 6000) to 0.85 (for ADT < 350). A rehabilitation and reconstruction index, UTI, determines whether to repair, rehabilitate, or reconstruct as follows:

$$\text{UTI} = k_p \times k_l \times (\text{condition} + \text{load capacity} + \text{functionality}) \quad (A41.7)$$

where k_p and k_l are factors reflecting the deck area and the ADT, respectively.

Kawamura et al. (Frangopol and Furuta, 2001)

The authors develop a bridge rating expert system (BREX) evaluating bridge members in terms of serviceability, load-carrying capacity, and durability (Appendix 47).

Taiwan Bridge Management

Table A41.2 shows a weight systems developed by Taiwan Central University for the rating system described in Table A40.4.

Table A41.2 Structural Elements and Weights for Overall Condition Ratings

Rating No.	Element/Component	Weight
1	Approach (roadway)	3
2	Approach guardrail	2
3	Channel	4
4	Approach bank protection	3
5	Abutment foundation	6
6	Abutment	6
7	Wingwall/retaining wall	5
8	Wearing surface	3
9	Bridge drainage system	4
10	Curb and sidewalk	2
11	Railing and barrier	3
12	Pier protection	6
13	Pier foundation	8
14	Pier (shaft)	7
15	Bearing/bearing plate	5
16	Bracing for earthquake	5
17	Expansion joint	6
18	(Main) girder	8
19	(Intermediate) diaphragm	6
20	Deck	7

APPENDIX 42.   EXPERT SYSTEMS (ESs) AND ARTIFICIAL INTELLIGENCE (AI)

At a relatively early stage of development, Adeli (1988) discussed ESs as an offspring of AI. For the latter he adopted the following (among many) definitions (p. 2):

- AI is a branch of computer science concerned with symbolic reasoning and problem solving.
- AI is a subfield of computer science concerned with the concepts and methods of symbolic inference by a computer and the symbolic representation of the knowledge to be used in making inferences.
- AI is a field pursuing the possibilities that a computer can be made to behave in ways that humans recognize as "intelligent" behavior in each other.

ESs are similarly overdefined (Adeli, 1988, p. 5). One of the numerous quoted definitions is as follows: "An interactive computer program, incorporating judgment, experience, rules of thumb, intuition, and other expertise to provide knowledgeable advice about a variety of tasks."

As in the case of bridge management systems, more representative are the essential components of an ES. Adeli (1988, p. 8) listed the following ES components:

1. Knowledge base: a repository of information available in a particular domain. The knowledge base may consist of well-established and documented definitions, facts, and rules, as well as judgmental information, rules of thumb, and heuristics.

2. Inference mechanism (also known as inference engine or reassign mechanism). It controls the reasoning strategy of the ES by making assertions, hypotheses, and conclusions.

3. Working memory (also context or global database): a temporary storage for the current state of the specific problem being solved. Its content changes dynamically and includes information provided by the user about the problem and the information derived by the system.

Desirable features include explanation, debugging and help facilities, intelligent interfaces, and knowledge base editors.

Adeli (1988, p. 14) identified two fundamental approaches to knowledge representation as follows:

Procedural Representation. It is used in traditional algorithmic programming and has the advantage of efficiency; however, knowledge is context dependent and embedded in the code. The result tends to fit the "black-box" description.

Declarative Representation. Knowledge is encoded as data and is therefore more understandable, easier to modify, and context independent.

The author pointed out that in engineering applications, where substantial numerical computation is involved, a hybrid procedural–declarative knowledge representation appears to be the best solution. Examples include the following:

1. Formal methods based on predicate calculus and mathematical logic
2. Semantic networks
3. Semantic (object–attribute–value) triplets
4. Rule-based or production systems
5. Frames of genetic data structures in predefined information slots

Singled out for improvements are the heuristic (e.g. learning) capability, the treatment of unique experiences, inductive problems, and analogies. Subsequent developments have exploited the two contrasted heuristic options, described in Appendix 43:

Combinatorial methods

Updating expertise

APPENDIX 43.   SIMULATED ANNEALING (SA) AND GENETIC ALGORITHMS (GAs)

Diwekar (2003, Chapter 4.4) presented SA and GAs as alternatives to the traditional mathematical programming techniques. They are probabilistic combinatorial methods based on ideas from the physical world.

Simulated annealing is a heuristic combinatorial optimization method derived from statistical mechanics. The analogy is to the behavior of physical systems in the absence of a heat bath. In physical annealing all atomic particles arrange themselves in a lattice formation that minimizes the amount of energy in the substance, provided the initial temperature is sufficiently high and the cooling is carried out slowly. At each temperature T the system is allowed to reach thermal equilibrium, characterized by the probability P_r of being in a state of energy E, given by the Boltzmann distribution:

$$P_r \text{ (energy state} = E) = \frac{1}{Z(t) \exp(-E/K_b T)} \tag{A43.1}$$

where K_b = Boltzmann's constant (1.3806×1023 J/K),
  $1/Z(t)$ = normalization factor

The objective function to be minimized (usually cost) becomes the energy of the system. The behavior of the system is simulated by generating a random perturbation that displaces a "particle," moving the system to another configuration. If the configuration that results from the move has a lower energy state, the move is accepted. If the move is to a higher energy state, the move is accepted with probability $\exp(-\Delta E/K_b T)$, according to the Metropolis criterion. This implies that at higher temperatures a larger percentage of uphill moves are accepted. After the system has evolved to thermal equilibrium at a given temperature, the temperature is lowered and the annealing process continues until the system reaches a temperature that represents "freezing." Thus SA combines both iterative improvements in local areas and random jumping to ensure that the system does not get stuck in a local optimum.

A major difficulty in the application of SA is defining the analogs to the entities in physical annealing. It is necessary to specify the configuration of the space, the cost function, the move generator (a method of randomly jumping from one configuration to another), the initial and final temperatures, the temperature decrement, and the equilibrium detection method. All of the listed parameters depend on the problem structure.

Genetic algorithms are search algorithms based on the mechanics of natural selection and genetics, particularly the survival of the fittest, rather than a simulated reasoning process. Domain knowledge is embedded in the abstract representation of a candidate solution, termed an *organism,* and organisms are grouped into sets called *populations.*

Successive populations are called *generations*. A general GA creates an initial generation (a population or a discrete set of decision variables) $G(0)$ and for each generation $G(t)$ generates a new one, $G(t+1)$, until a solution is found.

The GA may be terminated when an acceptable approximation is found by fixing the total number of generations or by some other special criterion. Key GA parameters are population size in each generation, the percentage of the population undergoing reproduction, crossover and mutation, and number of generations. The crossover operator randomly exchanges parts of the genes of two parent solution strings of generation $G(t)$ to generate two child solution strings of generation $G(t+1)$. Mutation is a secondary search operator increasing the variability of the population.

Furuta et al. (in Miyamoto and Frangopol, 2001, pp. 305–323) apply a virus evolutionary GA to the life-cycle cost analysis of an infrastructure system consisting of multiple facilities.

APPENDIX 44.   CONDITION DETERIORATION MODELS

Pennsylvania DOT

An early comprehensive bridge management system was developed by the Pennsylvania Department of Transportation (PennDOT) (FHWA-PA, 1987). It stressed the importance of maintenance as an inventory item. PennDOT models bridge useful life in terms of NBIS bridge condition ratings. Equivalent and estimated ages of the inspected components are defined. A normalized convex curve results, partially governed by the equation

$$\text{CNR} = 9 \left(1 - \frac{\text{EQA}}{\text{ESL}} \right)^{0.7} \tag{A44.1}$$

where EQA = equivalent age of bridge element (years)
     ESL = estimated life of bridge element (years)
    CNR = condition rating at equivalent age

Miyamoto (in Frangopol and Furuta, 2001)

The deterioration of reinforced-concrete bridge members is represented by an integrated convex graph. Bridge age appears on the abscissa. The ordinate represents *mean soundness* scores of load-carrying capacity and durability, defined respectively as

$$S_L(t) = f(t) = b_L - a_L t^4 \tag{A44.2a}$$

$$S_D(t) = g(t) = b_D - a_D t^3 \tag{A44.2b}$$

where a_L, b_L, a_D, b_D = constants
        t = bridge age (years)
  $f_0(t), g_0(t)$ = deterioration functions from beginning of service to first inspection

Mean soundness is scored from 100 to 0. The deterioration functions are adjusted according to inspection findings, maintenance, and repair. Bridge durability is assumed to reduce faster than load-carrying capacity, hence the lower order of the durability function. (*Note:* In this typical deterministic choice, durability signifies conditions that can be observed during inspections.)

Busa et al. (1985)

The authors propose the following models for NBI bridges:

$$\text{Deck} = 9 - 0.119 \,(\text{AGE}) - 2.158 \times 10^{-6} \,(\text{ADT} \times \text{AGE}) \quad \text{(A44.3a)}$$

$$\text{Super} = 9 - 0.103 \,(\text{AGE}) - 1.982 \times 10^{-6} \,(\text{ADT}) \quad \text{(A44.3b)}$$

$$\text{Sub} = 9 - 0.105 \,(\text{AGE}) - 2.051 \times 10^{-6} \,(\text{ADT}) \quad \text{(A44.3c)}$$

where AGE = age of bridge (years)
ADT = average daily traffic

Ellingwood (in Frangopol, 1998)

Concrete deterioration $X(t)$ comprises depth of penetration or section loss and is represented as function of time t by the following equation (referred to as kinetic or time order):

$$X(t) = C(t - t_i)^\alpha \quad \text{(A44.4)}$$

where C = rate parameter (random variable)
α = time-order parameter
t_i = induction or initiation period for deterioration process (random variable)

Here, C and α are determined by regression analysis of experimental data. Deterministic values recommended for α are $\alpha = \frac{1}{2}$ for diffusion-controlled processes, such as alkali–aggregate reaction, and $\alpha = 1$ for reaction-controlled ones, such as corrosion and sulfate attacks. The proposed method must be calibrated for various field conditions. It appears to correspond to the "durability" assessment of the preceding model. Probabilistic and deterministic aspects of the model are clearly stated.

Sørensen and Engelund (in Frangopol, 1998)

The authors model probabilistically the chloride ingress and progress of the carbonation front into concrete. Parameters are calibrated by measurements using Bayesian statistics. The probability of corrosion initiation is estimated as a function of time and used to optimize maintenance and repair strategies for a structure in marine environment.

Thoft-Christensen (in Frangopol, 1998)

The lifetime reliability of short bridge concrete slabs is estimated when subjected to chloride attacks, resulting corrosion and traffic. The failure models reflect shear and flexural limit states. The author introduces sensitivity factors estimating whether a parameter can be modeled deterministically or by stochastic analysis. The example suggests that the physical properties of the structure and the dynamic load factors are critical initially. Once critical chloride concentration is reached, the stochastic model of corrosion governs failure.

Barlow et al. (1965, Chapter 5)

The authors model complex systems stochastically by Markov chains for "at least two good reasons":

> If each component has an approximately exponential failure law, the complete system can be described approximately by a Markov process.

> The Markov process is the stochastic equivalent of a process in which past history has no predictive value.

In the forecasting of structural conditions, the latter property is considered a liability.

The authors proceed to define semi-Markov processes such that transitions between states take place only at certain times. The semi-Markov process is described as a "marriage of renewal theory and Markov chain theory" (p. 121). Renewal theory is based on assumptions of decreasing mean residual life and increasing failure rate (IFR). The authors show (p. 13) that the exponential distribution has constant failure rate, whereas the Weibull distribution (Eq. A9.10) has increasing failure rate for $\alpha > 1$. The Poisson distribution is a conservative estimate of the probability of n or more failures on $[0,t]$ when t exceeds the mean life of a single component (p. 49).

Ng and Moses (in Frangopol, 1999a, pp. 202–215)

The authors propose a *semi-Markov chain* model such that transitions from one state i to another state j are governed by the transition probabilities p_{ij} while the duration in state i prior to the transition is governed by the holding time distribution h_{ij}.

Frangopol and Liu (Miyamoto et al., 2005, pp. 57–70)

Bridge deterioration can be modeled by Markov chains, continuously and/or mechanistically.

Markov chain models can consider a limited number of previous states. The one-step Markov model considers only the immediately preceding state. In stationary models the essential transition matrix prescribing the probabilities of changing a state (of a component) to other states is constant throughout the specified time horizon.

Continuous computational models combine deterministically assigned deterioration functions with probabilistic distributions of the controlling parameters. Monte Carlo simulation can obtain statistical time-varying performance profiles of deteriorating structures under maintenance, assuming available data.

Mechanistic models account for known patterns and functional dependencies defining material behavior.

BRIDGIT (Hawk, in TRR 1490, 1995, pp. 19–22)

The author describes the Markovian chain process for modeling bridge element deterioration in BRIDGIT as the following sequence:

1. Elements cannot improve their condition unless some action has been effected.

2. An element quantity can transition to a lower condition by, at most, one state in a year.

3. For the total quantity of an element, TOTQUAN, the sum of the normalized quantities in each condition state must be equal to unity as follows:

$$\sum_{i=1}^{5} \frac{QUAN_i}{TOTQUAN} = 1 \qquad (A44.5a)$$

where $QUAN_i$ is the quantity of the element in state i ($1 \leq i \leq 5$) at the beginning of the year of the analysis.

4. The sum of the normalized quantities in each condition state in Y years must be unity:

$$\sum_{i=1}^{5} \frac{NEWQUAN_i}{TOTQUAN} = 1 \qquad (A44.5b)$$

where $NEWQUAN_i$ is the quantity of element in state i in Y years.

The following Markovian transition probability matrix results:

$$\frac{NEWQUAN}{TOTQUAN} = [P]^Y \left[\frac{QUAN}{TOTQUAN} \right] \qquad (A44.5c)$$

where P_{ii} is the probability that a quantity in state i will remain in it one year later.

Because of stipulations 1 and 2, $[P]$ has significant terms only on and immediately below the main diagonal. Stipulation 5 implies that $P_{55} \equiv 1$.

The index Y can vary according to conditions. The deterioration model of a bridge element requires the following information, to be obtained by inspections, stochastic or deterministic analysis:

Average number of years during which a specific percentage of an unprotected element quantity deteriorates from new condition to another condition state or worse

Corresponding fractional element quantity

Pontis (in Golabi et al., 1992)

The *Pontis Technical Manual* (Golabi et al., 1992) describes two condition state prediction models.

The *prior* model is used in the absence of two consecutive biennial inspection reports. The probability P_{ij} of transiting from state i to state j in two years is

$$P_{ij} = 1 - 0.5^{2/T_1} \qquad (A44.6)$$

where T_1 = engineer's estimate
$$1 \leq i \leq 4$$
$$1 \leq j \leq 4$$

The *posterior* model determines the transition probabilities in the Markov chain by linear regression once there are data from at least two consecutive inspection reports.

Brühwiler, Roelfstra, and Hajdin (in Miyamoto and Frangopol, 2001, p. 215)

The authors make the important observation that the Bayesian theorem (Section 5.2) allows for two interpretations of the terms in the Markovian transition probability matrix as follows:

> As the percentage of a segment that changes from condition state i to condition state j after one inspection period

> As the probability of a unit quantity of a segment to pass from condition state i to condition state j after one inspection period

Roelfstra, Adey, Hajdin, and Brühwiler (in TRC 498, 2000, C-2) divide concrete bridges into segments according to function. Deterioration parameters are exposure to chlorides and humidity, permeability, and cover thickness. The safety coefficient is modeled with respect to time, resulting in a partially linearized flat-S deterioration history.

Hearn (in TRC 498, 2000, C-1)

As in the preceding example, the author defines the transition probability P of an element moving to the next condition state as a function of the time spent in the current state $t_{Residence}$ as

$$P = 1 - \frac{1}{t_{Residence}} \qquad (A44.7a)$$

$$t_{Residence} = t_{State} + t_{Trans} \qquad (A44.7b)$$

where t_{State} = prior time in current state (years)
t_{Trans} = time left before transition (years)

Both times are obtained from the calibration of a deterioration function.

The condition states are the generic *protected, exposed, vulnerable, attacked,* and *damaged* (see Appendix 40), formulated by Hearn (in Frangopol, 1999b).

Madanat and Lin (in TRR 1697, 2000, pp. 14–18)

The authors propose a Bayesian method for modeling the deterioration of reinforced concrete. Measurements are subjected to sequential hypothesis testing and a probability is assigned to the conditions falling into one of the five previously defined levels.

Fitch, Weyers, and Johnson (in TRR 1490, 1995, pp. 60–66)

The proposed estimate of deterioration in reinforced-concrete decks is a typical combination of phenomenological, statistical, and deterministic input. The progress of steel corrosion is correlated to the cumulative damage observed on the surface area over time according to the model proposed by Cady and Weyers (1984). The main stages of corrosion are as follows:

1. Chloride ion diffusion through the cover concrete begins.
2. Corrosion of reinforcing steel begins (no surface evidence).
3. Cracking of concrete surrounding the reinforcing steel begins (linear increase in surface evidence).
4. Bridge component reaches the end of service life because of an accumulation of physical damage.

The authors combine this phenomenological model with statistical data from inspections, rehabilitation history, anti-icing needs, and average daily traffic (ADT) in order to estimate time to rehabilitation (TTR). The resulting models are of the form

$$y' = -10.3 + 14.0x - 11.4x^{1.05} \tag{A44.8a}$$

where y' = fitted time to rehabilitate decks based on applicable standards
x = percentage of deck delaminated, spalled, or patched

In the snowbelt:

$$y' = -11.2 + 5.34x - 3.41x^{1.1} \tag{A44.8b}$$

where x is the percentage of the worst traffic lane delaminated, spalled, or patched.

APPENDIX 45.   AASHTO LOAD RATING

AASHTO has adopted the following definitions:

Nominal resistance	Resistance of a component or connection to load effects, based on its geometry, permissible stresses, or specified strength of materials
Safe load capacity	Live load that can safely utilize a bridge repeatedly over the duration of a specified inspection cycle
Strength limit state	Safety limit state relating to strength and stability
Serviceability limit state	Collective term for service and fatigue limit states

For highway bridges the *Coding Guide* (FHWA, 1988) defines two load ratings as follows:

Operating (Maximum) Rating, NBI Item 64, Item L-5 in NBI Specifications (*FHWA, 2005a*). "A capacity rating resulting in the absolute maximum permissible load level to which the structure may be subjected for the vehicle type used in the rating."

A bridge can be loaded to the maximum permissible level under unique circumstances, but indefinite use at that level may shorten its useful life.

Inventory (Service) Rating, NBI Item 66, Item L-4 in NBI Specifications (FHWA, 2005a). "A capacity rating resulting in a load level, which can safely utilize an existing structure for an indefinite period of time."

The design live-loads bridges are rated for are defined by the AASHTO (Appendix 30 herein). In the most general terms, the bridge rating RT in tons is defined as

$$RT = (RF)W \tag{A45.1}$$

where W = weight (in tons) of nominal truck used to determine live-load effects (L)
RF = rating factor

The LRFD rating calculations of AASHTO (2003) were preceded by the AASHTO *Guide Specifications for Strength Evaluation of Existing Steel and Concrete Bridges* (AASHTO, 1989). The guide formulated the basic structural engineering requirement for supply (of structural strength) to exceed demand (of applied loads) as follows:

$$R \geq \sum_k Q_k \tag{A45.2}$$

where R = resistance
Q_k = effect of load k

Design specifications stipulate the *sufficient* supply of strength. The AASHTO (1998) LRFD bridge design code uses the following form (Eq. 1.3.2.1-1) (LRFD, 1998):

$$R_r = \varphi R_n \geq \sum_i \eta_i \gamma_i Q_i \tag{A45.2a}$$

where $\eta_i = \eta_D \eta_R \eta_I$ = load modifier for ductility, redundancy, and operating importance, $0.95 \leq \eta_i \leq 1.0$
γ_i = statistically based load factor applied to force effects
φ = statistically based resistance factor applied to nominal resistance

Based on Eqs. A45.2 and A45.2a, the guide (AASHTO, 1989) defines the rating factor as

$$RF = \frac{\varphi R_n - \sum_{i=1}^m \gamma_i^D D_i - \sum_{j=1}^n \gamma_j^L L_j (1 + I)}{\gamma_R^L L_R (1 + I)} \tag{A45.3}$$

where RF = rating factor (portion of rating vehicle allowed on bridge)
φ = resistance factor
m = number of elements included in dead load
R_n = nominal resistance
n = number of live loads other than rating vehicle
γ_i^D = dead-load factor for element i
D_i = minimal dead-load effect of element i
γ_j^L = live-load factor for live load j other than rating vehicle's
L_j = nominal traffic live-load effects for load j other than rating vehicle's
γ_R^L = live-load factor for rating vehicle

L_r = nominal live-load effect for rating vehicle

I = live-load impact factor

Equation A45.3 is paraphrased as follows:

Rating vehicle effects = capacity − dead-load effects − other live-load effects

If live-load effects are computed according to wheel line distribution factors, implying more than one vehicle, the "other than the rating vehicle's" load effects L_j can be ignored, resulting in a simplified rating factor expression:

$$RF = \frac{\varphi R_n - \gamma_D D}{\gamma_L L(1 + I)} \qquad (A45.3a)$$

"Load effects" can be any of the member forces entering design calculations, including bending moment and shear and axial load. The lowest rating factor to be obtained for any of the evaluated structural members governs the structure.

The dead- and live-load factors γ_D and γ_L depend on the structure and the traffic volume. These parameters are subject to design considerations and updates.

The nominal resistance R_n and the resistance factor φ pertain to existing bridges. A flowchart taking into account structural deterioration, redundancy, quality of inspection, and maintenance determines φ in the following range:

$$0.55 \leq \varphi \leq 0.95$$

The *Manual for Condition Evaluation* (AASHTO, 2000b) refers to Eq. A45.3 above in the following form:

$$RF = \frac{C - A_1 D}{A_2 L(1 + I)} \qquad (A45.3b)$$

where A_1, A_2 = dead- and live-load factors in lieu of γ_D and γ_L

C = capacity of rated structural member in lieu of φR_n

The altered notation in Eq. A45.3b is a reminder that the AASHTO allowable stress design (2002) is also an option, in which case both A_1 and A_2 are equal to 1.

For load factor design

$$A_1 = 1.3 \qquad A_2 = \begin{cases} 2.17 & \text{for inventory level} \\ 1.3 & \text{for operating level} \end{cases}$$

Capacity C is the same for operating and inventory rating (LRFD) but varies in allowable stress.

AASHTO (2003) computes member capacities for the limit states defined in the LRFD manual (AASHTO, 1998a). Three load rating levels are established as follows:

- Design load rating (first-level evaluation)
- Legal load rating (second-level evaluation)
- Permit load level (third-level evaluation)

The legal load rating is required when the first-level evaluation obtains an operating load rating RF < 1. Permit load rating cannot be considered unless bridges qualify for the legal one. Load factors are specifically calibrated for each of these levels.

The load rating factor RF for a component or connection under a single force effect (axial flexure or shear) assumes the form

$$\text{RF} = \frac{C - \gamma_{\text{DC}}\text{DC} - \gamma_{\text{DW}}\text{DW} \pm \gamma_p P}{\gamma_L(\text{LL} + \text{IM})} \qquad \text{(A45.3c)}$$

where
$C = f_R$ for service limit states
f_R = allowable stress specified in LRFD code
$C = \varphi_c\varphi_s\varphi R_n$ for strength limit states
R_n = nominal member resistance (as inspected)
$\varphi_C = 0.85 - 1.0$, condition factor
$\varphi_S = 0.85 - 1.0$, system factor (incorporating various η_i factors of Eq. A45.3a)
φ = LRFD resistance factor
$\varphi_C\varphi_S \geq 0.85$ for strength limit states
DC = dead-load effect due to structural components and attachments
DW = dead-load effect due to wearing surface and utilities
P = permanent loads other than dead loads
LL = live-load effects
IM = dynamic load allowance
$\gamma_{\text{DC}}, \gamma_{\text{DW}}, \gamma_P, \gamma_L$ = respective LRFD load factors

A variety of system factors φ_S were proposed (NCHRP Report 12-46, 2000), depending on the structural type and presumed redundancy. The live load factors γ_L generally range from 1.35 for operating to 1.75 for inventory level (1.85 for permit level). In order to permit greater live loads on a structure with RF < 1.0, AASHTO (2003) requires detailed evaluations, including level of reliability, refined analysis, load testing, site-specific load factors, and direct safety assessment. Each procedure is defined and illustrated by example.

Lichtenstein (in NCHRP Research Results Digest 234, 1993)

The report recommended procedures for bridge rating through nondestructive and proof load testing in cases where standard load rating formulas do not reflect the structural behavior realistically. Such cases include the following:

- Unintended composite action
- Load distribution effects
- Participation of parapets, railings, curbs, and utilities
- Material property differences
- Unintended continuity
- Participation of secondary members
- Skew

- Deterioration and damage
- Load carried by deck
- Unintended arching action because of frozen bearings

The preceding list is also quoted in NCHRP Synthesis 327 (2004).

The NBI specifications (FHWA, 2005a) propose the following coding of rating methods:

0 No rating analysis performed

1 Load factor (LF)

2 Allowable stress (AS), acceptable only for timber and masonry structures

3 Load and resistance factor (LRF)

4 Load testing

5 Field evaluation only (no plans available)

APPENDIX 46. FLAGS

NYS DOT engineering instruction EI 7.35-13 of February 22, 1994, superseded EI 88-39 and EI 85-38, designating potential hazards as "Flags" as follows:

1. *Red Structural Flag.* Report of a failure or potentially imminent failure of a critical primary structural component. Potentially imminent implies before the next inspection.

   Typical red-flag conditions include scour and movement of foundations, distortions, cracks, and significant loss of section in fracture-critical primary members. "Significant" requires expert judgment. Cracks and section loss may be considered a less than imminent hazard and are reported as yellow flags.

   Red flags are issued when the load rating of the bridge (according to inspection documentation) is more than 3 tons (or 50%) below the posted rating at the site, less than 22 tons when there is no actual posting, or less than 3 tons.

   Thus red flags are linked to both condition and load ratings.

2. *Yellow Structural Flag.* Report of a potentially hazardous condition which, if left unattended beyond the next anticipated inspection, would likely become a clear and present danger. To be used also for reporting actual or imminent failure of a noncritical structural component, where such failure may reduce the reserve capacity or redundancy of the bridge but would not result in a structural collapse or cause clear and present danger. A yellow flag is not used to draw attention to maintenance or routine repair needs.

3. *Safety Flag.* Report of a condition presenting clear and present danger to vehicle or pedestrian traffic but no danger of structural failure or collapse. Could be issued for closed bridges whose condition presents a threat to vehicular (or pedestrian) traffic underneath.

The failures of nonstructural elements, such as claddings, veneer, and parapets (Figs. E16.7–E16.9), utilities, railings (Figs. 14.17 and 14.18), sidewalks, and curbs (Fig. 4.40) are reported as safety flags. Despite their well-recognized structural implications, the failures of scuppers (Fig. E12.B) and expansion joints (Figs. 4.26, 4.28, 4.74, and 4.75) are flagged as high-priority safety hazards, because the potential of injury is the only urgency arising from such conditions.

Bridge deck failures do not entail the collapse of the primary structure but can entirely compromise service and cause a bridge closure. Therefore, they are more than just safety flags. In the case of monodecks it becomes difficult to distinguish between failures limited to the wearing surface and those extending through the deck (Figs. 4.14 and 4.15).

Flags are processed according to NYS DOT (1997, Appendix I). Flag "packets" include a report, notes, sketches, photos, and scour documentation (if available). Packets are transmitted to all responsible parties within five working days. Flags requiring "prompt interim action" (PIA) within 24 hours are transmitted immediately and the inspector remains at the site until the responding party arrives. Response to all other flags must not take longer than six weeks and may consist of temporary or permanent repair, engineering analysis, or closure.

Example EA46 describes the application of the flag method to the New York City bridge network.

Example EA46. Flag Forecasting (NYC DOT)

Between 1980 and 1994 the flags reported annually on the roughly 800 New York City bridges (with approximately 5500 spans) escalated from 30 to 3000, as shown in Fig. EA46.1. The change seemed to imply a transition of the network from a manageable stable equilibrium to one of overwhelming emergency repair demands. In response, after having lapsed into a branch of the Streets and Highways Division during the post–World War II years, a Bureau of Bridges was reestablished at the NYC DOT in 1988. Over the following 15 years annual investments exceeding U.S. $500 million in capital reconstruction and up to U.S. $50 million for emergency repairs reversed the trend of flag proliferation.

The following subjective and objective reasons contributed to the escalation of flags:

- More than half of the city bridges were older than 30 years and, after several decades of deferred maintenance, their conditions were reaching a critical point. Example 12 describes the distribution of bridge age and conditions during the period 1990–2005.

- The definitions in Appendix 46 evolved from a single flag signifying any potential hazard to safety, yellow, and red structural flags and ultimately prompt interim action (PIA) flags to be resovevd within 24 hours. Most significantly, after 1988, all structural elements involved in a reported condition had to be flagged separately. The condition shown in Fig. 4.24, for instance, would have constituted a single flag in the early 1980s but would have had to be flagged for primary member, secondary member, deck, and joint after 1988. This requirement facilitated flag-related data analysis but rendered the history of Fig. EA46.1 nonhomogeneous (as its pattern suggests).

Figure EA46.1 Flags issued for New York City bridges.

- Accidents influenced the perception of inspectors. The fatality of June 1, 1989 (Example 16), was followed by a pronounced rise in flags related to the spalling of concrete deck undersides. Fracture-critical and special-emphasis details were continually added to the list of "100% hands-on inspection" items (Section 14.2).

Once reported, flags must be addressed according to the prescribed schedule (Appendix 46). The average cost of a flag repair was estimated at U.S. $10,000 in 1990 and at U.S. $15,000 in 2002. Repairs include temporary traffic closures, specialized in-house and contracted emergency work, possible load postings, and bridge closures. In the long term the only resolution for a flag-ridden network is rehabilitation. The simultaneous response to emergency demands and strategic objectives is a managerial challenge which in this case was met by an organization structured as shown in Fig. E18.2. In the short term, a flag forecasting procedure had to be developed for annual budget, workload, and traffic estimates.

The forecast (Yanev, 1994) combined available flag and condition rating histories in the following steps:

- For every year on record, flags are linked to the bridge elements in potentially hazardous conditions. Only new flags are considered (reflagging is quite common).
- The 22 bridge elements in Table EA46.1 account for approximately 85% of the flags.
- The following 12 elements received approximately 70% of all flags: approach guard rail, primary member, deck, wearing surface, columns, abutment bearings, backwall, abutment stem, railing, sidewalk, utilities, abutment joint.
- The following 10 bridge elements contributed approximately 15% of the flags: abutment seat, wingwalls, curbs, scuppers, median, pier joints, pier bearings, pier pedestals, pier stem, cap beams.

The likelihood of the 22 elements of Table EA46.1 to be flagged was estimated as follows:

- The flagged elements were sorted according to their condition ratings (equating elements with spans rated).
- For every rating level, the ratio of flagged elements was assumed equal to the likelihood of flagging. If 50% of all steel primary members were flagged at their rating of 3, then an element rated 3 would be considered to have a 50% chance of being flagged.
- The history of condition ratings for the 22 elements under consideration was obtained deterministically. Worst-case deterioration rates for each type of element (Examples 9 and 12) were considered. Known emergency repairs were taken into account.

Based on past records and trends, the procedure assigned a ratio of flagged elements to every rating level of 22 elements. The result was assumed to represent roughly 85% of all anticipated flags as follows:

Table EA46.1 Flags Issued for Twenty-Two Bridge Elements in Selected Years (NYC DOT; Yanev, 1994)

Bridge Element	1982	1987	1988	1989	1990	1991 Projected
Abutment joint	0	0	3	2	2	1
Abutment bearings	0	0	0	1	6	4
Abutment seat	0	2	5	9	19	17
Abutment stem	3	0	7	12	28	11
Backwall	0	1	0	5	17	14
Wingwalls	0	4	5	15	25	15
Approach guide rail	5	16	35	28	74	69
Wearing surface	1	3	11	14	27	35
Curbs	2	3	8	16	25	51
Sidewalk	18	26	41	40	85	106
Railing	3	29	38	46	92	118
Scuppers	1	0	2	3	16	22
Median	0	1	1	6	7	23
Deck	10	26	44	87	264	463
Primary member	11	22	80	261	391	679
Pier joints	5	8	7	10	26	43
Pier bearings	0	2	7	3	11	12
Pier pedestals	0	1	3	8	7	11
Pier stem	0	0	1	4	9	4
Cap beam	0	5	4	6	33	45
Columns	1	6	12	16	45	118
Utilities	0	6	9	6	39	50
Total	60	161	323	598	1248	1909

$$0.85 \text{ flags} = \sum_{i=1}^{22} \sum_{R=1}^{7} (N_R^i \times \text{flags}_R^i) \qquad \text{(EA46.1)}$$

where i = each of 22 elements of Table EA46.1

R = each of NYS DOT condition ratings 1–7 (Appendix 40)

N^i = number of elements type i, rated 1–7, respectively, such that

$$N_R^i = \sum_{R=1}^{7} N_R^i$$

N^i = number of spans in network, where element i is present

flags_R^i = percentage of elements in each group N_R^i expected to be flagged during year of forecast (%)

The method is purely deterministic. The estimates would improve if they were made separately not only for every condition rating level but also for each year when that rating of the element under consideration was assigned. PONTIS and BRIDGIT use Markov chain models for such estimates, assuming adequate data. In this case the intent was to forecast the immediate needs (and the corresponding budget requests) for eradicating the conditions before their historic record could become statistically significant. The exercise yielded several observations of consequence:

- Comparable conditions attract more flags on essential bridges. This effect is not purely subjective, since essential bridges usually carry heavier traffic.

Thus, out of 800 bridges in New York City, 29 were responsible for 75% of all flags. These, however, were the multispan East River bridges (Figs. E1.1–E1.3; Example 3) and the 25 movable bridges, all of which are both essential and complex structures. All were aged approximately 100 years or more. Network-level management had already assigned top priority to the rehabilitation of the same bridges based on importance and overall condition.

- Flag and bridge condition data sets could not be automatically linked. All cross-referencing had to be done manually. Flag packets still do not indicate the condition ratings of flagged elements, nor do element condition rating files suggest the presence of flags. A correlation between the two types of assessment is an essential need addressed by current inspection software design (Example 27).
- Figures EA46.2–EA46.4 show that, below a certain level of flag occurrence, the lowest ratings attract the most flags. At higher levels of flag incidence, however, flag numbers begin to peak at the rating of 3 (which is designated as "not functioning as designed"). It can be speculated that under moderate deterioration condition ratings are consistent with hazard evaluations. As overall conditions deteriorate further, the findings of potential hazards precede condition ratings. In those cases, by the time the rating declines to 2, the newly flagged conditions decrease, because the bridge may have been closed or the

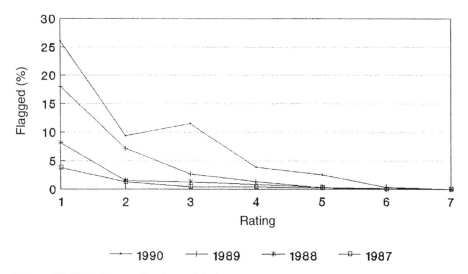

Figure EA46.2 Flags and ratings of decks.

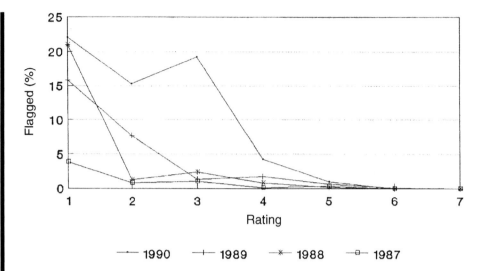

Figure EA46.3 Flags and ratings of primary members.

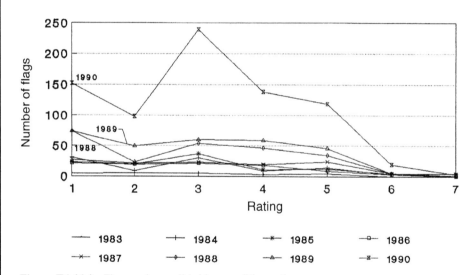

Figure EA46.4 Flags and overall bridge condition ratings.

condition repaired. Temporary repairs such as timber shorings or shielding eliminate the hazard, and hence close the flag, but do not improve the rating (see Section 11.4).

- Flags related to traffic accidents typically occur at a constant rate and are generally independent of bridge condition ratings.

- Beginning in 1982, low-rated primary members and decks were flagged increasingly. By 1990 all spans with decks rated 1 were flagged (Fig. EA46.2).

For primary members that ratio approached 300% (Fig. EA46.3); for example, the spans where primary members were rated 1 carried on the average three flags.

The obtained forecasts (e.g., those for 1991 in Table EA46.1) proved accurate, suggesting that condition ratings and potential hazards are indeed correlated. Implied is the conclusion that potential hazards are most effectively eliminated by maintaining condition ratings higher than 4 on the seven-grade scale.

APPENDIX 47.   EXPERT SYSTEMS FOR BRIDGE MANAGEMENT

Kawamura, Nakamura, and Miyamoto (in Frangopol and Furuta, 2001, pp. 161–178) developed a concrete bridge rating expert system (BREX). The concrete bridge serviceability is a function of the *capacity* and *durability* (e.g., rate of material degradation). The load-carrying capacity is determined by analysis and can be improved by strengthening. Durability is assessed by visual inspections and can be enhanced by repair.

The performance of a target bridge is evaluated in a diagnostic inference process simulating the assessment conducted by experts. A hierarchical neural network structure expresses the relationship between 12 judgment items and the input data in terms of if–then rules and fuzzy variables (hence the term neuro-fuzzy expert system). The system "learns" by back-propagation. An effort is made to render all hierarchical rules verifiable by the users in order to avoid a "black-box" effect. A correlation is sought between results obtained by the system and by experts.

Miyamoto (in Frangopol and Furuta, 2001, pp. 179–198) proposed to integrate BREX into a bridge management system (J-BMS). The system provides decision support in the selection of optimal rehabilitation and repair actions, minimizing cost or maximizing quality. A genetic algorithm is used for the selection of optimal candidates based on the convex deterioration curves generated by BREX. An example applies the algorithm to seven bridges, taking into account strictly structural conditions.

Mizuno et al. (in Miyamoto and Frangopol, 2002, pp. 111–126) extend decision support to field inspections by a Web-based interactive system (Fig. A47.1).

The system creates check-up lists for field inspections in two ways:

* Following rule-based reasoning deduced from maintenance standards, inspector's manuals, and so on (e.g., phenomenologically)
* Based on inference from past inspection records (e.g., based on experience)

Support information is provided in two forms:

* Lists of possible causes for the symptoms being reported, based on records of precedents
* Criteria for the identification of defects requiring urgent remediation

Various enhancements, such as speech and image recognition, are intended to simplify the inspector's tasks. The authors stress the future need for a capability to extract

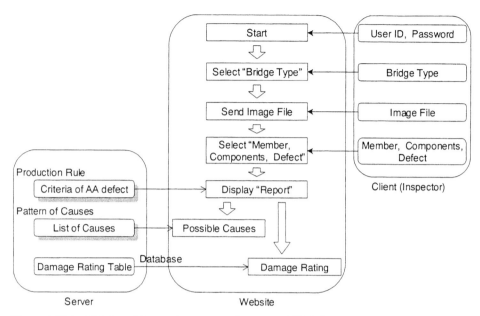

Figure A47.1 Web-based interactive system. Courtesy of Y. Fujino.

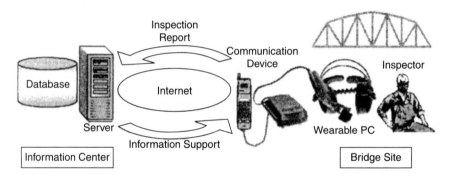

Figure A47.2 Flow of inspection support information. Courtesy of Y. Fujino.

information from speech and image input. The data flow between the server, the website, and the client is shown in Fig. A47.2.

Thoft-Christensen (in Frangopol 1999a, pp. 236–247) has developed the expert systems BRIDGE1 and BRIDGE2 for use during and after inspection for data acquisition and analysis, respectively. The systems are supplements to DANBRO (Appendix 40).

APPENDIX 48.   PREVENTIVE MAINTENANCE (PM)/MAINTAINABILITY

NCHRP Synthesis 327 (2004) surveyed PM practices employed at off-system and local (e.g., low-traffic-volume) bridges in the United States. The following generally applicable maintenance techniques are identified as the most common:

- Periodic application of boiled linseed oil for the first several years of the deck's life
- Use of penetrating concrete sealers
- Annual washing of the bridge deck to remove accumulated debris and road salts
- Maintenance of bridge scuppers and drainage systems

Also cited are the washing of bridge underside and debris removal, joint cleaning and repair, scour mitigation by cleaning of the channel, and adding of riprap.

Reference is made to the maintenance manuals of Florida and Iowa State transportation departments.

NCHRP Synthesis 327 (2004) compiled the following summary of maintainability or durability measures popular among local bridge owners in the United States:

Consistent with the ranking of bridge decks as the most significant maintenance concern, many . . . initial construction . . . techniques cited are aimed at prolonging the life of decks, including:

- Use of epoxy-coated reinforcing;
- High-performance or dense concrete mixes;
- Silica fume admixtures;
- Increased cover. . . .

With regard to details, the most common problem in bridge structures is the detrimental effect of water, whether as an undermining force on bridge foundations, a pressure behind abutments, or as a corrosive catalyst in bridge superstructures. . . . Several [respondents] indicated the use of continuous bridges, jointless bridges, and integral abutments.

Local bridge owners are reported to seek "maintenance-free" superstructures by the following design options:

- Weathering steel
- Galvanized steel
- Precast concrete components, including pipe and box culverts

The cost effectiveness of concrete deck overlays is another relevant factor. Llanos and Yanev (1991) observed that the useful life of decks with waterproofing and asphalt overlay has been up to twice longer than that of monodecks in New York City, where up to 300,000 tons of de-icing rock salt is applied during heavy winters.

NCHRP Synthesis 327 (2004) identified several popular measures against scour, including:

- Riprap protection, possibly combined with geotextile reinforcing
- Larger hydraulic openings
- Pier foundations outside water channels
- Piers on piles and pile bents

APPENDIX 49. USER COST ESTIMATES

BLCCA (NCHRP Report 483, 2003, p. 42) points out that some user costs "may be incurred as monetary expenses, e.g. increased vehicle fuel consumption, but most are not." The following types of delays are considered as potential causes of monetary losses:

Traffic Congestion and Closure Delays TDC_C

$$\text{TDC}_C = [\text{tdc}_1 v_1 + \text{tdc}_2 v_2 + \cdots + \text{tdc}_n v_n] \text{DT}_c \qquad (A49.1)$$

where $\text{tdc}_1, \text{tdc}_2, \ldots, \text{tdc}_n$ = delay cost per vehicle per unit time for vehicle types
$\qquad\qquad\qquad\qquad 1, 2, \ldots, n$
$\qquad v_1, v_2, v_n$ = number of affected vehicles of types $1, 2, \ldots, n$
$\qquad\qquad \text{DT}_c$ = average delay time per vehicle due to congestion and closure

Traffic flow and queing models may be employed to estimate the number of vehicles likely to be delayed and the likely lengths of delay during partial and complete temporary closures.

Average Delays Caused by Deck Surface Pavement Condition DT_p

$$\text{DT}_p = A \left(\frac{\text{CI}_t}{\text{CI}_F}\right)^Z \qquad (A49.2)$$

where CI_t = condition deterioration index (increasing with roughness) during period considered
$\quad \text{CI}_F$ = condition index level considered failure and warranting resurfacing
$\quad Z$ = empirical or judgmental exponent, typically > 2 (Purvis et al., in NCHRP Report 377, 1994, set $Z = 4$)
$\quad A$ = unit delay calibrating factor

Traffic Detours and Delay-Induced Diversions TDC_D

$$\text{TDC}_D = [\text{tde}_1 v_1 + \text{tdc}_2 v_2 + \cdots + \text{tdc}_n v_n] \text{DT}_D \qquad (A49.3a)$$

where DT_D is the average delay time per vehicle diverted. In case of a known distance D_D to the nearest alternate crossing:

$$\text{TDC}_D = 2\left(\frac{\text{FP}}{\text{FC}_n} + \frac{\text{TV}_n}{S_n}\right) D_D v_n \qquad (A49.3b)$$

where FP = prevailing average price of fuel
$\quad \text{FC}_n$ = average fuel consumption rate for vehicle type n
$\quad \text{TV}_n$ = unit value of time for vehicle type n, e.g., average wage rate
$\quad S_n$ = average speed of vehicle n

The report further recognizes the possibilities for highway vehicle damage and environmental, business, and other effects. Vulnerability costs are estimated separately.

Thompson et al. (in TRR 1697, 2000, pp. 6–13) adjust the model used in PONTIS and BRIDGIT to reflect more accurately the effects of road widening on accident risk estimates. The authors refer to the formula

$$BW_r = CA_c \ (R_r - R'_r) \tag{A49.4}$$

where BW_r = benefit of widening
CA_c = average cost per accident
R_r = estimate of current annual accident risk per average daily traffic (ADT)
R'_r = estimate current annual accident risk per ADT after improvement

The North Carolina Bridge Management System estimates R approximately by the following regression model:

$$R_r = 365 \times 200 \times (3.28084W_r)^{-6.5} \left(1 + \frac{9 - A_b}{14} \right) \tag{A49.5}$$

where W_r = roadway width (curb to curb) (m) (NBI item 51)
A_b = approach alignment rating (NBI item 72)

For the BMS of North Carolina, D. W. Johnston (in TRC 423, 1994, pp. 139–149) estimated user costs by the formula

$$AURC(t) = 365 \ ADT(t)(C_{WDA}U_{AC} + C_{ALA}U_{AC} + C_{CLA}U_{AC} + C_{CLD}U_{DC}DL +$$

$$+ \ C_{LCD}(t)U_{DL}DL) \tag{A49.6}$$

where $AURC(t)$ = annual user cost of bridge at year t (U.S. \$)
$ADT(t)$ = average daily traffic using the bridge at year t
C_{WDA} = coefficient for proportioning of vehicles incurring accidents due to width deficiency
C_{ALA} = incurring accidents due to poor alignment
C_{CLA} = incurring accidents due to vertical clearance deficiency
C_{CLD} = coefficient for proportioning of vehicles detoured due to vertical clearance deficiency
$C_{LCD}(t)$ = incurring accidents due to load capacity deficiency
U_{AC} = unit cost of vehicle accidents on bridges, [U.S. \$/accident)
U_{DC} = unit cost of average vehicle detours due to vertical clearance deficiency [U.S. \$/mile (km)]
U_{DL} = unit cost of average vehicle detours due to load capacity deficiency [U.S. \$/mile (km)]
DL = detour length [miles (km)]

The method assumes a linear relationship between vehicle operating costs and vehicle weight and empirical deterioration rates for the load capacity of timber, concrete, and steel bridges. The unit costs of accidents and the rates of accident caused by approach geometry and vertical and horizontal clearance restrictions are estimated based on state-wide surveys. Corrections are provided for future traffic growth.

NCHRP Synthesis 330 (2004) correlates the international roadway roughness index (IRI) and vehicle operating costs by the curve provided by Pavement Management Systems, 1987. Direct operating costs are expressed as U.S. dollars (1976) per 1000 km at 50 km/h. Surface conditions are grouped in five levels: very good, good, fair, poor, and very poor. The cost increases almost linearly from U.S. \$48 to \$60 as the pavement condition declines from 5 to 2 (e.g., from very good to poor). A decline of the surface condition from 4 to 0 corresponds to the steeper cost increase from U.S. \$60 to \$90.

APPENDIX 50. GLOSSARY OF HIGHWAY QUALITY ASSURANCE TERMS (TRC E-C037, 2002)

NCHRP Synthesis 346 (2005, p. 40) praised the glossary of terms provided by TRC E-C037. The terms quoted by NCHRP 346 are as pertinent for bridges as they are for highways:

Acceptance. Sampling and testing, or inspection, to determine the degree of compliance with contract requirements.

End-Result Specifications. Specifications that require the contractor to take the entire responsibility for supplying a product or an item of construction. The highway agency's responsibility is to either accept or reject the final product or to apply a price adjustment commensurate with the degree of compliance with the specifications.

Independence Assurance (IA). Management tool that requires a third party, not directly responsible for process control or acceptance, to provide an independent assessment of the product and/or the reliability of test results obtained from the process control and acceptance testing. (The results of IA tests are not to be used as the basis for product acceptance.)

Lot (or population). Specific quantity of similar material, construction, or units of product subjected to either an acceptance or process control decision. A lot, as a whole, is assumed to be produced by the same process.

Materials and Methods (or Method, Recipe, Prescriptive) Specifications. Specifications that direct the contractor to use specified materials in definite proportions and specific types of equipment and methods to place the material. Each step is directed by a representative of the highway agency.

Performance-Related Specifications. QA specifications that describe levels of key materials and construction quality characteristics that have been found to correlate with fundamental engineering properties that predict performance. These characteristics (e.g., air voids in asphalt concrete and compressive strength of Portland cement concrete) are amenable to acceptance testing at the time of construction.

Performance Specifications. Specifications that describe how the finished product should perform over time.

Quality Assurance (QA). All planned and systematic actions necessary to provide confidence that a product or facility will perform satisfactorily in service.

Quality Assurance Specifications. Combination of end-result specifications and materials and methods specifications.

Quality (or Process) Control (QC). QA actions and considerations necessary to assess and adjust production and construction processes as to control the level of quality being produced in the end product.

Statistically Based Specifications. Specifications based on random sampling and in which properties of the desired product or construction are described by appropriate statistical parameters.

Verification. Process of determining or testing the truth or accuracy of test results by examining the data and/or providing objective evidence.

APPENDIX 51.   DESIGN EXCEPTION PRACTICES

In NCHRP Synthesis 316 (2003), Mason and Mahoney summarized design exception practices prevalent among U.S. transportation agencies (STAs). Design exceptions are the process and the resulting documentation associated with a geometric feature created or perpetuated by a highway construction project that does not conform to the minimum criteria set forth in the standards and policies. Exceptions were found to arise in the following areas of project development:

- Location/system
- Funding source
- Scope/type
- Supplemental criteria (i.e., in addition to FHWA controlling criteria)
- STA criteria values higher than those of AASHTO
- Rehabilitation, restoration, or resurfacing criteria (known as the 3 R's)

The Federal-Aid Policy Guide (FHWA, 1997) has identified the following 13 criteria controlling where formal design exceptions can arise:

- Design speed
- Lane width
- Shoulder width
- Bridge width
- Structural capacity
- Horizontal alignment
- Vertical alignment
- Grade
- Stopping sight distance
- Cross slope
- Superelevation
- Vertical clearance
- Horizontal clearance (other than clear zone)

APPENDIX 52.   MAINTENANCE IMPLEMENTATION

AASHTO (1999a) The Maintenance and Management of Roadway Bridges

The manual (p. 1-4) recommends the *hub* and *spokes* maintenance management structure earlier advanced by NCHRP Report 363 (1994). The hub is the information center of an *integrated maintenance management system.* Spokes link hubs to the following modules or other systems:

- Central maintenance function
- Bridge management system (BMS)
- Pavement management
- Equipment management
- Transportation planning
- Contract management
- Program development/budgeting
- Payroll
- Financial accounting
- Permits
- Materials
- District satellite hubs

As in relational databases, the structure allows for integration with other management systems, such as the BMS.

NCHRP Report 511 (2004) Guide for Customer-Driven Benchmarking of Maintenance Activities

Benchmarking is presented as a cycle consisting of the following steps:

Select partners $\rightarrow$ Establish measures $\rightarrow$ Measure performance

$\uparrow$ $\downarrow$

Implement and continuously improve $\leftarrow$ Identify best performances and practices

The benchmarking, recommended for transportation asset maintenance is *customer driven*. Chapter 3 of the report (p. 45) identifies the following four types of performance measures:

- *Resources (Inputs).* Inputs are the resources used to deliver a product or service, perform an activity, or undertake a business process. Included are labor, equipment, and materials as well as the required funding. Land, water, and air can be considered.
- *Outputs.* Measures of production or accomplishment, or performance indicators.
- *Outcomes.* The results, effects, or changes that occur due to delivering a product or service, conducting an activity, or carrying out a business process. Value added can be measured in increase of customer satisfaction or economic value from saved travel time or avoided life-cycle cost. The three main measurable outcomes are:
  - Customer satisfaction
  - Condition of assets and other attributes of roads
  - Value received by customer

- *Hardship factors.* Factors outside the control of the maintenance organization, such as weather and terrain, that influence the outcomes and level of resources used.

The method recommends a shift from the internally oriented output measurements of productivity to the externally oriented measurements of user satisfaction. The latter are used for adjusting the performance.

APPENDIX 53. NBIS QUALIFICATIONS OF PERSONNEL

NBIS 23 CFR 650 subpart C stipulates the following qualifications for inspection personnel:

§650.307 (a) The individual in charge of the organizational unit that has been delegated the responsibilities for bridge inspection, reporting, and inventory shall possess the following minimum qualifications:

(1) Be a registered professional engineer; or

(2) Be qualified for registration as a professional engineer under the laws of the State; or

(3) Have a minimum of 10 years experience in bridge inspection assignments in a responsible capacity and have completed a comprehensive training course based on the "Bridge Inspector's Training Manual", which has been developed by a joint Federal–State task force, and subsequent additions to the manual.

(b) An individual in charge of a bridge inspection team shall possess the following minimum qualifications:

(1) Have the qualifications specified in paragraph (a) of this section; or

(2) Have a minimum of 5 years experience in bridge inspection assignments in a responsible capacity and have completed a comprehensive training course based on the "Bridge Inspector's Manual".

(3) Current certification as a Level III or IV Bridge Safety Inspector under the National Society of Professional Engineer's program for National Certification in Engineering Technologies (NICET) is an alternative acceptable means for establishing that a bridge inspection team leader is qualified.

The bridge inspector's training manual (FHWA, 1979) of the preceding quote has been repeatedly updated. The most recent edition is FHWA (2002a). The FHWA periodically offers training courses, as do the states.

(FHWA, 2005a) specifies *program manager, team leader, individual responsible for load ratings,* and *diver* as follows:

§650.309 (a) A program manager must, at a minimum:

(1) Be a registered professional engineer, or have ten years (of) bridge inspection experience; and,

(2) Successfully complete an FHWA approved bridge inspection training course.

(b) A team leader must, at minimum:

(1) Have the qualifications specified in paragraph (a) of this section, or

(2) Have five years bridge inspection experience and have successfully completed an FHWA approved comprehensive bridge inspection training course; or

(3) Be certified as a Level II or IV Bridge Safety Inspector under the National Society of Professional Engineer's program for National Certification in Engineering Technologies (NICET)and have successfully completed an FHWA approved comprehensive bridge inspection training course, or

(4) Have all of the following:

(i) A bachelor's degree in engineering from a college or university accredited by or determined as substantially equivalent by the Accreditation Board for Engineering and Technology;

(ii) Successfully passed the National Council of Examiners for Engineering and Surveying Fundamentals of Engineering examination;

(iii) Two years of bridge inspection experience, and

(iv) Successfully completed an FHWA approved comprehensive bridge inspection training course, or

(5) Have all of the following:

(i) An associate's degree in engineering or engineering technology from a college or university accredited by or determined as substantially equivalent by the Accreditation Board for Engineering and Technology;

(ii) Four years of bridge inspection experience

(iii) Successfully completed an FHWA approved comprehensive bridge inspection training course.

(*c*) *The individual charged with the overall responsibility for load rating bridges must be a registered professional engineer.*

(*d*) *An underwater bridge inspection diver must complete an FHWA approved comprehensive bridge inspection training course or other FHWA approved underwater bridge inspection training course.*

Index